# Handbook of Experimental Pharmacology

## Volume 207



For further volumes:
http://www.springer.com/series/164

Rebecca Lever • Barbara Mulloy • Clive P. Page
Editors

# Heparin - A Century of Progress

Springer

*Editors*
Rebecca Lever
School of Pharmacy
University of London
London
United Kingdom

Barbara Mulloy
National Institute for Biological
Standards and Control
Potters Bar, Hertfordshire
South Mimms
United Kingdom

Clive P. Page
Sackler Institute of Pulmonary
Pharmacology
King's College London
London
United Kingdom

ISSN 0171-2004 e-ISSN 1865-0325
ISBN 978-3-642-23055-4 e-ISBN 978-3-642-23056-1
DOI 10.1007/978-3-642-23056-1
Springer Heidelberg Dordrecht London New York

Library of Congress Control Number: 2012931938



Printed on acid-free paper

Springer is part of Springer Science+Business Media (www.springer.com)

# Preface

It will shortly be the 100th anniversary of the discovery of heparin by Jay McLean in 1916, so a volume dedicated to the enormous progress that has since been made in the understanding and clinical exploitation of this complex biomolecule is particularly timely. The present volume has the principal aim of recording the current state of affairs concerning the use and study of heparin; in addition, we have tried to record the progress made, particularly within the last 30 years, a period full of incident in this field. For this reason, we begin with a chapter on the history of heparin (Barrowcliffe), before going on to outline what is currently known about its biosynthesis (Carlsson and Kjellen) and the molecular basis for its anticoagulant and antithrombotic activity (Gray et al.)

The number and variety of heparin products in use rose dramatically with the development of the low-molecular-weight heparins. These are now among the most commonly used drugs and are commercially very important indeed, all of which has provided the impetus for development of the advanced techniques of standardisation (Gray) and characterisation (Mulloy) addressed in Part II of this volume. The commercial importance of heparin may also have its negative aspect, in motivating the extraordinary recent contamination incident described by Chess et al. The necessity to determine detailed structures of such complex molecules as low-molecular-weight heparins and heparan sulphate has led to recent progress in techniques for their analysis (Guerrini and Bisio; Shriver et al.)

The three chapters that make up Part III of this volume are concerned with the current clinical use of heparin in the treatment and prevention of thrombotic disease (Gresele et al.). Being a powerful drug with multiple biological activities, however, heparin has a number of undesirable side effects (Alban), sometimes necessitating the use of specific antidotes (Pai and Crowther).

We have been especially interested in looking to the future of heparin use. In Part IV of this volume an introduction to the biological activities of heparin other than those based on its anticoagulant and antithrombotic activity (Lever and Page) is followed by two chapters going into more detail on the subjects of chemokines and growth factors (Shute) and the growing field of research into the neuroprotective effects of heparin and other glycosaminoglycans (Dudas and Semeniken).

Finally, the wide range of heparin-like compounds and their biological activities is explored in Part V. Many of these had scarcely been described 30 years ago; some are naturally occurring compounds (Allegra et al.; Colliec-Jouault; Gallagher) and others are wholly or partially modified (Coombe and Kett; Oreste).

It seems arbitrary and still mysterious that a complex polysaccharide extracted from hogs' intestines should be a potent and useful anticoagulant and antithrombotic agent in human medicine. No possible application of "rational design" could have found heparin. Had it not been found and recognised so many years ago, it is hard to see how it could be discovered, much less accepted for human use, by the drug discovery paradigms of the present age. We should be grateful for the skill and acuity of the team which identified heparin so long ago, and look forward to an expanded range of applications of its use in the future.

London, UK

R. Lever
B. Mulloy
C.P. Page

# Contents

# Contributors

**S. Alban** Pharmazeutisches Institut, Abteilung Pharmazeutische Biologie, Christian-Albrechts-Universität zu Kiel, Kiel Gutenbergstr. 76, 24118 Kiel, Germany, salban@pharmazie.uni-kiel.de

**Luigi Allegra** Università degli Studi, IRCCS Fondazione Ca' Granda, Ospedale Policlinico, Via Francesco Sforza 35, 20122, Milano, Italy, luigi.allegra@unimi.it

**Shawn Bairstow** Baxter Healthcare Corporation, Round Lake, IL, USA

**T.W. Barrowcliffe** 3 White Point Court, Whitby, North Yorkshire, UK, twbarrowcliffe@yahoo.co.uk

**C. Bavington** Glycomar Ltd, European Centre for Marine Biotechnology, Dunstfaffnage Marine Laboratory, Dunbeg, Oban, Argyll, PA37 1QA, UK

**Antonella Bisio** Istituto di Ricerche Chimiche e Biochimiche G. Ronzoni, Milan, Italy

**Chiara Busti** Division of Internal and Cardiovascular Medicine, Department of Internal Medicine, University of Perugia, Via E. dal Pozzo, 06126, Perugia, Italy

**Ishan Capila** Momenta Pharmaceuticals Inc., 675 West Kendall Street, Cambridge, MA 02142, USA

**Pernilla Carlsson** Department of Medical Biochemistry and Microbiology, Uppsala University, Box 582, SE-751 23 Uppsala, Sweden

**Edward K. Chess** Baxter Healthcare Corporation, Round Lake, IL, USA, ed_k_chess@baxter.com

**S. Colliec-Jouault** Laboratoire de Biotechnologie et Molécules Marines, Ifremer, Rue de l'Ile d'Yeu, BP 21105, 44311, Nantes Cedex 3, France, Sylvia.Colliec.Jouault@ifremer.fr

**Deirdre R. Coombe** Molecular Immunology, School of Biomedical Sciences, Curtin Health Innovation Research Institute, Curtin University of Technology, Level 3 MRF Building, Rear 50 Murray Street, Perth, WA 6000, Australia, d.coombe@curtin.edu.au

**Mark A. Crowther** Department of Medicine, McMaster University, Hamilton, ON, Canada; Department of Pathology and Molecular Medicine, McMaster University, Hamilton, ON, Canada; Hamilton Regional Laboratory Medicine Program, McMaster University, Hamilton, ON, Canada; St. Joseph's Hospital, Room L-208, 50 Charlton Avenue East, Hamilton, ON, Canada L8N 4A6, crowthrm@mcmaster.ca

**C. Delbarre** Laboratoire de Biotechnologie et Molécules Marines, Ifremer, Rue de l'Ile d'Yeu, BP 21105, 44311, Nantes Cedex 3, France

**Sabrina Della Patrona** IRCCS Fondazione S. Maugeri, Istituto Scientifico di Riabilitazione, Via Roncaccio 16, 21049, Tradate (Varese), Italy, s.dellapatrona@libero.it

**Shane Donovan** Baxter Healthcare Corporation, Round Lake, IL, USA

**B. Dudas** Neuroendocrine Organization Laboratory, Lake Erie College of Osteopathic Medicine, 1858 West Grandview Blvd, Erie, PA1509, USA, bdudas@lecom.edu

**J.T. Gallagher** Paterson Institute for Cancer Research, Iduron Ltd., University of Manchester, Manchester, M20 4BX, UK, jgallagher@picr.man.ac.uk

**Elaine Gray** National Institute for Biological Standards and Control, Blanche Lane, South Mimms, Potter's Bar, Hertfordshire EN6 3QG, UK; Haemostasis Section, Biotherapeutics Group, National Institute for Biological Standards and Control, Blanche Lane, South Mimms, Potters Bar, Hertfordshire, EN6 3QG, UK Elaine.Gray@nibsc.hpa.org.uk

**Paolo Gresele** Division of Internal and Cardiovascular Medicine, Department of Internal Medicine, University of Perugia, Via E. dal Pozzo, 06126, Perugia, Italy, grespa@unipg.it

**Marco Guerrini** Istituto di Ricerche Chimiche e Biochimiche G. Ronzoni, Milan, Italy, guerrini@ronzoni.it

**Karalyn Havel** Baxter Healthcare Corporation, Round Lake, IL, USA

**John Hogwood** National Institute for Biological Standards and Control, Blanche Lane, South Mimms, Potter's Bar, Hertfordshire EN6 3QG, UK, John.Hogwood@nibsc.hpa.org.uk

**Peifeng Hu** Baxter Healthcare Corporation, Round Lake, IL, USA

**Richard J. Johnson** Baxter Healthcare Corporation, Round Lake, IL, USA

**Warren C. Kett** Glycan Biosciences Inc., 2 Buck Road, Suite D Mail Box 5, Hanover, NH03755, USA, wkett@glycanbio.com

**Lena Kjellén** Department of Medical Biochemistry and Microbiology, Uppsala University, Box 582, SE-751 23 Uppsala, Sweden, Lena.Kjellen@imbim.uu.se

**Sarah Lee** Baxter Healthcare Corporation, Round Lake, IL, USA

**Rebecca Lever** The School of Pharmacy, University of London, 29-39 Brunswick Square, London, WC1N 1AX, UK, rebecca.lever@pharmacy.ac.uk

**Jeff McKee** Baxter Healthcare Corporation, Round Lake, IL, USA

**Reagan Miller** Baxter Healthcare Corporation, Round Lake, IL, USA

**Edwin Moore** Baxter Healthcare Corporation, Round Lake, IL, USA

**Barbara Mulloy** National Institute for Biological Standards and Control, Blanche Lane, South Mimms, Potter's Bar, Hertfordshire EN6 3QG, UK, Barbara.Mulloy@nibsc.hpa.org.uk

**Mark Nordhaus** Baxter Healthcare Corporation, Round Lake, IL, USA

**P. Oreste** Glycores 2000 S.r.l., Milan, Italy, oreste@glycores.it

**Gloria Paganelli** Division of Internal and Cardiovascular Medicine, Department of Internal Medicine, University of Perugia, Via E. dal Pozzo, 06126, Perugia, Italy

**Clive P. Page** Sackler Institute of Pulmonary Pharmacology, Institute of Pharmaceutical Science, King's College London, Franklin-Wilkins Building, Waterloo Campus, London, SE1 9NH, UK, clive.page@kcl.ac.uk

**Menaka Pai** Department of Medicine, McMaster University, Hamilton, ON, Canada; Hamilton Regional Laboratory Medicine Program, McMaster University, Hamilton, ON, Canada

**Giuseppe Petrigni** Università degli Studi, IRCCS Fondazione Ca' Granda, Ospedale Policlinico, Via Francesco Sforza 35, 20122, Milano, Italy, Giuseppe.petrigni@unimi.it

**Joseph Ray** Baxter Healthcare Corporation, Round Lake, IL, USA

**Ram Sasisekharan** Harvard-MIT Division of Health Sciences & Technology, Cambridge, MA 02139, USA; Koch Institute for Integrative Cancer Research, Massachusetts Institute of Technology, Cambridge, MA 02139, USA; Department of Biological Engineering, Massachusetts Institute of Technology, Cambridge, MA 02139, USA, rams@mit.edu

**K. Semeniken** Millcreek Community Hospital, 5515 Peach Street, Erie, PA16509, USA

**Zachary Shriver** Harvard-MIT Division of Health Sciences & Technology, Cambridge, MA 02139, USA; Koch Institute for Integrative Cancer Research, Massachusetts Institute of Technology, Cambridge, MA 02139, USA; Department of Biological Engineering, Massachusetts Institute of Technology, Cambridge, MA 02139, USA

**Janis Shute** Institute of Biomedical and Biomolecular Sciences, University of Portsmouth, Portsmouth, PO1 2UP, UK, jan.shute@port.ac.uk

**Christina Szabo** Baxter Healthcare Corporation, Round Lake, IL, USA

**Ganesh Venkataraman** Momenta Pharmaceuticals Inc., 675 West Kendall Street, Cambridge, MA 02142, USA

**Todd Wielgos** Baxter Healthcare Corporation, Round Lake, IL, USA

**G. Zoppetti** Glycores 2000 S.r.l., Milan, Italy

# Part I
# Introduction

# History of Heparin

**T.W. Barrowcliffe**

## Contents



**Abstract** The history of heparin is described from its initial discovery in 1916 to recent developments in knowledge of its mechanism of action and clinical use. Commercial production started soon after its discovery, in the 1920s, and improved purification methods led to animal studies and the first clinical trials in the 1930s. Research into heparin's chemical structure proved difficult, with uncertainty about the uronic acid moiety and the N-acetyl content, but the structure of the basic disaccharide unit was established by the 1960s, though knowledge of the heterogeneity and fine structure of heparin chains continued to accumulate over the next 20 years. In 1976, it was found that only one third of heparin chains bound with high affinity to antithrombin, and subsequent studies identified a unique pentasaccharide sequence, which was essential for antithrombin binding and

T.W. Barrowcliffe (Formerly NIBSC) (✉)
3 White Point Court, Whitby, North Yorkshire, UK
e-mail: twbarrowcliffe@yahoo.co.uk

R. Lever et al. (eds.), *Heparin - A Century of Progress*,
Handbook of Experimental Pharmacology 207,
DOI 10.1007/978-3-642-23056-1_1, 

anticoagulant activity – this pentasaccharide was synthesised in 1983. Clinical usage of heparin continued to increase and two major developments were the use of low- dose heparin for prevention of deep vein thrombosis and pulmonary embolism, and the development of low-molecular-weight heparin as a separate drug.

**Keywords** Drug discovery • Heparin • History • LMW heparin • Standardisation

# 1 The First 50 Years (1911–1961)

## *1.1 Discovery and Early Development*

In 1916, Jay McLean, a medical student working in the laboratory of Professor Howell in Toronto, discovered heparin (McLean 1916). Or did he? The discovery of heparin is in many ways a classic example of a fairly frequent phenomenon in science, namely an unexpected observation during investigations of an apparently unrelated area. In McLean's case, he was not looking for an anticoagulant – he was actually investigating procoagulant phospholipid extracts from tissues, but found that two extracts, one from heart (cuorin) and the other from liver (heparphosphatid), had anticoagulant properties. The phospholipid materials (cephalins) from various tissues were prepared by extraction in ether, and these extracts were prepared by alcohol precipitation from the ethereal solutions at 60°C. This precipitation produced at least a partial separation of the anticoagulant substances from the ether-soluble cephalin, but it is noteworthy that the anticoagulant activity became most apparent after prolonged storage, around 3 months. This was attributed to the residual procoagulant cephalin in the preparation initially masking the anticoagulant activity, and subsequently being inactivated by oxidation.

From the known properties of heparin, it would be expected that hardly any heparin would appear in organic solvent extracts, and as pointed out by Jaques (1940), it seems likely that what McLean isolated consisted mainly of a phospholipid with anticoagulant properties. The increased anticoagulant activity on storage can be explained by oxidation of the polyunsaturated fatty acids in the phospholipids, which, as shown in the author's laboratory, yields substances which markedly prolong clotting times (Barrowcliffe et al. 1975), and thus McLean's extracts may have contained very little of what we now know as heparin.

McLean reported his findings to Howell, who was initially most sceptical; as described in subsequent reminiscences by McLean (1959), he had to give a practical demonstration of the anticoagulant properties of his extract by showing that it could completely prevent coagulation of fresh cat's blood (this later became the basis of the heparin unit). Nonetheless, Howell continued investigations into the new anticoagulant after McLean had completed his year of studies, naming it "heparin" in 1918 (Howell and Holt 1918). At this time, the ethereal extraction method was still being used, and as noted above, the extracts probably contained very little heparin. However, because of poor yields Howells crucially changed the extraction method to an aqueous extraction with acetone precipitation (Howell 1925), and these extracts were the first crude preparations of heparin as we know it today.

Quite apart from the nature of what McLean actually isolated, there had been a number of earlier studies of peptone shock in dogs which described how the blood became incoagulable – such a process is known to induce release of heparin by mast cells, and so these studies may have been the first description of heparin, though it was not recognised as such. During investigations of this phenomenon, a water soluble anticoagulant had been isolated from liver by Doyon et al. (1911), but the development of this early isolation of heparin was not pursued.

Thus, on two counts it can be said that McLean did not "discover" heparin in the strictest meaning of the term, since what he isolated was probably not the material we know today, and earlier observations had described the appearance of heparin in the blood and its isolation from liver. Nonetheless, as Jaques (1978) has pointed out in a perceptive article on the discovery of heparin, scientific discovery is generally ascribed to an individual whose observations lead to successful development and application. McLean's observations paved the way for Howell to develop the purification process of heparin and subsequently to the first in vivo studies, and in this sense, as Jaques noted, "Jay McLean discovered heparin (no more and no less)".

## *1.2 Commercial Production*

The commercial potential of heparin was recognised from the beginning, and the first commercial production of heparin outside Professor Howell's laboratory was by Hynson, Westcott and Dunning of Baltimore in the early 1920s. Dr. Dunning had been a student of Professor Howell's and was asked to make the material. This material, from dog liver, had a low potency, around 5 U/mg, and hence was less than 5% pure – it was used mainly for laboratory work as it was too toxic for animal or human studies.

The original tissue source, dog liver, was unsuitable for large-scale preparation, and back in Toronto the purification process was greatly improved during the 1930s by Charles and Scott, who also changed the tissue source from dog or beef liver to beef lung, the latter being cheaper and giving higher yields (Charles and Scott 1933a, b). Beef lung became the main source material for heparin production until the 1950s, when it was largely replaced by porcine mucosa.

Production was soon moved from the Department of Physiology at Toronto University to the Connaught Laboratories, which had already been set up to manufacture insulin for human use, under the supervision of Dr. Charles Best, a key figure in the development and early clinical use of heparin in Toronto. However, as described by Coyne in his review of heparin manufacture (Coyne 1981), the production process is lengthy, and since it involves large amounts of crude animal tissue can be somewhat odoriferous – so much so that at one point Charles and Scott were forced to move some stages of their heparin production to the more isolated

Connaught farm! Best worked closely with Charles and Scott to improve the manufacturing process, and by the 1940s heparin from the Connaught Laboratories was being sold quite widely; however, production ceased in 1950 as other manufacturers had patented improved methods of production giving higher yields.

Some of the material produced in Toronto found its way to Europe, and further chemical analyses were carried out in Denmark by Schmitz and Fischer (1933). However, the main centre for heparin research in Europe during the 1930s was the Karolinska Institute in Stockholm, in the laboratory of Dr. Erik Jorpes, who produced his own heparin from beef and horse liver, following a visit to Howell's laboratory in 1929. Subsequently, production of heparin was transferred to the Vitrum Company in Stockholm, which like the Connaught laboratories was already manufacturing insulin.

The first heparin marketed as a pharmaceutical product in the USA was in 1939. It was "Liquaemin", made by Roche Organon from beef lung tissues; the product under the same name was marketed by Organon for at least another 60 years. The manufacturing process was essentially similar to that of Charles and Scott, and comprised around 20 successive stages. Some of the processes involved are among the most extreme of any used in manufacture of biological drugs, and include heating at 95°C in alkaline solution and acidification down to pH 2, and precipitation with alcohol. Fortuitously these processes, though not designed as such, are also effective virucidal steps, and have ensured that heparin has been free of pathogenic virus transmissions which have been such a major concern with many biological drugs.

Beef lung continued as source material for heparin manufacture until the mid-to-late 1950s, when other tissues began to be investigated as alternatives. This was partly because of problems with handling, storage and extraction of the large amount of rotting lung material, and in addition some of the autolysis processes and other methods used for degradation of tissue also gave some degradation of the heparin. Another driving force was the availability of the source material; beef lung was being increasingly used in animal food, whereas porcine intestinal mucosa, which became the preferred alternative, was readily available as a by-product of manufacture of sausage skins. In fact during the Second World War, the Vitrum Company in Sweden had already used pig intestinal mucosa as an alternative source material, because almost all beef lung tissue in Sweden was used in animal (and possibly human?) food production. Thus, it can be seen that heparin manufacture has been intimately linked with the food industry.

Porcine intestinal mucosa was found to be a cleaner source tissue requiring less degradation than bovine lung, and together with improvements in the purification process, gave higher yields. One truck load of 40,000 pounds of pig "intestinal slime" gave around 5 pounds of the finished heparin product (Coyne 1981).

## 1.3 Chemical Analysis

The original material isolated by McLean was described as "heparphosphatide", and was thought to be a variety of phospholipid. However, following the change in extraction method introduced by Howells in the 1920s, it was realised that the phospholipid was an impurity, and by 1928 Howell had established two of the most important chemical characteristics of heparin, namely that it was a carbohydrate containing sulphur (Howell 1928). In 1959, knowledge of heparin chemistry was summarised by Charles and Best as follows. "Heparin is a complex polysaccharide. The carbohydrate moieties are glucuronic acid and glucosamine which are present in the molecular ratio of 1:1. The carbohydrate is highly sulphated. The amino group is not free and does not appear to be acetylated as in mucoitin or chondroitin sulphate. Evidence has been presented which indicates that the nitrogen is sulphated." However, the growth in knowledge of heparin's chemistry during these 30 years, and beyond, was not straightforward, as described in detail by Roden (1989). For instance although Howell had proposed in 1928 that heparin contained uronic acid, this was disputed by Charles and Scott in their papers of 1933 and 1936 (Charles and Scott 1933a, b), because heparin was negative in the test for uronic acid used at that time.

In Europe, the main centre for heparin research during the 1930s was the Karolinska Institute, in the laboratory of Dr. Erik Jorpes. In 1935, Jorpes published a landmark paper on the chemistry of heparin (Jorpes 1935), in which he identified the main carbohydrate constituents of heparin as uronic acid and hexosamine present in a 1:1 ratio. Jorpes also emphasised the importance of the sulphur content, found as magnesium sulphate in the ash, which constituted around 40% of the total weight. Jorpes was the first to recognise sulphate groups as an integral feature of heparin molecules, and calculated a sulphate content of 2.5 per disaccharide unit. However, as the identity of the uronic acid and hexosamine moieties could not be determined at this time, Jorpes assumed that the material isolated was a type of chondroitin sulphate. Further purification via the brucine salt yielded material with a somewhat higher sulphate content, and since it was also assumed that heparin was a homogenous chemical compound, its structure was assumed to be chondroitin trisulphuric acid.

Although Jorpes was mistaken with his final conclusion, his studies correctly identified the basic structural unit of heparin as a sulphated disaccharide consisting of alternate uronic acid and hexosamine groups; the hexosamine was subsequently identified by Jorpes as glucosamine (Jorpes and Bergström 1936). It would be some 20 years before the complexity and heterogeneity of heparin's fine structure would be understood, and a further 25 years before the molecular nature of heparin's anticoagulant action would be revealed.

One of the difficulties was that for many years heparin was regarded as a single chemical compound with a unique structure. Thus, although in the 1930s it was believed that the hexosamine residues were N-acetylated, studies by Jorpes et al. in 1950 showed that N-sulphate groups accounted for the majority of the hexosamine residues, N-acetyl groups accounting for only 10%. However because of the difficulty in accepting that both N-acetyl and N-sulphate groups could be present in the same molecule (and that heparin might consist of a variety of molecules with different structures), Jorpes assumed that the N-acetyl groups were due to a contaminant. Similar difficulties arose in explaining the ratio of 2.5 sulphates per disaccharide unit, although it was pointed out by Charles and Todd (1940) that the ratio could be equally well expressed as 5 sulphates per tetrasaccharide.

Identification of the uronic acid component of heparin proved particularly difficult. Because of the resistance of heparin to acid hydrolysis fairly severe conditions had to be used, which in turn destroyed most of the liberated uronic acid. In the 1930s, it had been assumed by both Howells and Jorpes that the uronic acid was glucuronic acid, but what appeared to be definitive proof was not provided until 1946, when Wolfrom and Rice (1946) devised new hydrolysis conditions which preserved the liberated uronic acid, and enabled them to identify it as D-glucuronic acid: this was subsequently confirmed in several laboratories. However, studies in the 1960s were to show that in fact the major uronic acid component is iduronic, rather than glucuronic acid (see Section 2.2).

## *1.4 Animal Studies and Early Clinical Use*

During the 1930s, the availability of increasing amount of material from the Connaught Laboratories in Toronto, and also from the Vitrum Company in Stockholm, led to in vivo studies in animals, and soon after, commencing around 1935, in patients. The move from animal experiments to the first clinical studies was remarkably rapid by today's standards, especially considering that the chemical structure and degree of purification of the heparin available was still uncertain. One reason for the rapid acceptance of heparin in the clinic was that it was thought to be a physiological substance, present in blood. At this time, Howell's theories of blood coagulation still held sway, and according to these theories blood contained a substance, "antiprothrombin", which prevented conversion of prothrombin to thrombin, and which was neutralised by the addition of procoagulant tissue extracts (thromboplastin). The concentration of this hypothetical anticoagulant was supposedly increased after peptone injections in dogs, and it was thought (correctly) that the anticoagulant present in dog blood after peptone shock was the same as that isolated from dog liver and other tissues. Thus, heparin was initially identified with the hypothetical "antiprothrombin" of blood, and as a physiological substance was considered unlikely to be harmful.

The initial animal and clinical studies in Toronto were described by Murray and Best in 1938 (Best 1959), although preliminary animal experiments and human use in transfusion had started in the 1920s. The heparin preparations of the late 1920s and early 1930s, which were probably only about 10–15% pure, were found to be toxic in animals, but following the efforts of Charles and Scott in the Connaught Laboratories heparin preparations with specific activity of around 250 units (Toronto units)/mg were produced (Charles and Scott 1933b), and these preparations were devoid of toxic effects in dogs. It was soon found that heparin was an effective anticoagulant in a variety of situations in dogs, preventing thrombosis induced by mechanical or chemical means in veins, and allowing the successful performance of a number of operative procedures. Studies in patients soon followed, although the initial injections of heparin were marked by reactions in about half the patients; these included headache, nausea, vomiting, chills, rapid pulse and a marked fall in blood pressure – similar reactions had been reported earlier by Mason (1924) and Howell (1928) in transfusion experiments, and were clearly due to use of impure material. Further work by Charles and Scott produced a crystalline barium salt of heparin which had about twice the potency of this material, and when this was given to patients there were no reactions. According to Murray and Best (Best 1959), the potency of this material was around 500 U/mg, which is apparently over twice as high as current heparin preparations, and four times higher than the material from the Toronto group subsequently established as a standard. However, the assay used at this stage was based on inhibition of clotting of citrated bovine plasma, and the unit was defined as the amount of heparin required to prevent coagulation of 1 mL of recalcified plasma for 4 h. This was substantially different from the original definition of the unit by Howell, and in practice 1 "Howell unit" was equivalent to about 5 "new units"; hence, the potency of this material would be around 100 U/mg in current units (see subsequent section).

In Stockholm, clinical studies started in 1935, around the same time as those in Toronto. The early clinical work was pioneered by Dr. Clarence Crafoord, using material purified by Jorpes and colleagues at the Vitrum Company, and reactions in patients similar to those observed in Toronto were not infrequent. Nonetheless, in the less rigorous regulatory environment that pertained at the time, this did not prevent further studies continuing, and Crafoord published the results of his clinical studies in 1939 (Crafoord 1939). In both Toronto and Stockholm, the initial clinical results were encouraging, and the effectiveness of heparin in preventing postoperative thrombosis was quickly established. Murray and Best (Best 1959) mention that heparin was given postoperatively to 315 patients, who showed no evidence of pulmonary embolism; at that time the rate of pulmonary embolism following major surgery was around 2–7.5%, and many patients died of this condition. They also described seven cases of existing pulmonary embolism treated successfully with heparin. These initial results, though of course unblinded and uncontrolled, anticipated one of the most important and lifesaving uses of heparin; the prevention and treatment of pulmonary embolism.

During the early 1940s, progress in heparin research was inevitably slow. Nonetheless, it was during this period that the first International Standard was established. Subsequently, an increasing number of publications documented the clinical effectiveness of heparin, though controlled clinical trials did not commence until 1960.

## 1.5 Units and Assays

The original unit of biological activity of heparin was defined by Howell in 1923 as the minimum quantity necessary to maintain the fluidity of 1 mL of cats' blood for 24 h at 0°C (Howell 1925). For measurement of activity, the blood was taken directly from an anaesthetised cat and placed in tubes containing various concentrations of heparin. Charles and Scott (1933a) used the same method but took readings after 2 h. Jorpes (1935) used fresh ox blood, initially reading at 24 h, but later taking readings at 2, 4, 8 and 16 h, as did Quick (Quick 1938a). Schutz (1941) used the original method of Howell, reading at 24 h, but with rabbit blood.

Although all these methods use non-anticoagulated whole blood, and can therefore be regarded as "physiological", it is impractical to collect fresh blood for large numbers of assays, and the assays cannot be repeated with the same substrate. The use of anticoagulated bovine plasma was described by Reinert and Winterstein (1939) and Foster and Nutley (1942); the anticoagulant was citrate. The anticoagulated plasma could be stored frozen in large batches and hence used repeatedly; an advantage in convenience and reproducibility over the whole blood assays. Kuizenga et al. (1943) used citrated sheep plasma, and this assay became the basis of the USP method in 1950.

In all of the above methods, no clot-promoting agents were present, except for the glass surface of the tubes, and hence coagulation times in the presence of heparin were very long. Therefore, in these early methods the degree of coagulation after a fixed incubation time was measured, rather than the clotting time. In 1938, Quick first described the prothrombin time (Quick 1935), in which tissue extracts (called thromboplastin, or thrombokinase) and $CaCl_2$ were added to anticoagulated whole blood or plasma; this accelerated clotting times to less than 30 s. This method was developed into a heparin assay by McIntosh (1941), who used oxalated horse plasma and a rabbit brain tissue extract (thrombokinase); clotting times in this assay were around 100–500 s. In 1950, Adams and Smith described a similar method using anticoagulated bovine blood and bovine brain thromboplastin (Adams and Smith 1950) – this subsequently became the basis of the British Pharmacopoeia (BP) heparin assay, first described in 1953. The anticoagulant used in this method was sodium sulphate, which prevented coagulation via its high ionic strength, rather than by calcium chelation.

A feature of these assays which use thromboplastin reagents is the relatively high concentration of heparin required to prolong clotting times, usually in the range 1–3 U/mL. This contrasts with assays involving thrombin or APTT reagents, which are sensitive to heparin concentrations as little as 0.1 U/mL. The reason for this is that tissue thromboplastin extracts inhibit heparin's anticoagulant activity – it was shown much later that this inhibition is due to the active principle of thromboplastin reagents, i.e. tissue factor itself (Gomperts and Zucker 1978). Thus, heparin assays which use thromboplastin are a balance between two mutually antagonistic substances, which could cause discrepancies if heparin preparations differ from the standard in their interaction with tissue factor.

It was recognised from the early days of heparin research that heparin enhanced the neutralisation of thrombin by plasma, and this was used as the basis of an assay method by a number of investigators. In 1941, Jaques and Charles described a method adding thrombin to bovine blood, but the degree of coagulation after 15 min was assessed rather than the clotting time (Jaques and Charles 1941). Kjems and

Wagner (1948) added bovine thrombin to bovine oxalated plasma, adjusting the amount of thrombin to give clotting times in the range of 20–60 s with a range of heparin concentrations. Although this method was not taken up by heparin manufacturers, it was the forerunner of the thrombin time method which is used in some clinical laboratories to control heparin therapy.

During the 1950s, most manufacturers of heparin used either the USP or BP methods. The USP assay was essentially that described by Reinert and Winterton (1939), with assessment of the degree of coagulation in citrated sheep plasma. A refinement was introduced by Foster and Nutley (1942), who described 11 grades of clotting, between zero (full fluidity) and 4+ (full clot). The first BP method in 1948 was based on the original Howell method with non-anticoagulated cat blood, but this was superseded in 1953 by the method of Adams and Smith (1950), using sulphated ox blood and thromboplastin. The Japanese Pharmacopoeia also adopted the BP method.

## *1.6 Standards*

It was realised from the early days of heparin assays that a standard was needed, especially as different methods were developed. The first commercial material produced from dogs' liver was assigned a potency of 5 U/mg by the original Howell method, and when the purified barium salt was produced by Charles and Scott this assayed at 110 U/mg against this material. For convenience, this was changed to 100 U/mg, and a sample of this material was adopted in 1933 as a provisional Standard in Toronto (Charles and Scott 1933b; Best 1959). A similar material was used as a provisional Swedish Standard by Jorpes and colleagues, but no unitage was assigned and test samples were measured in terms of mgs of the Standard.

The increasing international interest in heparin and especially the first clinical studies enhanced the need for an International Standard. As noted by McIntosh (1941) the barium salt produced by Charles and Scott was not ideal as a standard as it was not very stable and when in solution it was highly acidic. However, this was subsequently converted into the sodium salt by Charles and Scott and a sample of this was sent to the Division of Biological Standards at the National Institute of Medical Research in the UK; at that time, this was one of the two centres with responsibility for International Standards (the other was the State Serum Institute in Copenhagen). Because of the Second World War, it was not possible to organise an international collaborative study, and this material was adopted as a provisional International Standard in 1942 by the League of Nations, the predecessor of the World Health Organisation (WHO) (League of Nations 1943/1944). The potency of this standard was assigned as 130 U/mg, as assayed in Toronto, and after the war it was officially established by WHO as the first International Standard (IS) for heparin with the same potency (WHO Expert Committee on Biological Standardisation 1947/1948). Thus, the International Unit as defined by this Standard can be traced back to the Toronto unit as originally defined by Howell (1925).

By the late 1950s, stocks of the first IS were running low and a collaborative study was organised to calibrate its replacement. The material for the second IS was of the same origin as the first IS, i.e. bovine lung and eight laboratories took part in the calibration. A diverse range of methods was used, with the USP method predominating (4 labs), but with the exception of one laboratory, the results were very similar, and the overall mean potency turned out to be 130 IU/mg, i.e. identical to that of the first Standard. It is interesting that two sets of data from the one laboratory whose results were discrepant came from methods involving human plasma, which appeared to give lower potencies than the methods using animal plasma.

# 2 The Last 50 Years (1961–2011)

In the early 1960s, heparin was a well-established anticoagulant drug with an increasing therapeutic profile. However, during the last 50 years there have been several major developments, which have considerably expanded both our theoretical knowledge of heparin and its therapeutic role. The fine structure of heparin's chemistry has been unravelled and the molecular nature of its mechanism of action revealed. This has been followed by chemical synthesis of the antithrombin binding region and the development of low-molecular-weight (LMW) heparin as a separate drug.

These developments during the last 50 years are mostly covered in detail in subsequent chapters of this book, and will only be mentioned briefly in this section.

## *2.1 Assays and Standards*

The methods developed by the US and British Pharmacopoeias continued to be used with no substantial changes by manufacturers up till the mid-1980s, when the EP assay replaced the BP method (Barrowcliffe 1989). The EP assay used the same substrate as the USP assay, i.e. anticoagulated sheep plasma, but whereas the USP method measured the degree of coagulation the EP method was based on measurement of clotting times after adding an accelerator, i.e. similar to the APTT in human plasma, used for clinical control of heparin therapy. Both the USP and EP assays are still used by manufacturers to this day.

A major development in the 1970s was the introduction of the anti-Xa method, by Yin et al. (1973) and Denson and Bonnar (1973). Initially, the residual FXa was measured by a clotting method, but subsequently a chromogenic version was developed (see Barrowcliffe 1989 for review). The introduction of the anti-Xa method coincided with clinical studies of low-dose heparin and with the early studies of LMW heparin (see Sect. 2.5), and it was particularly useful in measurement of ex vivo samples in these studies. Although not adopted for assay of UFH, it became the mainstay of standardisation of assay of LMW heparin (see Gray 2011).

The 1960s saw a gradual transition from bovine lung to porcine mucosa as the source material for heparin, and in a collaborative study to compare samples of the new material with the second IS (lung), the results showed considerable differences between methods, as well as high variability within each method, (Bangham and Woodward 1970). This is an example of the importance of the "like vs like" principle; the differences between lung and mucosal heparin were highlighted in a study in our laboratory in 1978 (Barrowcliffe et al. 1978). A sample of one of the mucosal heparins was eventually established by WHO as the third IS in 1973, and these method differences had important consequences, leading to a discrepancy of around 7% between the USP unit, as defined by the USP Standard, and the International Unit, as defined by the International Standard (see Gray 2011). The third IS has been replaced three times since 1973, each time with similar mucosal material.

The development of several different low-molecular-weight (LMW) heparin products in the late 1970s and early 1980s was quite a challenge for standardisation. In addition to the large differences between anti-Xa and APTT assays, which were recognised as a feature of LMW heparin (see later section), there was substantial variability between labs using the same method when assaying LMW heparins against the unfractionated heparin (UFH) Standard (Barrowcliffe et al. 1985). It became apparent that a separate standard was needed for LMW heparin, and from the results of the first collaborative study (Barrowcliffe et al. 1985), two LMW heparins were selected and subjected to a larger international collaborative study (Barrowcliffe et al. 1988); one of the candidates was established by WHO as the first IS for LMW heparin in 1986.

As described in a later section, the establishment of the first IS was crucial for eradicating the confusion associated with the unitage of LMW heparin during the early development of the drug. The first IS for LMW heparin has recently been replaced with the second IS, following a comprehensive collaborative study of eight different candidate materials (see Gray 2011).

## 2.2 *Chemistry and Biochemistry*

As already mentioned, the uronic acid component of heparin was mistakenly thought to be only D-glucuronic for many years, particularly after the studies of Wolfrom and colleagues (Wolfrom and Rice 1946). However in 1962, Cifonelli and Dorfman found L-iduronic acid in an acid hydrolysate of heparin (Cifonelli and Dorfman 1962), and other studies by the groups of Lindahl and Axelsson (1971), and Perlin et al. (1971) confirmed this. Wolfrom et al. (1969) had accepted that his earlier results had led to an incorrect conclusion; he had repeated the earlier studies and found that under the hydrolysis conditions used only glucuronic acid was produced. With the development of improved analytical methods, it was eventually realised that iduronic acid was in fact the major uronic acid component.

Further work on the fine chemical structure of heparin is described in detail by Roden (1989), and by Lindahl et al. (1980). Important findings include the configuration of the glusosaminidic and uronidic linkages, and the presence and location of O-sulphate groups, in particular a 3-O-sulphate substituent, which is essential for heparin's anticoagulant activity.

The heterogeneity of heparin's chemical structure was increasingly appreciated, including its MW range, from 5,000 to 35,000 (Johnson and Mulloy 1976). However, the most important, and surprising, finding was described in 1976 by the groups of Lindahl (Höök et al. 1976) and Rosenberg (Lam et al. 1976), who used immobilised antithrombin to fractionate heparin into high-affinity and low-affinity components. Both groups found that only one third of the heparin bound with high affinity to antithrombin and this high-affinity material accounted for nearly all the anticoagulant activity; the low affinity fraction had very little activity. Although there were no gross differences in chemical composition between high and low affinity components, this finding implied that there must be a specific sequence in heparin molecules, which confers high affinity binding to antithrombin. Further studies by Lindahl and colleagues isolated an octasaccharide, which appeared to be the smallest fragment capable of binding to antithrombin (Lindahl et al. 1979), and within this fragment it was shown that there was a unique pentasaccharide sequence which conferred high affinity to antithrombin. The fine structure of this pentasaccharide was elucidated by Lindahl's group (Thunberg et al. 1982); a 3-O-sulphate group on the middle glucosamine residue was found to be crucial for anticoagulant activity, although other sulphate groups were also found to be important.

One of the most important landmarks in heparin chemistry was the synthesis of this unique pentasaccharide by Petitou and colleagues from the group of Choay – this was first announced in 1983 (Choay et al. 1983). This and subsequent developments in synthesis of heparin oligosaccharides are described in detail in Alban (2011).

## 2.3 *Mechanism of Action*

Already by 1925 Howell had recognised that heparin required a plasma co-factor for its anticoagulant action (Howell 1928), but its nature was not known. In the late 1930s, Quick (1938b) and Brinkhous et al. (1939) related the heparin co-factor to the naturally occurring antithrombin in plasma, but it was not until the 1960s that Abildgaard purified antithrombin (then called antithrombin III) and showed it to be identical with heparin co-factor (Abildgaard 1968, 1975). As noted above, only one third of heparin chains bind with high affinity to antithrombin, implying considerable structural specificity in both molecules, though the exact nature of the heparin binding site in the antithrombin molecule would not be revealed until the advent of information from structural studies and molecular genetics, including studies of naturally occurring mutants [for reviews see Björk et al. (1989) and Carrell et al. (1995)].

It was shown by Damus et al. (1973) that binding of heparin to antithrombin increased its rate of inhibition of thrombin by up to 1,000 times; they also found a similar mechanism for potentiation of the inhibition of other proteases, namely Factors Xa, IXa, and XIa by antithrombin. It was proposed by Rosenberg and others that heparin induced a conformational change in the antithrombin molecule, which greatly enhanced its reaction rate with thrombin and other serine proteases. However during the late 1970s and early 1980s, there was a competing hypothesis, namely that heparin bound to thrombin and thereby increased its reaction rate with antithrombin (Griffith 1979; Longas et al. 1980). Although heparin does bind to thrombin, this binding was shown not to require a specific structure in the heparin molecule, and to be of much lower affinity than that of the heparin antithrombin interaction (Björk et al. 1989). After many detailed biochemical studies in the 1980s, it was eventually realised that the accelerating effect of heparin on the reaction between thrombin and antithrombin requires both proteins to bind in close proximity to the same heparin chain (Björk et al. 1989; Danielsson et al. 1986). Another important feature of heparin's action, which was elucidated in this period, was its catalytic nature; it was shown that the affinity of the thrombin – antithrombin complex for heparin was much lower than that of antithrombin itself; hence after the complex is formed, heparin is free to dissociate for further interactions (Carlström et al. 1977).

In another line of investigation, studies in our laboratory in the mid-1970s focused on the activity of heparin fractions of different molecular weights in various assay systems. It was already known that anticoagulant activity decreased with decreasing MW in pharmacopoeial assays, and this was confirmed using the APTT assay with human plasma. However, a surprising finding was that the low MW fractions retained their activity in what was then a new method, the anti-Xa assay (Anderson et al. 1976). Subsequent studies in collaboration with colleagues at Kabi in Stockholm, using high-affinity heparin and purified reagents, confirmed this dichotomy, with the low MW fractions having very little ability to potentiate thrombin inhibition but retaining high anti-Xa activity (Andersson et al. 1979). It became clear that the mechanism of heparin's action was quite different for the two enzymes, with a requirement for a certain chain length for thrombin inhibition. Further studies by Lane, in collaboration with the group of Lindahl (Lane et al. 1984) pinpointed the precise chain length required for heparin to potentiate thrombin inhibition – a minimum of 18 saccharides was required, though activity on a molar basis continued to increase with increasing MW.

These studies confirmed the template mechanism previously proposed for thrombin inhibition, with thrombin and antithrombin binding to the same heparin chain, hence the requirement for a minimum MW higher than that of the pentasaccharide. The fact that anti-Xa activity was maintained down to the lowest MW oligosaccharides (Lane et al. 1984) showed that this template mechanism did not apply to inhibition of Factor Xa, and focused attention on the possible therapeutic properties of LMW heparin.

## 2.4 Development of LMW Heparin

The origins of LMW heparin as a drug go back to the mid-1970s in the laboratory of Dr Edward Johnson at NIBSC. Dr Johnson had been investigating the MW range of Dextran, which was known to affect its clinical properties. He decided to carry out similar investigations on heparin, which was known to be at least as heterogeneous as Dextran, if not more so, and prepared three broadly cut fractions of heparin, of high, medium and low MW, by gel filtration. With the help of a medical colleague, Dr Milica Brozovic, a volunteer study was organised in which samples of the high and low MW fractions, as well as sodium and calcium salts of UFH were injected subcutaneously into medical students. The results, published in 1976, showed clearly that the LMW fraction gave much higher and more prolonged blood levels by anti-Xa assay than either the HMW fraction or UFH (Johnson et al. 1976).

As noted by Dr Johnson (Johnson 1992), the fractions were prepared in an ordinary chemistry laboratory with no GMP validation, and though they were sterilised and tested for pyrogens by a pharmaceutical company, no further animal studies were done. Under today's regulatory environment, it is very doubtful whether the studies would have been allowed to go ahead. Nonetheless, this pioneering study, together with the in vitro observations of anticoagulant activities mentioned earlier, focused attention on the LMW fractions of heparin, which up to that time had attracted little interest, and in the late 1970s several pharmaceutical companies started to prepare and investigate LMW heparin as a potential therapeutic agent.

The fractions used by Johnson and colleagues were prepared by gel filtration, and initially one French manufacturer, Choay Laboratories (later part of Sanofi) used the same method to prepare a fraction of about 6,000 average MW (Fraxiparine). However as this meant wasting two-thirds of the heparin, this and other manufacturers subsequently used various methods to depolymerise the heparin, improving the yield. Because of patent restrictions, each manufacturer used a different method of depolymerisation, the main methods being nitrous acid, enzymatic cleavage, and β-elimination. By 1980, four main products were being developed; these were Fraxiparin (Choay, later Sanofi ), Logiparin (Leo, later Novo), Enoxaparin (Pharmuka, later Sanofi) and Fragmin (Kabi, later Pfizer) though four others were added by the mid-1980s. Unusually for pharmaceutical development, all manufacturers were in mainland Europe. A LMW heparin was prepared in the UK by Glaxo Ltd., but despite promising results in animals, Glaxo decided to stop all work on heparin and LMW heparin, thereby missing out on a multi-billion dollar market. These products were soon shown to be effective antithrombotic agents in animal studies, and their antithrombotic activity appeared to correlate with their anti-Xa activity rather than their activity in clotting assays (Thomas et al. 1981; Thomas 1992).

At the time of development of LMW heparins, low-dose UFH had become established for prophylaxis of venous thromboembolism, and the theory which held sway was that the anti-Xa activity of heparin (and LMW heparin) was responsible for its ability to *prevent* thrombosis, by virtue of preventing prothrombin activation. When it became apparent that LMW heparin retained anti-Xa activity but had much less effect than UFH on clotting times, it was also proposed that LMW heparin might have less effect on bleeding. Although some animal studies did support this hypothesis, others did not, and the picture was far from clear-cut (Thomas 1992). It became evident from initial clinical studies that the hope of separating the antithrombotic properties of heparin from its propensity to enhance bleeding had not been realised, and as reviewed in detail elsewhere, the concept of anti-Xa = antithrombotic and anti-IIa = bleeding was overly simplistic (Barrowcliffe 1995).

## 2.5 *Clinical Use of Heparin and LMW Heparin*

By the early 1960s, heparin was a well-established therapeutic drug for the prevention and treatment of venous thrombosis, being given mainly intravenously. In 1960, Barritt and Jordan published a landmark clinical trial comparing heparin to placebo for the treatment of pulmonary embolism (Barritt and Jordan 1960). The results were so overwhelmingly positive in favour of heparin that since then heparin has been standard treatment for pulmonary embolism, further placebo controlled trials being considered unethical.

A new concept introduced by Kakkar and colleagues during the 1970s was that of "low-dose" heparin, given subcutaneously for prevention of post-operative DVT (Kakkar et al. 1972), though this approach had previously been suggested by Sharnoff some 10 years earlier (Sharnoff et al. 1962). The theoretical basis for this was the findings of Rosenberg and other workers that low concentrations of heparin could enhance the inhibition of FXa by antithrombin and hence prevent the generation of thrombin. Dosage was 5,000 units two or three times daily, and several randomised controlled trials demonstrated the effectiveness of this approach (Gallus et al. 1973; Nicolaides et al. 1972) in preventing DVT; subsequent larger multi-centre studies showed that it was also effective in preventing pulmonary embolism (Anonymous 1975).

Collins et al. (1988) reviewed more than 70 randomised trials involving over 16,000 patients, and found that low-dose subcutaneous heparin given twice daily could prevent about half of all pulmonary emboli and about two-thirds of all DVTs. Advantages of subcutaneous low-dose heparin compared to standard dose intravenous administration were the longer lasting anticoagulant effect, the reduced incidence of haemorrhage, and less need for monitoring. These advantages were extended when LMW heparin began to be used in the same way.

The first clinical study of LMW heparin was published by Kakkar et al. (1982), using “Fraxiparine” (Choay Laboratories, prepared at that time by gel filtration). The drug was given either once or twice daily to groups of 100 and fifty patients, respectively. The results showed that both regimes were effective; in particular, the once-daily regimen prevented DVT in 97 of the 100 patients and was not associated with any increase in bleeding. Although uncontrolled, this pioneering trial paved the way for many comparative trials of LMW heparin with low-dose UFH during the 1980s, and was the first to show that effective prophylaxis could be achieved with a single daily dose of LMW heparin – a significant advantage over UFH in terms of patient convenience. Although in this first trial, there were no major problems with haemorrhage this was not the case with two other trials published shortly after. Both Schmitz-Huebner et al. (1984) and Koller et al. (1986) found an unacceptably high incidence of bleeding in groups of patients given two different LMW heparin preparations, and it became clear that LMW heparin was not the “holy grail” of anticoagulants, providing antithrombotic effectiveness with no risk of bleeding. Koller et al. (1986) showed that the bleeding risk was dose related by repeating the study with the same LMW heparin at one-third of the dose – the bleeding risk was reduced to the same as that of UFH. It was clear in retrospect that the doses used in both studies were too high – the main reason for this was the lack of standardisation of units of activity of LMW heparin at the time, and these clinical results emphasised the importance of the first International Standard for LMW heparin (see earlier section).

Since these early studies, the number of trials of the various different LMW heparin products for prevention of DVT has mushroomed, and meta-analyses have been published. Overall, it has been shown that LMW heparin is at least as effective as UFH in general surgery, and may be more effective in orthopaedic surgery; however except for a few studies, there is no reduction of haemorrhagic tendency [for review, see Gray (2008)].

Some studies in the mid-1980s investigated the use of LMW heparin, in higher doses but still given subcutaneously, to treat established venous thrombosis (Bratt et al. 1985). Several trials in the 1990s demonstrated the effectiveness of LMW heparin in this indication; a major advantage is the ability to continue treatment on an outpatient basis, without the need for monitoring as required with UFH [for review, see Hull and Pineo (2000)].

Probably the most significant recent clinical development, and one which provides a fitting conclusion to this chapter, is the demonstration of the clinical effectiveness of the synthetic pentasaccharide and its derivatives; this is covered in detail in Alban (2011). Thus, nearly 100 years after heparin was first described in Howell’s laboratory, the small sequence of the heparin chains which represents its active principle, having been identified and synthesised, has now been shown to be an effective antithrombotic drug.

## References


Abildgaard U (1968) Highly purified antithrombin 3 with heparin cofactor activity prepared by disc electrophoresis. Scand J Clin Lab Invest 21(1):89–91

Abildgaard U (1975) Heparin cofactor and antithrombin. Thromb Diath Haemorrh 33(1):38–42

Adams SS, Smith KL (1950) The use of sulphated whole blood in the assay of heparin. J Pharm Pharmacol 2(11):836–846

Alban S (2011) Adverse effects of heparin. In: Lever R, Mulloy B, Page CP (eds) Heparin – a century of progress. Springer, Heidelberg 211–263

Anderson L-O, Barrowcliffe TW, Holmer E, Johnson EA, Sims GEC (1976) Anticoagulant properties of heparin fractionated by affinity chromatography on matrix-bound antithrombin III and by gel filtration. Thromb Res 9:575–583

Andersson L-O, Barrowcliffe TW, Holmer E, Söderström G (1979) Molecular weight dependency of the heparin potentiated inhibition of thrombin and activated Factor X. Effect of heparin neutralisation in plasma. Thromb Res 15:531–541

Anonymous (1975) Prevention of fatal postoperative pulmonary embolism by low doses of heparin. An international multicentre trial. Lancet 2(7924):45–51

Bangham DR, Woodward PM (1970) A collaborative study of heparins from different sources. Bull World Health Organ 42(1):129–149

Barritt D, Jordan S (1960) Anticoagulant drugs in the treatment of pulmonary embolism: a controlled trial. Lancet 1:1309–1312

Barrowcliffe TW (1989) Heparin assays and standardisation. In: Lane DA, Lindahl U (eds) Heparin, chemical and biological properties, clinical applications. Edward Arnold, London

Barrowcliffe TW (1995) Low molecular weight heparin(s). Br J Haematol 90(1):1–7

Barrowcliffe TW, Gutteridge JMC, Dormandy TL (1975) The effect of fatty acid autoxidation products on blood coagulation. Thromb Diath Haemorrh 33:271–277

Barrowcliffe TW, Johnson EA, Eggleton CA, Thomas DP (1978) Anticoagulant activities of lung and mucosal heparins. Thromb Res 12:27–36

Barrowcliffe TW, Curtis AD, Tomlinson TP, Hubbard AR, Johnson EA, Thomas DP (1985) Standardisation of low molecular weight heparins: a collaborative study. Thromb Haemost 54:675–679

Barrowcliffe TW, Curtis AD, Johnson EA, Thomas DP (1988) An International Standard for LMW heparin. Thromb Haemost 60:1–7

Best CH (1959) Preparation of heparin and its use in the first clinical cases. Circulation 19:79–86

Björk I, Olson ST, Shore JD (1989) Molecular mechanisms of the accelerating effect of heparin on the reactions between antithrombin and clotting preteinases. In: Lane DA, Lindahl U (eds) Heparin, chemical and biological properties, clinical applications. Edward Arnold, London

Bratt G, Törnebohm E, Granqvist S, Aberg W, Lockner D (1985) A comparison between low molecular weight heparin (KABI 2165) and standard heparin in the intravenous treatment of deep venous thrombosis. Thromb Haemost 54(4):813–817

Brinkhous KM, Smith HP Jr, Warner ED, Seegers WH (1939) The inhibition of blood clotting: an unidentified substance which acts in conjunction with heparin to prevent the conversion of prothrombin into thrombin. Am J Physiol 125:683–687

Carlström AS, Liedén K, Björk I (1977) Decreased binding of heparin to antithrombin following the interaction between antithrombin and thrombin. Thromb Res 11(6):785–797

Carrell R, Skinner R, Warden M, Whisstock J (1995) Antithrombin and heparin. Mol Med Today 1(5):226–231

Charles AF, Scott DA (1933a) Studies on heparin I: the preparation of heparin. J Biol Chem 102:425–429

Charles AF, Scott DA (1933b) Studies on heparin II: heparin in various tissues. J Biol Chem 102:431–435

Charles AF, Todd AR (1940) Observations on the structure of the barium salt of heparin. Biochem J 34(1):112–118

Choay J, Petitou M, Lormeau JC, Sinaÿ P, Casu B, Gatti G (1983) Structure-activity relationship in heparin: a synthetic pentasaccharide with high affinity for antithrombin III and eliciting high anti-factor Xa activity. Biochem Biophys Res Commun 116(2):492–499

Cifonelli JA, Dorfman A (1962) The uronic acid of heparin. Biochem Biophys Res Commun 7:41–45

Collins R, Scrimgeour A, Yusef S, Peto R (1988) Reductions in fatal pulmonary embolism and venous thrombosis by perioperative administration of subcutaneous heparin. Overview of results of randomized trials in general, orthopedic and urological surgery. N Engl J Med 318:1162–1173

Coyne E (1981) Heparin – past, present and future. In: Lundblad RL, Brown WV, Mann KG, Roberts HR (eds) Chemistry and biology of heparin. Elsevier, New York

Crafoord C (1939) Heparin and post-operative thrombosis. Acta Chir Scand 82:319–333

Damus PS, Hicks M, Rosenberg RD (1973) Anticoagulant action of heparin. Nature 246 (5432):355–357

Danielsson A, Raub E, Lindahl U, Björk I (1986) Role of ternary complexes, in which heparin binds both antithrombin and proteinase, in the acceleration of the reactions between antithrombin and thrombin or factor Xa. J Biol Chem 261(33):15467–15473

Denson KW, Bonnar J (1973) The measurement of heparin. A method based on the potentiation of anti-factor Xa. Thromb Diath Haemorrh 30(3):471–479

Doyon M, Morel A, Policard A (1911) Estraition directe de l'antithrombine du foie. Influence de la congelation du sang. C R Soc Biol Paris 70:341–346

Foster RHK, Nutley NJ (1942) The assay of heparin. J Lab Clin Med 27:820–827

Gallus AS, Hirsh J, Tutle RJ, Trebilcock R, O'Brien SE, Carroll JJ, Minden JH, Hudecki SM (1973) Small subcutaneous doses of heparin in prevention of venous thrombosis. N Engl J Med 288(11):545–551

Gomperts ED, Zucker ML (1978) Heparin, brain thromboplastin and the insensitivity of the prothrombin time to heparin activity. Thromb Res 12(1):105–117

Gray E, Mulloy B, Barrowcliffe TW (2008) Heparin and low-molecular-weight heparin. Thromb Haemost 99:807–818

Gray E (2011) Standardisation of Unfractionated and Low-Molecular-Weight Heparin. In: Lever R, Mulloy B, Page CP (eds) Heparin – a century of progress. Springer, Heidelberg 65–76

Griffith MJ (1979) Kinetic analysis of the heparin-enhanced antithrombin III/thrombin reaction. Reaction rate enhancement by heparin-thrombin association. J Biol Chem 254(23):12044–12049

Höök M, Björk I, Hopwood J, Lindahl U (1976) Anticoagulant activity of heparin: separation of high-activity and low-activity heparin species by affinity chromatography on immobilized antithrombin. FEBS Lett 66(1):90–93

Howell WH (1925) The purification of heparin and its presence in blood. Am J Physiol 17:553–562

Howell WH (1928) The purification of heparin and its chemical and physiological reactions. Bull John Hopkins Hosp 42:199–206

Howell WH, Holt E (1918) Two new factors in blood coagulation – heparin and pro-antithrombin. Am J Physiol 47:328–341

Hull RD, Pineo GF (2000) Low-molecular-weight heparin in the treatment of venous thromboembolism. Semin Thromb Hemost 26(Suppl 1):61–67

Jaques LB (1940) The heparins of various mammalian species and their relative anti-coagulant potency. Science 92(2395):488–489

Jaques LB (1978) Addendum: the discovery of heparin. Semin Thromb Hemost 4:350–353

Jaques LB, Charles AF (1941) The assay of heparin. Quart J Pharm Pharmacol 14:1–15

Johnson EA (1992) Historical note. In: Barrowcliffe TW, Johnson EA, Thomas DP (eds) Low molecular weight heparin. Wiley, Chichester

Johnson EA, Mulloy B (1976) The molecular-weight range of mucosal-heparin preparations. Carbohydr Res 51(1):119–127

Johnson EA, Kirkwood TB, Stirling Y, Perez-Requejo JL, Ingram GI, Bangham DR, Brozović M (1976) Four heparin preparations: anti-Xa potentiating effect of heparin after subcutaneous injection. Thromb Haemost 35(3):586–591

Jorpes E (1935) The chemistry of heparin. Biochem J 29(8):1817–1830

Jorpes E, Bergström S (1936) Der aminozucker des heparins. Hoppe Seylers Z Physiol Chem 244:253–259

Kakkar VV, Corrigan T, Spindler J, Fossard DP, Flute PT, Crellin RQ, Wessler S, Yin ET (1972) Efficacy of low doses of heparin in prevention of deep-vein thrombosis after major surgery. A double-blind, randomised trial. Lancet 2(7768):101–106

Kakkar VV, Djazaeri B, Fok J, Fletcher M, Scully MF, Westwick J (1982) Low-molecular-weight heparin and prevention of postoperative deep vein thrombosis. Br Med J 284:375–379

Kjems H, Wagner H (1948) A new method for the assay of heparin. Acta Pharmacol 4:155–163

Koller M, Schoch U, Buchmann P, Largiader F, von Felten A, Frick PG (1986) Low molecular weight heparin (KABI 2165) as thromboprophylaxis in elective visceral surgery. A randomized double-blind study versus unfractionated heparin. Thromb Haemost 56:243–246

Kuizenga MH, Nelson JW, Cartland GF (1943) The bioassay of heparin preparations. Am J Physiol 139:612–616

Lam LH, Silbert JE, Rosenberg RD (1976) The separation of active and inactive forms of heparin. Biochem Biophys Res Commun 69(2):570–577

Lane DA, Denton J, Flynn AM, Thunberg L, Lindahl U (1984) Anticoagulant activities of heparin oligosaccharides and their neutralization by platelet factor 4. Biochem J 218(3):725–732

League of Nations (1943/1944) Memorandum on a provisional International Standard for heparin. Bull Health Organ League Nations 10:51

Lindahl U, Axelsson OJ (1971) Identification of iduronic acid as the major sulfated uronic acid of heparin. J Biol Chem 246(1):74–82

Lindahl U, Bäckström G, Höök M, Thunberg L, Fransson LA, Linker A (1979) Proc Natl Acad Sci USA 76(7):3198–3202

Lindahl U, Bäckström G, Thunberg L, Leder IG (1980) Evidence for a 3-O-sulfated D-glucosamine residue in the antithrombin-binding sequence of heparin. Proc Natl Acad Sci USA 77(11):6551–6555

Longas MO, Ferguson WS, Finlay TH (1980) Studies on the interaction of heparin with thrombin, antithrombin, and other plasma proteins. Arch Biochem Biophys 200(2):505–602

Mason EC (1924) A note on the use of heparin in blood transfusion. J Lab Clin Med 10:203–206

McIntosh FC (1941) A method for estimating the potency of heparin preparations. Biochem J 35:770–775

McLean J (1916) The thromboplastic action of cephalin. Am J Physiol 41:250–257

McLean J (1959) The discovery of heparin. Circulation 19(1):75–78

Nicolaides AN, Dupont PA, Desai S, Lewis JD, Douglas JN, Dodsworth H, Fourides G, Luck RJ, Jamieson CW (1972) Small doses of subcutaneous sodium heparin in preventing deep venous thrombosis after major surgery. Lancet 2(7783):890–893

Perlin AS, Mackie DM, Dietrich CP (1971) Evidence for a (1→4)-linked 4-O-(-L-idopyranosyluronic acid 2-sulfate)-(2-deoxy-2-sulfoamino-D-glucopyranosyl 6-sulfate) sequence in heparin. Long-range H-H coupling in 4-deoxy-hex-4-enopyranosides. Carbohydr Res 18 (2):185–194

Quick AJ (1935) The prothrombin in haemophilia and in obstructive jaundice. J Biol Chem 109:lxiii–lxiv

Quick AJ (1938a) On the action of heparin and its relation to thromboplastin. Am J Physiol 115:317–333

Quick AJ (1938b) The normal antithrombin of the blood and its relation to heparin. Am J Physiol 123:712–719

Reinert M, Winterstein A (1939) Contribution to the study of heparin. Arch Int Pharmacodyn Ther 62:47–48

Roden L (1989) Highlights in the history of heparin. In: Lane DA, Lindahl U (eds) Heparin, chemical and biological properties, clinical applications. Edward Arnold, London
Schmitz A, Fischer A (1933) Uber die chemische natur des heparins. III. Einige untersuchungen zur constitution des heparins. Hoppe Seylers Z Physiol Chem 216:274–280
Schmitz-Huebner U, Bünte H, Freise G, Reers B, Rüschemeyer C, Scherer R, Schulte H, van de Loo J (1984) Clinical efficacy of low molecular weight heparin in postoperative thrombosis prophylaxis. Klin Wochenschr 62(8):349–353
Schutz F (1941) Biological standardisation of anticoagulants. Quart J Pharm Pharmacol 14:45–48
Sharnoff JG, Kass HH, Mistica BA (1962) A plan of heparinization of the surgical patient to prevent postoperative thromboembolism. Surg Gynecol Obstet 115:75–79
Thomas DP (1992) Experimental studies in animals. In: Barrowcliffe TW, Johnson EA, Thomas DP (eds) Low molecular weight heparin. Wiley, Chichester
Thomas DP, Merton RE, Lewis WE, Barrowcliffe TW (1981) Studies in man and experimental animals of a low molecular weight heparin fraction. Thromb Haemost 45(3):214–218
Thunberg L, Bäckström G, Lindahl U (1982) Further characterization of the antithrombin-binding sequence in heparin. Carbohydr Res 100:393–410
WHO Expert Committee on Biological Standardisation (1947/1948) Bull World Health Organ 1:9
Wolfrom ML, Rice FAH (1946) The uronic acid component of heparin. J Am Med Assoc 68:532–537
Wolfrom ML, Wang PY, Honda S (1969) On the distribution of sulphate in heparin. Carbohydr Res 11:179–185
Yin ET, Wessler S, Butler JV (1973) Plasma heparin: a unique, practical, submicrogram-sensitive assay. J Lab Clin Med 81(2):298–310

# Heparin Biosynthesis

**Pernilla Carlsson and Lena Kjellén**

## Contents



**Abstract** Heparin and heparan sulfate share the same polysaccharide backbone structure but differ in sulfation degree and expression pattern. Whereas heparan sulfate is found in virtually all cells of the human body, heparin expression is restricted to mast cells, where it has a function in storage of granular components such as histamine and mast cell specific proteases. Although differing in charge and sulfation pattern, current knowledge indicates that the same pathway is used for synthesis of heparin and heparan sulfate, with a large number of different enzymes

L. Kjellén (✉)
Department of Medical Biochemistry and Microbiology, Uppsala University, Box 582, SE-751 23 Uppsala, Sweden
e-mail: Lena.Kjellen@imbim.uu.se

R. Lever et al. (eds.), *Heparin - A Century of Progress*,
Handbook of Experimental Pharmacology 207,
DOI 10.1007/978-3-642-23056-1_2, © Springer-Verlag Berlin Heidelberg 2012

taking part in the process. At present, little is known about how the individual enzymes are coordinated and how biosynthesis is regulated. These questions are addressed in this chapter together with a review of the basic enzymatic steps involved in initiation, elongation, and modification of the polysaccharides.

**Keywords** GAGosome • Golgi • Heparan sulfate • Heparin • Heparin biosynthesis • Mast cell

## 1 Introduction

Heparin is found in mast cell granules where it interacts with histamine, proteases, and inflammatory mediators. The negative charge of the polysaccharide, due to a high degree of sulfate substitution, is important for its ability to bind the other granule constituents. Storage, retention, and in some cases activation of these components are most likely the main functions of mast cell heparin. The sulfate groups of the heparin molecule are added during biosynthesis, which occurs in the Golgi compartment of the mast cell. Here, the heparin chains are simultaneously elongated and modified by a large number of enzymes.

Heparin chains are attached to serglycin. In addition to mast cells, this protein is found also in other hematopoietic cells as well as in endothelial cells. However, in these cells the serglycin core protein is substituted with chondroitin sulfate instead of heparin (Kolset and Tveit 2008). The physiological role of heparin was long believed to be control of blood coagulation, since this is a potent pharmacological effect of the polysaccharide (Petitou et al. 2003). However, the localization of endogenous heparin in mast cells, and not in blood, makes it unlikey to fulfill this function.

Heparan sulfate (HS) contains the same polysaccharide backbone as heparin and is also sulfated. The major difference between the two is the degree of modification, heparin being more heavily sulfated than HS (Fig. 1b). Also, HS isolated from different tissues and cell types differ in structure when compared to each other (Ledin et al. 2004), but the overall sulfation degree of all these HS species is much lower than that of heparin (Gallagher and Walker 1985). In contrast to heparin, HS is produced by virtually all cells of the human body. HS chains are attached to a variety of core proteins that are either secreted to the extracellular space (e.g., perlecan, agrin, and collagen XVIII) or associated with the cell surface (syndecans and glypicans). The four members of the syndecan core protein family are transmembrane proteins, while the six glypicans are linked with the plasma membrane through a GPI-anchor (Fig. 1a).

Current knowledge indicates that the same biosynthesis pathway is used for heparin and HS biosynthesis. Why then, do mast cells make heparin while other cells synthesize HS?

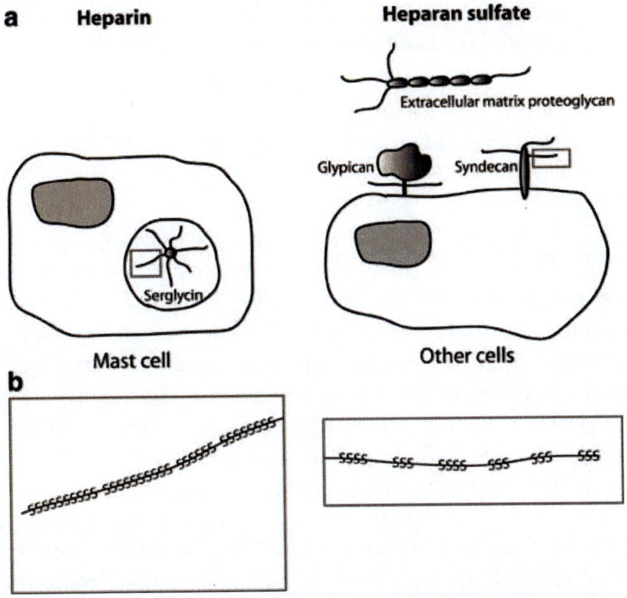


**Fig. 1** (**a**) Mast cells produce highly sulfated heparin attached to the serglycin core protein. The heparin proteoglycan is stored in mast cell granules. Other cells synthesize less negatively charged heparan sulfate, which is found at the cell surface as glypican and syndecan proteoglycans, and in the extracellular matrix. (**b**) A blow-up of the boxes in panel a, illustrating the difference in sulfation pattern between heparin and heparan sulfate

## 2 Heparin and Heparan Sulfate Biosynthesis

As mentioned above, heparin and HS share the same basic structure, consisting of N-acetyl-D-glucosamine (GlcNAc) and glucuronic acid (GlcA) units that are partly modified by epimerization of GlcA to iduronic acid (IdoA) and by sulfation at different positions of mainly GlcNAc and IdoA residues. The synthesis is a rapid process, as shown for heparin produced by mouse mastocytoma microsomes. Polymerization and modification of a polysaccharide chain was estimated to be completed in ~1 min (Hook et al. 1975; Lidholt et al. 1989).

Below follows an overview of the different steps in heparin/HS synthesis, followed by a section in which the enzymes responsible for the different reactions are described (Fig. 3).

### 2.1 Initiation

The biosynthesis of heparin and HS begins with the formation of a tetrasaccharide linkage region (-GlcA-Gal-Gal-Xyl-). This process is catalyzed by four enzymes

```
MMQKLLKCSRLVLALALILVLESSVQGYPTQRARYQWVRC  40
NPDSNSANCLEEKGPMFELLPGESNKIPRLRTDLFPKTRI  80
QDLNRIFPLSEDYSGSGFGSGSGSGSGSGSGFLTEMEQDY 120
QLVDESDAFHDNLRSLDRNLPSDSQDLGQHGLEEDFML   158
```

**Fig. 2** Human serglycin amino acid sequence (UniProtKB accession P10124). The glycosaminoglycan attachment site is *underlined*

adding individual monosaccharides sequentially to the growing glycosaminoglycan (GAG) chain. A xylose residue from UDP-xylose is first transferred to the hydroxyl group of a serine residue on the core protein (Fig. 2). Only certain serines, defined by Ser–Gly residues flanked by acidic residues, are selected for GAG attachment (Esko and Zhang 1996). The serglycin protein contains a stretch of repetitive Ser–Gly residues that can be decorated with heparin chains, making the protein densely glycosylated. This repetitive sequence is conserved in different species, although the number of Ser–Gly units varies. Notably, this sequence is more than twice as long in rat as in, for example, mouse and human.

The attachment of xylose is followed by a stepwise transfer of two galactose and one glucuronic acid residues. These units may be modified by phosphorylation (xylose) and/or sulfation (galactose units). In vitro studies have shown that the linkage region enzymes are sensitive to these modifications (Gulberti et al. 2005; Tone et al. 2008). Phosphorylation/sulfation of the tetrasaccharide linker may thus be a way of regulating GAG synthesis.

Formation of the linkage region is identical in heparin/HS and chondroitin sulfate (CS) synthesis. The crucial point in determining whether a heparin/HS chain or a CS chain will be attached to the linkage tetrasaccharide is the addition of the next monosaccharide. While a N-acetylgalactosamine residue will initiate CS elongation, addition of GlcNAc results in HS/heparin formation. The enzymes responsible for transferring this first hexosamine to the tetrasaccharide linkage region are unique to this reaction step and do not take part in the subsequent polymerization process. It has been suggested that O-sulfation of the Gal residues may lead preferentially to CS synthesis (Ueno et al. 2001).

## 2.2 *Elongation*

Upon addition of the first GlcNAc residue, the heparin/HS chain is elongated by addition of alternating glucuronate and N-acetyl-glucosamine residues from their respective UDP-sugars. The final products are extended polysaccharides, heparin chains from different sources being in the range of $M_r$ = 60,000–100,000 Da (Robinson et al. 1978). Commercially available heparins are processed and their molecular weights range between ~7,000 and 25,000 Da. In comparison with newly synthesized heparin, heparan sulfate chains are generally shorter ($M_r$ = 22,000–45,000 Da) (Lyon et al. 1994).

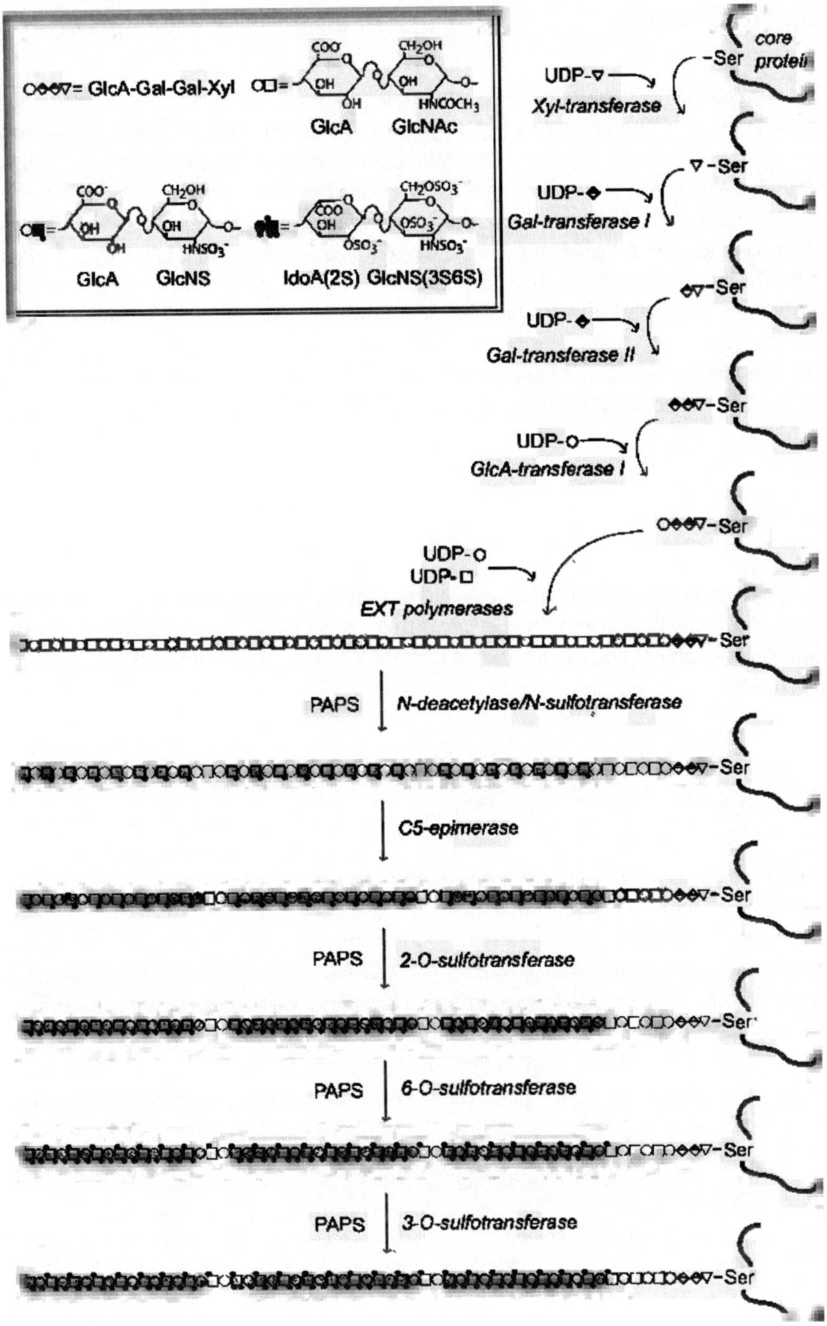


**Fig. 3** Heparin/heparan sulfate biosynthesis. The growing polysaccharide is attached to a serine residue in a core protein. Different UDP-sugars and PAPS are used as substrates and each

## 2.3 *Modification*

As the HS/heparin chain grows, it is modified by a set of various enzymes (Esko and Lindahl 2001). HS chains are only partly modified, with the modifications occurring in clusters, resulting in polysaccharide chains having regions that are highly sulfated interspersed with unmodified regions. Heparin is more heavily sulfated, containing 80–90% N-sulfated glucosamine, whereas about 30–60% of the glucosamine residues in HS are N-sulfated (Gallagher and Walker 1985; Lyon et al. 1994). D-Glucuronic acid residues adjacent to N-sulfated glucosamine can be epimerized to L-iduronic acid followed by 6-O-sulfation of GlcNAc and 2-O-sulfation of IdoA and, more rarely, of GlcA. Sulfate groups can also be found at the C3-position of GlcNAc, although this modification is not very common. The occurrence of $GlcNH_2$ residues has also been reported (Westling and Lindahl 2002).

# 3 The Enzymes

All enzymes taking part in HS/heparin biosynthesis have been cloned. They are transmembrane proteins (with the exception of 3-O-sulfotransferase-1) with the enzymatically active domain located in the Golgi lumen, where HS/heparin synthesis takes place.

## 3.1 *Linkage Region Enzymes*

Transfer of the first xylose residue to the core protein is performed by a xylosyltransferase. Two highly similar isoforms, XylT1 and XylT2, with tissue-specific expression patterns exist in mammals (Gotting et al. 2007). Although the XylT2 was cloned already in 2000 (Gotting et al. 2000), its enzyme activity was not demonstrated until several years later when three independent papers on the subject were published (Schon et al. 2006; Cuellar et al. 2007; Voglmeir et al. 2007). It has long been discussed whether xylosylation takes place in the endoplasmic reticulum or in the Golgi compartment. However, by using fluorescently tagged

---

**Fig. 3** (continued) individual reaction step is catalyzed by a specific enzyme. Symbols used for individual monosaccharides are explained in the box to the *upper left*. After synthesis of the linkage region, the polymerase complex composed of EXT1 and EXT2 add alternating units of glucuronic acid and N-acetylglucosamine to the nonreducing end of the chain. In the presence of the sulfate donor 3′-phosphoadenosine 5′-phosphosulfate (PAPS), a series of modifications takes place, beginning with N-deacetylation and N-sulfation of the original N-acetylglucosamine units. N-Deacetylation/N-sulfation is followed by epimerization of glucuronic acid to iduronic acid and finally ending with stepwise O-sulfation of the sugars, including 2-O-sulfation of the uronic acid and 6-O- and 3-O-sulfation of the glucosamine xylosyltransferases, Schon and colleagues could determine localization of both enzymes to the early cisternae of the Golgi apparatus (Schon et al. 2006).

Although the xylosyltransferases are Golgi-located transmembrane proteins, xylosyltransferase enzyme activity can be detected also in cell culture supernatants and in human body fluids (see references in Gotting et al. 2007). However, the substrate for these enzymes, UDP-xylose, is not found extracellularly, and the function of secreted enzyme is not known. Also, other glycosaminoglycan biosynthesis enzymes are secreted (Nagai et al. 2007). Could proteolytic cleavage of the enzymes be a way of regulating GAG biosynthesis by rapidly downregulating enzyme activity in the Golgi compartment?

The enzymes responsible for the subsequent steps of the linkage region formation, galactosyltransferase-I (GalT1), galactosyltransferase-II (GalT2), and glucuronyl-transferase (GlcAT1), occur as single isoforms and have been shown to be located to the medial Golgi (Bai et al. 2001; Pinhal et al. 2001).

Addition of the first GlcNAc residue to the GlcA of the heparin/HS linkage region is performed by enzymes with GlcNAcT-I activity, in contrast to polymerization of the HS chain, which is dependent on enzymes with GlcNAcT-II activity. Two members of the exostosin gene family, Exostosin-like 2 (EXTL2) and EXTL3, have been shown to possess GlcNAc-TI activity, thus being capable of adding the first GlcNAc unit to the heparin/HS chain (Kitagawa et al. 1999; Kim et al. 2001). While EXTL2 possesses only GlcNAcT-I activity, EXTL3 is capable of transferring GlcNAc also in the polymerization reaction. Another member of the exostosin-like gene family, EXTL1, could possibly take part in elongation of heparin/HS chains since it has GlcNAc-TII activity (Kim et al. 2001). However, little is known about the individual contribution of exostosin-like enzymes to HS/heparin biosynthesis in vivo.

### *3.2 Polymerases*

HS/heparin polymerization is carried out by EXT1 and EXT2 (Zak et al. 2002). These enzymes were first characterized as tumor suppressors since heterozygous mutations in the genes encoding the enzymes are responsible for the development of benign skeletal tumors in patients with hereditary multiple exostoses (HME) (McCormick et al. 1999). Both enzymes have been shown to have dual enzyme activities in vitro, GlcA-TII and GlcNAc-TII (Lind et al. 1998; McCormick et al. 2000; Busse and Kusche-Gullberg 2003), but the EXT2 polymerizing activity is weak. While EXT1 alone is able to polymerize HS chains in vitro, EXT2 does not seem to have this capacity (Busse and Kusche-Gullberg 2003). It has been suggested that the role of EXT2 in HS biosynthesis is to act as a chaperone for EXT1 (Wei et al. 2000). Consistent with this idea, there is evidence suggesting that the localization of EXT1 in the Golgi compartment depends on expression of EXT2 (McCormick et al. 2000). It has been demonstrated that the two enzymes form a hetero-oligomeric complex and that this dimer probably represents the biologically relevant form of the heparin/HS polymerization unit (Kobayashi et al. 2000; McCormick et al. 2000; Senay et al. 2000; Busse and Kusche-Gullberg 2003; Kim et al. 2003).The EXT enzymes may also be part of larger enzyme complexes, sometimes referred to as GAGosomes (Esko and Selleck 2002), see below.

## *3.3 Modification Enzymes*

Mammalian HS/heparin modification enzymes, with the exception of C5-epimerase and 2-O-sulfotransferase, exist in several isoforms with four NDSTs, three 6-O-sulfotransferases and seven 3-O-sulfotransferases reported. For an overview of substrate preferences for the different enzymes, see Lindahl and Li (2009).

NDSTs have a key role in HS biosynthesis as further modifications occur mainly in N-sulfated regions. However, 6-O-sulfate groups can be found also in cells lacking N-sulfated HS, showing that N-sulfation is not an absolute prerequisite for other modifications at all occasions (Holmborn et al. 2004). The NDSTs are bifunctional enzymes, responsible for both N-deacetylation and N-sulfation of GlcNAc residues in HS chains (Grobe et al. 2002). The N-sulfotransferase domain has been located to the carboxyl part of the protein and has been crystallized (Berninsone and Hirschberg 1998; Kakuta et al. 1999). The three-dimensional structure of the deacetylase domain is not known, but expression of a truncated form of NDST2 (A66-P604) resulted in a protein with retained N-deacetylase activity, indicating that the domain is located closer to the N-terminal than is the N-sulfotransferase active site (Duncan et al. 2006). Studies on cells expressing NDST1 mutants lacking either N-deacetylase or N-sulfotransferase activity have shown that the two reaction steps do not have to be performed by the same NDST molecule. Instead, two separate NDST molecules can work together, one performing the N-deacetylation reaction and the other transferring the sulfate group (Bengtsson et al. 2003).

Four vertebrate NDST isoforms have been identified and cloned, NDST1–NDST4 (Hashimoto et al. 1992; Eriksson et al. 1994; Orellana et al. 1994; Kusche-Gullberg et al. 1998; Aikawa and Esko 1999; Aikawa et al. 2001). NDST1 and NDST2 transcripts are found in most tissues both during the embryonic stage and in adult mice, whereas NDST3 and NDST4 show more restricted mRNA expression (Kusche-Gullberg et al. 1998; Aikawa et al. 2001; Pallerla et al. 2008). However, transcription levels may not necessarily correlate with translation levels. In fact, as discussed further below, expression of the NDST isoforms may be both translationally and posttranslationally regulated (Grobe and Esko 2002).

In addition to N-sulfated and N-acetylated glucosamine residues, low levels of N-unsubstituted glucosamine can also be detected in HS/heparin preparations (Westling and Lindahl, 2002). Such residues have been suggested to be generated by the N-deacetylase activity of NDST enzymes without subsequent N-sulfation. $GlcNH_2$ residues have been identified in mouse embryonic stem cells lacking both NDST1 and NDST2 (Holmborn et al. 2004) as well as in tissues of NDST3 knockout mice (Pallerla et al. 2008). It is possible that all NDST isoforms under certain conditions have the capacity to generate $GlcNH_2$. Alternatively, such residues are formed through other, so far unknown, mechanisms.

After N-sulfation, the C5-epimerase acts to transform some of the GlcA residues into IdoA by epimerization of the C5 carboxyl group. GlcA residues may be recognized as substrates if they are linked to an N-sulfated unit at the nonreducing end (Jacobsson et al. 1984). IdoA residues are thus confined to N-sulfated domains.

2-O-Sulfation is closely associated with epimerization. Most IdoA units are sulfated at the C2 position, whereas 2-O-sulfated GlcA residues are rare (Rong et al. 2001). In cerebral cortex, however, this type of modification is more abundant (Lindahl et al. 1995). 2-O-Sulfation of IdoA units is largely confined to contiguous N-sulfated domains, although 2-O-sulfated IdoA is occasionally found also adjacent to N-acetylated GlcN residues (Rong et al. 2001). The 2-O-sulfotransferase, like the C5-epimerase, only occurs in one single isoform.

There are three enzymes catalyzing the 6-O-sulfotransferase reaction, 6OST1-3 (Habuchi et al. 2000). The occurrence of a 6-OST-2 splice variant has also been reported (Habuchi et al. 2003). The three isoforms differ slightly regarding expression pattern and substrate preferences, but are all able to modify both GlcNAc and GlcNS residues in different sequence settings (Jemth et al. 2003; Smeds et al. 2003).

A 3-O-sulfate group is a crucial component of the antithrombin-binding pentasaccharide of heparin. The high affinity binding of this pentasaccharide to antithrombin results in a conformational change of the protein and enhanced interaction with thrombin leading to inhibition of blood coagulation (Petitou et al. 2003). The involvement of 3-O-sulfate groups in this interaction, as well as in HS binding to the herpes simplex gD protein, where the uncommon 3-O-sulfated $GlcNH_2$ residues are recognized (Shukla et al. 1999), indicates that 3-O-sulfation probably is dedicated to interactions involving very specific HS structures. Notably, there are seven isoforms of the enzyme responsible for 3-O-sulfation, also suggesting a crucial role for this type of HS modification. Although the 3-O-sulfotransferase-1 has been suggested to be the most critical isoform for producing the antithrombin-binding HS sequence, ablation of the gene in mice does not result in a procoagulant phenotype (Shworak et al. 2002). Instead, genetic-background-dependent lethality and intrauterine growth retardation are observed, but the cause of these abnormalities is not yet known.

### 3.4 *Postbiosynthesis Endosulfatases*

Postsynthetic modification of HS also occurs, performed by two endosulfatases, Sulf1 and Sulf2, located at the cell surface (Lamanna et al. 2007). These sulfatases act after extracellular translocation of the GAG chains, by cleaving off 6-O-sulfate groups in the internal part of the polysaccharide chain. Whether heparin, exposed in the extracellular space after mast cell degranulation, is modified by the Sulfs is not known. However, the preferred substrate of the Sulfs is an internal trisulfated disaccharide, abundant in heparin.

### 3.5 *Which Enzyme Isoforms are Present in Mast Cells?*

So far, only a few studies have dealt with mast cell expression of heparin/HS biosynthesis enzymes. Obviously, all enzymes responsible for the formation of the linkage region and polymerization of the polysaccharide chain can be expected to be expressed by mast cells. However, it is not known whether xylosyltransferase-1 or -2 (or both) are responsible for the initiation of the heparin chain and which of the EXTL enzymes that participate in heparin formation. From studies of mouse mastocytoma NDSTs, it is clear that NDST2 is the dominating NDST isoform, present at high concentration, while NDST1 transcript is barely detected (Kusche-Gullberg et al. 1998). Accordingly, connective tissue type mast cells from mice deficient in NDST2 lack sulfated heparin, show abnormal morphology and contain reduced amounts of histamine and mast cell proteases (Forsberg et al. 1999; Humphries et al. 1999). The altered morphology and decreased levels of inflammatory mediators are also seen in mice deficient in serglycin, where no proteoglycans are found in the intracellular granules (Abrink et al. 2004).

Heparin contains both iduronic acid and 2-O-sulfate groups and hence the single isoform enzymes C5-epimerase and 2-O-sulfotransferase must be expressed by mast cells. Analysis of glycosaminoglycans isolated from mouse ears, where mast cells are abundant, showed no difference in heparin composition when mice deficient in 6-OST-1 were compared to wild-type mice. Thus, heparin 6-O-sulfation preferentially relies on 6-OST-2 and/or 6-OST-3, at least in skin mast cells (Habuchi et al. 2007). An immortalized mouse mast cell line has been shown to express 3-O-sulfotransferase-1 (Shworak et al. 1997), but it has not been studied whether also other 3-O-sulfotransferase isoforms are expressed.

## 4 Why Do Mast Cells Form Heparin When Other Cells Synthesize Heparan Sulfate?

As mentioned above, modifications of HS chains are made in clusters resulting in regions that are highly sulfated (NS-domains) interspersed with nonsulfated regions (NA-domains) and with intermediately sulfated regions (NA/NS-domains) usually surrounding the NS-domains (Gallagher 2001). The sulfated regions are not completely modified at all possible sites, resulting in specific patterns depending on cell type and developmental stage (David et al. 1992; Lindahl et al. 1995; Maccarana et al. 1996; Brickman et al. 1998; van Kuppevelt et al. 1998; Jenniskens et al. 2000, 2002; Allen and Rapraeger 2003; ten Dam et al. 2003; Ledin et al. 2004; Warda et al. 2006). Notably, very little structural variation is seen when HS from the same tissue but from different individuals is compared (Lindahl et al. 1995; Ledin et al. 2004), suggesting that HS biosynthesis is a highly regulated process. Heparin, on the contrary, lacks the pattern of alternating NA- and NS-domains and can be characterized as a more or less continuous NS-domain.

### *4.1 Enzyme Abundance*

How is the domain pattern in HS formed, and why do not all positions available for sulfation become modified in HS biosynthesis? In contrast to DNA in protein synthesis, there is no template to determine the design of the final glycosaminoglycan product. Instead, expression levels of the individual enzymes are obviously an important factor. By regulating the abundance of the enzymes/isoenzymes, at transcriptional or translational levels, or by changing their turnover, different HS/heparin modification patterns may be obtained. As mentioned above, when the NDST2 gene is knocked out in mice, the mast cells are abnormal and lack heparin (Forsberg et al. 1999), although HS from different other tissues of these animals appears unaffected (Ledin et al. 2004). It can therefore be concluded that NDST2 is the NDST isoform mainly responsible for heparin synthesis. When instead mice devoid of NDST1 are analyzed for HS structure, a dramatic reduction of N-sulfation in various tissues, including liver, is observed (Ledin et al. 2006), indicating that N-sulfation in HS producing cells relies mostly on NDST1. However, the question of how the difference in expression levels of the isoforms is regulated remains to be answered.

### *4.2 Regulation of Enzyme Expression*

The knowledge of transcriptional control of HS biosynthesis enzymes is scarce, but the NDST2 gene has been shown to be under regulation of the GA-binding protein, which is a transcription factor (Morii et al. 2001). Mice carrying a mutation in the *mi* allele express abnormal *mi* transcription factor. This results in decreased amounts of NSDT2 protein in skin mast cells as a consequence of disturbed nuclear localization of the GA-binding protein, which normally binds to a GGAA motif in the 5′-untranslated region of NDST2.

Regulation of the NDST proteins probably also occurs at the translational level as suggested by differential expression of constructs in which the different NDST 5′-untranslated regions were ligated to a reporter gene (Grobe and Esko 2002). Evidence for translational regulation of HS biosynthetic enzymes in *Drosophila melanogaster* has also been presented (Bornemann et al. 2008). In addition, post-translational modifications such as glycosylation could play a role as has been shown for one of the chondroitin sulfate sulfotransferases (Yusa et al. 2005). In fact, glycosylation of NDST protein affects its enzyme activity (Carlsson and Kjellén, unpublished).

## 4.3 GAGosome Composition

The GAGosome model (Esko and Selleck 2002), suggesting close proximity of the enzymes in a physical complex, also offers a tentative explanation for regulation of HS/heparin modification. Here, the ability of the individual enzymes to associate with the other enzymes/components of the GAGosome as well as their relative concentration will be important (Fig. 4). Supporting the hypothesis, the HS polymerases EXT1 and EXT2 are known to function as a complex (Kobayashi et al. 2000; McCormick et al. 2000; Senay et al. 2000). Moreover, physical association has been observed between XylT and GalT (Schwartz 1975) and between GlcA C5-epimerase and IdoA 2-O-sulfotransferase (Pinhal et al. 2001). Recently, interaction between EXT2 and NDST1 was also reported (Presto et al. 2008). In this study, it was also shown that the expression levels of EXT polymerases can affect the amount of NDST1 protein in cells, in turn influencing HS structure.

Based on our studies of liver HS structure in NDST1- and NDST2-deficient mouse embryos, we previously suggested that NDST1 is preferentially incorporated into the GAGosomes (Ledin et al., 2006). In a control liver, where similar amounts of NDST1 and NDST2 are expressed, the GAGosomes will contain NDST1 and HS will be produced (Fig. 4). In mast cells which express lots of NDST2 compared to NDST1 (Kusche-Gullberg et al. 1998), NDST2 will be the dominating NDST isoform incorporated into the GAGosomes resulting in heparin production (Fig. 4).

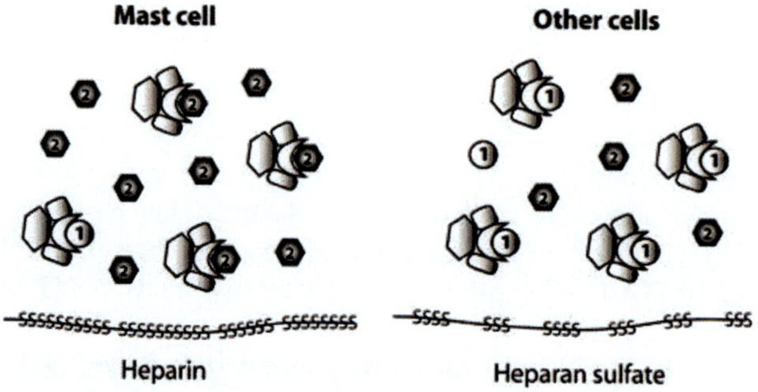


**Fig. 4** A tentative model to explain why mast cells make heparin while other cells synthesize heparan sulfate. In most cells, NDST1 and NDST2 are expressed at similar levels, but NDST1 is more readily incorporated into the GAGosome. The NDST1 containing enzyme complexes synthesize heparan sulfate. In heparin-producing mast cells, NDST2 expression is massive, whereas NDST1 transcript is barely detected. Despite its lower affinity for the GAGosome, NDST2 can now be incorporated into GAGosomes, resulting in heparin production

## 4.4 Control of Substrate Levels

Access to the different substrates needed for HS/heparin biosynthesis, i.e. UDP-sugars (UDP-GlcNAc and UDP-GlcA) and the sulfate donor PAPS is also an obvious point of regulation.

### 4.4.1 UDP-Sugars

Formation of glycosidic linkages between monosaccharides is an energetically unfavorable process, which requires coupling of the monosaccharides to high-energy nucleotides through reaction with UTP. The resulting UDP-sugars can then be used in glycoconjugate synthesis such as GAG formation. While synthesis of UDP-sugars takes place in the cytosol, GAG synthesis occurs in the lumen of ER and Golgi compartments. The activated sugar donors must thus be translocated into these organelles. This is mediated through the action of membrane energy-independent nucleotide sugar antiporters, which shuffle nucleotide sugars into the organelles while simultaneously transporting nucleotide monophosphates back to the cytosol (Berninsone and Hirschberg 2000; Caffaro and Hirschberg 2006). Obviously, altering the concentration of available UDP-sugars by regulating either synthesis or transport of the UDP-sugars may influence HS/heparin biosynthesis.

### 4.4.2 PAPS: The Sulfate Donor

3′-phosphoadenosine 5′-phosphosulphate (PAPS) is the universal sulfate donor for all biochemical sulfotransferase reactions. For sulfation reactions taking place in the Golgi compartment, as in heparin/HS biosynthesis, inorganic sulfate must be transported into the cell, transformed into its activated form, PAPS, and be translocated into the Golgi where it is used as a substrate. Cellular uptake of sulfate is performed by a number of transmembrane antiporter and symporter molecules (ul Haque et al. 1998). A spectrum of recessively inherited disorders affecting bone and cartilage development have been related to mutations of genes encoding such sulfate transporters (Superti-Furga et al. 1996).

In the cytosol, PAPS is synthesized from inorganic sulfate and ATP in a two-step reaction by PAPS synthase (PAPSS), a bifunctional enzyme containing both an ATP sulfurylase domain and an APS kinase domain needed for the reaction. Two isoforms exist in vertebrates, PAPSS1 and PAPSS2 (Fuda et al. 2002; Strott 2002; Venkatachalam 2003). The expression of the two isoforms differs, with PAPSS1 being found ubiquitously, while PAPSS2 has a more restricted expression pattern (Fuda et al. 2002).

The PAPS synthases are localized to the cytosol and, unexpectedly, to the nucleus. After synthesis, PAPS utilized in HS and heparin sulfation is transported into the Golgi compartment by PAPS transporters, of which two have been identified in humans (Kamiyama et al. 2003, 2006). PAPS transporters probably act by an antiport mechanism, but the antiporter molecule has so far not been identified. PAP or 5′-AMP are possible candidates (Frederick et al. 2008).

Recent results in our laboratory indicate that the PAPS concentration may be a critical factor for regulation of NS-domain length (Carlsson et al. 2008). In the absence of PAPS, NDST catalyzes limited and seemingly random N-deacetylation of GlcNAc residues. In the presence of PAPS, the NDST enzymes work in a processive manner adding sulfate groups to contiguous disaccharides creating NS-domains, the length of which depends on the concentration of PAPS.

## 5 Past, Present, and Future

Studies of heparin biosynthesis have been ongoing since 1960s, when Jeremiah Silbert started to study how mast cell granule fraction incorporated radioactively labeled UDP-sugars into a polysaccharide that was degradable with heparinase (see refs. 9 and 10 in Silbert 2009). Many more important articles on heparin biosynthesis have come from this lab (see Silbert 2009). In the seventies, Ulf Lindahl began his investigations of heparin biosynthesis using microsomal fractions prepared from a transplantable mouse mastocytoma (see Lindahl 2000). In this system, the order of the modification reactions was worked out, and it was demonstrated that the substrate specificities of the modification enzymes to a large extent regulated the final structure of the polysaccharide. The enzymes taking part in the biosynthesis reactions of both heparin and HS biosynthesis are now all known and have been cloned through the efforts of several labs, including that of Ulf Lindahl (Lindahl and Li 2009). Much work still remains, e.g. to understand how the enzymes are assembled in the Golgi compartment, the nature of the potential GAGosome, and how enzyme expression is regulated. With this knowledge, we may in the future be able to influence both heparin and HS biosyntheses in vivo. In addition, it may be possible to construct biosynthetic machineries consisting of selected recombinant enzymes able to synthesize heparin or HS oligosaccharides of desired structure.

## References

Abrink M, Grujic M, Pejler G (2004) Serglycin is essential for maturation of mast cell secretory granule. J Biol Chem 279:40897–40905

Aikawa J, Esko JD (1999) Molecular cloning and expression of a third member of the heparan sulfate/heparin GlcNAc N-deacetylase/N-sulfotransferase family. J Biol Chem 274:2690–2695

Aikawa J, Grobe K, Tsujimoto M, Esko JD (2001) Multiple isozymes of heparan sulfate/heparin GlcNAc N-deacetylase/GlcN N-sulfotransferase. Structure and activity of the fourth member, NDST4. J Biol Chem 276:5876–5882

Allen BL, Rapraeger AC (2003) Spatial and temporal expression of heparan sulfate in mouse development regulates FGF and FGF receptor assembly. J Cell Biol 163:637–648

Bai X, Zhou D, Brown JR, Crawford BE, Hennet T, Esko JD (2001) Biosynthesis of the linkage region of glycosaminoglycans: cloning and activity of galactosyltransferase II, the sixth member of the beta 1,3-galactosyltransferase family (beta 3GalT6). J Biol Chem 276:48189–48195

Bengtsson J, Eriksson I, Kjellen L (2003) Distinct effects on heparan sulfate structure by different active site mutations in NDST-1. Biochemistry 42:2110–2115

Berninsone P, Hirschberg CB (1998) Heparan sulfate/heparin N-deacetylase/N-sulfotransferase. The N-sulfotransferase activity domain is at the carboxyl half of the holoenzyme. J Biol Chem 273:25556–25559

Berninsone PM, Hirschberg CB (2000) Nucleotide sugar transporters of the Golgi apparatus. Curr Opin Struct Biol 10:542–547

Bornemann DJ, Park S, Phin S, Warrior R (2008) A translational block to HSPG synthesis permits BMP signaling in the early Drosophila embryo. Development 135:1039–1047

Brickman YG, Ford MD, Gallagher JT, Nurcombe V, Bartlett PF, Turnbull JE (1998) Structural modification of fibroblast growth factor-binding heparan sulfate at a determinative stage of neural development. J Biol Chem 273:4350–4359

Busse M, Kusche-Gullberg M (2003) In vitro polymerization of heparan sulfate backbone by the EXT proteins. J Biol Chem 278:41333–41337

Caffaro CE, Hirschberg CB (2006) Nucleotide sugar transporters of the Golgi apparatus: from basic science to diseases. Acc Chem Res 39:805–812

Carlsson P, Presto J, Spillmann D, Lindahl U, Kjellen L (2008) Heparin/heparan sulfate biosynthesis: processive formation of N-sulfated domains. J Biol Chem 283(29):20008–20014

Cuellar K, Chuong H, Hubbell SM, Hinsdale ME (2007) Biosynthesis of chondroitin and heparan sulfate in chinese hamster ovary cells depends on xylosyltransferase II. J Biol Chem 282:5195–5200

David G, Bai XM, Van der Schueren B, Cassiman JJ, Van den Berghe H (1992) Developmental changes in heparan sulfate expression: in situ detection with mAbs. J Cell Biol 119:961–975

Duncan MB, Liu M, Fox C, Liu J (2006) Characterization of the N-deacetylase domain from the heparan sulfate N-deacetylase/N-sulfotransferase 2. Biochem Biophys Res Commun 339:1232–1237

Eriksson I, Sandback D, Ek B, Lindahl U, Kjellen L (1994) cDNA cloning and sequencing of mouse mastocytoma glucosaminyl N-deacetylase/N-sulfotransferase, an enzyme involved in the biosynthesis of heparin. J Biol Chem 269:10438–10443

Esko JD, Lindahl U (2001) Molecular diversity of heparan sulfate. J Clin Invest 108:169–173

Esko JD, Selleck SB (2002) Order out of chaos: assembly of ligand binding sites in heparan sulfate. Annu Rev Biochem 71:435–471

Esko JD, Zhang L (1996) Influence of core protein sequence on glycosaminoglycan assembly. Curr Opin Struct Biol 6:663–670

Forsberg E, Pejler G, Ringvall M, Lunderius C, Tomasini-Johansson B, Kusche-Gullberg M, Eriksson I, Ledin J, Hellman L, Kjellen L (1999) Abnormal mast cells in mice deficient in a heparin-synthesizing enzyme. Nature 400:773–776

Frederick JP, Tafari AT, Wu SM, Megosh LC, Chiou ST, Irving RP, York JD (2008) A role for a lithium-inhibited Golgi nucleotidase in skeletal development and sulfation. Proc Natl Acad Sci USA 105:11605–11612

Fuda H, Shimizu C, Lee YC, Akita H, Strott CA (2002) Characterization and expression of human bifunctional 3'-phosphoadenosine 5'-phosphosulphate synthase isoforms. Biochem J 365:497–504

Gallagher JT (2001) Heparan sulfate: growth control with a restricted sequence menu. J Clin Invest 108:357–361

Gallagher JT, Walker A (1985) Molecular distinctions between heparan sulphate and heparin. Analysis of sulphation patterns indicates that heparan sulphate and heparin are separate families of N-sulphated polysaccharides. Biochem J 230:665–674

Gotting C, Kuhn J, Kleesiek K (2007) Human xylosyltransferases in health and disease. Cell Mol Life Sci 64:1498–1517

Gotting C, Kuhn J, Zahn R, Brinkmann T, Kleesiek K (2000) Molecular cloning and expression of human UDP-d-Xylose: proteoglycan core protein beta-d-xylosyltransferase and its first isoform XT-II. J Mol Biol 304:517–528

Grobe K, Esko JD (2002) Regulated translation of heparan sulfate N-acetylglucosamine N-deacetylase/n-sulfotransferase isozymes by structured 5'-untranslated regions and internal ribosome entry sites. J Biol Chem 277:30699–30706

Grobe K, Ledin J, Ringvall M, Holmborn K, Forsberg E, Esko JD, Kjellen L (2002) Heparan sulfate and development: differential roles of the N-acetylglucosamine N-deacetylase/N-sulfotransferase isozymes. Biochim Biophys Acta 1573:209–215

Gulberti S, Lattard V, Fondeur M, Jacquinet JC, Mulliert G, Netter P, Magdalou J, Ouzzine M, Fournel-Gigleux S (2005) Phosphorylation and sulfation of oligosaccharide substrates critically influence the activity of human beta1,4-galactosyltransferase 7 (GalT-I) and beta1,3-glucuronosyltransferase I (GlcAT-I) involved in the biosynthesis of the glycosaminoglycan-protein linkage region of proteoglycans. J Biol Chem 280:1417–1425

Habuchi H, Miyake G, Nogami K, Kuroiwa A, Matsuda Y, Kusche-Gullberg M, Habuchi O, Tanaka M, Kimata K (2003) Biosynthesis of heparan sulphate with diverse structures and functions: two alternatively spliced forms of human heparan sulphate 6-O-sulphotransferase-2 having different expression patterns and properties. Biochem J 371:131–142

Habuchi H, Nagai N, Sugaya N, Atsumi F, Stevens RL, Kimata K (2007) Mice deficient in heparan sulfate 6-O-sulfotransferase-1 exhibit defective heparan sulfate biosynthesis, abnormal placentation, and late embryonic lethality. J Biol Chem 282:15578–15588

Habuchi H, Tanaka M, Habuchi O, Yoshida K, Suzuki H, Ban K, Kimata K (2000) The occurrence of three isoforms of heparan sulfate 6-O-sulfotransferase having different specificities for hexuronic acid adjacent to the targeted N-sulfoglucosamine. J Biol Chem 275:2859–2868

Hashimoto Y, Orellana A, Gil G, Hirschberg CB (1992) Molecular cloning and expression of rat liver N-heparan sulfate sulfotransferase. J Biol Chem 267:15744–15750

Holmborn K, Ledin J, Smeds E, Eriksson I, Kusche-Gullberg M, Kjellen L (2004) Heparan sulfate synthesized by mouse embryonic stem cells deficient in NDST1 and NDST2 is 6-O-sulfated but contains no N-sulfate groups. J Biol Chem 279:42355–42358

Hook M, Lindahl U, Hallen A, Backstrom G (1975) Biosynthesis of heparin. Studies on the microsomal sulfation process. J Biol Chem 250:6065–6071

Humphries DE, Wong GW, Friend DS, Gurish MF, Qiu WT, Huang C, Sharpe AH, Stevens RL (1999) Heparin is essential for the storage of specific granule proteases in mast cells. Nature 400:769–772

Jacobsson I, Lindahl U, Jensen JW, Roden L, Prihar H, Feingold DS (1984) Biosynthesis of heparin. Substrate specificity of heparosan N-sulfate D-glucuronosyl 5-epimerase. J Biol Chem 259:1056–1063

Jemth P, Smeds E, Do AT, Habuchi H, Kimata K, Lindahl U, Kusche-Gullberg M (2003) Oligosaccharide library-based assessment of heparan sulfate 6-O-sulfotransferase substrate specificity. J Biol Chem 278:24371–24376

Jenniskens GJ, Hafmans T, Veerkamp JH, van Kuppevelt TH (2002) Spatiotemporal distribution of heparan sulfate epitopes during myogenesis and synaptogenesis: a study in developing mouse intercostal muscle. Dev Dyn 225:70–79

Jenniskens GJ, Oosterhof A, Brandwijk R, Veerkamp JH, van Kuppevelt TH (2000) Heparan sulfate heterogeneity in skeletal muscle basal lamina: demonstration by phage display-derived antibodies. J Neurosci 20:4099–4111

Kakuta Y, Sueyoshi T, Negishi M, Pedersen LC (1999) Crystal structure of the sulfotransferase domain of human heparan sulfate N-deacetylase/ N-sulfotransferase 1. J Biol Chem 274:10673–10676

Kamiyama S, Sasaki N, Goda E, Ui-Tei K, Saigo K, Narimatsu H, Jigami Y, Kannagi R, Irimura T, Nishihara S (2006) Molecular cloning and characterization of a novel 3'-phosphoadenosine 5'-phosphosulfate transporter, PAPST2. J Biol Chem 281:10945–10953

Kamiyama S, Suda T, Ueda R, Suzuki M, Okubo R, Kikuchi N, Chiba Y, Goto S, Toyoda H, Saigo K, Watanabe M, Narimatsu H, Jigami Y, Nishihara S (2003) Molecular cloning and identification of 3'-phosphoadenosine 5'-phosphosulfate transporter. J Biol Chem 278:25958–25963

Kim BT, Kitagawa H, Tamura J, Saito T, Kusche-Gullberg M, Lindahl U, Sugahara K (2001) Human tumor suppressor EXT gene family members EXTL1 and EXTL3 encode alpha 1,4- N-acetylglucosaminyltransferases that likely are involved in heparan sulfate/ heparin biosynthesis. Proc Natl Acad Sci USA 98:7176–7181

Kim BT, Kitagawa H, Tanaka J, Tamura J, Sugahara K (2003) In vitro heparan sulfate polymerization: crucial roles of core protein moieties of primer substrates in addition to the EXT1-EXT2 interaction. J Biol Chem 278:41618–41623

Kitagawa H, Shimakawa H, Sugahara K (1999) The tumor suppressor EXT-like gene EXTL2 encodes an alpha1, 4-N-acetylhexosaminyltransferase that transfers N-acetylgalactosamine and N-acetylglucosamine to the common glycosaminoglycan-protein linkage region. The key enzyme for the chain initiation of heparan sulfate. J Biol Chem 274:13933–13937

Kobayashi S, Morimoto K, Shimizu T, Takahashi M, Kurosawa H, Shirasawa T (2000) Association of EXT1 and EXT2, hereditary multiple exostoses gene products, in Golgi apparatus. Biochem Biophys Res Commun 268:860–867

Kolset SO, Tveit H (2008) Serglycin – structure and biology. Cell Mol Life Sci 65:1073–1085

Kusche-Gullberg M, Eriksson I, Pikas DS, Kjellen L (1998) Identification and expression in mouse of two heparan sulfate glucosaminyl N-deacetylase/N-sulfotransferase genes. J Biol Chem 273:11902–11907

Lamanna WC, Kalus I, Padva M, Baldwin RJ, Merry CL, Dierks T (2007) The heparanome – the enigma of encoding and decoding heparan sulfate sulfation. J Biotechnol 129:290–307

Ledin J, Ringvall M, Thuveson M, Eriksson I, Wilen M, Kusche-Gullberg M, Forsberg E, Kjellen L (2006) Enzymatically active N-deacetylase/N-sulfotransferase-2 is present in liver but does not contribute to heparan sulfate N-sulfation. J Biol Chem 281(47):35727–35734

Ledin J, Staatz W, Li JP, Gotte M, Selleck S, Kjellen L, Spillmann D (2004) Heparan sulfate structure in mice with genetically modified heparan sulfate production. J Biol Chem 279 (41):42732–42741

Lidholt K, Kjellen L, Lindahl U (1989) Biosynthesis of heparin. Relationship between the polymerization and sulphation processes. Biochem J 261:999–1007

Lind T, Tufaro F, McCormick C, Lindahl U, Lidholt K (1998) The putative tumor suppressors EXT1 and EXT2 are glycosyltransferases required for the biosynthesis of heparan sulfate. J Biol Chem 273:26265–26268

Lindahl B, Eriksson L, Lindahl U (1995) Structure of heparan sulphate from human brain, with special regard to Alzheimer's disease. Biochem J 306(Pt 1):177–184

Lindahl U (2000) 'Heparin' – from anticoagulant drug into the new biology. Glycoconj J 17:597–605

Lindahl U, Li JP (2009) Interactions between heparan sulfate and proteins-design and functional implications. Int Rev Cell Mol Biol 276:105–159

Lyon M, Deakin JA, Gallagher JT (1994) Liver heparan sulfate structure. A novel molecular design. J Biol Chem 269:11208–11215

Maccarana M, Sakura Y, Tawada A, Yoshida K, Lindahl U (1996) Domain structure of heparan sulfates from bovine organs. J Biol Chem 271:17804–17810

McCormick C, Duncan G, Goutsos KT, Tufaro F (2000) The putative tumor suppressors EXT1 and EXT2 form a stable complex that accumulates in the Golgi apparatus and catalyzes the synthesis of heparan sulfate. Proc Natl Acad Sci USA 97:668–673

McCormick C, Duncan G, Tufaro F (1999) New perspectives on the molecular basis of hereditary bone tumours. Mol Med Today 5:481–486

Morii E, Ogihara H, Oboki K, Sawa C, Sakuma T, Nomura S, Esko JD, Handa H, Kitamura Y (2001) Inhibitory effect of the mi transcription factor encoded by the mutant mi allele on GA binding protein-mediated transcript expression in mouse mast cells. Blood 97:3032–3039

Nagai N, Habuchi H, Kitazume S, Toyoda H, Hashimoto Y, Kimata K (2007) Regulation of heparan sulfate 6-O-sulfation by beta-secretase activity. J Biol Chem 282:14942–14951

Orellana A, Hirschberg CB, Wei Z, Swiedler SJ, Ishihara M (1994) Molecular cloning and expression of a glycosaminoglycan N-acetylglucosaminyl N-deacetylase/N-sulfotransferase from a heparin-producing cell line. J Biol Chem 269:2270–2276

Pallerla SR, Lawrence R, Lewejohann L, Pan Y, Fischer T, Schlomann U, Zhang X, Esko JD, Grobe K (2008) Altered heparan sulfate structure in mice with deleted NDST3 gene function. J Biol Chem 283(24):16885–16894

Petitou M, Casu B, Lindahl U (2003) 1976–1983, a critical period in the history of heparin: the discovery of the antithrombin binding site. Biochimie 85:83–89

Pinhal MA, Smith B, Olson S, Aikawa J, Kimata K, Esko JD (2001) Enzyme interactions in heparan sulfate biosynthesis: uronosyl 5-epimerase and 2-O-sulfotransferase interact in vivo. Proc Natl Acad Sci USA 98:12984–12989

Presto J, Thuveson M, Carlsson P, Busse M, Wilen M, Eriksson I, Kusche-Gullberg M, Kjellen L (2008) Heparan sulfate biosynthesis enzymes EXT1 and EXT2 affect NDST1 expression and heparan sulfate sulfation. Proc Natl Acad Sci USA 105:4751–4756

Robinson HC, Horner AA, Hook M, Ogren S, Lindahl U (1978) A proteoglycan form of heparin and its degradation to single-chain molecules. J Biol Chem 253:6687–6693

Rong J, Habuchi H, Kimata K, Lindahl U, Kusche-Gullberg M (2001) Substrate specificity of the heparan sulfate hexuronic acid 2-O-sulfotransferase. Biochemistry 40:5548–5555

Schon S, Prante C, Bahr C, Kuhn J, Kleesiek K, Gotting C (2006) Cloning and recombinant expression of active full-length xylosyltransferase I (XT-I) and characterization of subcellular localization of XT-I and XT-II. J Biol Chem 281:14224–14231

Schwartz NB (1975) Biosynthesis of chondroitin sulfate: immunoprecipitation of interacting xylosyltransferase and galactosyltransferase. FEBS Lett 49:342–345

Senay C, Lind T, Muguruma K, Tone Y, Kitagawa H, Sugahara K, Lidholt K, Lindahl U, Kusche-Gullberg M (2000) The EXT1/EXT2 tumor suppressors: catalytic activities and role in heparan sulfate biosynthesis. EMBO Rep 1:282–286

Shukla D, Liu J, Blaiklock P, Shworak NW, Bai X, Esko JD, Cohen GH, Eisenberg RJ, Rosenberg RD, Spear PG (1999) A novel role for 3-O-sulfated heparan sulfate in herpes simplex virus 1 entry. Cell 99:13–22

Shworak NW, HajMohammadi S, De Agostini AI, Rosenberg RD (2002) Mice deficient in heparan sulfate 3-O-sulfotransferase-1: normal hemostasis with unexpected perinatal phenotypes. Glycoconj J 19:355–361

Shworak NW, Liu J, Fritze LM, Schwartz JJ, Zhang L, Logeart D, Rosenberg RD (1997) Molecular cloning and expression of mouse and human cDNAs encoding heparan sulfate D-glucosaminyl 3-O-sulfotransferase. J Biol Chem 272:28008–28019

Silbert JE (2009) Glycosaminoglycan metabolism before molecular biology: reminiscences of our early work. Glycocon J 27(2):201–209

Smeds E, Habuchi H, Do AT, Hjertson E, Grundberg H, Kimata K, Lindahl U, Kusche-Gullberg M (2003) Substrate specificities of mouse heparan sulphate glucosaminyl 6-O-sulphotransferases. Biochem J 372:371–380

Strott CA (2002) Sulfonation and molecular action. Endocr Rev 23:703–732

Superti-Furga A, Hastbacka J, Rossi A, van der Harten JJ, Wilcox WR, Cohn DH, Rimoin DL, Steinmann B, Lander ES, Gitzelmann R (1996) A family of chondrodysplasias caused by mutations in the diastrophic dysplasia sulfate transporter gene and associated with impaired sulfation of proteoglycans. Ann N Y Acad Sci 785:195–201

ten Dam GB, Hafmans T, Veerkamp JH, van Kuppevelt TH (2003) Differential expression of heparan sulfate domains in rat spleen. J Histochem Cytochem 51:727–739

Tone Y, Pedersen LC, Yamamoto T, Izumikawa T, Kitagawa H, Nishihara J, Tamura J, Negishi M, Sugahara K (2008) 2-o-phosphorylation of xylose and 6-o-sulfation of galactose in the protein linkage region of glycosaminoglycans influence the glucuronyltransferase-I activity involved in the linkage region synthesis. J Biol Chem 283:16801–16807

Ueno M, Yamada S, Zako M, Bernfield M, Sugahara K (2001) Structural characterization of heparan sulfate and chondroitin sulfate of syndecan-1 purified from normal murine mammary

gland epithelial cells. Common phosphorylation of xylose and differential sulfation of galactose in the protein linkage region tetrasaccharide sequence. J Biol Chem 276:29134–29140
ul Haque MF, King LM, Krakow D, Cantor RM, Rusiniak ME, Swank RT, Superti-Furga A, Haque S, Abbas H, Ahmad W, Ahmad M, Cohn DH (1998) Mutations in orthologous genes in human spondyloepimetaphyseal dysplasia and the brachymorphic mouse. Nat Genet 20:157–162
van Kuppevelt TH, Dennissen MA, van Venrooij WJ, Hoet RM, Veerkamp JH (1998) Generation and application of type-specific anti-heparan sulfate antibodies using phage display technology. Further evidence for heparan sulfate heterogeneity in the kidney. J Biol Chem 273:12960–12966
Venkatachalam KV (2003) Human 3'-phosphoadenosine 5'-phosphosulfate (PAPS) synthase: biochemistry, molecular biology and genetic deficiency. IUBMB Life 55:1–11
Voglmeir J, Voglauer R, Wilson IB (2007) XT-II, the second isoform of human peptide-O-xylosyltransferase, displays enzymatic activity. J Biol Chem 282:5984–5990
Warda M, Toida T, Zhang F, Sun P, Munoz E, Xie J, Linhardt RJ (2006) Isolation and characterization of heparan sulfate from various murine tissues. Glycoconj J 23:555–563
Wei G, Bai X, Gabb MM, Bame KJ, Koshy TI, Spear PG, Esko JD (2000) Location of the glucuronosyltransferase domain in the heparan sulfate copolymerase EXT1 by analysis of Chinese hamster ovary cell mutants. J Biol Chem 275:27733–27740
Westling C, Lindahl U (2002) Location of N-unsubstituted glucosamine residues in heparan sulfate. J Biol Chem 277:49247–49255
Yusa A, Kitajima K, Habuchi O (2005) N-linked oligosaccharides are required to produce and stabilize the active form of chondroitin 4-sulphotransferase-1. Biochem J 388:115–121
Zak BM, Crawford BE, Esko JD (2002) Hereditary multiple exostoses and heparan sulfate polymerization. Biochim Biophys Acta 1573:346–355

# The Anticoagulant and Antithrombotic Mechanisms of Heparin

**Elaine Gray, John Hogwood, and Barbara Mulloy**

## Contents



**Abstract** The molecular basis for the anticoagulant action of heparin lies in its ability to bind to and enhance the inhibitory activity of the plasma protein antithrombin against several serine proteases of the coagulation system, most importantly factors IIa (thrombin), Xa and IXa. Two major mechanisms underlie heparin's potentiation of antithrombin. The conformational changes induced by heparin binding cause both expulsion of the reactive loop and exposure of exosites of the surface of antithrombin, which bind directly to the enzyme target; and a template mechanism exists in which both inhibitor and enzyme bind to the same heparin molecule. The relative importance of these two modes of action varies between enzymes. In addition, heparin can act through other serine protease inhibitors such as heparin co-factor II, protein C inhibitor and tissue factor plasminogen inhibitor. The antithrombotic action of heparin in vivo, though dominated by anticoagulant mechanisms, is more complex, and interactions with other plasma proteins and cells play significant roles in the living vasculature.

---

E. Gray (✉) • J. Hogwood (✉) • B. Mulloy (✉)
National Institute for Biological Standards and Control, Blanche Lane, South Mimms, Potter's Bar, Hertfordshire EN6 3QG, UK
e-mail: Elaine.Gray@nibsc.hpa.org.uk; John.Hogwood@nibsc.hpa.org.uk; Barbara.Mulloy@nibsc.hpa.org.uk

R. Lever et al. (eds.), *Heparin - A Century of Progress*,
Handbook of Experimental Pharmacology 207,
DOI 10.1007/978-3-642-23056-1_3, 

**Keywords** Antithrombin • Factor Xa • Heparin cofactor II • Protein C inhibitor • Thrombin • Tissue factor plasminogen activator

## Abbreviations

| | |
|---|---|
| APC | Activated protein C |
| AT | Antithrombin |
| DS | Dermatan sulphate |
| EPCR | Endothelial protein C receptor |
| fIIa | Factor IIa (thrombin) |
| fVIIIa | Factor VIIIa |
| fIXa | Factor IXa |
| fVa | Factor Va |
| fXa | Factor Xa |
| fXIa | Factor XIa |
| HCII | Heparin cofactor II |
| HRG | Histidine-rich glycoprotein |
| HS | Heparan sulphate |
| MW | Molecular weight |
| PCI | Protein C Inhibitor |
| PF4 | Platelet factor 4 |
| RCL | Reactive centre loop |
| TFPI | Tissue Factor Pathway Inhibitor |

## 1 Introduction

In vivo, free flowing of blood depends on the balance of pro-coagulant and anticoagulant processes. A complex network of serine proteases, acting in an amplification cascade, converts pro-enzymes to their active form (Fig. 1). Factor IIa (fIIa, commonly known as thrombin) is the final serine protease that cleaves fibrinogen to form fibrin which, together with a platelet plug, is the basis of a clot. This coagulation cascade is activated when there is an injury to the vasculature, so that the serine proteases are exposed to pro-coagulant stimuli such as tissue factor and collagen. Several natural or endogenous anticoagulant proteins, which include antithrombin (AT), heparin co-factor II (HCII), protein C inhibitor (PCI) and tissue factor pathway inhibitor (TFPI), are also in place to regulate the formation of thrombin. These inhibitors are found at a higher total concentration than the proteases and under normal physiological condition act to keep the clot local to the wound by mopping up any proteases straying into the rest of the vasculature.

Antithrombin, HCII and PCI are members of the structural class of proteins known as "serpins" (short for serine protease inhibitors). There are two common features in the inhibitory mechanism of serpins; first, there is usually a requirement for conformational change in both the protease and the protease inhibitor and

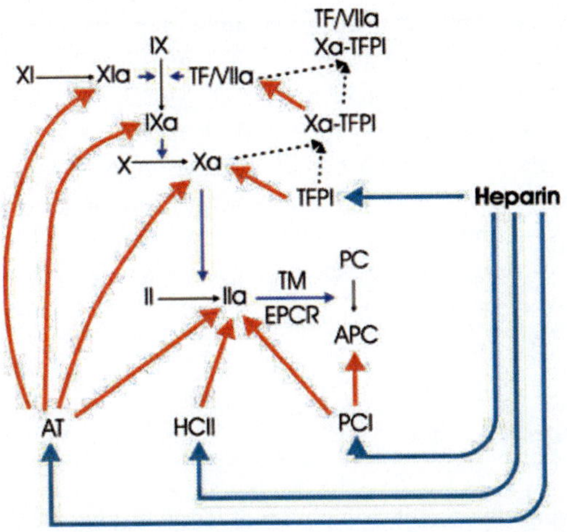


**Fig. 1** A simplified diagrammatic representation of the interactions of heparin with the natural anticoagulant systems. *Blue arrows* represent the potentiating interactions of heparin with coagulation inhibitors; *red arrows* represent the inhibition of coagulation enzymes; *thin black arrows* represent the conversion of pro-enzymes to their active forms, and *purple arrows* represent the effects of the enzymes. *Dotted arrows* indicate the formation of complexes. Abbreviations not defined in the text are: *TM* thrombomodulin, *PC* protein C, *APC* activated protein C, *TF* tissue factor

second the reaction is often greatly enhanced by the presence of glycosaminoglycans such as heparin, heparan sulphate (HS) and dermatan sulphate (DS) (Huntington 2003). Unlike direct enzyme inhibitors, heparin acts in a catalytic fashion to activate the inhibitor and to stabilise the inhibitor–enzyme complex; after the permanent, stoichiometric inactivation of the enzyme by the serpin, heparin is released and can act again, as long as free serpin is available.

## 2 Heparin and Antithrombin

Heparin has anticoagulant action in several ways, but by far the most significant of these is through its potentiating action on the serpin antithrombin. The serpins are a structurally related group of proteins with a reactive loop that mimics a serine protease substrate sequence (Silverman et al. 2010; Whisstock et al. 2010). When this loop is cleaved, the protease is trapped in a covalent, inactive complex (Huntington et al. 2000). Antithrombin is one of the best-studied examples of this group, and though it is active in other biological systems, antithrombin is best known as the major heparin co-factor in the inhibition of the coagulation proteases, particularly factors Xa and IIa (Fig. 1). The molecular mechanisms by which heparin activates antithrombin have recently been reviewed (Olson et al. 2010) and will be briefly summarised here.

It has long been understood that antithrombin binds with high affinity to a specific, unusual pentasaccharide sequence in heparin; the structure of this sequence, illustrated in fig. 1c of Mulloy (2011), is αDGlcNAc(6S)-βDGlcA-αDGlcNS(3S, 6S)-α LIdoA-αDGlcNS(6S) (Lindahl et al. 1984). Heparin lacking this sequence is capable of activating antithrombin in vitro, but very much higher concentrations are needed (Streusand et al. 1995). Over the past few decades, a series of studies in the kinetics and structural biology of this interaction and its consequences has provided a clear picture of the molecular mechanisms of heparin/antithrombin-mediated inhibition of coagulation.

## 2.1 Thrombin and Factor Xa

The heparin-binding site of antithrombin involves an arrangement of basic residues distant from each other in sequence, but which form an approximately linear basic patch on the protein surface over several helices and in the N-terminal region (Fig. 2a) (Jin et al. 1997). As the high-affinity pentasaccharide sequence in a heparin molecule binds to antithrombin, it induces local conformational changes which improve the fit between protein and ligand; in turn, further conformational changes propagate through the protein structure, eventually leading to expulsion of the reactive centre loop (RCL) (Huntington 2003). This loop contains the apparent protease substrate sequence, and its expulsion increases its exposure to the protease. On interaction with the protease (thrombin, for example) active site, the loop is cleaved, and thrombin is trapped by a covalent linkage. At this point, the RCL is incorporated as an extra strand into a beta sheet, pulling with it the thrombin, to the opposite end of the antithrombin molecule, both inactivating the thrombin and disrupting its structure (Huntington et al. 2000).

The same mechanism holds true for factor Xa (fXa), by and large, but there are substantial differences between the way thrombin and fXa engage with heparin-activated antithrombin, as was initially indicated by detailed kinetic studies (Craig et al. 1989) . More recently, it has been found that fXa interacts with antithrombin not only through the RCL, but in addition at a specific exosite. This direct protein–protein interaction is the basis for the specificity of antithrombin for fXa (Gettins and Olson 2009). Thrombin does not bind to the same exosite; instead it binds, rather non-specifically, to the same heparin molecule as is bound to antithrombin. This extra interaction means that a heparin molecule needs to be long enough to reach both proteins; in practice, this needs a chain of about 13 monosaccharide units attached at the non-reducing end of the AT-binding pentasaccharide sequence. This minimum motif in heparin for inhibition of thrombin through antithrombin has been called the C-domain (Al Dieri et al. 2003), and has a molecular weight (MW) of about 5,400. Although the inhibition of fXa by antithrombin and heparin is also somewhat MW dependent, and fXa also interacts with both heparin and antithrombin (Gettins and Olson 2009; Wagenvoord et al. 2008), this is not an essential feature of the process and heparin molecules of MW below 5,400 can still inhibit fXa. For this

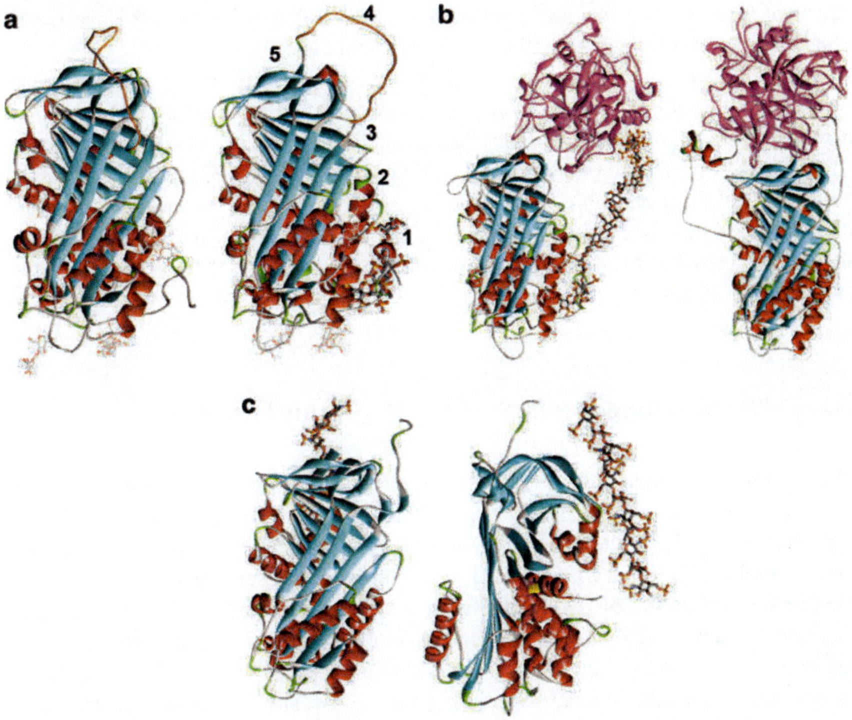


**Fig. 2** Serpin structures, shown as ribbon diagrams coloured according to secondary structure (beta strands in *blue*, helix *red*, turns *green*). Bound heparin molecules are depicted in stick representation coloured by element (carbon *grey*, oxygen *red*, sulphur *yellow*). (**a**) Structures of antithrombin in its native form (left: from the crystal structure 1TIF.pdb) and its heparin-activated form (*right*: from the high-resolution crystal structure in complex with fIXa, 3KCG.pdb). The bound heparin pentasaccharide (1) causes helix D to lengthen (2), expelling the minor central strand segment (3) allowing the full extension of the reactive centre loop, coloured orange (4); the exosite that binds to fXa is exposed (5). (**b**) Crystal structures of, on the left, the ternary complex between antithrombin, a heparin mimetic, and thrombin (coloured *magenta*) and on the right the complex between heparin cofactor II and thrombin. Note that the orientation of thrombin with respect to the serpin is different in the two cases. In addition to the interaction between the serpin reactive centre loop and thrombin active site, thrombin exosite interactions are seen; in the case of antithrombin, the essential extended heparin-like molecule interacts with exosite II of thrombin, and in the case of HCII the long, hirudin-like N-terminal tail interacts with exosite I of thrombin. (**c**) A molecular model of the complex between protein C inhibitor (PCI) and heparin, based on the crystal structure of PCI (2HI9.pdb, chain A) with a heparin oligosaccharide model (based on the NMR structure 1HPN.pdb) in the bound position predicted by docking calculations (Mulloy and Forster 2008). Two orientations are shown; on the left a view for comparison with the antithrombin-heparin and HCII-heparin complexes shown in Fig. 2a, b, and on the right rotated through about 90° from that orientation. Heparin binds to the H-helix, rather than the D-helix as for antithrombin, and is positioned much closer to the RCL, well placed to interact also with protein C in a ternary complex, as has been proposed on the basis of the cleaved PCI/heparin crystal structure (Li and Huntington 2008)

reason, low-molecular-weight heparins have ratios of anti-Xa to anti-IIa activity greater than 1 (for unfractionated heparin, this value is 1 – see Gray 2011).

Antithrombin is a glycoprotein, with several glycosylation sites. One of these lies within the heparin-binding site and interferes with heparin binding. It is naturally absent in a proportion of antithrombin; this material is called β-antithrombin (the fully glycosylated version is called α-antithrombin) (Swedenborg 1998). Not surprisingly, β-antithrombin has higher affinity for heparin than α-antithrombin, and is more readily activated by heparin. In plasma, only about 10% of antithrombin is in the β-form, but there is evidence that in other tissues the β-form predominates (Kamp et al. 2001).

## 2.2 *Other Clotting Factors Affected by Antithrombin*

The antithrombin-mediated anticoagulant effects of heparin are best described for factors Xa and thrombin, but a third factor, IXa, is an important contributor (Olson et al. 2004). In the high-resolution crystal structure of the pentasaccharide-antithrombin-fIXa ternary complex (Johnson et al. 2010), the similarities between the mode of inhibition of fIXa and fXa are evident; fIXa binds to the same exosite of antithrombin, though in a somewhat different orientation, and as for fXa the heparin-induced unfolding of the reactive loop exposes the exosite as well as presenting the pseudo-substrate to the enzyme's active site. In this crystal structure, the pentasaccharide is found not only in its site on antithrombin, but also at a site on fIXa previously identified as a heparin-binding site by mutagenesis studies (Yang et al. 2002). This site is located at the corresponding position on the protein surface to the heparin-binding sites of fXa and thrombin. This is in agreement with earlier descriptions of a calcium-dependent template mechanism for the enhancement of antithrombin's inhibition of fIXa by full-length heparin (Wiebe et al. 2003).

Factor XIa, a less significant contributor to the anticoagulant action of heparin/antithrombin (Olson et al. 2004), is structurally different from other coagulation proteases, but like factors IXa, Xa and thrombin it is formed by cleavage of a pro-enzyme and is subject to inhibition by heparin-bound antithrombin (Emsley et al. 2010). Heparin accelerates the inhibition of fXIa by both antithrombin and protein C inhibitor (Yang et al. 2009), and though heparin binding to the enzyme is involved, this seems to be more complex than is the case for other coagulation proteases and is not only a straightforward template mechanism. There are two distinct heparin-binding sites on fXIa, one of which is in the catalytic domain. Direct interaction of heparin with this domain can enhance inhibition by a mutant of antithrombin with no heparin-binding site; presumably by an allosteric or charge-related mechanism (Yang et al. 2010).

## 3 Heparin Cofactor II

In addition to antithrombin, there is second plasma cofactor that binds to heparin and can influence plasma coagulation, its presence first reported in 1974 (Briginshaw and Shanberge 1974). Heparin Cofactor II is a single chain glycoprotein that is a member of the serine protease inhibitor family, with structural similarities to antithrombin (Griffith et al. 1985). In coagulation heparin cofactor II selectively inhibits thrombin (Parker and Tollefsen 1985; Griffith 1983) by formation of a stoichiometric 1:1 covalent bimolecular complex in the absence or presence of heparin. The binding of heparin to heparin cofactor II is analogous to heparin and antithrombin binding (O'Keeffe et al. 2004), and both serpins have similarities in their reactive site peptide structure (Griffith et al. 1985). The crystal structures of HCII alone and in complex with thrombin bring out an interesting difference between antithrombin and HCII (Fig. 2b) (Baglin et al. 2002). Both serpins require an extra interaction with a thrombin exosite; for antithrombin, this takes place via the heparin molecule, which binds to thrombin's exosite II, but HCII has a long N-terminal tail, which interacts with the hirudin-binding thrombin exosite I.

The presence of heparin potentiates thrombin inhibition by HCII, though this is an order of magnitude less than the action of antithrombin and heparin on thrombin (Tollefsen and Blank 1981). At low concentrations of heparin, thrombin is preferentially inhibited by antithrombin; at higher therapeutic concentrations, it is also able to act through HCII (Tollefsen et al. 1982). Physiologically, both HCII and antithrombin are able to also bind to heparan sulphate; HCII is less discriminating than antithrombin, requiring no highly specific structural sequence, and is also able to bind to a range of other polyanions (Pratt et al. 1989). Within the vascular system, HCII primarily acts in an anticoagulant capacity with the glycosaminoglycan DS (McGuire and Tollefsen 1987). DS is selective for only HCII among the serpins, and is found on the surface of cell walls underlining the vascular endothelium (He et al. 2002; Tollefsen 2002). HCII deficiency has no effect on plasma coagulation but causes an increase in the formation of occlusive thrombi at the site of vascular damage (He et al. 2002; Tovar et al. 2005). It is thus viewed that whilst HCII can bind to heparin and inhibit coagulation, the role it plays when bound to DS in extravascular tissue is the one where its thrombin regulation function is most physiologically important (Vicente et al. 2004). Although HCII has no strong preference for specific sequences in heparin, it requires specific disulphated sequences in DS (Maimone and Tollefsen 1991); DS-like compounds of marine invertebrate origin have been used to explore the structural specificity of HCII for DS (Pavao et al. 1998). The heparin contaminant over-sulphated chondroitin sulphate exerts its modest anticoagulant activity through HCII rather than antithrombin (Li et al. 2009).

## 4 Protein C Inhibitor

Protein C is a vitamin K-dependent factor, produced by the liver (Stenflo 1976), that circulates as a zymogen until its activation by thrombin to an activated form (APC) (Comp et al. 1982). It is now well recognised that thrombin activation of protein C is accelerated by the binding of thrombin to thrombomodulin and the binding of protein C to endothelial protein C receptor (EPCR) (Esmon 2010). Thrombomodulin-bound thrombin is also less able to convert fibrinogen to fibrin and activate factor V. Activated protein C (APC) has been described to have two physiological roles. It acts as an anticoagulant by inactivating factors Va and VIIIa in the presence of a co-factor, protein S. When complexed with EPCR and PAR-1, APC elicits a cytoprotective effect through modulation of expression of genes related to anti-inflammatory and anti-apoptotic pathways (Mosnier et al. 2007). The activity of APC is mainly regulated by protein C inhibitor (PCI), an inhibitor from the serpin family.

Heparin influences the activity of the protein C pathway not only by potentiation of antithrombin inhibition of thrombin, but also through its interaction with PCI. It is interesting that the inhibition of thrombin would lead to a decrease in activation of protein C, thereby limiting the inactivation of fVa and fVIIIa. In addition, heparin has been shown to inhibit the inactivation of factor Va by APC (Nicolaes et al. 2004). Similarly, heparin binding to PCI accelerates the rate of inhibition of APC shifting the balance towards coagulation. However, PCI can also act as anticoagulant as it has the ability to inhibit thrombin, fXa and fXIa (Van Walderveen et al. 2010; Sun et al. 2009). The rate of inhibition is MW or size and concentration dependent. In contrast to the potentiation of antithrombin inhibition of thrombin where a threshold of 18 saccharides unit is critical for any significant inhibition to be observed (Hemker and Beguin 1992), the binding of APC to PCI has been shown to require only a minimum of 7 saccharides (Aznar et al. 1996) and the rate of inhibition of APC and factor Xa increases linearly with increasing saccharide length (Pratt et al. 1992). Pratt and Church (1992) indicated that different optimal concentrations of heparin are needed to induce maximal rate of inhibition of different proteases by PCI. The concentrations reported are relatively high, 10, 30 and 100 μg/ml equating to approximately 2, 6 and 10 IU/ml of heparin for thrombin, factor Xa and APC, respectively, and suggests both PCI and the serine protease need to bind simultaneously to the same heparin molecule. This fits well with the concept of ternary complex template model established for antithrombin and heparin (Fig. 2b). The physiological significance of the interaction between heparin and PCI is still unclear.

The crystal structures of both native (Li et al. 2007) and cleaved (Huntington et al. 2003) PCI have been elucidated. By comparison with the RCL of the other serpins, the RCL for PCI is relatively long and flexible. Unlike antithrombin and heparin co-factor II, where the heparin-binding site involves helix D, the PCI heparin-binding site is located at helix H, closer to the reactive loop (Li et al. 2008). The only crystal structure of the PCI-heparin complex to date involves the cleaved form of PCI (Li and Huntington 2008), but docking calculations by an established method (Mulloy and Forster 2008) allow the construction of a model illustrating the interaction for the active form of PCI; the orientation of heparin in the docked complex is similar to that of the crystal structure (Fig. 2c).

## 5 Tissue Factor Pathway Inhibitor

Tissue factor pathway inhibitor is a pleiotropic serine protease inhibitor. Apart from being the main physiological inhibitor of the extrinsic coagulation pathway, it also has important influence in lipid metabolism, innate immunity and angiogenesis (Holroyd and Simari 2010). Publications in 1947 gave the first descriptions of an endogenous inhibitor of tissue factor and the extrinsic pathway (Thomas 1947; Schneider 1947); however, TFPI was not isolated and cloned until 1987 (Broze and Miletich 1987). TFPI is a single chain polypeptide with 276 amino acids, forming an acidic N-terminal region, three kringle or Kunitz tandem domains and a highly basic carboxy-terminal end (Bajaj et al. 2001). It is now well established and documented that the Kunitz domains are important for the anticoagulant activity of TFPI. The first and second Kunitz domains bind and inhibit factor VIIa/tissue factor complex and fXa, respectively (Broze et al. 1988). NMR and crystal structures have been published for the second Kunitz domain (Burgering et al. 1997). Although no inhibitory activity has been ascribed to the third domain, the amino acid residues Gly-212 to Phe-243 have been identified as a binding site for heparin and binding of the third domain to heparin potentiates the inhibitory activity of the second domain (Wesselschmidt et al. 1993; Enjyoji et al. 1995). Other studies have now demonstrated that the C-terminus has higher affinity for heparin than the third Kunitz domain and is essential for the expression of anticoagulant activity (Petersen et al. 1993; Nordfang et al. 1991). Ye et al. subsequently determined that binding to heparin is facilitated by 12 amino acid residues that include Arg-257 and Arg-259 at the C-terminus region, and although they have not elucidated any specific heparin sequence, they have ascertained that 6-O, N- and 2-O sulphate groups are necessary for binding to TFPI (Ye et al. 1998). Binding and fXa inhibition studies of heparin oligosaccharides and recombinant TFPI indicated that a dodecasaccharide is the minimum sugar chain length that will enhance the inhibitory activity of TFPI and that saccharide units greater than an octadecasaccharide are needed to achieve full potentiation, as observed with unfractionated heparin (Xu et al. 2002).

TFPI is predominantly expressed by endothelial cells, with small quantities synthesised by other cell types such as monocytes, macrophages, lung fibroblasts and smooth muscle cells (Bajaj et al. 1990; Werling et al. 1993). Interestingly, TFPI is not produced by normal adult hepatocytes, erythrocytes, neutrophils and lymphocytes (Osterud et al. 1995). In vivo, there are three distinct pools of TFPI: ~80–85% endothelium bound probably via cell surface heparan sulphate proteoglycan involving glypican-3 and syndecan 4, ~10% complexed with circulating lipoproteins (TFPI is also known as lipoprotein-associated coagulation inhibitor – LACI) and ~3% is associated with platelets (Broze et al. 1994; Osterud et al. 1995; Mast et al. 1997; Kojima et al. 1996). Circulating lipoprotein associated TFPI is C-terminal truncated and is not as active an anticoagulant as TFPI from the endothelium (Broze et al. 1994). The endothelial pool can be released into circulation as full-length TFPI by intravenous injection of heparin and this can be observed as an apparent potentiation of the anticoagulant activity of the heparin administered (Sandset et al. 1988). Continuous infusion of unfractionated, but not low-molecular-weight heparin depletes cell associated TFPI (Hansen et al. 1996, 1998; Hansen and Sandset 1998), and this may help to partly explain rebound activation of the coagulation system following heparin treatment (Hansen et al. 2000; Naumnik et al. 2003). The structural requirement of heparin for release of TFPI is not yet clear. The differential effect of unfractionated and low-molecular-weight heparin suggests a molecular weight-dependent mechanism, and this is supported by a report from Ma et al. (2007) in which higher molecular weight heparin fractions induced higher concentration of releasable TFPI in a primate model. There is also a suggestion that both the total sulphate content and the localisation of the charge groups of heparin are also important for release of TFPI (Valentin et al. 1994).

Both unfractionated and low-molecular-weight heparin have been found to increase TFPI mRNA expression by endothelial cells in a time and dose-dependent manner, leading to de novo synthesis of functionally active TFPI (Thyzel et al. 2007; Lupu et al. 1999).

## 6 Effects of Heparin Structure and Heterogeneity on Anticoagulant Activity

The picture of heparin/antithrombin-mediated anticoagulation which has been presented so far is simplistic. It is a convenient approximation to imagine the antithrombin binding sequence as a single, simple pentasaccharide (the "A" domain) sometimes with a sulphated, non-specific thrombin binding extension in addition (the "C" domain) (Al Dieri et al. 2003). Things are in fact more complex than this. The antithrombin-binding sequence is not completely specific, and is longer than a pentasaccharide in native heparin; only a subset of heparin molecules contains this sequence in any case; and the length of the heparin chain affects more than simply its ability to bind thrombin. These factors are further complicated for the low-molecular-weight heparins by subtle structural changes introduced by depolymerization.

The heparin sequence associated with high affinity for antithrombin is longer than the minimal pentasaccharide; a high-affinity octasaccharide was defined some time ago (Lindahl et al. 1984) with a characteristic unsulphated iduronate at the non-reducing side of the pentasaccharide, signs of which are also evident in NMR spectra of high-affinity heparin fractions (Mulloy and Johnson 1987).

Very much more recently, an octasaccharide in which the unsulphated iduronate residue is replaced by a glucuronic acid has been found (Guerrini et al. 2008), and, surprisingly, this variant has about 10 times higher affinity for AT than its iduronate-containing equivalent.

Species differences in heparin are found in the AT-binding region; for example, the N-acetylated glucosamine residue in porcine mucosal heparin is largely replaced by N-sulphated glucosamine in bovine lung heparin (Loganathan et al. 1990).

The content of high-affinity sequence may vary between heparin samples; for unfractionated heparin, this can lead to variations in specific anticoagulant activity between different manufacturer's products, especially when measured by the anti-Xa assay. High-specific activity seems to be correlated with high molecular weight and high degree of sulphation (Mulloy et al. 2000), within a limited scope. For low-molecular-weight heparins, the method of depolymerization may affect the high-affinity sequence in different ways. The GlcNS3S6S-IdoA2S linkage is easily cleaved by heparinase I (Shriver et al. 2000; Xiao et al. 2011), leaving partial high-affinity sequences at the reducing ends of heparin molecules with much reduced affinity for AT and lower anti-Xa activity (Shriver et al. 2000). It has been found that isolated monodisperse oligosaccharides of tinzaparin, the LMWH made using this enzyme, have lower anti-Xa activity than oligosaccharides of the same length isolated from enoxaparin, produced by beta-elimination of the benzyl ester of heparin (Schroeder et al. 2011); this is also consistent with the findings of Bisio et al. (2009) that tinzaparin fractions have a lower content of the intact high-affinity pentasaccharide than equivalent MW fractions of enoxaparin or dalteparin. Modification of the chemical beta-elimination-based depolymerisation of heparin using a strong organic phosphazene base has been used to generate an ultra-LMWH with particularly high-specific activity (about 160 IU/mg) (Viskov et al. 2009).

## 7 The Relationship Between Anticoagulant and Antithrombotic Activity of Heparin

The ability of heparin to prevent clotting in patients is related to its anticoagulant activity, but the relationship is not simple. An antithrombotic agent is a therapeutic that possesses the ability to prevent and resolve thrombus formation in vivo, while an anticoagulant is a substance that will prevent the clotting of blood ex vivo or prolongs clotting times or inhibits activated clotting factors in vitro. It has been demonstrated that anticoagulants can be antithrombotics, but efficient antithrombotics do not necessary behave as anticoagulants. For example, heparan sulphate prepared from bovine pancreas and fucan derived from brown algae have negligible anticoagulant activity, and yet they were found to be efficacious in arterial and venous experimental models (Nader et al. 2004).

Heparin and low-molecular-weight heparins have been used successfully for the treatment of thrombotic diseases for decades, and it is well known that the administration of heparin leads to a measureable anticoagulant effect in plasma samples (Dougherty et al. 1992; Andrassy et al. 1988). Therefore, it is reasonable to assume that there is a close relationship between in vivo antithrombotic effect and ex vivo or in vitro anticoagulant activity of heparin. However, results from experimental and clinical studies over the last decades have indicated that the in vitro anticoagulant action of unfractionated and low-molecular-weight heparin does not always correlate with their in vivo antithrombotic activity (Morris 2000; Zancan and Mourao 2004). The lack of correlation could be partly explained by the heterogeneity of heparin structure and partly due to the promiscuous interaction of the highly negative charged heparin with a wide variety of endothelial bound and plasma proteins. The binding of heparin to these proteins has a profound effect on its overall antithrombotic action that cannot be predicted by measurement of anticoagulant activity. Heparin-binding proteins have been subjects of thorough reviews (Conrad 1998; Capila and Linhardt 2002).

As has been discussed above, the anticoagulant effect of heparin is primarily dependent on potentiation of the inhibitory activity of antithrombin; just as different heparin preparations contain different proportions of material that have high affinity to antithrombin, so the proportions of low affinity material also differs for these heparin preparations. This is the main reason why heparin cannot be dosed by weight/mass as different batches from the same manufacturer and preparations from different manufacturers have different specific activities (Mulloy et al. 2000). The ability of these fractions to activate HCII, PCI, TFPI and other antithrombin-independent anticoagulant mechanisms do not give rise to strong anticoagulant activity in vitro. However, antithrombotic action can be enhanced by the release of TFPI and as mentioned earlier, the release of endothelial TFPI could be MW and sulphate content related, so that different heparin preparations may have different TFPI releasing capacity. This idea is supported by a healthy volunteer study, which showed the molecular weight dependency of ex vivo anti-Xa activity exhibited by different commercial low-molecular-weight heparins (Alban and Gastpar 2001).

The low affinity fraction lacking the essential antithrombin binding pentasaccharide sequence can interact with other heparin-binding proteins thereby "freeing" the high-affinity material to bind to antithrombin leading to the enhancement of the anticoagulant and antithrombotic activity of "whole" heparin (Merton et al. 1984; Gray et al. 1994; Cosmi et al. 1997). However, there are heparin-binding proteins that neutralise the anticoagulant activity of heparin. Platelet factor 4 (PF4), a protein released upon activation of platelets and histidine rich glycoprotein (HRG), a circulating plasma protein have been studied extensively (Eslin et al. 2004; Lambert et al. 2007; Jones et al. 2005; Blank and Shoenfeld 2008). The neutralisation of anticoagulant activity by both PF4 and HRG is MW dependent and less than 16 saccharide units were found to be resistant to the action of these two proteins (Lane et al. 1986). In addition, pro-coagulant properties of sulphated polysaccharides have been described; over-sulphated glycosaminoglycans can trigger the contact system in a non-specific, charge related way, and heparin itself shares this property (Pan et al. 2010). The pro-coagulant properties of heparin in antithrombin and HCII free plasma have also been attributed to accelerated fV activation (Smith and Morrissey 2008), again in common with other sulphated polysaccharides with limited anticoagulant activity (Liu et al. 2006).

Unfractionated heparin is cleared from the circulation largely by the endothelium (Boneu et al. 1990), by a mechanism not yet thoroughly understood. Administration of heparin, particularly in the substantial doses required for haemodialysis, competes with endothelial heparan sulphate for several proteins, such as TFPI (see above) and lipoprotein lipase (Stegmayr et al. 2009).

The complexity of heparin's interactions with proteins in circulation and on the endothelial surface means that patient response to heparin treatment is variable. Indeed, it is well known that the unpredictable pharmacodynamics and pharmacokinetics of unfractionated heparin led to the recent guidelines being set up for monitoring of heparin ex vivo (Hirsh et al. 2008).

## References


Al Dieri R, Wagenvoord R, van Dedem GW, Beguin S, Hemker HC (2003) The inhibition of blood coagulation by heparins of different molecular weight is caused by a common functional motif – the C-domain. J Thromb Haemost 1:907–914

Alban S, Gastpar R (2001) Plasma levels of total and free tissue factor pathway inhibitor (TFPI) as individual pharmacological parameters of various heparins. Thromb Haemost 85:824–829

Andrassy K, Morike K, Koderisch J, Weber E (1988) Human pharmacological studies of a defined low molecular weight heparin fraction (Fragmin) evidence for a simultaneous inhibition of factor Xa and IIa (thrombin). Thromb Res 49:601–611

Aznar J, Espana F, Estelles A, Royo M (1996) Heparin stimulation of the inhibition of activated protein C and other enzymes by human protein C inhibitor – influence of the molecular weightof heparin and ionic strength. Thromb Haemost 76:983–988

Baglin TP, Carrell RW, Church FC, Esmon CT, Huntington JA (2002) Crystal structures of native and thrombin-complexed heparin cofactor II reveal a multistep allosteric mechanism. Proc Natl Acad Sci USA 99:11079–11084

Bajaj MS, Kuppuswamy MN, Saito H, Spitzer SG, Bajaj SP (1990) Cultured normal human hepatocytes do not synthesize lipoprotein-associated coagulation inhibitor: evidence that endothelium is the principal site of its synthesis. Proc Natl Acad Sci USA 87:8869–8873

Bajaj MS, Birktoft JJ, Steer SA, Bajaj SP (2001) Structure and biology of tissue factor pathway inhibitor. Thromb Haemost 86:959–972

Bisio A, Vecchietti D, Citterio L, Guerrini M, Raman R, Bertini S, Eisele G, Naggi A, Sasisekharan R, Torri G (2009) Structural features of low-molecular-weight heparins affecting their affinity to antithrombin. Thromb Haemost 102:865–873

Blank M, Shoenfeld Y (2008) Histidine-rich glycoprotein modulation of immune/autoimmune, vascular, and coagulation systems. Clin Rev Allergy Immunol 34:307–312

Boneu B, Caranobe C, Sie P (1990) Pharmacokinetics of heparin and low molecular weight heparin. Baillieres Clin Haematol 3:531–544

Briginshaw GF, Shanberge JN (1974) Identification of two distinct heparin cofactors in human plasma. II. Inhibition of thrombin and activated factor X. Thromb Res 4:463–477

Broze GJ Jr, Miletich JP (1987) Isolation of the tissue factor inhibitor produced by HepG2 hepatoma cells. Proc Natl Acad Sci USA 84:1886–1890

Broze GJ Jr, Warren LA, Novotny WF, Higuchi DA, Girard JJ, Miletich JP (1988) The lipoprotein-associated coagulation inhibitor that inhibits the factor VII-tissue factor complex also inhibits factor Xa: insight into its possible mechanism of action. Blood 71:335–343

Broze GJ Jr, Lange GW, Duffin KL, MacPhail L (1994) Heterogeneity of plasma tissue factor pathway inhibitor. Blood Coagul Fibrinolysis 5:551–559

Burgering MJ, Orbons LP, van der DA, Mulders J, Theunissen HJ, Grootenhuis PD, Bode W, Huber R, Stubbs MT (1997) The second Kunitz domain of human tissue factor pathway inhibitor: cloning, structure determination and interaction with factor Xa. J Mol Biol 269:395–407

Capila I, Linhardt RJ (2002) Heparin-protein interactions. Angew Chem Int Ed Engl 41:391–412

Comp PC, Jacocks RM, Ferrell GL, Esmon CT (1982) Activation of protein C in vivo. J Clin Invest 70:127–134

Conrad HE (1998) Heparin-binding proteins. Academic, San Diego

Cosmi B, Fredenburgh JC, Rischke J, Hirsh J, Young E, Weitz JI (1997) Effect of nonspecific binding to plasma proteins on the antithrombin activities of unfractionated heparin, low-molecular-weight heparin, and dermatan sulfate. Circulation 95:118–124

Craig PA, Olson ST, Shore JD (1989) Transient kinetics of heparin-catalyzed protease inactivation by antithrombin III. Characterization of assembly, product formation, and heparin dissociation steps in the factor Xa reaction. J Biol Chem 264:5452–5461

Dougherty KG, Gaos CM, Bush HS, Leachman DR, Ferguson JJ (1992) Activated clotting times and activated partial thromboplastin times in patients undergoing coronary angioplasty who receive bolus doses of heparin. Cathet Cardiovasc Diagn 26:260–263

Emsley J, McEwan PA, Gailani D (2010) Structure and function of factor XI. Blood 115:2569–2577

Enjyoji K, Miyata T, Kamikubo Y, Kato H (1995) Effect of heparin on the inhibition of factor Xa by tissue factor pathway inhibitor: a segment, Gly212-Phe243, of the third Kunitz domain is a heparin-binding site. Biochemistry 34:5725–5735

Eslin DE, Zhang C, Samuels KJ, Rauova L, Zhai L, Niewiarowski S, Cines DB, Poncz M, Kowalska MA (2004) Transgenic mice studies demonstrate a role for platelet factor 4 in thrombosis: dissociation between anticoagulant and antithrombotic effect of heparin. Blood 104:3173–3180

Esmon CT (2010) The discovery of the endothelial cell protein C receptor. J Thromb Haemost 8:2–5

Gettins PGW, Olson ST (2009) Exosite determinants of serpin specificity. J Biol Chem 284:20441–20445

Gray E (2011) Standardisation of unfractionated and low-molecular-weight heparin. In: Lever R, Mulloy B, Page CP (eds) Heparin – a century of progress. Springer, Heidelberg 65–76

Gray E, Cesmeli S, Lormeau JC, Davies AB, Lane DA (1994) Low affinity heparin is an antithrombotic agent. Thromb Haemost 71:203–207

Griffith MJ (1983) Heparin-catalyzed inhibitor/protease reactions: kinetic evidence for a common mechanism of action of heparin. Proc Natl Acad Sci USA 80:5460–5464

Griffith MJ, Noyes CM, Church FC (1985) Reactive site peptide structural similarity between heparin cofactor II and antithrombin III. J Biol Chem 260:2218–2225

Guerrini M, Guglieri S, Casu B, Torri G, Mourier P, Boudier C, Viskov C (2008) Antithrombin-binding octasaccharides and role of extensions of the active pentasaccharide sequence in the specificity and strength of interaction. Evidence for very high affinity induced by an unusual glucuronic acid residue. J Biol Chem 283:26662–26675

Hansen JB, Sandset PM (1998) Differential effects of low molecular weight heparin and unfractionated heparin on circulating levels of antithrombin and tissue factor pathway inhibitor (TFPI): a possible mechanism for difference in therapeutic efficacy. Thromb Res 91:177–181

Hansen JB, Sandset PM, Huseby KR, Huseby NE, Nordoy A (1996) Depletion of intravascular pools of tissue factor pathway inhibitor (TFPI) during repeated or continuous intravenous infusion of heparin in man. Thromb Haemost 76:703–709

Hansen JB, Sandset PM, Huseby KR, Huseby NE, Bendz B, Ostergaard P, Nordoy A (1998) Differential effect of unfractionated heparin and low molecular weight heparin on intravascular tissue factor pathway inhibitor: evidence for a difference in antithrombotic action. Br J Haematol 101:638–646

Hansen JB, Naalsund T, Sandset PM, Svensson B (2000) Rebound activation of coagulation after treatment with unfractionated heparin and not with low molecular weight heparin is associated with partial depletion of tissue factor pathway inhibitor and antithrombin. Thromb Res 100:413–417

He L, Vicente CP, Westrick RJ, Eitzman DT, Tollefsen DM (2002) Heparin cofactor II inhibits arterial thrombosis after endothelial injury. J Clin Invest 109:213–219

Hemker HC, Beguin S (1992) A rational approach to heparins. Nouv Rev Fr Hematol 34:5–9

Hirsh J, Bauer KA, Donati MB, Gould M, Samama MM, Weitz JI (2008) Parenteral anticoagulants: American College of Chest Physicians Evidence-Based Clinical Practice Guidelines (8th Edition). Chest 133:141S–159S

Holroyd EW, Simari RD (2010) Interdependent biological systems, multi-functional molecules: the evolving role of tissue factor pathway inhibitor beyond anti-coagulation. Thromb Res 125 (Suppl 1):S57–S59

Huntington JA (2003) Mechanisms of glycosaminoglycan activation of the serpins in hemostasis. J Thromb Haemost 1:1535–1549

Huntington JA, Read RJ, Carrell RW (2000) Structure of a serpin-protease complex shows inhibition by deformation. Nature 407:923–926

Huntington JA, Kjellberg M, Stenflo J (2003) Crystal structure of protein C inhibitor provides insights into hormone binding and heparin activation. Structure 11:205–215

Jin L, Abrahams JP, Skinner R, Petitou M, Pike RN, Carrell RW (1997) The anticoagulant activation of antithrombin by heparin. Proc Natl Acad Sci USA 94:14683–14688

Johnson DJ, Langdown J, Huntington JA (2010) Molecular basis of factor IXa recognition by heparin-activated antithrombin revealed by a 1.7-A structure of the ternary complex. Proc Natl Acad Sci USA 107:645–650

Jones AL, Hulett MD, Parish CR (2005) Histidine-rich glycoprotein: a novel adaptor protein in plasma that modulates the immune, vascular and coagulation systems. Immunol Cell Biol 83:106–118

Kamp P, Strathmann A, Ragg H (2001) Heparin cofactor II, antithrombin-beta and their complexes with thrombin in human tissues. Thromb Res 101:483–491

Kojima T, Katsumi A, Yamazaki T, Muramatsu T, Nagasaka T, Ohsumi K, Saito H (1996) Human ryudocan from endothelium-like cells binds basic fibroblast growth factor, midkine, and tissue factor pathway inhibitor. J Biol Chem 271:5914–5920

Lambert MP, Rauova L, Bailey M, Sola-Visner MC, Kowalska MA, Poncz M (2007) Platelet factor 4 is a negative autocrine in vivo regulator of megakaryopoiesis: clinical and therapeutic implications. Blood 110:1153–1160

Lane DA, Pejler G, Flynn AM, Thompson EA, Lindahl U (1986) Neutralization of heparin-related saccharides by histidine-rich glycoprotein and platelet factor 4. J Biol Chem 261:3980–3986

Li W, Huntington JA (2008) The heparin binding site of protein C inhibitor is protease-dependent. J Biol Chem 283:36039–36045

Li W, Adams TE, Kjellberg M, Stenflo J, Huntington JA (2007) Structure of native protein C inhibitor provides insight into its multiple functions. J Biol Chem 282:13759–13768

Li W, Adams TE, Nangalia J, Esmon CT, Huntington JA (2008) Molecular basis of thrombin recognition by protein C inhibitor revealed by the 1.6-A structure of the heparin-bridged complex. Proc Natl Acad Sci USA 105:4661–4666

Li B, Suwan J, Martin JG, Zhang F, Zhang Z, Hoppensteadt D, Clark M, Fareed J, Linhardt RJ (2009) Oversulfated chondroitin sulfate interaction with heparin-binding proteins: new insights into adverse reactions from contaminated heparins. Biochem Pharmacol 78:292–300

Lindahl U, Thunberg L, Backstrom G, Riesenfeld J, Nordling K, Bjork I (1984) Extension and structural variability of the antithrombin-binding sequence in heparin. J Biol Chem 259:12368–12376

Liu T, Scallan CD, Broze GJ Jr, Patarroyo-White S, Pierce GF, Johnson KW (2006) Improved coagulation in bleeding disorders by Non-Anticoagulant Sulfated Polysaccharides (NASP). Thromb Haemost 95:68–76

Loganathan D, Wang HM, Mallis LM, Linhardt RJ (1990) Structural variation in the antithrombin III binding site region and its occurrence in heparin from different sources. Biochemistry 29:4362–4368

Lupu C, Poulsen E, Roquefeuil S, Westmuckett AD, Kakkar VV, Lupu F (1999) Cellular effects of heparin on the production and release of tissue factor pathway inhibitor in human endothelial cells in culture. Arterioscler Thromb Vasc Biol 19:2251–2262

Ma Q, Tobu M, Schultz C, Jeske W, Hoppensteadt D, Walenga J, Cornelli U, Lee J, Linhardt R, Hanin I, Fareed J (2007) Molecular weight dependent tissue factor pathway inhibitor release by heparin and heparin oligosaccharides. Thromb Res 119:653–661

Maimone MM, Tollefsen DM (1991) Structure of a dermatan sulfate hexasaccharide that binds to heparin cofactor II with high affinity. J Biol Chem 266:14830

Mast AE, Higuchi DA, Huang ZF, Warshawsky I, Schwartz AL, Broze GJ Jr (1997) Glypican-3 is a binding protein on the HepG2 cell surface for tissue factor pathway inhibitor. Biochem J 327(Pt 2):577–583

McGuire EA, Tollefsen DM (1987) Activation of heparin cofactor II by fibroblasts and vascular smooth muscle cells. J Biol Chem 262:169–175

Merton RE, Thomas DP, Havercroft SJ, Barrowcliffe TW, Lindahl U (1984) High and low affinity heparin compared with unfractionated heparin as antithrombotic drugs. Thromb Haemost 51:254–256

Morris TA (2000) Low molecular weight heparins: background and pharmacology. Semin Respir Crit Care Med 21:537–546

Mosnier LO, Zlokovic BV, Griffin JH (2007) The cytoprotective protein C pathway. Blood 109:3161–3172

Mulloy B (2011) Structure and physicochemical characterisation of heparin. In: Lever R, Mulloy B, Page CP (eds) Heparin – a century of progress. Springer, Heidelberg 77–98

Mulloy B, Forster MJ (2008) Application of drug discovery software to the identification of heparin-binding sites on protein surfaces: a computational survey of the 4-helix cytokines. Mol Simulat 34:481–489

Mulloy B, Johnson EA (1987) Assignment of the 1H-N.M.R. spectra of heparin and heparan sulphate. Carbohydr Res 170:151–165

Mulloy B, Gray E, Barrowcliffe TW (2000) Characterization of unfractionated heparin: comparison of materials from the last 50 years. Thromb Haemost 84:1052–1056

Nader HB, Lopes CC, Rocha HA, Santos EA, Dietrich CP (2004) Heparins and heparinoids: occurrence, structure and mechanism of antithrombotic and hemorrhagic activities. Curr Pharm Des 10:951–966

Naumnik B, Borawski J, Mysliwiec M (2003) Different effects of enoxaparin and unfractionated heparin on extrinsic blood coagulation during haemodialysis: a prospective study. Nephrol Dial Transplant 18:1376–1382

Nicolaes GA, Sorensen KW, Friedrich U, Tans G, Rosing J, Autin L, Dahlback B, Villoutreix BO (2004) Altered inactivation pathway of factor Va by activated protein C in the presence of heparin. Eur J Biochem 271:2724–2736

Nordfang O, Bjorn SE, Valentin S, Nielsen LS, Wildgoose P, Beck TC, Hedner U (1991) The C-terminus of tissue factor pathway inhibitor is essential to its anticoagulant activity. Biochemistry 30:10371–10376

O'Keeffe D, Olson ST, Gasiunas N, Gallagher J, Baglin TP, Huntington JA (2004) The heparin binding properties of heparin cofactor II suggest an antithrombin-like activation mechanism. J Biol Chem 279:50267–50273

Olson ST, Swanson R, Raub-Segall E, Bedsted T, Sadri M, Petitou M, Herault JP, Herbert JM, Bjork I (2004) Accelerating ability of synthetic oligosaccharides on antithrombin inhibition of proteinases of the clotting and fibrinolytic systems. Comparison with heparin and low-molecular-weight heparin. Thromb Haemost 92:929–939

Olson ST, Richard B, Izaguirre G, Schedin-Weiss S, Gettins PG (2010) Molecular mechanisms of antithrombin-heparin regulation of blood coagulation proteinases. A paradigm for understanding proteinase regulation by serpin family protein proteinase inhibitors. Biochimie 92(11):1587–1596

Osterud B, Bajaj MS, Bajaj SP (1995) Sites of tissue factor pathway inhibitor (TFPI) and tissue factor expression under physiologic and pathologic conditions. On behalf of the Subcommittee on Tissue factor Pathway Inhibitor (TFPI) of the Scientific and Standardization Committee of the ISTH. Thromb Haemost 73:873–875

Pan J, Qian Y, Zhou X, Lu H, Ramacciotti E, Zhang L (2010) Chemically oversulfated glycosaminoglycans are potent modulators of contact system activation and different cell signaling pathways. J Biol Chem 285:22966–22975

Parker KA, Tollefsen DM (1985) The protease specificity of heparin cofactor II. Inhibition of thrombin generated during coagulation. J Biol Chem 260:3501–3505

Pavao MS, Aiello KR, Werneck CC, Silva LC, Valente AP, Mulloy B, Colwell NS, Tollefsen DM, Mourao PA (1998) Highly sulfated dermatan sulfates from Ascidians. Structure versus anticoagulant activity of these glycosaminoglycans. J Biol Chem 273:27848–27857

Petersen JG, Meyn G, Rasmussen JS, Petersen J, Bjorn SE, Jonassen I, Christiansen L, Nordfang O (1993) Characterization of human tissue factor pathway inhibitor variants expressed in *Saccharomyces cerevisiae*. J Biol Chem 268:13344–13351

Pratt CW, Church FC (1992) Heparin binding to protein C inhibitor. J Biol Chem 267:8789–8794

Pratt CW, Whinna HC, Meade JB, Treanor RE, Church FC (1989) Physicochemical aspects of heparin cofactor II. Ann N Y Acad Sci 556:104–115

Pratt CW, Whinna HC, Church FC (1992) A comparison of three heparin-binding serine proteinase inhibitors. J Biol Chem 267:8795–8801

Sandset PM, Abildgaard U, Larsen ML (1988) Heparin induces release of extrinsic coagulation pathway inhibitor (EPI). Thromb Res 50:803–813

Schneider CL (1947) The active principle of placental toxin; thromboplastin; its inactivator in blood; antithromboplastin. Am J Physiol 149:123–129

Schroeder M, Hogwood J, Gray E, Mulloy B, Hackett AM, Johansen KB (2011) Protamine neutralisation of low molecular weight heparins and their oligosaccharide components. Anal Bioanal Chem 399:763–771

Shriver Z, Sundaram M, Venkataraman G, Fareed J, Linhardt R, Biemann K, Sasisekharan R (2000) Cleavage of the antithrombin III binding site in heparin by heparinases and its implication in the generation of low molecular weight heparin. Proc Natl Acad Sci USA 97:10365–10370

Silverman GA, Whisstock JC, Bottomley SP, Huntington JA, Kaiserman D, Luke CJ, Pak SC, Reichhart JM, Bird PI (2010) Serpins flex their muscle: I. Putting the clamps on proteolysis in diverse biological systems. J Biol Chem 285:24299–24305

Smith SA, Morrissey JH (2008) Heparin is procoagulant in the absence of antithrombin. Thromb Haemost 100:160–162

Stegmayr B, Olivecrona T, Olivecrona G (2009) Lipoprotein lipase disturbances induced by uremia and hemodialysis. Semin Dial 22:442–444

Stenflo J (1976) A new vitamin K-dependent protein. Purification from bovine plasma and preliminary characterization. J Biol Chem 251:355–363

Streusand VJ, Bjork I, Gettins PG, Petitou M, Olson ST (1995) Mechanism of acceleration of antithrombin-proteinase reactions by low affinity heparin. Role of the antithrombin binding pentasaccharide in heparin rate enhancement. J Biol Chem 270:9043–9051

Sun W, Eriksson AS, Schedin-Weiss S (2009) Heparin enhances the inhibition of factor Xa by protein C inhibitor in the presence but not in the absence of $Ca^{2+}$. Biochemistry 48:1094–1098

Swedenborg J (1998) The mechanisms of action of alpha- and beta-isoforms of antithrombin. Blood Coagul Fibrinolysis 9(Suppl 3):S7–S10

Thomas L (1947) Studies on the intravascular thromboplastin effect of tissue suspensions in mice: II. A factor in normal rabbit serum which inhibits the thromboplastin effect of the sedimentable tissue component. Bull Johns Hopkins Hosp 81:26–42

Thyzel E, Kohli S, Siegling S, Prante C, Kleesiek K, Gotting C (2007) Relative quantification of glycosaminoglycan-induced upregulation of TFPI-mRNA expression in vitro. Thromb Res 119:785–791

Tollefsen DM (2002) Heparin cofactor II deficiency. Arch Pathol Lab Med 126:1394–1400

Tollefsen DM, Blank MK (1981) Detection of a new heparin-dependent inhibitor of thrombin in human plasma. J Clin Invest 68:589–596

Tollefsen DM, Majerus DW, Blank MK (1982) Heparin cofactor II. Purification and properties of a heparin-dependent inhibitor of thrombin in human plasma. J Biol Chem 257:2162–2169

Tovar AM, de Mattos DA, Stelling MP, Sarcinelli-Luz BS, Nazareth RA, Mourao PA (2005) Dermatan sulfate is the predominant antithrombotic glycosaminoglycan in vessel walls: implications for a possible physiological function of heparin cofactor II. Biochim Biophys Acta 1740:45–53

Valentin S, Larnkjer A, Ostergaard P, Nielsen JI, Nordfang O (1994) Characterization of the binding between tissue factor pathway inhibitor and glycosaminoglycans. Thromb Res 75:173–183

Van Walderveen MC, Berry LR, Chan AK (2010) Effect of covalent antithrombin-heparin on activated protein C inactivation by protein C inhibitor. J Biochem 148:255–260

Vicente CP, He L, Pavao MS, Tollefsen DM (2004) Antithrombotic activity of dermatan sulfate in heparin cofactor II-deficient mice. Blood 104:3965–3970

Viskov C, Just M, Laux V, Mourier P, Lorenz M (2009) Description of the chemical and pharmacological characteristics of a new hemisynthetic ultra-low-molecular-weight heparin, AVE5026. J Thromb Haemost 7:1143–1151

Wagenvoord R, Al Dieri R, Van DG, Beguin S, Hemker HC (2008) Linear diffusion of thrombin and factor Xa along the heparin molecule explains the effects of extended heparin chain lengths. Thromb Res 122:237–245

Werling RW, Zacharski LR, Kisiel W, Bajaj SP, Memoli VA, Rousseau SM (1993) Distribution of tissue factor pathway inhibitor in normal and malignant human tissues. Thromb Haemost 69:366–369

Wesselschmidt R, Likert K, Huang Z, MacPhail L, Broze GJ Jr (1993) Structural requirements for tissue factor pathway inhibitor interactions with factor Xa and heparin. Blood Coagul Fibrinolysis 4:661–669

Whisstock JC, Silverman GA, Bird PI, Bottomley SP, Kaiserman D, Luke CJ, Pak SC, Reichhart JM, Huntington JA (2010) Serpins flex their muscle: II. Structural insights into target peptidase recognition, polymerization, and transport functions. J Biol Chem 285:24307–24312

Wiebe EM, Stafford AR, Fredenburgh JC, Weitz JI (2003) Mechanism of catalysis of inhibition of factor IXa by antithrombin in the presence of heparin or pentasaccharide. J Biol Chem 278:35767–35774

Xiao Z, Zhao W, Yang B, Zhang Z, Guan H, Linhardt RJ (2011) Heparinase 1 selectivity for the 3,6-di-O-sulfo-2-deoxy-2-sulfamido-alpha-D-glucopyranose (1,4) 2-O-sulfo-alpha-L-idopyranosyluronic acid (GlcNS3S6S-IdoA2S) linkages. Glycobiology 21:13–22

Xu X, Takano R, Nagai Y, Yanagida T, Kamei K, Kato H, Kamikubo Y, Nakahara Y, Kumeda K, Hara S (2002) Effect of heparin chain length on the interaction with tissue factor pathway inhibitor (TFPI). Int J Biol Macromol 30:151–160

Yang L, Manithody C, Rezaie AR (2002) Localization of the heparin binding exosite of factor IXa. J Biol Chem 277:50756–50760

Yang L, Sun MF, Gailani D, Rezaie AR (2009) Characterization of a heparin-binding site on the catalytic domain of factor XIa: mechanism of heparin acceleration of factor XIa inhibition by the serpins antithrombin and C1-inhibitor. Biochemistry 48:1517–1524

Yang L, Manithody C, Qureshi SH, Rezaie AR (2010) Contribution of exosite occupancy by heparin to the regulation of coagulation proteases by antithrombin. Thromb Haemost 103:277–283
Ye Z, Takano R, Hayashi K, Ta TV, Kato H, Kamikubo Y, Nakahara Y, Kumeda K, Hara S (1998) Structural requirements of human tissue factor pathway inhibitor (TFPI) and heparin for TFPI-heparin interaction. Thromb Res 89:263–270
Zancan P, Mourao PA (2004) Venous and arterial thrombosis in rat models: dissociation of the antithrombotic effects of glycosaminoglycans. Blood Coagul Fibrinolysis 15:45–54

# Part II
# Unfractionated and Low-Molecular-Weight Heparins

# Standardisation of Unfractionated and Low-Molecular-Weight Heparin

**Elaine Gray**

## Contents



**Abstract** Unfractionated and low-molecular-weight heparins are complex biologicals. Standardisation and global harmonisation of units and methods of measurement are essential for safety and efficacy of this important class of anticoagulants. This chapter describes the traceability of the international unit and current status of the relationship between the international and pharmacopoeial standards, together with a review on current pharmacopoeial assay methods.

E. Gray (✉)
Haemostasis Section, Biotherapeutics Group, National Institute for Biological Standards and Control, Blanche Lane, South Mimms, Potters Bar, Hertfordshire EN6 3QG, UK
e-mail: elaine.gray@nibsc.hpa.org.uk

R. Lever et al. (eds.), *Heparin - A Century of Progress*,
Handbook of Experimental Pharmacology 207,
DOI 10.1007/978-3-642-23056-1_4, 

**Keywords** Standardisation • Anticoagulant activity • Unfractionated heparin • Low-molecular-weight heparin

## Abbreviations

| | |
|---|---|
| EP | European Pharmacopoeia |
| IS | International Standard |
| IU | International Unit |
| JP | Japanese Pharmacopoeia |
| USP | United States Pharmacopeia |
| WHO | World Health Organisation |

## 1 Introduction

For clinical efficacy and safety, it is important that unfractionated and low-molecular-weight heparins are clearly and accurately labelled with their respective concentrations and units. Although heparin is a polysaccharide and the mechanism by which it functions as an anticoagulant has been well established (see Gray et al. 2011), its complex structure means that its activity cannot be represented simply in gravitimetric mass units. Similar to macromolecular proteins such as blood coagulation factors and cytokines, the activity of heparin is defined in arbitrary units and determined by biological assays relative to a reference standard. For unfractionated heparin, the unitage of different products should be interchangeable. For low-molecular-weight heparins, it is still debatable whether the unitage of the different low-molecular-weight heparins are interchangeable (Prandoni 2003; Lopez 2001; Planes et al. 1999; Nenci 2003; van der Heijden et al. 2000; Simonneau et al. 2006). This is due to the differences in manufacturing processes; the ratio of the anti-Xa and anti-IIa activity is distinct for each product and therefore 100 anti-Xa units of enoxaparin may not have the same in vivo activity as 100 anti-Xa units of tinzaparin. However, the unitage of biosimilar or biogeneric low-molecular-weight heparin products should be interchangeable with the reference product. Standardisation of the unit of activity is essential to ensure patients will receive the same dosage of heparin wherever they are in the world. The evolution of the unit of anticoagulant activity and assay methods for unfractionated and low-molecular-weight heparins has been reviewed in detail by Barrowcliffe (1989). This chapter will briefly summarise the early history and bring up-to-date developments in standardisation, detailing the importance of different types of biological or functional assays, including those that are defined in pharmacopoeial monographs for the determination of units of heparin activity.

## 2 Unfractionated Heparin

### 2.1 *History of Standardisation of Unfractionated Heparin*

In 1923, Howell first defined a unit of activity as the amount of heparin, which just prevented the clotting of 1 ml of cat's blood for 24 h at 0°C (Howell 1922). This unit of activity was related to the specific assay method described by Howell. With modifications introduced by other groups (Charles and Scott 1936; Jorpes 1935), the definition of the unit became somewhat ambiguous. Murray and Best (1938) applied the comparative bioassay principle developed by Dale for the measurement of insulin and developed a reference material for heparin (Bangham 1999). In 1942, the League of Nations recognised the clinical importance of heparin by the acceptance of a batch of reference material as a provisional International Standard (IS) with the International Unit (IU) assigned traceable to the "Toronto unit" defined by the reference material produced by Murray and Best (1943). This standard was officially established after the Second World War as the 1st International Standard for Heparin in 1947 (WHO Expert Committee on Biological Standardization 1947). For more than 30 years, the 1st IS and its successor, the 2nd IS (WHO Expert Committee on Biological Standardization 1957) served well to harmonise the global measurement of heparin, and they were used to calibrate pharmacopoeial reference standards, which were used extensively for the potency labelling of therapeutic products. The replacement of the 2nd IS, a bovine lung preparation with a porcine mucosal heparin in 1973 as the 3rd IS (WHO Expert Committee on Biological Standardization 1969) introduced the divergence of the IU and the United States Pharmacopeial (USP) Unit for heparin. The major reason for this disparity was that a 7% lower potency was obtained when the 3rd IS was assayed against the 2nd IS using the USP method when compared with the potency estimate attained with the British Pharmacopoeial (BP) method (Bangham and Woodward 1970). This was a classical issue of the need to assay "like against like" for comparative biological assays and will be discussed in later section of this chapter. The USP at that time decided against using the WHO assigned value, which was the mean potency estimates by all methods used by participants of the collaborative study. Instead, they used the potency estimate obtained using the USP method only for calibration of the subsequent USP reference standard for heparin. Since then, the next few generations of USP reference standard for heparin were calibrated against the previous lots of USP reference standard. The WHO replacement standards, the 4th (Thomas et al. 1984) and the 5th (Gray et al. 2000) IS, were value assigned relative to the previous IS using potency estimates from all methods. This meant that there was a discontinuity between the IU and the USP unit for heparin. In October 2009, this situation was resolved when the USP calibrated their current reference standard for potency of heparin sodium (Lot F) against the 5th IS in the same collaborative study that established the 6th IS. Although the potency of Lot F is still given in USP units, it is harmonised with the IU and therefore 1 USP unit is now equivalent to 1 IU of heparin.

## 2.2 International and Pharmacopoeial Standards

International standards for biologicals are prepared in accordance with WHO recommendations (WHO Expert Committee on Biological Standardization 2006). The following quotes the four general principles described:

1. That the reference standard should be assigned a value in arbitrary rather than absolute units, but there can be exceptions, where justified
2. That the unit is defined by a reference standard with a physical existence
3. That in the establishment of the standard a variety of methods is usually used and that the value assignment of the standard, and therefore the definition of the unit, is not necessarily dependent on a specific method of determination
4. That the behaviour of the reference standard should resemble as closely as possible the behaviour of test samples in the assay systems used to test them

The production of international standards for heparin follows these principles in that a new IS is value assigned in arbitrary units, as defined by the current IS (the 6th IS at the time of writing of this chapter). The value assigned is not dependent on a specific method but taken as the mean of potency estimates from all methods used by the participants of the collaborative study. The IS for heparin is typically a batch of clinical grade active pharmaceutical ingredient for heparin. As the IU is defined by the current IS, and the methods used for value assignment may change over time, there is a danger of discontinuity of the IU when an IS is replaced. To minimise this risk, every effort is made to ensure that the previous and replacement standards have similar characteristics, so that the IU defined by a replacement standard is close to equivalent to the IU defined by the previous standard. This allows the continuity of the IU over generations of international standards. However, another issue may arise when the properties, such as the physiochemical characteristics or purity profile of clinical products change over time. The replacement standard will have to reflect this change, which means it will not have similar characteristics to the previous standard against which it will be value assigned. It is clear that the source and the physicochemical characteristics of heparin have changed since the first human clinical trial carried out in 1935. Mulloy et al. (2000) surveyed samples of heparin produced between 1950 and 2000 and showed that the mean molecular weight and specific activity of therapeutic heparin products have both increased over time. Notably, the tissue source of heparin was changed in the 1960s from lung to mucosa. The WHO therefore recommended switching from bovine lung material to a porcine mucosal reference standard, so that the IS would be closer to the characteristics of therapeutic heparin at that time. This resulted in the disparity of potency estimates in different assay methods for the porcine mucosa 3rd IS, when it was assayed against the bovine lung 2nd IS. Bioassays tend to be complex, and the end-point or final readout may be the result of a cascade of events within the assay. As most biologicals have multiple targets, if the test samples are different to the reference standard, they may induce different responses at the varying steps within the assay. This illustrates the importance of assaying like against like.

Pharmacopoeial standards for heparin are widely available, and in accordance with the WHO recommendation for establishment of secondary standards, the unitage of these standards should be directly traceable to the IU and therefore calibrated against the current IS. This is certainly true for the European Pharmacopoeial (EP) and Japanese Pharmacopoeial (JP) standards and is also the case for the USP standard, Lot F which was established in October 2009. One major difference between the value assignment for the IS and for pharmacopoeial reference standards is that the calibration of the pharmacopoeial standard employs only the specified pharmacopoeial method. The continuity of the unit for the pharmacopoeial standards is also reassessed whenever the IS is replaced. This is usually carried out by the inclusion of the pharmacopoeial reference standards in the collaborative study that value assigned the replacement standard, though it could be carried out as a separate exercise.

## 2.3 In Vitro Assay Methods for Unfractionated Heparin

In vitro assay methods for the determination of potency of heparin products should be precise, accurate, reflect the mode of its anticoagulant action and not be easily influenced by the presence of impurities or contaminants. Pharmacopoeial methods used for potency labelling should evolve with new understandings of the mechanism of action of heparin and incorporate new technologies to ensure the robustness of these assays. However, pharmacopoeial monographs for heparin have been in existence for more than 70 years and revisions of these monographs have been infrequent. In the absence of any pressure to change, the pharmacopoeias did not adopt more advanced assay methods as soon as they became available. In addition, the changes in manufacturing processes over the years that lead to progressively higher purity heparin products have not been taken into account. Contaminants such as over-sulphated chondroitin sulphate (OSCS) that could give an apparent increase of the potency of heparin in pharmacopoeial assays have caused adverse drug reaction and fatalities (Guerrini et al. 2008; Kishimoto et al. 2008).

Potency assays for heparin can be divided into two groups: (a) plasma/blood and clot based and (b) purified reagents and chromogenic substrate based. Until the 1980s, when chromogenic substrates specific for thrombin (IIa) and activated factor X (Xa) became widely available, all heparin assays were clot based methods.

The clot-based methods use fibrin formation as the end-point and therefore can be readily influenced by the quality of the blood or plasma, impurities and contaminants. Pre October 2009, the USP and EP potency methods were clot based and utilised sheep plasma as the substrate. It has been shown that sheep plasmas from different producers have varying amounts of platelet factor 4 (PF4) that can neutralise the anticoagulant activity of heparin preparations (van Dedem et al. 1996). It is also known that the neutralisation of heparin by PF4 is molecular weight dependent and that higher molecular weight materials are more easily neutralisable than lower molecular weight fractions (Lane et al. 1984). So if the heparin test sample has a significantly different molecular weight profile to the reference

**Table 1** Potency estimates of unfractionated heparin relative to the 5th IS

| Samples | Mn | Potency relative to the 5th IS Iu/ampoule | | Potency ratio sheep plasma: anti-IIa |
|---|---|---|---|---|
| | | USP sheep plasma assay | USP anti-IIa assay | |
| 5th IS | 14,910 | – | – | – |
| T | 11,992 | 1,935 | 1,736 | 1.11 |
| Y | 11,231 | 2,026 | 1,866 | 1.09 |
| W | 12,772 | 2,177 | 2,112 | 1.03 |
| X | 13,016 | 2,154 | 2,144 | 1.00 |

standard, the potency obtained using the sheep plasma methods may be under- or over-estimated. Data from the 6th IS collaborative study illustrate this problem (Table 1). Candidate T and Y had lower number average molecular weight (Mn) than the 5th IS, and the potencies obtained using the USP sheep plasma assay were higher than potencies from anti-IIa chromogenic assay, using purified reagents. Candidates X and W had Mn closer to the 5th IS, and the potencies by USP sheep plasma assay were similar to those obtained by anti-IIa chromogenic assay. This illustrates the importance of assaying like against like, the basis of comparative bioassay. If the test sample has the same properties as the reference standard, similar potency estimates will be obtained regardless of assay methods used; this justifies the WHO recommendation that the potency assignment of IS should be independent of methods.

The anti-IIa chromogenic assay uses purified proteins and reagents and the absence of PF4 and other interfering plasma proteins means that the difference in molecular weight profile of the standard and test unfractionated heparin will not influence the end results of the assay. The anti-Xa chromogenic assay has a similar principle to the anti-IIa assay except the protease in the assay is Xa rather than thrombin. Both chromogenic assays are highly specific for unfractionated and low-molecular-weight heparins as they are based on the rationale that only heparin preparations have the prerequisite pentasaccharide sequence that potentiates the inhibition of thrombin or Xa by antithrombin. Potency estimates for a purified unfractionated heparin preparation obtained from anti-IIa and anti-Xa assays against a purified heparin reference standard should be similar and therefore the potency ratio should be close to 1.

Another issue with clot-based assays, especially assays using sheep plasma, came to light when in 2007 some heparin products were found to be contaminated with OSCS (Guerrini et al. 2008; Kishimoto et al. 2008). It has been speculated that in 2006 the incidence of porcine reproductive and respiratory syndrome in China (Ma et al. 2009), commonly known as Blue Ear Disease of pigs, caused a shortage of raw material for heparin production. Oversulphated chondroitin sulphate was used as an adulterant as it was relatively cheap to produce and it could not be differentiated from heparin in a sheep plasma-based assay. Like heparin, OSCS is also a glycosaminoglycan; unlike heparin, its anticoagulant action can only be exerted via the potentiation of the inhibition of thrombin by heparin co-factor II. The presence of OSCS in heparin preparation will enhance the anticoagulant

**Table 2** Effect of OSCS on activity of unfractionated heparin

| Assay type | IU/mg (95% confidence limits) | |
|---|---|---|
| | Heparin | Heparin + 15% OSCS |
| EP Sheep Plasma | 165.9 (160.4–171.4) | 200.1 (193.7–206.5) |
| Anti-Xa | 172.8 (169.1–176.5) | 177.5 (173.7–181.3) |
| Anti-IIa | 168.4 (154.4–182.4) | 179.0 (162.9–186.1) |

activity of heparin in a plasma clot-based assay as plasma contains heparin co-factor II. Oversulphated chondroitin sulphate has minimal activity in anti-IIa or anti-Xa chromogenic assays as it cannot potentiate the activity of antithrombin. Table 2 shows that the addition of OSCS to pure heparin gave higher potency by the EP sheep plasma assay, but it had no effect on the potency estimates by anti-Xa or anti-IIa assays. It is also possible that OSCS having higher proportions of high-molecular-weight material than heparin, PF4 may preferentially binds to OSCS, thereby freeing heparin to act. This hypothesis is currently being investigated in the author's own laboratory. So if the antithrombin-dependent anti-IIa and anti-Xa assays were used for potency estimation of heparin products, there would be no economic motive to adulterate heparin with OSCS. Indeed, the USP revised heparin monographs, effective from October 2009, specified the use of the antithrombin-dependent anti-IIa chromogenic assay for potency estimation and in addition a specification of 0.9–1.1 for the anti-IIa to anti-Xa ratio was also included to ensure the quality of heparin products. The EP is also in the process of considering similar revisions to its heparin monographs.

### 2.4 *Analysis of Assay Results*

The potency estimates obtained using pharmacopoeial methods are only valid if the data are analysed and have passed the acceptance criteria for statistical analysis of bioassays. The analysis is based on the assumption of assaying like against like and the test dilutions should behave as if they were dilutions of the reference standard. The potency of a test sample can be calculated when its dose response curve is compared with the dose response of the reference standard with a known potency. These principles and analysis have been described for general bioassays (Finney 1978) and more specifically for assays of coagulation factors (Kirkwood and Snape 1980). The parallel line and slope ratio models are the two most useful analyses for heparin assays. For both models, it is important to have an appropriate assay design:

(a) More than three dilutions each for standard and test preparations should be included.
(b) The treatments or dilutions for standard and test preparations should be randomised.
(c) The responses to each treatment are normally distributed

(d) The standard deviations of the responses within each treatment group of both standard and unknown preparations do not differ significantly from one another.

For parallel line analysis, the logarithm of the dose response relationships for both standard and test preparations should be linear and parallel to each other. For slope ratio analysis, the untransformed dose response relationships for both standard and test preparation should be linear, and both the standard and test response should have the same intercept at the *y*-axis; i.e. standard and test dilution curves should have a common intercept. The validity of linearity and parallelism is assessed by analysis of variance. Non-parallelism or different intercept suggests the test preparation and the reference standard do not behave in the same way in the assay and that they may have different mechanisms of action. To obtain a reliable and valid potency estimate for a batch of product, the results from several independent assays should be statistically combined to give a weighted mean potency estimate with confidence limits. The EP and USP provide detailed guidance for the design and analysis of bioassays by both models and combination of potency estimates(European Pharmacopoeia 2010; United States Pharmacopeia 2009a). Commercial software packages for statistical analysis which have been validated for analysis according to the EP and USP are available. In Europe, most regulatory laboratories and manufacturers use the "CombiStat" software package provided by the EP.

## 3 Low-Molecular-Weight Heparin

### *3.1 History of Standardisation of Low-Molecular-Weight Heparin*

Since the discovery of low-molecular-weight heparin in 1976 (Johnson et al. 1976), there are at least eight different low-molecular-weight heparin products licenced globally for clinical use and they each have their own individual WHO International Non-proprietary Names (INNs), indicating that these products have different active ingredients http://www.who.int/medicines/services/inn/en/. Barrowcliffe et al. (1985) investigated the use of unfractionated heparin as a reference standard for the anti-Xa and thrombin inhibitory activities of low-molecular-weight heparins and found not only high intra- and inter-laboratory variability in the potency estimates, but also large numbers of invalid assays due to non-parellelism. They concluded that a low-molecular-weight heparin standard was required, and the 1st International Standard for Low Molecular Weight Heparin was established in 1986 (Barrowcliffe et al. 1988). When the 1st IS was replaced, the suitability of unfractionated heparin was again investigated, and in addition, the study also addressed whether one low-molecular-weight heparin can serve as a valid comparator to all existing low-molecular-weight heparins (Gray et al. 2001). This study found that when assayed against the 5th IS for Unfractionated Heparin, all 8 low-molecular-weight heparins gave non-parallel anti-IIa assays and 2 out

8 preparations were non-parallel by the anti-Xa method. This confirmed the earlier finding that unfractionated heparin should not be used as a reference for anticoagulant activities of low-molecular-weight heparins (Barrowcliffe et al. 1985). When assayed against the 1st IS for Low Molecular Weight Heparin, only 1 out of 8 and 2 out of 8 samples were assayed non-parallel by the anti-IIa and anti-Xa assays, respectively, and with the exception of two samples, which gave non-parallel assays to most of the other samples, all the low-molecular-weight heparins compared well against each other. This indicated that not all low-molecular-weight heparins can serve as a reference standard for this diverse range of products. The 2nd IS was subsequently established in 2003 (Gray et al. 2003). The stock of the 2nd IS is now close to depletion and since new generation of ultra-low low-molecular-weight heparins are in clinical trials, a study is now ongoing to investigate the feasibility of one low-molecular-weight heparin replacement standard to cover all products.

## 3.2 *In Vitro Assay Methods for Low-Molecular-Weight Heparin*

A wide range of assays are available for laboratory measurement of low-molecular-weight heparins (Barrowcliffe 1989), but for potency labelling, the European Pharmacopoeia has established the antithrombin-dependent anti-Xa and anti-IIa chromogenic assays. The principle of these assays is the same as that earlier described in the unfractionated heparin section of this chapter. The USP assays as described in the monograph for enoxaparin sodium are similar to the EP assays. The assay design and analysis of data are the same as those detailed for unfractionated heparin.

## 3.3 *International and Pharmacopoeial Standards*

The European Pharmacopoeia provides a general monograph for low-molecular-weight heparin (European Pharmacopoeia 2008a) (note the EP refers low-molecular-weight heparin as low molecular mass heparin) and specific monographs for five of these individual products (European Pharmacopoeia 2008b, c, d, e, f). There is a USP monograph for enoxaparin (United States Pharmacopeia 2009b) and currently, a monograph for dalteparin and two general chapters, one on potency assays and the other on molecular weight profiling, are also being drafted. It is clear that these low-molecular-weight heparins are produced by different manufacturing process with different physiochemical characteristics and anticoagulant profiles. Patent protection for some of these low-molecular-weight heparins is coming to an end and a biosimilar low-molecular-weight heparin has already been approved by the US Food and Drugs Administration (FDA) this year, although biosimilar products have been in use in India, China and some South American countries for quite some time. Reliable reference standards and robust assay methods are essential to control not only the innovator products, but also the biosimilars. The current WHO international standard for low-molecular-weight heparin (2nd IS) was produced from a clinical batch of low-molecular-weight heparin. Since the ratio of anti-Xa to anti-IIa activities defines a particular low-molecular-weight heparin, the WHO IS has to be assigned with two values: an anti-Xa and an anti-IIa IU. The anti-Xa and anti-IIa activity units for the 1st IS were value assigned relative to the 4th IS for Unfractionated Heparin (Barrowcliffe et al. 1985); the 2nd IS Low Molecular Weight Heparin was value assigned against the 1st IS Low Molecular Weight Heparin (Gray et al. 2003). The EP Biological Reference Preparation (BRP) for Low Molecular Mass Heparin was calibrated against the WHO IS for low-molecular-weight heparin, using the EP monograph methods for low-molecular-mass heparin and therefore directly traceable to the IU (Gray et al. 2004). The EP recommends the use of this BRP as the reference standard for the monograph potency assays of all eight preparations of low-molecular-weight heparins. The USP has produced a product specific reference standard for the assay of enoxaparin calibrated against the IS and unlike the unfractionated heparin (Heparin Sodium) reference standard, it is labelled in IU. It is envisaged that once the general chapter for potency measurement of low-molecular-weight heparin becomes official, a USP reference standard for potency assay will be available for the other low-molecular-weight heparins.

## 4 Conclusion

Clinically, unfractionated and low-molecular-weight heparins have been in use for more than 70 years while low-molecular-weight heparins have been important therapeutics for close to 30 years. New generations of ultra-low molecular weight heparin are now in clinical trials. The standardisation of these two anticoagulants is matured, the units of activities are globally harmonised and providing the pharmacopoeias and the WHO continues to work together, keeping up with the changes in the properties of these products, safety and efficacy of heparins will be ensured.

**Acknowledgement** Dr. Barbara Mulloy for the molecular weight analysis of heparin samples and John Hogwood for the assays of heparin and OSCS samples.

## References

Bangham DR (1999) Biological standardization: contributions from the UK 1900–1995. Society for Endocrinology, Bristol

Bangham DR, Woodward PM (1970) A collaborative study of heparins from different sources. Bull World Health Organ 42:129–149

Barrowcliffe TW (1989) Heparin assays and standardisation. In: Lane DA, Lindahl U (eds) Heparin: chemical and biological properties, clinical applications. Edward Arnold, London
Barrowcliffe TW, Curtis AD, Tomlinson TP, Hubbard AR, Johnson EA, Thomas DP (1985) Standardization of low molecular weight heparins: a collaborative study. Thromb Haemost 54:675–679
Barrowcliffe TW, Curtis AD, Johnson EA, Thomas DP (1988) An international standard for low molecular weight heparin. Thromb Haemost 60:1–7
Charles AF, Scott DA (1936) Studies on heparin: observations on the chemistry of heparin. Biochem J 30:1927–1933
European Pharmacopoeia (2008a) Heparins, low-molecular mass, monograph 0828. Council of Europe, Strasbourg
European Pharmacopoeia (2008b) Enoxaparin sodium, monograph 1097. Council of Europe, Strasbourg
European Pharmacopoeia (2008c) Dalteparin sodium, monograph 1195. Council of Europe, Strasbourg
European Pharmacopoeia (2008d) Tinzaparin sodium, monograph 1271. Council of Europe, Strasbourg
European Pharmacopoeia (2008e) Parnaparin sodium, monograph 1252. Council of Europe, Strasbourg
European Pharmacopoeia (2008f) Nadroparin sodium, monograph 1134. Council of Europe, Strasbourg
European Pharmacopoeia (2010) Statistical analysis of results of biological assays and tests, monograph 50300. Council of Europe, Strasbourg
Finney DJ (1978) Statistical methods in biological assay. Charles Griffin, London
Gray E, Walker AD, Mulloy B, Barrowcliffe TW (2000) A collaborative study to establish the 5th international standard for unfractionated heparin. Thromb Haemost 84:1017–1022
Gray E, Sands DA, Barrowcliffe TW (2001) Report on the pilot study on proposed candidate materials for the 2nd international standard for low molecular weight heparin. World Health Organ Tech Rep Ser
Gray E, Rigsby P, Behr-Gross ME (2003) Collaborative study to establish the 2nd international standard for low molecular weight heparin and the low molecular mass heparin for assay European pharmacopoeia biological reference preparation. World Health Organ Tech Rep Ser
Gray E, Rigsby P, Behr-Gross ME (2004) Collaborative study to establish the Low-molecular-mass heparin for assay – European pharmacopoeia biological reference preparation. Pharmeuropa Bio 2004:59–76
Gray E et al (2011) The anticoagulant and antithrombotic mechanisms of heparin. In: Lever R, Mulloy B, Page CP (eds) Heparin – a century of progress. Springer, Heidelberg, pp 43–61
Guerrini M, Beccati D, Shriver Z, Naggi A, Viswanathan K, Bisio A, Capila I, Lansing JC, Guglieri S, Fraser B, Al-Hakim A, Gunay NS, Zhang Z, Robinson L, Buhse L, Nasr M, Woodcock J, Langer R, Venkataraman G, Linhardt RJ, Casu B, Torri G, Sasisekharan R (2008) Oversulfated chondroitin sulfate is a contaminant in heparin associated with adverse clinical events. Nat Biotechnol 26:669–675
Howell WH (1922) Heparin: an anticoagulant. Am J Physiol 63:434–435
Johnson EA, Kirkwood TB, Stirling Y, Perez-Requejo JL, Ingram GI, Bangham DR, Brozovic M (1976) Four heparin preparations: anti-Xa potentiating effect of heparin after subcutaneous injection. Thromb Haemost 35:586–591
Jorpes E (1935) The chemistry of heparin. Biochem J 29:1817–1830
Kirkwood TB, Snape TJ (1980) Biometric principles in clotting and clot lysis assays. Clin Lab Haematol 2:155–167
Kishimoto TK, Viswanathan K, Ganguly T, Elankumaran S, Smith S, Pelzer K, Lansing JC, Sriranganathan N, Zhao G, Galcheva-Gargova Z, Al-Hakim A, Bailey GS, Fraser B, Roy S, Rogers-Cotrone T, Buhse L, Whary M, Fox J, Nasr M, Dal Pan GJ, Shriver Z, Langer RS, Venkataraman G, Austen KF, Woodcock J, Sasisekharan R (2008) Contaminated heparin

associated with adverse clinical events and activation of the contact system. N Engl J Med 358:2457–2467
Lane DA, Denton J, Flynn AM, Thunberg L, Lindahl U (1984) Anticoagulant activities of heparin oligosaccharides and their neutralization by platelet factor 4. Biochem J 218:725–732
Lopez LM (2001) Low-molecular-weight heparins are essentially the same for treatment and prevention of venous thromboembolism. Pharmacotherapy 21:56S–61S
Ma Y, Feng Y, Liu D, Gao GF (2009) Avian influenza virus, *Streptococcus suis* serotype 2, severe acute respiratory syndrome-coronavirus and beyond: molecular epidemiology, ecology and the situation in China. Philos Trans R Soc Lond B Biol Sci 364:2725–2737
Mulloy B, Gray E, Barrowcliffe TW (2000) Characterization of unfractionated heparin: comparison of materials from the last 50 years. Thromb Haemost 84:1052–1056
Murray DWG, Best CH (1938) Heparin and thrombosis: present situation. J Am Med Assoc 110:118–119
World Health Organisation (1943) Memorandum on a provisional International Standard for Heparin. Bull Health Organ League Nations
Nenci GG (2003) Low molecular weight heparins: are they interchangeable? No. J Thromb Haemost 1:12–13
Planes A, Samama MM, Lensing AW, Buller HR, Barre J, Vochelle N, Beau B (1999) Prevention of deep vein thrombosis after hip replacement – comparison between two low-molecular heparins, tinzaparin and enoxaparin. Thromb Haemost 81:22–25
Prandoni P (2003) Low molecular weight heparins: are they interchangeable? Yes. J Thromb Haemost 1:10–11
Simonneau G, Laporte S, Mismetti P, Derlon A, Samii K, Samama CM, Bergman JF (2006) A randomized study comparing the efficacy and safety of nadroparin 2850 IU (0.3 mL) vs. enoxaparin 4000 IU (40 mg) in the prevention of venous thromboembolism after colorectal surgery for cancer. J Thromb Haemost 4:1693–1700
Thomas DP, Curtis AD, Barrowcliffe TW (1984) A collaborative study designed to establish the 4th international standard for heparin. Thromb Haemost 52:148–153
United States Pharmacopeia (2009) Design and analysis of biological assays. USP 32-NF 27:103–115
United States Pharmacopeia (2009) Enoxaparin Sodium. USP 32-NF 27: 2253–2256
van Dedem G, Maat A, Dollevoet G (1996) Pharmacopeial heparin assays-proposals for improvement. Pharmacopoeial Forum 22:2938–2945
van der Heijden JF, Prins MH, Buller HR (2000) Low-molecular-weight heparins: are they interchangeable? Haemostasis 30(Suppl 2):148–157
WHO Expert Committee on Biological Standardization (1947) Bull World Health Organ 1:9
WHO Expert Committee on Biological Standardization (1957) World Health Organ Tech Rep Ser 127:17
WHO Expert Committee on Biological Standardization (1969) World Health Organ Tech Rep Ser 413:17
WHO Expert Committee on Biological Standardization (2006) Recommendations for the preparation, characterization and establishment of international and other standards (revised 2004). World Health Organ Tech Rep Ser 932:75–130

# Structure and Physicochemical Characterisation of Heparin

**Barbara Mulloy**

## Contents



**Abstract** Heparin is a member of the heparan sulphate family of glycosaminoglycans, a linear polysaccharide with a complex sequence resulting from the action of post-polymerisation enzymes on a regular repeating disaccharide background. Its overall conformation is rod-like in solution as well as in the solid state, but the conformational fluctuations of iduronate residues give rise to considerable internal motion and variation in local three-dimensional structure. Structure/

B. Mulloy (✉)
National Institute for Biological Standards and Control, Blanche Lane, South Mimms, Potter's Bar, Hertfordshire EN6 3QG, UK
e-mail: Barbara.Mulloy@nibsc.hpa.org.uk

R. Lever et al. (eds.), *Heparin - A Century of Progress*,
Handbook of Experimental Pharmacology 207,
DOI 10.1007/978-3-642-23056-1_5, 

function relationships and their relation to sequence are still the subject of argument, but new methodologies to tackle the subject are emerging.

Heparin as a therapeutic agent and as the object of research may be characterised by numerous physico-chemical techniques. These include chromatographic methods for measurement of molecular weight; a variety of spectroscopic techniques; separation methods for whole polysaccharides, as well as for oligo- and monosaccharides; and mass spectrometric methods for mapping and sequence analysis. The impetus provided by the discovery of heparin contamination with oversulphated chondroitin sulphate has been influential in bringing combinations of many old and new techniques into use to ensure that heparin is sufficiently consistent and pure to be used safely. Synthetic and semi-synthetic heparins are in development and may become reality in the relatively near future.

**Keywords** Heparin · Structure · Analysis · Conformation · Spectroscopy · Chromatography

## Abbreviations

| | |
|---|---|
| CE | Capillary electrophoresis |
| CID | Collisionally induced dissociation |
| GAG | Glycosaminoglycan |
| GlcA | β-D-glucuronic acid |
| GlcNAc | N-acetylated α-D-glucosamine |
| GlcNS | N-sulphated α-D-glucosamine |
| GlcNS6S | N-sulphated 6-O-sulphated α-D-glucosamine |
| GPC | Gel permeation chromatography |
| HS | Heparan sulphate |
| IdoA | Alpha-L-iduronic acid |
| IdoA2S | 2-O-sulphated α-L-iduronic acid |
| IR | Infra-red |
| LC-MS | Liquid chromatography-Mass spectrometry |
| LMWH | Low Molecular Weight Heparin |
| MS | Mass spectrometry |
| NMR | Nuclear magnetic resonance |
| OSCS | Oversulphated chondroitin sulphate |
| PAGE | Polyacrylamide gel electrophoresis |
| RPIP | Reverse-phase ion pair |
| SAS | Strong Anion Exchange |
| UFH | Unfractionated Heparin |

# 1 Introduction: The Structure of Heparin

Heparin is a linear polysaccharide in all senses of the word. Like all mammalian glycosaminoglycans (GAGs), it is absolutely linear in sequence, with no branches. In addition, heparin has the unusual property for a polysaccharide of having a linear, rod-like conformation (Mulloy et al. 1993). In spite of this, as a member of the heparan sulphate (HS) family of GAGs, its structure is extremely complex through heterogeneity in both sequence and size.

The biosynthesis of heparin is well described by Carlsson and Kjellen (2011). The repeating disaccharide of the heparosan precursor sets the typical GAG pattern of alternating uronic acid (in this case, β-D-glucuronic acid (GlcA)) and hexosamine (in this case, N-acetyl α-D-glucosamine (GlcNAc))(Fig. 1a). In heparin, this disaccharide is mostly converted to a trisulphated form (Fig. 1b) in which the GlcA has been epimerised to α-L-iduronic acid (IdoA), unusual otherwise in nature but widespread through the animal kingdom within the glycosaminoglycan family (Nader et al. 1999).

Besides the uniform, highly sulphated domains which make up the larger part of heparin, there exist short unsulphated domains and more complex sequences, the most significant of which is the sequence with high affinity for the plasma protein antithrombin (Fig. 1c). This serine protease inhibitor inhibits not only thrombin, as its name implies, but also other proteases of the coagulation system as described by Gray (2011).

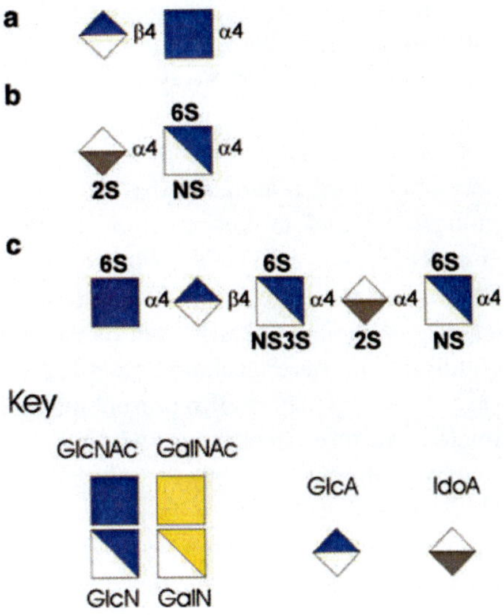


**Fig. 1** Structures in heparin, shown using recommended symbols and conventions for glycan structures (Varki et al. 2009). (**a**)The unsulphated disaccharide GlcA-GlcNAc, which forms the precursor polysaccharide of both heparin and heparan sulphate (Carlsson and Kjellen 2011) and is present in minor amounts in heparan. (**b**)The trisulphated disaccharide, which is the most abundant repeating unit in heparin. (**c**)The pentasaccharide, which is the minimum structure with high affinity for antithrombin (Gray et al. 2011)

## 1.1 *Heterogeneity of Heparin and Heparan Sulphate: Fine Structure and Domains, and Relationship to Function*

Heparin has been described as an extreme form of heparan sulphate, and is distinguished from HS both by its origin in mast cell granules, and by the use of an extracted subfraction of heparin in the prevention and treatment of thrombosis (see other chapters in this book). Methods for characterisation of heparins and heparan sulphates may be suitable for academic studies of the vast range of possible structures in these polysaccharides, or they may be suitable for screening the more limited range of clinical-grade heparins; or both. This chapter describes first of all the way in which the sequence of heparin determines its three-dimensional structure, and therefore its mode of interaction with the extracellular proteins such as those of the coagulation cascade and of inflammatory pathways, the immune system, and developmental morphogens. It then goes on to outline the many ways in which the size, shape and sequence of heparin samples, and of fragments of heparin, can be characterized.

## 1.2 *Three-Dimensional Structure and Dependence on Sequence*

Like heparan sulphate, heparin has two main domain types; those in which glucosamine residues are N-sulphated, alternating with IdoA [the "S" domains (Merry et al. 1999)], and those in which glucosamine residues are N-acetylated and alternate with GlcA, having been unaltered by enzymes of post-polymerisation transformation. In heparin, especially in commercial heparins, the S-domains overwhelmingly predominate, and are highly substituted with sulphate at C2 of iduronate and C6 of glucosamine. Unlike most polysaccharides, heparin behaves in solution as a rod-like molecule (Mulloy and Forster 2000); a recent study using ultracentrifugation and X-ray scattering showed that relatively small fragments of heparin are almost rigid, and longer lengths a little more flexible (Khan et al. 2010). Conformational studies of heparin by NMR gave the same result; data on distances between hydrogen atoms in the structure could not be interpreted on the basis of a globular or random-coil structure but fitted very well for a linear, rod-like shape (Fig. 2) (Forster et al. 1989; Mulloy et al. 1993). That is not to say, however, that heparin is a static molecule. The iduronate residue is not, as are many hexopyranose sugars, stable in a "chair" form of its six-membered ring, but exists in a dynamic equilibrium between a chair form and a twisted skew-boat form (Ferro et al. 1986) (Fig. 3), which may itself represent the average of a rapidly fluctuating ensemble of related structures (Forster and Mulloy 1993). While variations in sulphate substitution were found to affect the overall conformation of heparin only moderately (Mulloy et al. 1994), they have a considerable effect on the conformational equilibrium of the iduronate residue (Mulloy et al. 1994; Ferro et al. 1990). Comparison between the sequence depiction of heparin in Fig. 2a with the three-dimensional models in Fig. 2b shows the patterns of sulphate groups, in clusters of three along each side of the heparin chain. These clusters consist of sulphate substituents from

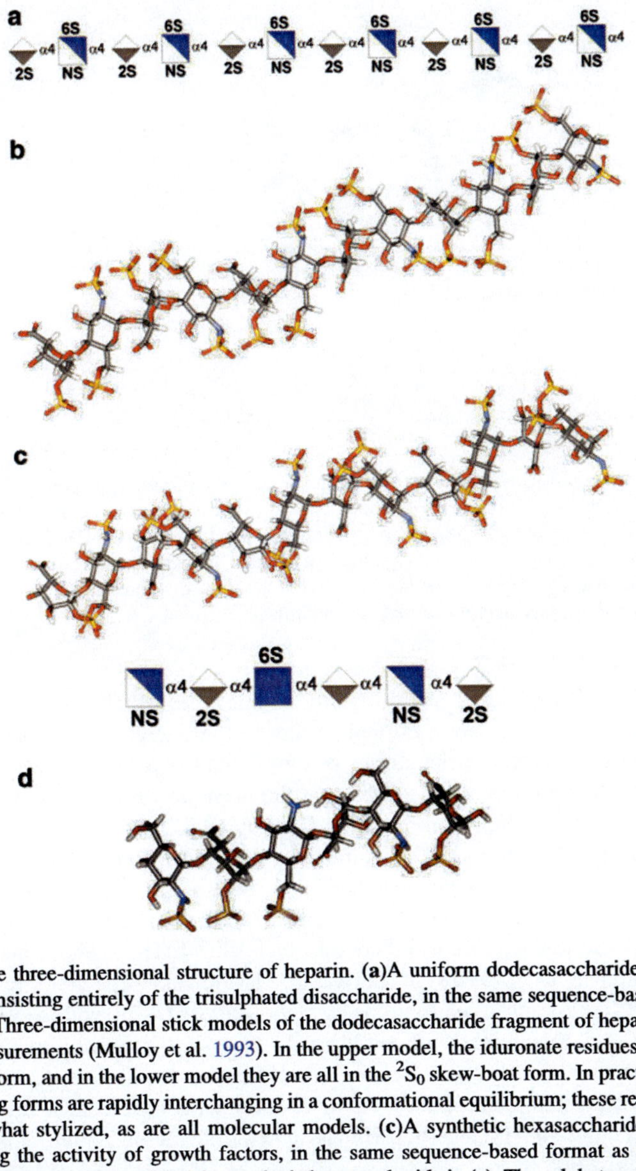


**Fig. 2** The three-dimensional structure of heparin. **(a)**A uniform dodecasaccharide fragment of heparin consisting entirely of the trisulphated disaccharide, in the same sequence-based format as Fig. 1. **(b)**Three-dimensional stick models of the dodecasaccharide fragment of heparin, based on NMR measurements (Mulloy et al. 1993). In the upper model, the iduronate residues are all in the $^{1}C_4$ chair form, and in the lower model they are all in the $^{2}S_0$ skew-boat form. In practice of course the two ring forms are rapidly interchanging in a conformational equilibrium; these representations are somewhat stylized, as are all molecular models. **(c)**A synthetic hexasaccharide, capable of potentiating the activity of growth factors, in the same sequence-based format as Fig. 1. **(d)** A three-dimensional stick model of the synthetic hexasaccharide in **(c)**. The sulphate substituents are arranged down one side of the molecule; this is not obvious from the sequence representation (Carbon atoms are *black*, oxygen *red*, nitrogen *blue*, sulphur *yellow* and hydrogen *white*).The heparin models were visualised using Discovery v. 2.6 from Accelrys

three contiguous monosaccharide residues, brought together by the geometry of the glycosidic linkages. The conformational flexibility of iduronate alters the detailed geometry of these sulphate clusters, but does not change the overall shape of the

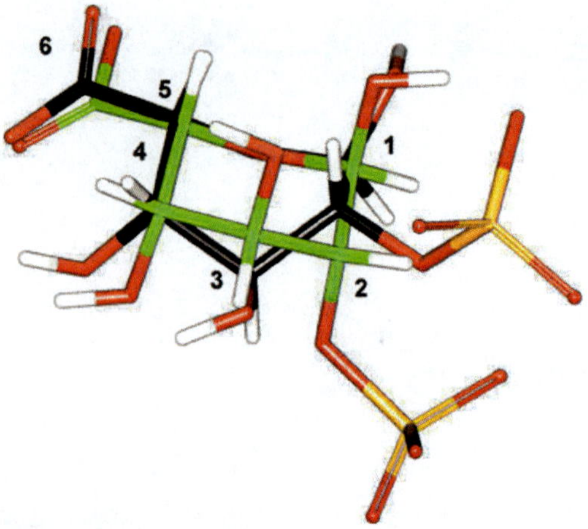


**Fig. 3** The two conformational forms of iduronic acid in equilibrium in heparin. The $^{1}C_{4}$ chair form (carbon atoms in green) and the $^{2}S_{0}$ skew boat form (carbon atoms in black)of α-L-iduronic acid (IdoA). The two molecules have been superimposed so that the C4-C5-O5-C1 part of the six-membered pyranose ring coincides as closely as possible in the two forms. The orientation of the sulphate attached at C2 of IdoA is the largest effect of the conformational change (Sulphur atoms are shown in *yellow*, oxygen atoms in *red*, and hydrogens in *white*)

polysaccharide. Three-dimensional features of this type, which cannot be inferred from the sequence alone, are a critical feature in the interactions of heparin with basic peptides and proteins; the cluster pattern repeats approximately every 17 Å, close enough to the pitch of an α-helix that the presence of heparin induces poly-lysine to adopt an α-helical conformation (Mulloy et al. 1996).

The conformation of the intervening GlcA-GlcNAc domains is not so well understood, and less thoroughly studied, and there is some evidence that these domains are more flexible than the S-domains (Mobli et al. 2008) [though there is also an opposite point of view (Hricovini et al. 1997)]. The view of heparin and heparan sulphate as a series of fairly stiff S-domains separated by flexible, unsulphated links is an attractive one, which would fit well with the ability of heparin and HS to interact with multiple proteins in, for example, chemokine oligomers (Stringer et al. 2002; Stringer and Gallagher 1997). It is also consistent with recent ideas that the size and spacing of S-domains is at least as important as the fine structure of HS and heparin as a determinant of protein interactions (Kreuger et al. 2006).

The functional significance of the iduronate ring flexibility is less obvious, and has to lie, if anywhere, in the details of sugar–protein interaction at the atomic level. It may be that specific, fixed iduronate conformations are required for particular specific interactions with proteins; or it may be that the flexibility is itself advantageous in terms of entropic contributions to binding processes. The conformational equilibrium is sensitive to surrounding sequence (Mulloy et al. 1994; Murphy et al. 2008). For proteins in complex with ligands, X-ray crystallography is still unrivalled in its ability to provide descriptions of modes of interaction at the atomic level. Crystal structures of a number of heparin oligosaccharides in complex with proteins are available in the database, though these are not always of high enough resolution to be sure about fine details of saccharide geometry. The linkage geometries of the heparin fragments in these structures, which determine the overall rod-like shape of the molecule, are almost always found to be similar to those in the NMR structure (Mulloy and Linhardt 2001; Khan et al. 2010). Neither heparin nor any of its fragments have been crystallised alone, though a solid-state heparin structure based on fibre diffraction has been published (Nieduszynski and Atkins 1973).

The dependence of function on fine structure for heparin has been studied most closely for its interaction with antithrombin, a rare instance of a heparin–protein interaction involving a highly specific heparin sequence (Fig. 1c). This is discussed in more detail in the mechanism of action chapter. The pentasaccharide which forms the core of the antithrombin-binding sequence has an unusual glucosamine residue, not only 2- and 6-O-sulphated but also 3-O-sulphated. The same residue, in a different context, is involved in the interaction between *Herpes simplex* virus type 1 (HSV1) and its host cell (Shukla et al. 1999). A heparin oligosaccharide containing the triply sulphated glucosamine is capable of inhibiting infection in vitro (Copeland et al. 2008).

However, the vast majority of heparin–protein interactions do not involve unusual residues, merely the more or less sulphated S-domains of heparin or heparan sulphate. Fully sulphated heparin seems to be able to stand in for less completely sulphated sequences, and its uniformity is of course useful to the crystallographer. The disadvantage of the common use of heparin as a model compound for HS is that any fine structure specificity is swamped; it is in fact remarkably difficult to tell from structural studies whether or not a protein has a preference for any particular pattern of sulphation, as extra sulphates increase affinity (Kreuger et al. 2006; Mulloy 2005). Some use has been made of systematically modified heparins, in which the pattern of sulphate substitution is altered uniformly throughout the heparin molecule; for example in the study of growth factor interactions (Ostrovsky et al. 2002). These compounds have given some interesting insights, but are still unrealistic as models for natural sequences in HS S-domains.

Other standard approaches for the study of protein ligands are difficult to apply to heparin/HS; for example, the synthesis of pure heparin mimetic oligosaccharides is still challenging. The contribution of synthetic chemistry to structure/function relationships of glycosaminoglycans has, however, been considerable (Gama and Hsieh-Wilson 2005). The heparin hexasaccharide illustrated in Fig. 2c, d is a synthetic product, designed to probe the structural requirements for binding to, and potentiating the action of, growth factors (Angulo et al. 2004; Lucas et al. 2003). It was designed to display a specific three-dimensional feature; the sulphates are ranged on one side of the chain only (Fig. 2d). This hexasaccharide is capable of inducing the mitogenic activity of FGF-1 more effectively than other hexasaccharides with similar, or even greater, degree of overall sulphation, arranged in different patterns.

More recently, a different synthetic strategy has been used to generate a small hexasaccharide library (Wang et al. 2010) with varying sulphation patterns. Experiments with this group of compounds have indicated that a trisaccharide sequence in which a 2-sulphated iduronate residue is flanked by two N-sulphated glucosamines is a minimal structural motif for binding to FGF-2. However, no 6-sulphated glucosamines are present in these hexasaccharides.

It is necessary to treat conclusions reached from studies on such restricted libraries of small oligosaccharides with care. The number of potential heparin/HS sequences, even for hexasaccharides, is large. Even considering only two glucosamine types (GlcNS and GlcNS6S) and two iduronate types (IdoA and IdoA2S) there are 48 different hexasaccharides with glucosamine at the non-reducing end and another 48 with iduronate at the non-reducing end. Including further monosaccharide types (such as GlcA and GlcNAc) or going up to the octasaccharide level would stretch the limits of any combinatorial chemistry approach.

# 2 Methods for Characterisation and Analysis of Heparin Samples

Physico-chemical characterisation of heparin as a therapeutic agent has been developed with a view to ensuring its consistency and safety. Although the biological activity of heparin is a key indicator of its suitability for human use, and is therefore the key measurand in heparin standardisation (see Gray 2011), the use of physicochemical methods has recently proved invaluable in assurance of purity and identity (see Chess et al. 2011).

## 2.1 *Compositional Analysis of Heparin*

### 2.1.1 Monosaccharide Analysis

Monosaccharide analysis for a polysaccharide with a repeating disaccharide structure should be simple, but of course the alternation of amino sugars with uronic acids means that it is not practicable to perform a simple analysis based on hydrolysis and chromatography. Conditions for release of amino sugars are too extreme for the survival of labile iduronic acid. In practice, amino sugar analysis is performed to determine the proportion of galactosamine-containing GAGs in a heparin samples (see below), and glucuronate/iduronate ratios either measured by methanolysis and gas chromatography (Inoue and Miyawaki 1975) or estimated by NMR.

### 2.1.2 Sulphate and Carboxylate

Several analytical methods are available for the measurement of sulphate in a glycosaminoglycan sample (Ruiz-Calero et al. 2001). For heparin and other glycosaminoglycans, the ratio of sulphate groups to carboxylate groups is a useful parameter, as there is a single carboxyl per disaccharide. It is possible to use conductimetric titration to determine this ratio (Casu and Gennaro 1975), a classic method still in use.

## *2.2 The Size of Heparin Molecules: Molecular Weight Determinations*

Heparin is a heterogeneous polysaccharide, in terms of both its sequence and its degree of polymerisation. This means that its molecular weight cannot be described by a single number. The measurement of molecular weight for heparin is usually, in fact, based on an experimental estimate of molecular volume in solution (usually by gel permeation chromatography (GPC) or occasionally by electrophoretic methods). The use of GPC in molecular weight determinations for heparin has been reviewed (Mulloy 2002).

Because heparin is a linear molecule, behaving like a slightly bendy rod (Khan et al. 2010), its hydrodynamic properties do not resemble those of either globular proteins (Volpi and Bolognani 1993) or more flexible polysaccharides; calibrants for heparin molecular weight measurement should therefore be made of heparin itself.

## *2.3 Molecular Weight of Unfractionated Heparin*

Given the complexity of the starting material and the variety of manufacturing procedures, it was a mild surprise to note the high degree of consistency in molecular weight displayed by porcine mucosal heparin samples produced between the 1940s and 1990s (Mulloy et al. 2000). The same study also noted a slight increase in molecular weight of some products over the previous 30 years, possibly associated with the use of unfractionated heparin (UFH) as a starting material for the production of low-molecular-weight-heparin (LMWH). Unfractionated heparin contains a little material below 3,000 and above 100,000 in molecular weight, with a mean somewhere between 10,000 and 20,000 depending on the tissue and species of origin and the method of preparation (Fig. 4). At present, the most useful method for the measurement of UFH molecular weight is GPC with some form of light scattering detector (Bertini et al. 2005), as at the time of writing there are no universally available and accepted calibrants.

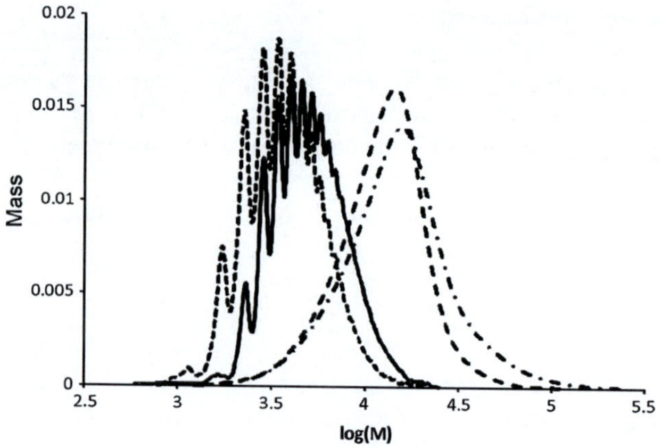


**Fig. 4** The molecular weight distributions of UFH and LMWH. Two UFH (*long dashes, dot-dashes*) and two LMWH (*solid line*: dalteparin, short dashes: reviparin) reproduced with permission (Mulloy et al. 2010). The two UFH products are similar but not identical, and the two LMWH products are easy to differentiate; however, the difference between UFH and LMWH is very much greater

## 2.4 *Molecular Weight of Low-Molecular-Weight Heparin*

The first study of the effects of low-molecular-weight heparin fractions in a group of volunteers (Johnson et al. 1976) indicated that they were able to induce strong and long-lasting anti-Xa activity in plasma. In time, this observation led to the widespread clinical use of low-molecular-weight heparin products. Each of these products has a characteristic molecular weight distribution, which is a major factor in determining its anti-Xa and anti-IIa activities and the ratio between them. The molecular weight specification for all low-molecular-weight heparin is found in a general monograph, and profiles of several products are defined in individual monographs of the European Pharmacopoeia (2010a).

The low molecular weight heparins licensed for clinical use may contain fragments as small as the tetrasaccharide (Mr about 1200) and a small amount of material over the 20,000 mark; though each product has its own characteristic molecular weight profile, they are all more similar to each other than they are to UFH (Fig. 4).

The same techniques of light scattering as for UFH are successful for LMWH (Knobloch and Shaklee 1997), but it is technically easier to use a calibrant material such as the European Pharmacopoeia CRS LMM Heparin for Calibration or the WHO second IS LMW Heparin for MW Calibration. These are closely related calibrants, though the EP material is characterised for use with the specific monograph method, requiring both RI and UV detection, as the ratio between RI (a measure of mass) and UV (in this case, a measure of molar concentration) responses is required for the calibrant. The second IS LMW Heparin for MW Calibration may be used successfully as a Broad Standard, and results using either method and calibrant are broadly comparable (Mulloy et al. 2007).

## 2.5 *Separations Techniques: Chromatography and Electrophoresis*

### 2.5.1 Chromatography of Heparin

Apart from the use of GPC to estimate the molecular weight of heparin, the same technique plays a part in the preparation and analysis of heparin oligosaccharides (Ziegler and Zaia 2006). Other chromatographic techniques have been developed which can both separate heparin from other similar compounds, and separate heparin fragments from each other with high resolution. Such highly anionic compounds as glycosaminoglycans are readily separated by charge. Strong Anion Exchange (SAX) chromatography has been in use for some time now for separation of heparin oligosaccharides (Rice et al. 1985). The technique has been improved and refined over the years (Mourier and Viskov 2004) and used in sequencing strategies, though the high concentration of salt required to elute sulphated oligosaccharides makes direct interfacing with mass spectrometry equipment difficult.

SAX chromatography can be used to separate heparin from dermatan sulphate and chondroitin sulphate, and, because the level of sulphation is different, from oversulphated chondroitin sulphate (OSCS) (Keire et al. 2010a).

### 2.5.2 Affinity Chromatography and Heparin

The large number of proteins that can bind to heparin in a manner which is dependent on ionic strength has been used in numerous applications for their separation and preparation (Xiong et al. 2008). This methodology has even been extended to the purification of viruses (Segura et al. 2010). As heparin and heparan sulphate are heterogeneous, individual species can also be separated on the basis of their affinity for a given protein. The classic example is that of antithrombin; fractions of heparin retained on immobilized antithrombin were found to have very much higher activity in anticoagulant assays than unbound heparin (Holmer et al. 1979), and oligosaccharides prepared in this way allowed the identification of the high-affinity sequence (Fig. 1c) (Casu et al. 1981; Thunberg et al. 1982).

### 2.5.3 Electrophoresis of Heparin

Electrophoretic techniques on gels or membranes were once extensively used for GAGs and their fragments, but in recent years have been superseded to some extent by capillary electrophoresis (CE). The advantage of the "flat" techniques is that they offer the chance to run several samples side-by-side for comparison, whereas CE, like chromatography, is a serial technique. On the other hand, quantification and electronic storage of data is easier for CE.

Whole glycosaminoglycans can be separated by electrophoresis on agarose (Volpi 1993) or cellulose acetate (Burlingame et al. 1981). Agarose separations, together with specific enzyme digestions, can be used to identify GAGs isolated from new sources (Gandra et al. 2000). Cellulose acetate electrophoresis is one of the techniques, which may be used to identify heparin lots contaminated with OSCS (Domanig et al. 2009). Polyacrylamide gel electrophoresis (PAGE) is more often used to separate oligomers of heparin by size, and has been used in the past to give estimates of molecular weight of heparin samples, as well as early applications of the technique of 'oligosaccharide mapping' to heparin/heparan sulphate samples (Turnbull and Gallagher 1988; Linhardt et al. 1990).

The use of capillary electrophoresis for glycosaminoglycans and their fragments is particularly appropriate given their high negative charge; this field has recently been reviewed (Volpi et al. 2008). Capillary electrophoresis has also recently been adapted successfully to the identification of heparin lots contaminated with OSCS (Somsen et al. 2009). CE is especially useful as a separation technique to deliver heparin/heparan subfragments for on-line MS analysis.

## 2.6 *Spectroscopic Techniques*

One advantage of spectroscopic techniques is of course that the sample is not degraded. Optical methods may be used for the quantitation and composition of heparin, as well as its three-dimensional structure, with varying degrees of resolution.

### 2.6.1 Optical Spectroscopy of Heparin

Heparin does not absorb light in the visible and near UV, but its effect on the absorbance spectra of other compounds has been used to develop quantitative methods for the measurement of heparin concentration. Binding to heparin changes the wavelength of maximum absorbance ($\lambda_{max}$) of some dyes; this phenomenon can be used in a "metachromatic" assay for heparin (Templeton 1988), measuring either the decrease in absorbance at one wavelength or the increase at another. Dimethylmethylene blue is perhaps the most frequently used dye (Farndale et al. 1986). Metachromatic staining can also reveal glycosaminoglycan containing granules in mast cells and related structures (Cavalcante et al. 2000).

The far UV absorbance of heparin is not directly useful (with the exception of the band at about 234 nm arising from the unsaturated uronic acid of β-elimination derived fragments of heparin, such as the LMW heparins tinzaparin and enoxaparin. However, the far-UV circular dichroism (difference in absorbance of left-handed and right-handed polarised light) of heparin and related compounds has been studied (Mulloy et al. 1994; Stone 1985; Stevens et al. 1985), most recently using synchrotron radiation, capable of reaching further into the high-energy far UV (Rudd et al. 2009a; Matsuo et al. 2009).

Infra-red (IR) spectroscopy and the related technique of Raman spectroscopy are often used in the characterisation of organic molecules. Even if bands in the IR spectrum are not completely assigned, a "fingerprint" spectrum can be useful as an identity test when compared with an authentic reference, or as a consistency test for a particular product. For hydrophilic compounds such as heparin, a major problem is the IR absorbance of water. However, bands due to sulphate vibration occur in the 1,400–950 $cm^{-1}$ "window" of high water transmittance, and these bands may therefore be observed for heparin in aqueous solution (Cabassi et al. 1978). Further bands, arising from carboxylate and acetamido groups, can be studied in deuterium oxide solution (Casu et al. 1978). The influence of cations and of chemical modifications of heparin on the IR spectrum has been studied (Grant et al. 1987, 1989). IR is not, like NMR, a high-resolution technique, but can be used to monitor gross changes in heparin, for example to assess the extent of chemical desulphation (Mulloy et al. 1994).

The complementary technique of Raman spectroscopy has been very rarely applied to heparin; the Raman spectrum of heparin is simple. A major envelope of bands around 1,050 $cm^{-1}$ arising from symmetric sulphate vibrations has the potential for use as a quantitative indication of the sulphate content of the sample (Cabassi et al. 1978). The weak optical activity of Raman scattering has now been recorded for glycosaminoglycans including heparin (Rudd et al. 2010) and interpreted in conformational terms.

## 3 NMR Spectroscopy

Both $^{1}$H and $^{13}$C NMR spectroscopy may be used to give a detailed "fingerprint" of an individual sample of heparin or heparan sulfate. Each monosaccharide residue gives a set of signals with characteristic chemical shifts depending on the state of sulphate substitution and the adjoining monosaccharide residues. The NMR spectra obtained, whether 1D or 2D, give a good reflection of the heterogeneity and domain structure of the GAG sample. In proton NMR, the anomeric (H1) signals, falling between 4.5 and 5.6 ppm, are informative (Mulloy and Johnson 1987). Signals from the regular trisulphated domains dominate the spectrum, and smaller signals from residues in irregular sequences are numerous; GlcA anomeric signals appear to be especially sensitive to their sequence context.

The information in the $^{1}$H and $^{13}$C spectra can be combined in 2D spectra with $^{1}$H chemical shift as one axis and $^{13}$C shift as the second. These heteronuclear correlated spectra are quite complex for heparin, but comparison of an unknown and a reference spectrum is just as straightforward as it is for 1D spectra, if not more so, and their acquisition is no longer limited to the research laboratory but accessible as part of the basic programmimg of any high-field NMR spectrometer. This type of spectroscopy has been used to discriminate between LMW heparins, in terms of minor components resulting from manufacturing processes (Guerrini et al. 2007). It has also been suggested as a robust method of ensuring the identity and purity of heparin for screening of heparin products (Guerrini et al. 2009a).

## *3.1 Systematic Analysis of Spectroscopic Data*

The interpretation of complex spectroscopic data for compounds as complex as heparin is not easy to systematise, and it is a common complaint that spectroscopic identity tests, whether by NMR, IR or any other fingerprinting technique, are subjective, with criteria such as "similarity" which seem vague, particularly to a non-specialist with no experience of what features are significant and which may safely be ignored. Modern methods for the mathematical analysis of multiple complex datasets can provide objective, systematic interpretations of spectra, and have the advantage that more than two or more types of spectra can be combined. Principal component analysis (PCA) has been applied to NMR and CD of heparin and other glycosaminoglycans (Rudd et al. 2009b).

## *3.2 Sequence Determinations of Oligosaccharides*

In order to find out exactly which structural features in heparin (or heparan sulphate) are critical for interactions with particular proteins, the classic strategy is to fragment a heparin sample into smaller oligosaccharides, separate the oligosaccharides by their affinity for the protein in question, and determine the structural differences between high-and low-affinity oligosaccharides (Casu et al. 1981). Where the heparin fragments and the protein involved are available in generous quantities, there are choices to be made between several suitable techniques for structural characterisation of the oligosaccharides. However, it is usually the case that sample is scarce, so that insensitive techniques such as NMR spectroscopy are not ideal. Mass spectrometry (MS) is sensitive, rapid, and can be linked with separations techniques as well as used in tandem MS-MS experiments (Zaia 2009). The combination of size separation, fractionation by binding to antithrombin, and LC-MS has been used to differentiate between high-and low-affinity hexasaccharides (Naimy et al. 2008) and recently extended to octasaccharides (Naimy et al. 2010) Even so, there are technical difficulties to be overcome; obtaining a molecular ion from a fully sulphated heparin oligosaccharide is challenging, and the results do not give a complete sequence analysis. More commonly, the oligosaccharides are broken down by enzymic digestion with bacterial lyases into disaccharides, which are then analysed by LC-MS; collisionally induced dissociation (CID) has proved useful at producing fragmentation patterns, which can distinguish between sites of sulphation of the disaccharides

(Meissen et al. 2009); UV photodissociation has also been applied to generate informative fragments from heparin disaccharides (Racaud et al. 2009).

Separations methods useful for oligosaccharide sequencing by MS have recently been reviewed (Zaia 2009). One recently discussed method is reverse-phase ion-pair chromatography (RPIP) (Jones et al. 2010) which can be used in conjunction with fluorescent labelling (Skidmore et al. 2009). A protocol for oligosaccharide sequencing using RPIP with volatile reagents, coupled to ESI MS has recently been published (Volpi and Linhardt 2010).

Recently, the use of collisionally induced dissociation (CID) to produce fragments capable of distinguishing between isomers with differing sulphation patterns has been applied to the structural study of an octasaccharide with high affinity for the chemokine CCL2 (Meissen et al. 2009)

## 3.3 *Identity and Purity of Heparin for Therapeutic use*

The concept of purity for a highly heterogeneous polysaccharide such as heparin is not easy to define. The structural differences between heparin and heparan sulphate are a question of degree, and the distinction between them, particularly in the mixtures of mucosal glycosaminoglycans from which commercial heparin is prepared, is one of the conventions. In addition, the structure of heparin differs between species and tissues; for example, the structural differences between bovine and porcine mucosal heparin are marked (Casu et al. 1995; Aquino et al. 2010). However, it is easier to define galactosaminoglycans as impurities, whether they are the naturally occurring mucosal chondroitin sulphates or the semi-synthetic oversulphated chondroitin sulphate which caused many adverse effects in patients in 2007–2008 (see Chess et al. 2011). The most recent compendial methods for the assessment of purity and identity of heparin are weighted towards recognising samples which are contaminated with OSCS, even at low quantities, but at the same time the use of combinations of spectroscopic and separations techniques, taken together with well thought out specifications, can be organised so as to find almost any kind of gross contamination. This orthogonal approach has been described (Guerrini et al. 2009b). The possibility that other contaminants might be present, such as oversulphated heparan sulphate, which would affect physico-chemical assays in more subtle ways than OSCS (Keire et al. 2010b), has recently been raised (Pan et al. 2010). In this case, the use of a combination of physico-chemical characterisation and anticoagulant assays with high specificity for heparin can identify the presence of such contaminants (Keire et al. 2010b). None of these oversulphated GAGs are naturally occurring compounds; they are semi-synthetic. Current tests and specifications should make deliberate adulteration of heparin with other sulphated polysaccharides detectable at a level, which would make the practice unprofitable.

The two pharmacopoeial compendia most used internationally, the United States Pharmacopeia (USP) and the European Pharmacopoeia (EP), both use the orthogonal strategy in their most recent revisions. The monographs for Heparin Sodium and Heparin Calcium in the EP (2010b, 2010c), and that for Heparin sodium in the USP (2009), have seen rapid change in response to the OSCS incident and have reached similar, but not entirely identical, repertoires of testing. The differences are due in part to the differences in regulatory regimes between the USA and Europe; the FDA in the USA has a zero-tolerance approach to OSCS contamination, whereas in Europe decisions on these matters are taken at the national level rather than centrally, leading to some variation in the level of OSCS deemed tolerable, and therefore necessitating a quantitative method for OSCS. Although NMR spectroscopy has been used quantitatively (McEwen et al. 2008), the current EP monograph uses SAX chromatography.

## 3.4 *Synthetic and Semi-Synthetic Heparin*

The molecular heterogeneity of heparin does not in any way hinder its usefulness as an anticoagulant and antithrombotic drug, but it does mean that no two preparations are exactly alike, and it means that although we know a great deal about the structure and chemistry of heparin, its potency is still necessarily expressed in more or less arbitrary units. Each molecule, containing the antithrombin-binding motif at a specific position in its sequence, will have a subtly different effect on proteins of the coagulation cascade to a heparin molecule with a different sequence. The reduction of complexity obtained by constructing a synthetic, homogeneous heparin would mean that its activity could be expressed simply in terms of its molar concentration, as is the case for small molecule drugs. This goal is currently being pursued in more than one way.

One approach is that of ab initio synthesis; this has already given rise to the synthetic pentasaccharide fondaparinux, now in clinical use (Petitou and van Boeckel 2004) and has been extended to more complex constructs, such as the compound currently designated EP217609, in which the pentasaccharide is attached to a direct thrombin inhibitor (Petitou et al. 2009). The attachment of a biotin moiety to the molecule also allows the efficient neutralisation by avidin of this potent anticoagulant.

A second initiative involves chemoenzymatic synthesis, on the basis of a bacterial polysaccharide with the same structure as heparosan, using recombinant versions of the enzymes of heparin biosynthesis. These may be re-engineered to change substrate specificity, so as to produce novel heparin-like compounds with predictable spectra of biological activity (Laremore et al. 2010). When such reactions take place using microfluidic devices, this strategy amounts to the construction of an "artificial Golgi" (Martin et al. 2009).

## 4 Conclusions

Although heparin is a relatively simple form of heparan sulphate, lacking the full range of diversity of this class of glycosaminoglycans, a complete chemical description of any naturally occurring heparin sample is beyond the grasp of physicochemical analysis, for reasons of its sheer complexity. The numerous techniques discussed briefly above all have their specific uses; whether for the challenging task of sequencing a heparin/heparan oligosaccharide, in a research environment, or for checking that a heparin product for clinical use meets the specifications that allow it to be distributed for that purpose.

This complexity has some interesting consequences. How are we to define the concentration of a heparin solution? We can measure, to good enough accuracy, the molecular weight distribution of a heparin sample, and use that information to tell us what weight and molar concentrations we have, but if we then go on to measure the thermodynamics and kinetics of an interaction of that material with one or other of the many heparin-binding proteins, or indeed to measure the potency of the sample in a biological assay, we still have not taken into account that the property of heparin we are measuring may be molecular weight dependent. Then, if we simply extract (with much labour) a single monodisperse heparin fragment, we find that the property may very well be sequence dependent. What is more, there is every possibility that the sequence dependence may not be all-or-nothing, but the "epitope" which carries affinity for a particular protein may be a conformationally defined pattern of sulphates, displayed by more than one sequence in either direction along the heparin molecule. It is not surprising that synthetic heparin substitutes and small molecule mimetics are available and under development. The needs of good regulation of medicines, and of quality control and assurance, are not readily compatible with such a degree of complexity; however, in spite of everything, heparin and its derived products will continue to be used in the clinic and in the research laboratory for a great deal longer than its first hundred years.

## References

Angulo J, Ojeda R, de Paz JL, Lucas R, Nieto PM, Lozano RM, Redondo-Horcajo M, Gimenez-Gallego G, Martin-Lomas M (2004) The activation of fibroblast growth factors (FGFs) by glycosaminoglycans: influence of the sulfation pattern on the biological activity of FGF-1. Chembiochem 5:55–61

Aquino RS, Pereira MS, Vairo BC, Cinelli LP, Santos GR, Fonseca RJ, Mourao PA (2010) Heparins from porcine and bovine intestinal mucosa: are they similar drugs? Thromb Haemost 103:1005–1015

Bertini S, Bisio A, Torri G, Bensi D, Terbojevich M (2005) Molecular weight determination of heparin and dermatan sulfate by size exclusion chromatography with a triple detector array. Biomacromolecules 6:168–173

Burlingame RW, Thomas GH, Stevens RL, Schmid K, Moser HW (1981) Direct quantitation of glycosaminoglycans in 2 mL of urine from patients with mucopolysaccharidoses. Clin Chem 27:124–128

Cabassi F, Casu B, Perlin AS (1978) Infrared absorption and raman scattering of sulfate groups of heparin and related glycosaminoglycans in aqueous solution. Carbohydr Res 63:1–11

Carlsson P, Kjellen L (2011) Heparin biosynthesis. In: Lever R, Mulloy B, Page CP (eds) Heparin – a century of progress. Springer, Heidelberg, pp 23–41

Casu B, Gennaro U (1975) A conductimetric method for the determination of sulphate and carboxyl groups in heparin and other mucopolysaccharides. Carbohydr Res 39:168–176

Casu B, Scovenna G, Cifonelli AJ, Perlin AS (1978) Infrared spectra of glycosaminoglycans in deuterium oxide and deuterium chloride solution: quantitative evaluation of uronic acid and acetamidodeoxyhexose moieties. Carbohydr Res 63:13–27

Casu B, Oreste P, Torri G, Zoppetti G, Choay J, Lormeau JC, Petitou M, Sinay P (1981) The structure of heparin oligosaccharide fragments with high anti-(factor Xa) activity containing the minimal antithrombin III-binding sequence. Chemical and $^{13}$C nuclear-magnetic-resonance studies. Biochem J 197:599–609

Casu B, Guerrini M, Naggi A, Torri G, De-Ambrosi L, Boveri G, Gonella S, Ronzoni G (1995) Differentiation of beef and pig mucosal heparins by NMR spectroscopy. Thromb Haemost 74:1205

Cavalcante MCM, Allodi S, Valente AP, Straus AH, Takahashi HK, Mourao PAS, Pavao MSG (2000) Occurrence of heparin in the invertebrate *Styela plicata* (Tunicata) is restricted to cell layers facing the outside environment: an ancient role in defense? J Biol Chem 275:36189–36196

Chess ED et al (2011) Case study: contamination of heparin with oversulfated chondroitin sulfate. In: Lever R, Mulloy B, Page CP (eds) Heparin – a century of progress. Springer, Heidelberg, pp 99–125

Copeland R, Balasubramaniam A, Tiwari V, Zhang F, Bridges A, Linhardt RJ, Shukla D, Liu J (2008) Using a 3-O-sulfated heparin octasaccharide to inhibit the entry of herpes simplex virus type 1. Biochemistry 47:5774–5783

Domanig R, Jobstl W, Gruber S, Freudemann T (2009) One-dimensional cellulose acetate plate electrophoresis-A feasible method for analysis of dermatan sulfate and other glycosaminoglycan impurities in pharmaceutical heparin. J Pharm Biomed Anal 49:151–155

European Pharmacopoeia (2010a) European Directorate on the Quality of Medicines and HealthCare.

European Pharmacopoeia (2010b) Heparin Sodium, monograph 0333. In: Council of Europe, Strasbourg

European Pharmacopoeia (2010c) Heparin Calcium, monograph 0332. In: Council of Europe, Strasbourg

Farndale RW, Buttle DJ, Barrett AJ (1986) Improved quantitation and discrimination of sulphated glycosaminoglycans by use of dimethylmethylene blue. Biochim Biophys Acta 883:173–177

Ferro DR, Provasoli A, Ragazzi M, Torri G, Casu B, Gatti G, Jacquinet JC, Sinay P, Petitou M, Choay J (1986) Evidence for conformational equilibrium of the sulfated L-iduronate residue in heparin and in synthetic heparin mono- and oligosaccharides: NMR and force-field studies. J Am Chem Soc 108:6773–6778

Ferro DR, Provasoli A, Ragazzi M, Casu B, Torri G, Bossennec V, Perly B, Sinay P, Petitou M, Choay J (1990) Conformer populations of L-iduronic acid residues in glycosaminoglycan sequences. Carbohydr Res 195:157–167

Forster MJ, Mulloy B (1993) Molecular dynamics study of iduronate ring conformation. Biopolymers 33:575–588

Forster M, Jones C, Mulloy B (1989) NOEMOL: integrated molecular graphics and the simulation of nuclear overhauser effects in NMR spectroscopy. J Mol Graph 7(196–201):217

Gama CI, Hsieh-Wilson LC (2005) Chemical approaches to deciphering the glycosaminoglycan code. Curr Opin Chem Biol 9:609–619

Gandra M, Cavalcante M, Pavao M (2000) Anticoagulant sulfated glycosaminoglycans in the tissues of the primitive chordate Styela plicata (Tunicata). Glycobiology 10:1333–1340
Grant D, Long WF, Williamson FB (1987) Infrared spectroscopy of heparin-cation complexes. Biochem J 244:143–149
Grant D, Long WF, Moffat CF, Williamson FB (1989) Infrared spectroscopy of chemically modified heparins. Biochem J 261:1035–1038
Gray E (2011) Standardisation of unfractionated and low molecular weight heparin. In: Lever R, Mulloy B, Page CP (eds) Heparin – a century of progress. Springer, Heidelberg, pp 65–76
Gray E et al (2011) The anticoagulant and antithrombotic mechanisms of heparin. In: Lever R, Mulloy B, Page CP (eds) Heparin – a century of progress. Springer, Heidelberg, pp 43–61
Guerrini M, Guglieri S, Naggi A, Sasisekharan R, Torri G (2007) Low molecular weight heparins: structural differentiation by bidimensional nuclear magnetic resonance spectroscopy. Semin Thromb Hemost 33:478–487
Guerrini M, Shriver Z, Bisio A, Naggi A, Casu B, Sasisekharan R, Torri G (2009a) The tainted heparin story: an update. Thromb Haemost 102:907–911
Guerrini M, Zhang Z, Shriver Z, Naggi A, Masuko S, Langer R, Casu B, Linhardt RJ, Torri G, Sasisekharan R (2009b) Orthogonal analytical approaches to detect potential contaminants in heparin. Proc Natl Acad Sci USA 106:16956–16961
Holmer E, Soderstrom G, Andersson LO (1979) Studies on the mechanism of the rate-enhancing effect of heparin on the thrombin-antithrombin III reaction. Eur J Biochem 93:1–5
Hricovini M, Guerrini M, Torri G, Casu B (1997) Motional properties of *E. coli* polysaccharide K5 in aqueous solution analyzed by NMR relaxation measurements. Carbohydr Res 300:69–76
Inoue S, Miyawaki M (1975) Quantitative analysis of iduronic acid and glucuronic acid in sulfated galactosaminoglycuronans by gas chromatography. Anal Biochem 65:164–174
Johnson EA, Kirkwood TB, Stirling Y, Perez-Requejo JL, Ingram GI, Bangham DR, Brozovic M (1976) Four heparin preparations: anti-Xa potentiating effect of heparin after subcutaneous injection. Thromb Haemost 35:586–591
Jones CJ, Membreno N, Larive CK (2010) Insights into the mechanism of separation of heparin and heparan sulfate disaccharides by reverse-phase ion-pair chromatography. J Chromatogr A 1217:479–488
Keire DA, Trehy ML, Reepmeyer JC, Kolinski RE, Ye W, Dunn J, Westenberger BJ, Buhse LF (2010a) Analysis of crude heparin by $^{1}$H NMR, capillary electrophoresis, and strong-anion-exchange-HPLC for contamination by over sulfated chondroitin sulfate. J Pharm Biomed Anal 51:921–926
Keire DA, Mans DJ, Ye H, Kolinski RE, Buhse LF (2010b) Assay of possible economically motivated additives or native impurities levels in heparin by $^{1}$H NMR, SAX-HPLC, and anticoagulation time approaches. J Pharm Biomed Anal 52:656–664
Khan S, Gor J, Mulloy B, Perkins SJ (2010) Semi-rigid solution structures of heparin by constrained x-ray scattering modelling: new insight into heparin-protein complexes. J Mol Biol 395:504–521
Knobloch JE, Shaklee PN (1997) Absolute molecular weight distribution of low-molecular-weight heparins by size-exclusion chromatography with multiangle laser light scattering detection. Anal Biochem 245:231–241
Kreuger J, Spillmann D, Li JP, Lindahl U (2006) Interactions between heparan sulfate and proteins: the concept of specificity. J Cell Biol 174:323–327
Laremore TN, Ly M, Solakyildirim K, Zagorevski DV, Linhardt RJ (2010) High-resolution preparative separation of glycosaminoglycan oligosaccharides by polyacrylamide gel electrophoresis. Anal Biochem 401:236–241
Linhardt RJ, Loganathan D, Al-Hakim A, Wang HM, Walenga JM, Hoppensteadt D, Fareed J (1990) Oligosaccharide mapping of low molecular weight heparins: structure and activity differences. J Med Chem 33:1639–1645
Lucas R, Angulo J, Nieto PM, Martin-Lomas M (2003) Synthesis and structural study of two new heparin-like hexasaccharides. Org Biomol Chem 1:2253–2266

Martin JG, Gupta M, Xu Y, Akella S, Liu J, Dordick JS, Linhardt RJ (2009) Toward an artificial Golgi: redesigning the biological activities of heparan sulfate on a digital microfluidic chip. J Am Chem Soc 131:11041–11048

Matsuo K, Namatame H, Taniguchi M, Gekko K (2009) Vacuum-ultraviolet circular dichroism analysis of glycosaminoglycans by synchrotron-radiation spectroscopy. Biosci Biotechnol Biochem 73:557–561

McEwen I, Mulloy B, Hellwig E, Kozerski L, Beyer T, Holzgrabe U, Wanko R, Spieser JM, Rodomonte A (2008) Determination of oversulphated chondroitin sulphate and dermatan sulphate in unfractionated heparin by $^{1}$H-NMR – collaborative study for quantification and analytical determination of LoD. Pharmeuropa Bio 2008:31–39

Meissen JK, Sweeney MD, Girardi M, Lawrence R, Esko JD, Leary JA (2009) Differentiation of 3-O-sulfated heparin disaccharide isomers: identification of structural aspects of the heparin CCL2 binding motif. J Am Soc Mass Spectrom 20:652–657

Merry CL, Lyon M, Deakin JA, Hopwood JJ, Gallagher JT (1999) Highly sensitive sequencing of the sulfated domains of heparan sulfate. J Biol Chem 274:18455–18462

Mobli M, Nilsson M, Almond A (2008) The structural plasticity of heparan sulfate NA-domains and hence their role in mediating multivalent interactions is confirmed by high-accuracy $^{15}$N-NMR relaxation studies. Glycoconj J 25:401–414

Mourier PA, Viskov C (2004) Chromatographic analysis and sequencing approach of heparin oligosaccharides using cetyltrimethylammonium dynamically coated stationary phases. Anal Biochem 332:299–313

Mulloy B (2002) Gel permeation chromatography of heparin. In: Volpi N (ed) Glycosaminoglycans. Research Signpost, Trivandrum

Mulloy B (2005) The specificity of interactions between proteins and sulfated polysaccharides. An Acad Bras Cienc 77:651–664

Mulloy B, Forster MJ (2000) Conformation and dynamics of heparin and heparan sulfate. Glycobiology 10:1147–1156

Mulloy B, Johnson EA (1987) Assignment of the 1H-N.M.R. spectra of heparin and heparan sulphate. Carbohydr Res 170:151–165

Mulloy B, Linhardt RJ (2001) Order out of complexity – protein structures that interact with heparin. Curr Opin Struct Biol 11:623–628

Mulloy B, Forster MJ, Jones C, Davies DB (1993) N.m.r. and molecular-modelling studies of the solution conformation of heparin. Biochem J 293:849–858

Mulloy B, Forster MJ, Jones C, Drake AF, Johnson EA, Davies DB (1994) The effect of variation of substitution on the solution conformation of heparin: a spectroscopic and molecular modelling study. Carbohydr Res 255:1–26

Mulloy B, Crane DT, Drake AF, Davies DB (1996) The interaction between heparin and polylysine: a circular dichroism and molecular modelling study. Braz J Med Biol Res 29:721–729

Mulloy B, Gray E, Barrowcliffe TW (2000) Characterization of unfractionated heparin: comparison of materials from the last 50 years. Thromb Haemost 84:1052–1056

Mulloy B, Heath A, Behr-Gross ME (2007) Establishment of replacement batches for heparin low-molecular-mass for calibration CRS, and the International Standard Low Molecular Weight Heparin for Calibration. Pharmeuropa Bio 2007:29–48

Mulloy B, Hogwood J, Gray E (2010) Assays and reference materials for current and future applications of heparins. Biologicals 38:459–466

Murphy KJ, McLay N, Pye DA (2008) Structural studies of heparan sulfate hexasaccharides: new insights into iduronate conformational behavior. J Am Chem Soc 130:12435–12444

Nader HB, Chavante SF, dos-Santos EA (1999) Heparan sulfates and heparins: similar compounds performing the same functions in vertebrates and invertebrates? Braz J Med Biol Res 32:529–538

Naimy H, Leymarie N, Bowman MJ, Zaia J (2008) Characterization of heparin oligosaccharides binding specifically to antithrombin III using mass spectrometry. Biochemistry 47:3155–3161

Naimy H, Leymarie N, Zaia J (2010) Screening for anticoagulant heparan sulfate octasaccharides and fine structure characterization using tandem mass spectrometry. Biochemistry 49:3743–3752

Nieduszynski IA, Atkins ED (1973) Conformation of the mucopolysaccharides. X-ray fibre diffraction of heparin. Biochem J 135:729–733

Ostrovsky O, Berman B, Gallagher J, Mulloy B, Fernig DG, Delehedde M, Ron D (2002) Differential effects of heparin saccharides on the formation of specific fibroblast growth factor (FGF) and FGF receptor complexes. J Biol Chem 277:2444–2453

Pan J, Qian Y, Zhou X, Pazandak A, Frazier SB, Weiser P, Lu H, Zhang L (2010) Oversulfated chondroitin sulfate is not the sole contaminant in heparin. Nat Biotechnol 28:203–207

Petitou M, van Boeckel CA (2004) A synthetic antithrombin III binding pentasaccharide is now a drug! What comes next? Angew Chem Int Ed Engl 43:3118–3133

Petitou M, Nancy-Portebois V, Dubreucq G, Motte V, Meuleman D, de KM, van Boeckel CA, Vogel GM, Wisse JA (2009) From heparin to EP217609: the long way to a new pentasaccharide-based neutralisable anticoagulant with an unprecedented pharmacological profile. Thromb Haemost 102:804–810

Racaud A, Antoine R, Joly L, Mesplet N, Dugourd P, Lemoine J (2009) Wavelength-tunable ultraviolet photodissociation (UVPD) of heparin-derived disaccharides in a linear ion trap. J Am Soc Mass Spectrom 20:1645–1651

Rice KG, Kim YS, Grant AC, Merchant ZM, Linhardt RJ (1985) High-performance liquid chromatographic separation of heparin-derived oligosaccharides. Anal Biochem 150:325–331

Rudd TR, Skidmore MA, Guimond SE, Holman J, Turnbull JE, Lauder RM, Fernig DG, Yates EA (2009a) The potential for circular dichroism as an additional facile and sensitive method of monitoring low-molecular-weight heparins and heparinoids. Thromb Haemost 102:874–878

Rudd TR, Skidmore MA, Guimond SE, Cosentino C, Torri G, Fernig DG, Lauder RM, Guerrini M, Yates EA (2009b) Glycosaminoglycan origin and structure revealed by multivariate analysis of NMR and CD spectra. Glycobiology 19:52–67

Rudd TR, Hussain R, Siligardi G, Yates EA (2010) Raman and Raman optical activity of glycosaminoglycans. Chem Commun (Camb) 46:4124–4126

Ruiz-Calero V, Puignou L, Diez M, Galceran MT (2001) Determination of free and total sulfate and phosphate in glycosaminoglycans by column-switching high-performance size-exclusion and ion chromatography and single-column ion chromatography. Analyst 126:169–174

Segura MM, Garnier A, Durocher Y, Ansorge S, Kamen A (2010) New protocol for lentiviral vector mass production. Methods Mol Biol 614:39–52

Shukla D, Liu J, Blaiklock P, Shworak NW, Bai X, Esko JD, Cohen GH, Eisenberg RJ, Rosenberg RD, Spear PG (1999) A novel role for 3-O-sulfated heparan sulfate in herpes simplex virus 1 entry. Cell 99:13–22

Skidmore M, Atrih A, Yates E, Turnbull JE (2009) Labelling heparan sulphate saccharides with chromophore, fluorescence and mass tags for HPLC and MS separations. Methods Mol Biol 534:157–169

Somsen GW, Tak YH, Torano JS, Jongen PMJM, de Jong GJ (2009) Determination of oversulfated chondroitin sulfate and dermatan sulfate impurities in heparin by capillary electrophoresis. J Chromatogr A 1216:4107–4112

Stevens ES, Morris ER, Rees B, Sutherland JC (1985) Synchroron light source applied to measuring the vacuum ultraviolet circular dichroism of heparin. J Am Chem Soc 107:2982–2983

Stone AL (1985) Far-ultraviolet circular dichroism and uronic acid components of anticoagulant deca-, dodeca-, tetradeca-, and octadecasaccharide heparin fractions. Arch Biochem Biophys 236:342–353

Stringer SE, Gallagher JT (1997) Specific binding of the chemokine platelet factor 4 to heparan sulfate. J Biol Chem 272:20508–20514

Stringer SE, Forster MJ, Mulloy B, Bishop CR, Graham GJ, Gallagher JT (2002) Characterization of the binding site on heparan sulfate for macrophage inflammatory protein 1α. Blood 100:1543–1550

Templeton DM (1988) General occurrence of isosbestic points in the metachromatic dye complexes of sulphated glycosaminoglycans. Int J Biol Macromol 10:131–136

Thunberg L, Backstrom G, Lindahl U (1982) Further characterization of the antithrombin-binding sequence in heparin. Carbohydr Res 100:393–410

Turnbull JE, Gallagher JT (1988) Oligosaccharide mapping of heparan sulphate by polyacrylamide-gradient-gel electrophoresis and electrotransfer to nylon membrane. Biochem J 251:597–608

United States Pharmacopeia (2009) Heparin sodium. USP 32-NF 27, 2552-2554. , USP, Rockville

Varki A, Cummings RD, Esko JD, Freeze HH, Stanley P, Marth JD, Bertozzi CR, Hart GW, Etzler ME (2009) Symbol nomenclature for glycan representation. Proteomics 9:5398–5399

Volpi N (1993) 'Fast moving' and 'slow moving' heparins, dermatan sulfate, and chondroitin sulfate: Qualitative and quantitative analysis by agarose-gel electrophoresis. Carbohydr Res 247:263–278

Volpi N, Bolognani L (1993) Glycosaminoglycans and proteins: different behaviours in high-performance size-exclusion chromatography. J Chromatogr 630:390–396

Volpi N, Linhardt RJ (2010) High-performance liquid chromatography-mass spectrometry for mapping and sequencing glycosaminoglycan-derived oligosaccharides. Nat Protoc 5:993–1004

Volpi N, Maccari F, Linhardt RJ (2008) Capillary electrophoresis of complex natural polysaccharides. Electrophoresis 29:3095–3106

Wang Z, Xu Y, Yang B, Tiruchinapally G, Sun B, Liu R, Dulaney S, Liu J, Huang X (2010) Preactivation-based, one-pot combinatorial synthesis of heparin-like hexasaccharides for the analysis of heparin-protein interactions. Chemistry 16:8365–8375

Xiong S, Zhang L, He QY (2008) Fractionation of proteins by heparin chromatography. Methods Mol Biol 424:213–221

Zaia J (2009) On-line separations combined with MS for analysis of glycosaminoglycans. Mass Spectrom Rev 28:254–272

Ziegler A, Zaia J (2006) Size-exclusion chromatography of heparin oligosaccharides at high and low pressure. J Chromatogr B Analyt Technol Biomed Life Sci 837:76–86

# Case Study: Contamination of Heparin with Oversulfated Chondroitin Sulfate

**Edward K. Chess, Shawn Bairstow, Shane Donovan, Karalyn Havel, Peifeng Hu, Richard J. Johnson, Sarah Lee, Jeff McKee, Reagan Miller, Edwin Moore, Mark Nordhaus, Joseph Ray, Christina Szabo, and Todd Wielgos**

## Contents



**Abstract** In late 2007 and early 2008, a cluster of adverse events in patients receiving Heparin Sodium Injection occurred in the United States and in some countries in Europe. The adverse events were reported as being "allergic type" reactions, chiefly characterized by acute hypotension, nausea, and shortness of breath. The root cause of the cluster of adverse events was determined to be a contamination of the heparin by oversulfated chondroitin sulfate. The isolation and structure determination of this contaminant was accomplished by an FDA-led consortium of academic and government laboratories and independently by Baxter

E.K. Chess (✉)
Baxter Healthcare Corporation, Round Lake, IL, USA
e-mail: ed_k_chess@baxter.com

R. Lever et al. (eds.), *Heparin - A Century of Progress*,
Handbook of Experimental Pharmacology 207,
DOI 10.1007/978-3-642-23056-1_6, © Springer-Verlag Berlin Heidelberg 2012

Healthcare, whose vial products were first identified in the USA as being associated with the adverse events. Oversulfated chondroitin sulfate was shown to produce acute hypotension in animal models, demonstrating that it was most likely the causative agent responsible for certain of the reported adverse events in patients receiving the contaminated heparin products.

**Keywords** Oversulfated chondroitin sulfate • Contamination • Heparin Sodium Injection USP • Hypotension

## 1 Introduction

Now that the Global Heparin Crisis has been resolved one can look back upon the record of reported adverse events and note that the first indication that there may be an issue with some lots of Baxter's Heparin Sodium Injection, USP vial products came in mid-November 2007, with an increase in the reports of "allergic type" reactions in pediatric dialysis patients. By late December 2007 and early January 2008, Baxter's Pharmacovigilance group had picked up a potential signal of increased adverse events associated with an "allergic type" reaction in multidose heparin vial products and initiated a review of manufacturing and quality control records at the manufacturing facility in Cherry Hill, New Jersey. This review showed no discrepancies in batch records, including results from all incoming active pharmaceutical ingredient (API) and in process testing. The Baxter Quality group continued their manufacturing investigation, and, even though no issues with quality control records were found, placed certain multidose heparin vial product inventory on hold at the plant and at all distribution centers. Baxter investigative teams conducted on-site visits of dialysis centers, collecting information about the dialysis protocols and materials used. After discussions with the FDA, Baxter then issued an urgent voluntary recall on January 17, 2008, for the nine lots of multidose heparin vials that had been associated with the cluster of adverse events. Baxter then began assembling teams to investigate different aspects of the problem: a Manufacturing Team reviewed Baxter's manufacturing process from receipt of the API to release and distribution of the finished vial products; the API Team was charged with investigating the manufacture of the API; the Analytical Team was given the responsibility to determine whether there was a contaminant in the finished product; and the Biological/Medical Team was to focus on what was the root cause of the biological response.

During January, the teams were assembled and drafted work plans, and samples of recalled product and appropriate controls were shipped to Baxter's Round Lake facility for the investigations. In a parallel effort, the US FDA assembled an international academic team to support its own laboratory investigations. Baxter continued to monitor adverse event reporting and the status of the investigation. As the investigations unfolded, it became apparent the adverse events were associated with specific heparin finished product vendors, and other vendors, such as APP, were not as affected and were able to increase production to fill the critical needs of the USA. When APP's ability to fill these needs was apparent, the FDA gave Baxter approval to withdraw all products, and on February 28, 2008, Baxter promptly expanded its recall to all heparin products manufactured at Cherry Hill, including single dose and Hep-Lock product offerings. By March 5, 2008, sufficient information about the identity of the contaminant was known by Baxter and the FDA consortium to allow the FDA to report the contaminant as a heparin-like compound. A day later, the FDA posted nuclear magnetic resonance (NMR) and capillary electrophoresis (CE) methods on their website that could be used to screen for the contaminant, even though the full identity was not yet known, and by March 14, 2008, all heparin manufacturers had committed to test all heparin APIs using these methods and report the results to the FDA. This case study will focus on the activities of the Baxter Analytical and Biological/Medical Teams, and their contributions to the resolution of the heparin crisis.

## 2 Analytical Team Plan

The Analytical and Biological Teams reviewed the information contained in the adverse event descriptions and created a list of chemical and material classes that were known to cause "allergic type" reactions in patients. This list included transition metals, variability in magnesium or calcium content, presence of a microbial product, such as peptidoglycan, presence of EDTA, citrate or histamine, presence of a complement activator (that would produce C3a or C5a in vivo), presence of an unknown peptide or protein (antibody), presence of a carbohydrate, and presence of an arachidonic acid metabolite. The Analytical Team next assessed analytical methodologies that could be applied to screen for these materials in products and made some risk-based decisions on which to try first and which to delay until the initial results were known. We also wanted to establish whether the issue originated with the API heparin, or whether the "allergic type" reactions were being caused by the final product manufacturing process (batching, excipients, filling, container closure system, sterilization). Thus, a work flow was established to investigate as many of these different pathways as possible in parallel until one or more of them found a difference between analytical results from the lots associated with the cluster of adverse events (test lots) and results from lots that were not (control lots). A tree of the techniques used to provide initial screening of API and Finished Product (FP) is given in Fig. 1.

Quality Assurance personnel, working with input from the latest data from the Pharmacovigilance group and batch records from the manufacturing facility, created a "Heparin Product Tree" that mapped the lot numbers for all lots of heparin API, sodium chloride, and benzyl alcohol (the latter two are the only formulation excipients) used over the past several quarters to all lots of all finished product codes (different sizes and concentrations), and then indicated the specific codes and lots that were associated with the cluster of adverse events. This analysis allowed rapid determination of which lots of FP, API, and excipients to evaluate in the

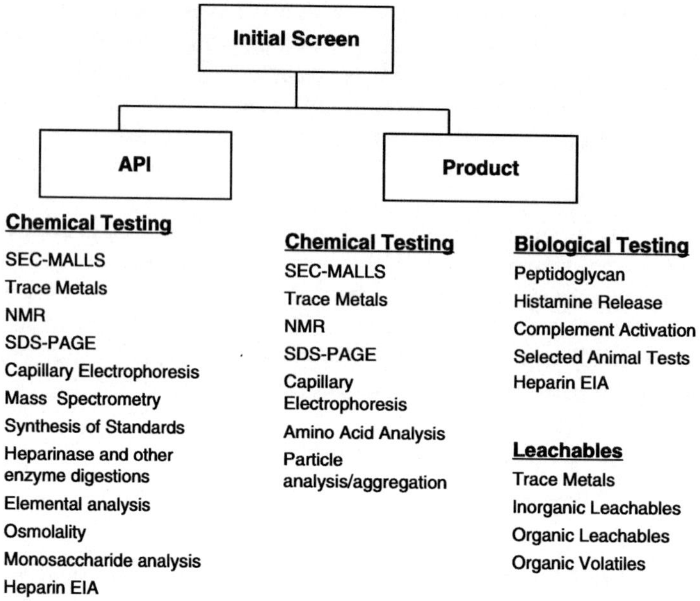


**Fig. 1** Initial screening assays for heparin API and finished product

screening process, and materials were shipped from the field or pulled from retained manufacturing samples and sent to the Baxter's analytical testing facilities. Representative analytical testing roadmaps for the FP and heparin API are shown in Figs. 2 and 3, respectively. These roadmaps are versions from late in the process, showing that differences among test and control articles were initially observed for certain analytical techniques, such as capillary electrophoresis (CE) and nuclear magnetic resonance (NMR). These techniques, and the results derived from them, will be examined in further detail. Also of great importance, however, were the "negative" results from the evaluations of the container/closure and excipients: no differences could be found in the extractables profiles from lots of container/closure materials or in the impurity profiles for the excipients used in test and control lots of finished product.

# 3 Analytical Team Results

## *3.1 General Screening Results for API and Finished Product*

When the heparin test lots and control lots of finished products were initially screened by the techniques shown in Fig. 1, SDS-PAGE, and particulate analysis did not show differences between the test and control units. As a further test for

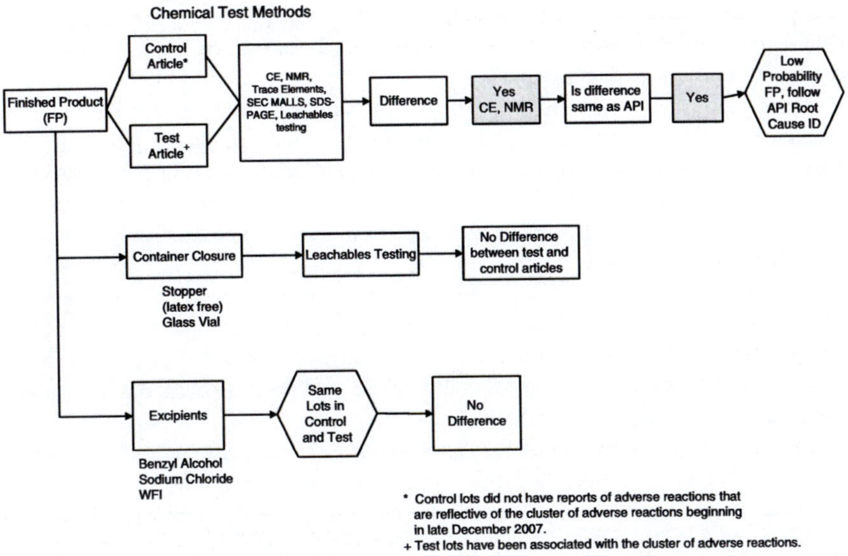


**Fig. 2** Roadmap for finished product testing

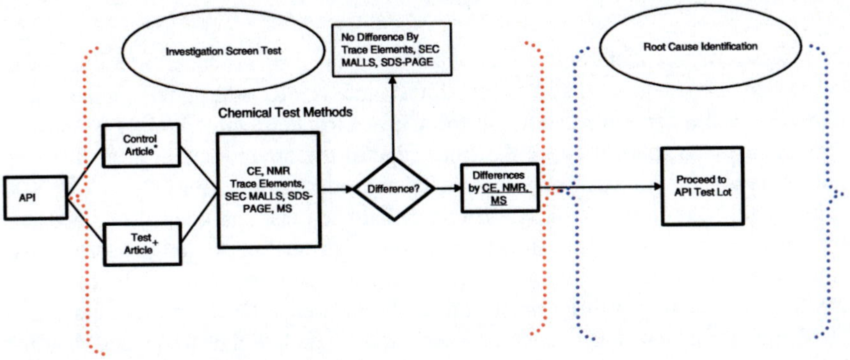


* Control lots did not have reports of adverse reactions that are reflective of the cluster of adverse reactions beginning in late December 2007.
+ Test lots have been associated with the cluster of adverse reactions.

**Fig. 3** Roadmap for heparin API testing

aggregation, the finished product lots were evaluated for their propensity to form aggregates by storage at 4°C for 6 weeks with the result that no observable aggregates formed in any samples. A substantial evaluation of the extractable and leachable chemical species from container/closure components was conducted. The results from this study conclusively demonstrated that there were no chemically relevant differences in the extractable and leachable profiles for the different lots of container closure components used in the manufacture of finished products (test or control lots). This result clearly demonstrated that any differences found between test and control lots must be a result of either the API or post-manufacture product manipulation (i.e., product tampering).

Results from trace metals analysis on the finished products by inductively coupled plasma atomic emission spectroscopy showed a slightly higher level of sulfur in test versus control articles. Analyses of finished products by SEC-MALLS produced results indicating a slightly higher average molecular weight in the control articles, and this difference appeared to be due to increased polydispersity in the low-molecular-weight materials. The differences observed between test and control finished products for both the ICP-AES and SEC-MALLS results were judged to be too small to be effective criteria for a screening method. Clear differences were observed in the amino acid analyses of test and control finished products, although these differences could not be explained by normal amino acids and could not initially be interpreted.

The APIs were also evaluated for the presence of citrate, EDTA, and histamine, chemicals known to cause "allergic type" reactions upon infusion, but these chemicals were not observed above the detection limits of the methods used.

## 3.2 *Initial CE and NMR Results*

In analyses of test and control lots of API and finished products by an in-house developed CE method (Patel et al. 2008) and NMR, additional signals were observed in the heparin test lots. In the CE electropherogram (Fig. 4) of the test lots, there was a sharp peak on the front edge of the broad heparin peak that was not observed in the electropherogram of the control lots. Initially, $^{13}$C NMR (Fig. 5) gave the most definitive difference between test and control lots because the spread in chemical shifts showed multiple additional peaks in the test lots that were free from overlap with the peaks due to heparin in the control lots. In the $^{1}$H NMR spectrum (Fig. 6), the most noticeable difference between the control and test heparin lots was a resonance observed at 0.11 ppm higher frequency than the heparin acetamide resonance. After comparison of about 20 test and control lots, it was decided that time would be saved by acquiring only the $^{1}$H NMR data with a focus on the presence or absence of the 2.16 ppm (unknown contaminant peak). The differences in the CE and $^{1}$H NMR data were judged as being significant enough that these two techniques could be used for screening.

Within days of determining that the same additional peaks were consistently observed in spectra of the test lots, difference spectra (i.e., the spectra that resulted by subtracting a spectrum of a control lot from that of a test lot) were calculated. Both $^{1}$H and $^{13}$C difference spectra (Figs. 5c and 6c) indicated that the contaminant was a saccharide. In addition, the difference spectra were used quantitatively to count the number of hydrogens and carbons. When these spectra indicated that there were about 12 hydrogens and 12 carbons per repeat unit, and the chemical shifts were consistent with a saccharide structure, the contaminant was suspected of being a polydisaccharide.

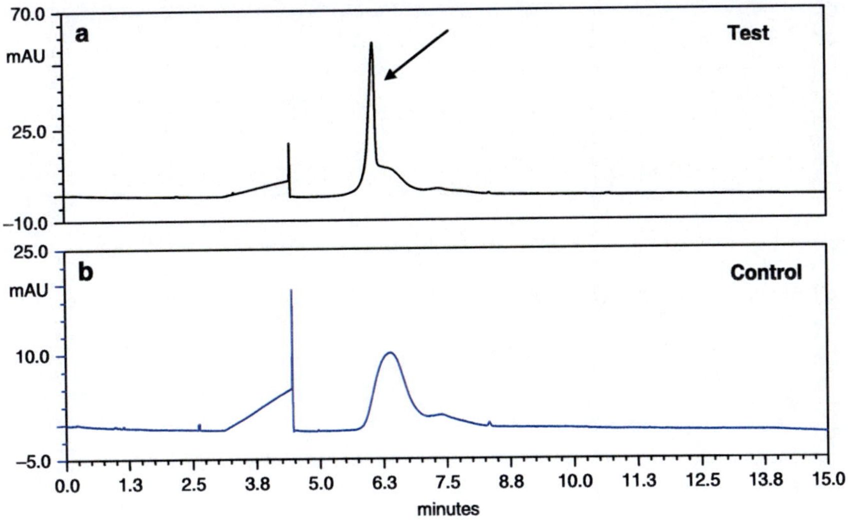


**Fig. 4** Capillary electropherograms of (**a**) Baxter heparin FP test lot and (**b**) Baxter heparin FP control lot

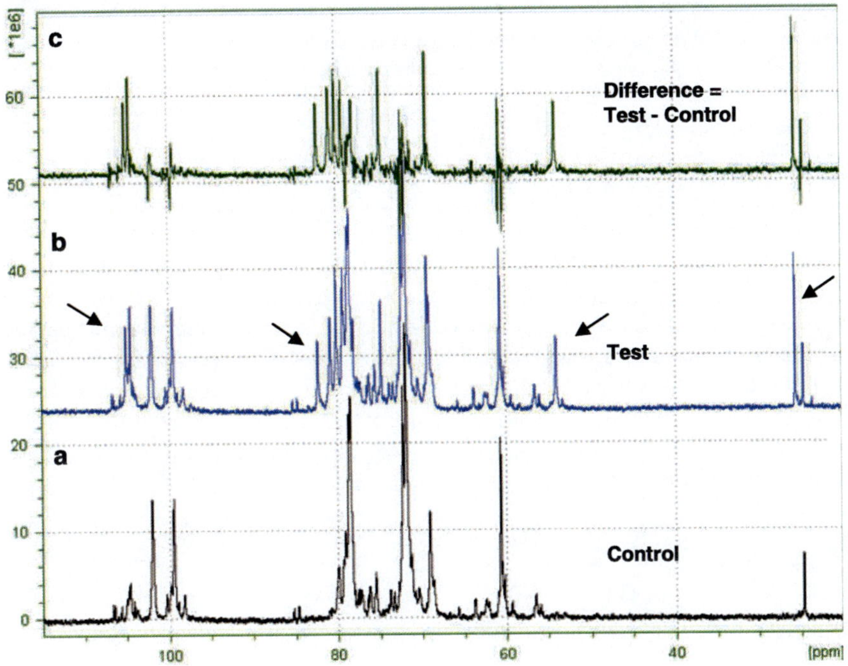


**Fig. 5** (**a**) $^{13}$C NMR spectra of a Baxter heparin control lot, (**b**) a Baxter heparin test lot, and (**c**) the difference spectrum [obtained by subtracting the control heparin spectrum from that of the test (contaminated) heparin spectrum]

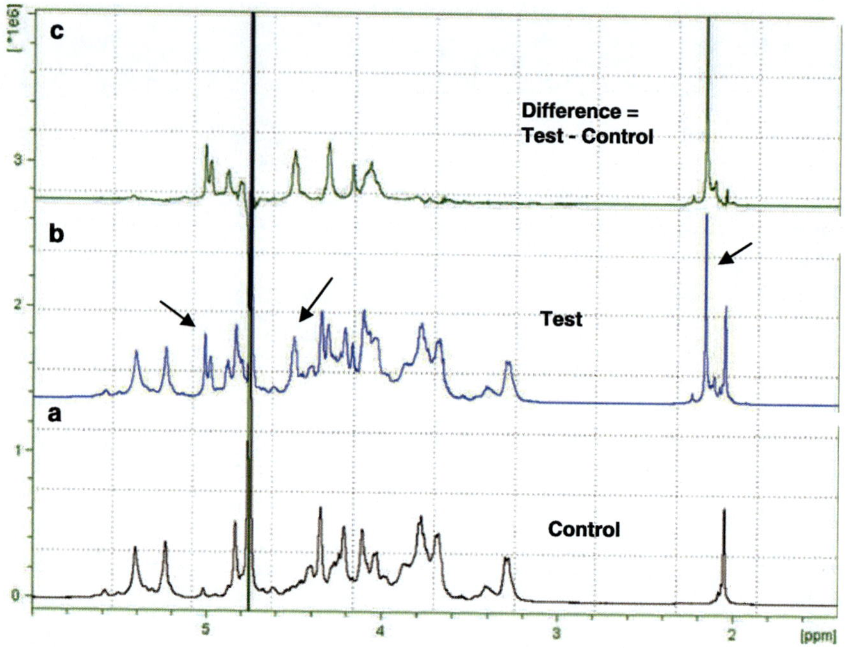


**Fig. 6** Proton NMR spectra of (**a**) Baxter heparin control lot, (**b**) Baxter heparin test lot, and (**c**) the difference spectrum [obtained by subtracting the control heparin spectrum from that of the test (contaminated) heparin spectrum]

## 3.3 *Initial MS Results*

Further evidence of a glycosaminoglycan (GAG)-like contaminant was obtained by direct infusion electrospray ionization mass spectrometric (ESI-MS) analysis of desalted heparin samples at high cone voltage (55 V). From the control lots of heparin, a prominent pair of ions was observed at *m/z* 240 and 416, corresponding to glucosamine (Structure **1**) and a disaccharide with a sulfate group (Structure **2**), respectively (Fig. 7). A second pair of ions with *m/z* values of 282, and 458, respectively, was also present. These two ions corresponded to an acetylated glucosamine (Structure **3**) and a disaccharide containing an acetylated glucosamine (Structure **4**).

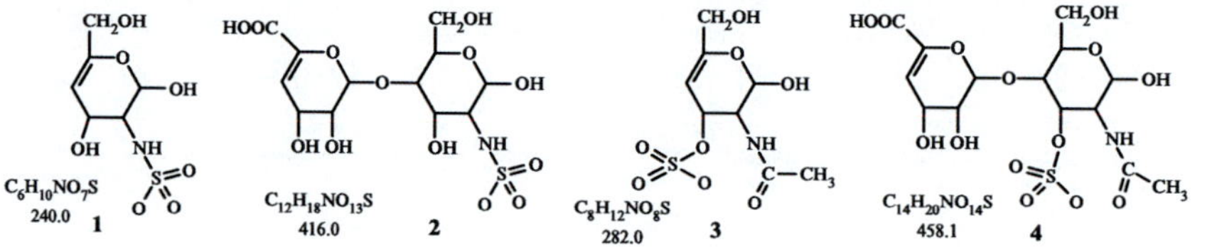

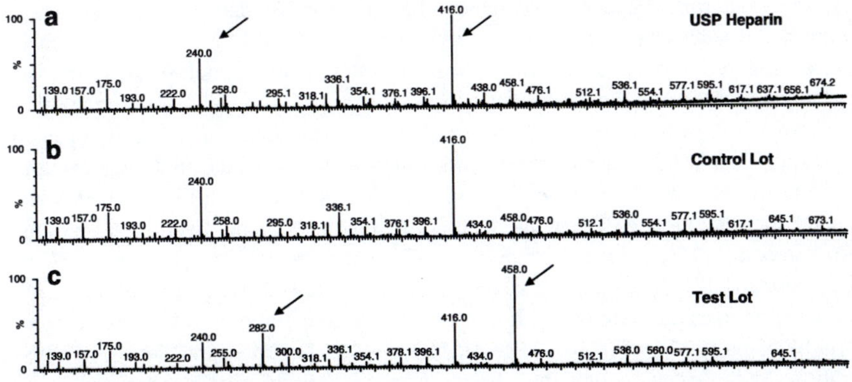


**Fig. 7** Electrospray mass spectra acquired at cone voltage 55 V of (**a**) USP heparin, (**b**) a heparin control lot, and (**c**) a heparin test lot

The observation of these ions suggested that the ESI-MS experiments, under high cone voltage conditions, desulfated heparin and broke the glycosidic linkages of the heparin backbone. The low relative abundance of the 282/458 ion pair qualitatively reflected the fact that ~15% of the glucosamine units in heparin structure were acetylated. These data showed that major structure features of heparin were represented despite the fact that the fragments were produced from a wide range of heparin components with different chain lengths and degree of sulfation (Hu et al. 2009).

In contrast, the test lots showed a much enhanced level of the *m/z* 282/458 ion pair, suggesting a higher degree of acetylation (O- or N-). There were two ways to interpret the data. One scenario was that the increased acetylation was due to additional acetylation of heparin. In this scenario, the acetylation can occur nearly exclusively on free hydroxyl groups (the presence of free amine in heparin is very low). The other interpretation was contamination by a GAG with higher degree of N-acetylation, such as a chondroitin or dermatan sulfate. The resonance observed at ~0.11 ppm higher frequency than the main heparin acetamide resonance (2.04 ppm) in the proton NMR spectra of the test lots (2.15 ppm) was in agreement with both the O-acetylation and N-acetylation theories.

### *3.4 Initial Conclusions for API and Finished Products*

At this point, it was clear that there was some type of contamination in the heparin API lots that was associated with the cluster of adverse events. The CE migration behavior indicated that the contaminant species had a higher negative charge density than did heparin, thus migrating faster and eluting sooner than the bulk of the heparin molecules. The NMR data indicated that the contaminant was a polydisaccharide that contained an acetyl group. The MS data was also consistent with the presence of either N- or O-acetylation of a GAG-like structure. Therefore, we considered modified heparin, dermatan sulfate, and chondroitin sulfate as potential sources of the contaminant.

The same results were found for the API as for the finished products for each analytical test used: if a difference was found in the finished product, the same difference was found in the API. Both $^{1}$H-NMR and CE were confirmed to be useful for screening API, and the results of finding these relatively fast methods for screening either finished product or API were communicated to the FDA in late February, 2008. By March 6, the FDA had reviewed the two methods, made minor modifications, and had published these methods to be used for screening all lots of heparin API prior to use in US drug products (Food and Drug Administration 2008). Even though the identity of the chemical species responsible for the difference between test and control lots had not yet been established, these methods, or variants of these methods, were adopted across the world to ensure that no additional lots of heparin that had $^{1}$H-NMR and CE data consistent with the lots of heparin associated with the adverse events would be used in further production.

### *3.5 Results from Enzyme Treatments*

Lots of heparin API containing the contaminant were treated with a cocktail of heparinases (lyase I, II, and III) or with chondroitinase IIa in an attempt to distinguish between the two potential contaminants, and these digests were examined by SEC with UV detection. The heparinase cocktail did digest a major portion of the sample, but left a portion of the sample undigested. This undigested material had a molecular weight comparable to that of unfractionated heparin as measured by SEC. This was substantiated by NMR diffusion experiments on a contaminated sample that indicated the contaminant had a hydrodynamic radius similar to that of heparin or that the contaminant was strongly associated with the heparin molecules. In addition, the isolated contaminant had NMR spectra that were identical to the $^{1}$H and $^{13}$C difference spectra obtained earlier. The chondroitinase IIa had no effect on the samples at all. This led us to conclude that the contaminant was not a naturally occurring chondroitin sulfate.

The chondroitinase results were confirmed with preliminary NMR spiking experiments with assorted chondroitins, including dermatan. Spiking experiments were also performed with other GAGs including heparan sulfate and oversulfated heparin, and over-acetylated heparin. In all cases, no positive identifications were made.

### 3.6 2-D NMR Results

Both one and two-dimensional NMR experiments had been initiated on a test lot of heparin where the contaminant level was especially high (later estimated to be about 20%). However, the broadness of the peaks as well as overlap with the heparin peaks made these data intractable. The isolation of the contaminant using the heparinase cocktail was one of the key steps to determining its structure. Both $^{1}H$ and $^{13}C$ NMR data (Figs. 8 and 9) confirmed that the isolated material had spectra identical to the one-dimensional difference spectra obtained earlier.

A series of two-dimensional NMR spectra were obtained, which included COSY (Correlation Spectroscopy), TOCSY (Total Correlation Spectroscopy), $^{1}H$ edited HSQC (Heteronuclear Single Quantum Coherence), and HMBC (Heteronuclear Multiple Bond Coherence) experiments. The analysis began with the edited HSQC, which allowed assignment of the $CH_2$ group that only occurs in the hexosamine ring A, assumed initially to be a glucosamine (Glc); assignments of the anomerics of ring A and ring B (GlcA) were also obvious from this spectrum (See Structures **5** and **6**). Once these assignments were made, it was straightforward to determine

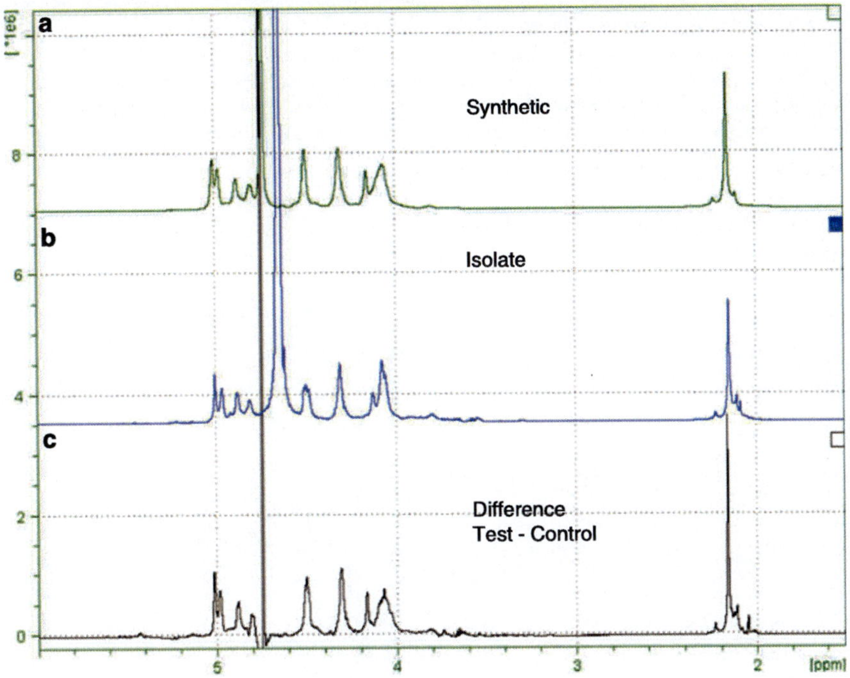


**Fig. 8** Proton NMR spectra of **(a)** a synthesized fully sulfated chondroitin, **(b)** the impurity isolated from heparin test samples, and **(c)** the difference spectrum [obtained by subtracting the normal heparin spectrum from that of the contaminated material (Fig. 6)]

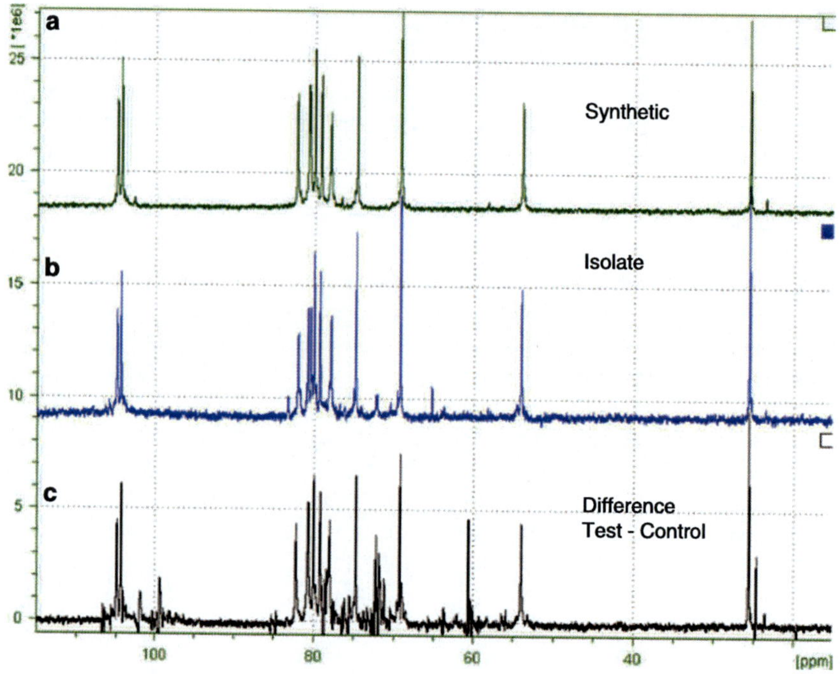


**Fig. 9** $^{13}$C NMRSpectra of (**a**) a synthesized fully sulfated chondroitin, (**b**) the impurity isolated from heparin test samples, and (**c**) the difference spectrum [obtained by subtracting the normal heparin spectrum from that of the contaminated material (Fig. 5)]

the assignments of all of the hydrogens in both rings using COSY, TOCSY, and HMBC.

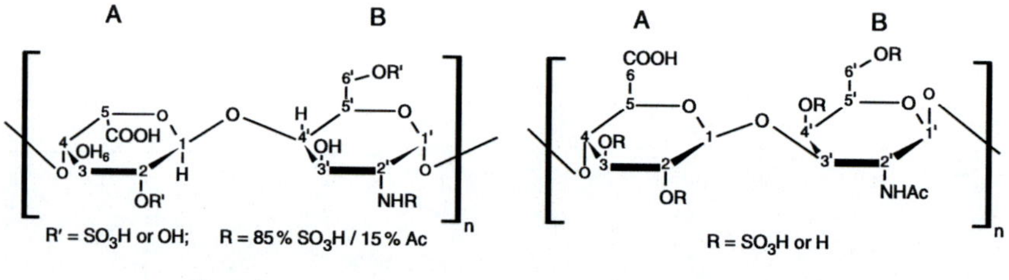


Heparin
**5**

Chondroitin
**6**

The HMBC spectra showed correlations from the anomeric proton on the GlcA ring to position 3 on the A ring, while the anomeric proton on the A ring showed correlation with the 4-position of the GlcA ring. This observation of both 1,3- and 1,4-linkages turned our focus back to chondroitins. We had obtained spectra of several chondroitin sulfates and none of these spectra matched those of the contaminant. Furthermore, the spectrum of a commercially available oversulfated chondroitin also failed to match our data.

However, there was reasonable agreement between chemical shifts of the isolated contaminant and those reported in the literature for a sample of fully sulfated chondroitin sulfate (FSCS), but the order of the assignments differed (Barzu et al. 1993). Assignments of the chemical shifts for our sample were determined primarily using a combination of COSY and edited HSQC spectra. The assignments of the anomeric carbons (A1 and B1) led to a clear assignment of the proton anomeric resonances as well as those assigned to the B3 and A4 positions as shown in Fig. 10a.

A second key contribution to this analysis was the synthesis of FSCS, which had a spectrum identical to that of the isolated contaminant, Fig. 10b. Also, Prof. Robert Linhardt graciously provided a sample of the material on which the literature reference was based (Barzu et al. 1993), and its $^{1}$H NMR spectrum had not changed in ten years of storage, Fig. 10c. The difference seen in the spectra of the isolated and synthetic contaminant, compared to the spectrum of Linhardt's sample, could not be explained by variations in temperature, concentration, or pH. However, when we converted our synthetic sample of Na FSCS to Ca FSCS, we obtained a spectrum Fig. 10d that was identical to that reported in the literature as the FSCS

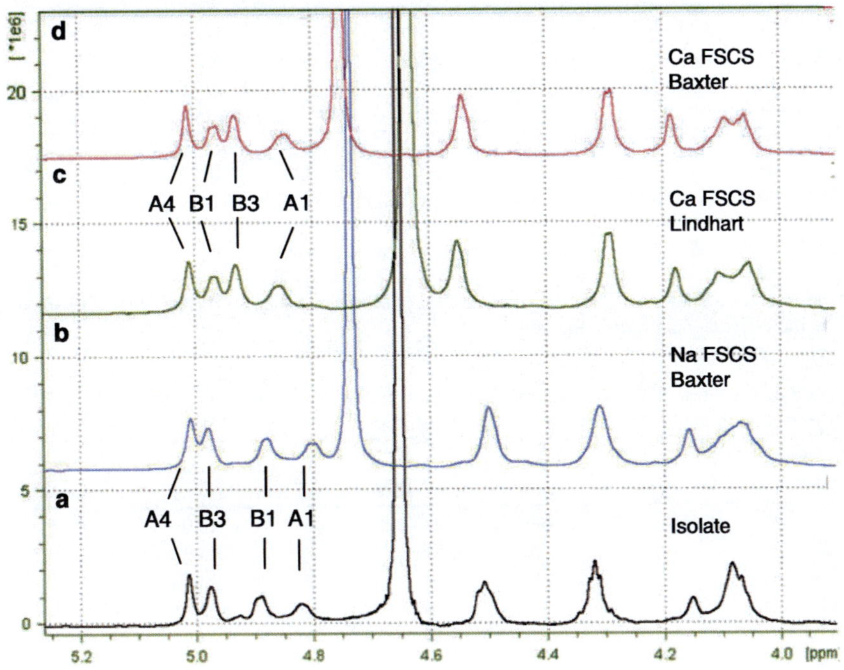


**Fig. 10** $^{1}$H NMR spectra of the impurity isolated from heparin (**a**), the Na (**b**) and Ca (**d**) salts of FSCS prepared at Baxter and the FSCS supplied by Linhardt (**c**)

sodium salt. Apparently, the NMR spectrum published in the literature was that of an FSCS calcium salt, not a sodium salt.

Chemical shift assignments for the contaminant are given in Tables 1 and 2 and are consistent with the literature Barzu et al. (1993) and Pavão et al. (1995) including the seminal paper defining the structure of the contaminant (Guerrini et al. 2008), coauthored by members of the FDA and an international consortium of heparin experts, mostly drawn from academia and government laboratories, and published on March 19, 2008. Subsequently, the structure of the contaminant has been confirmed by other groups as OSCS (Trehy et al. 2009; Viskov et al. 2009).

Other NMR experiments performed included the analysis of in-house synthesized FSCS and commercially available oversulfated chondroitin sulfate

**Table 1** Chemical shift assignments (ppm), referencing based on setting the expected $^{1}H/^{13}C$ NAc methyl in control heparin to 2.00/25.00 ppm

| Position on GalNAc (A) Ring | $^{1}H$ | | | $^{13}C$ | | |
|---|---|---|---|---|---|---|
| | Bax | Lit:GalNAc-4S6S[a] | Lit:GalNAc-4S6S[b] | Bax | Lit: GalNAc-4S[c] | Lit: GalNAc-6S[d] |
| A1 | 4.82 | 4.70 | 4.86 | 104.93 | 104.8 | 104 |
| A2 | 4.08 | 4.02 | 4.10 | 54.04 | 54.8 | 54.2 |
| A3[e] | 4.08 | 4.0–4.05 | 4.10 | 80.51 | 78.2 | 82.5 |
| A4[e] | 5.01 | 4.76 | 5.02 | 77.92 | 78.9 | 70.4 |
| A5 | 4.07 | 4.07 | 4.06 | 74.84 | 77.4 | 76.5 |
| A6 | 4.32 | 4.25, 4.21 | 4.29 | 69.27 | 63.8 | 69.7 |
| A7 CO | – | – | – | 177.64 | 177.8 | Not given |
| A7 Me | 2.16 | 2.06 | 2.16 | 25.59 | 25.4 | 25.6 |

[a]From Pavão et al. (1995) on the GalNAc-(4,6-di$SO_4$) unit of oversulfated mammalian dermatan sulfate (DS)
[b]From Barzu et al. (1993) on the GalNAc-(4,6-di$SO_4$) unit of fully sulfated chondroitin sulfate. 30°C
[c]From Pavão et al. (1995) on the GalNAc-(4-$S0_4$) unit of mammalian DS
[d]From Pavão et al. (1995) on the GalNAc-(6$SO_4$) unit of Ascidian DS
[e]Based only on Baxter data, these assignments may be interchanged, but the literature supports the specific assignments given in the table

**Table 2** Chemical Shift Assignments (ppm), referencing based on setting the expected $^{1}H/^{13}C$ NAc methyl in the control heparin to 2.046/24.64 ppm

| Position on GlcA (B) Ring | $^{1}H$ | | $^{13}C$ | |
|---|---|---|---|---|
| | Bax | Lit:GalNAc-4S6S[a] | Bax | Literature |
| B1 | 4.89 | 4.97 | 104.40 | N/A |
| B2 | 4.51 | 4.53 | 80.09 | N/A |
| B3 | 4.97 | 4.94 | 79.34 | N/A |
| B4 | 4.49 | 4.55 | 80.83 | N/A |
| B5 | 4.14 | 4.20 | 82.02 | N/A |
| B6 | – | – | 177.64 | N/A |

[a]From Barzu et al. (1993) on the GalNAc-(4,6-di$SO_4$) unit of fully sulfated chondroitin sulfate. 30°C

(OSCS). To clarify the nomenclature, FSCS refers to a chondroitin molecule that has sulfate groups on all four available sites of each disaccharide unit. Commonly occurring chondroitins have one or two sulfates per disaccharide unit, and oversulfated chondroitin sulfate (OSCS) has, on average, more than two sulfates per disaccharide. The spectrum of OSCS was very different from that of the contaminant, which showed that the commercial OSCS was not a FSCS and, therefore, not the contaminant in question. When a heparin control was spiked with synthetic FSCS, a new peak was observed with the same CE migration time as the contaminant peak. When a heparin test sample was spiked, the contaminant peak increased in intensity. The results of NMR experiments on similarly spiked samples were in agreement with CE and provided strong evidence that the contaminant in the heparin test samples was primarily FSCS. The complexity of the methyl regions of both the $^{1}H$ and $^{13}C$ NMR spectra of the contaminant indicates that the contaminant is a mixture of FSCS and OSCS, but FSCS is the major component.

## 3.7 *Confirmation of Structure by MS*

Confirmation of structure was achieved by MS analysis. Upon analysis by high performance liquid chromatography with mass spectrometric detection (HPLC-MS) (Korir et al. 2008), the test heparin API (Fig. 11b) showed a shoulder on the tail end of the heparin peak that was absent from the control lots (Fig. 11c) that eluted at the same time as the portion of the heparin sample that was not digested by the cocktail of heparinases I, II, and III (Fig. 11a). The presence of an additional component in the test lots, but not in the control lots of heparin, provided strong support to the contamination theory. The data, however, could also be explained by a different theory that the extra components in test heparin were heparin chemically modified by O-acetylation and that have slightly different chromatographic properties. This theory was called into question by comparison of the mass spectrum of the undigested component, remaining after heparinase digestion of a contaminated (test) sample (Fig. 12a), and mass spectrum of the trailing shoulder peak of heparin found in a nondigested test sample (Fig. 12b). Both spectra showed exclusively a 282/458 ion pair not observed in the mass spectrum of the heparin (nondigested control in Fig. 12c), which can be explained by a structure in which all amino groups were acetylated. Chemical (nonenzymatic) acetylation of the hydroxyl groups inevitably introduces some level of statistical distribution of acetyl groups among all sugar units, i.e., we reasoned that if the heparin had been chemically acetylated and the 282/458 ion pair was abundant, then the spectrum should also contain observable levels of the 240/416 and 324/500 ion pairs containing two acetyl groups. In order to test this hypothesis, we O-acetylated heparin following the procedure of Petitou (Petitou et al. 1992; see also Barzu et al. 1993). The ESI mass spectrum of the synthetically O-acetylated heparin illustrated the predicted pattern, lending further support to the contamination theory, and dismissing the O-acetylation theory.

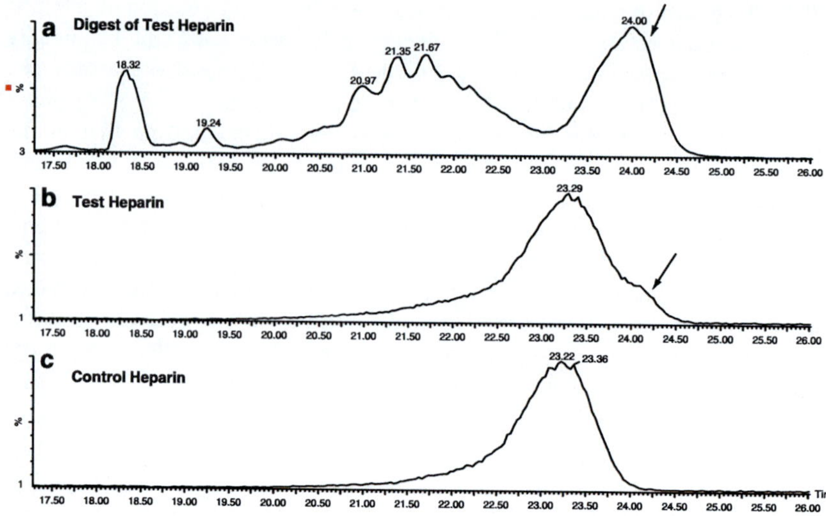


**Fig. 11** LC-MS chromatograms of (**a**) heparinase digest of Baxter heparin test lot, (**b**) Baxter heparin test lot, and (**c**) Baxter heparin control lot

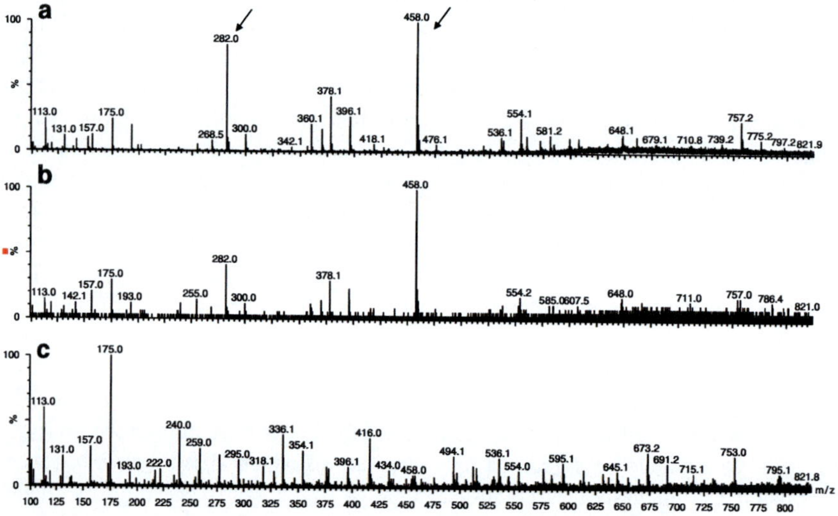


**Fig. 12** ESI spectra acquired at 55 V of (**a**) undigested compound(s) eluting at ~24.0 min in Fig. 11a; (**b**) the shoulder peak of heparin at ~24.0 min in Fig. 11b, and (**c**) heparin (Fig. 11c)

$^{1}$H NMR results indicated that the contaminant was possibly an over-acetylated heparin based on the chemical shift of the acetyl peak, but these results could not be used to distinguish between O- and N-acetylation. However, the $^{13}$C NMR data was not in agreement with O-acetylation (Petitou et al. 1992).

As suggested by the ESI-MS spectrum acquired at high cone voltage, the contaminant isolated from heparinase digest appeared to be fully N-acetylated. Chondroitin sulfates and dermatan sulfate are GAGs that have fully N-acetylated structures that can explain the mass spectrometric data. The ESI-MS spectra of chondroitin A and chondroitin C sulfates were very similar to that of the contaminant, including the feature of exclusive 282/458 ion pair. Commercial chondroitin and dermatan sulfates, however, did not have NMR spectra that were good matches with the contaminant, and, as they are less sulfated than heparin, should have longer CE elution times than heparin. Thus, unmodified chondroitin and dermatan sulfates were ruled out as the contaminants.

### 3.8 Results from Nitrous Acid Treatment

To obtain further information on the contaminant, nitrous acid digestion of the isolated contaminant was performed with and without hydrazine hydrolysis. Nitrous acid causes deaminative cleavage between GlcNS-GlcA and GlcNS-IdoA, regardless of the O-sulfation carried by either monosaccharide unit. This deaminative cleavage is blocked by N-acetyl substitution on the hexosamine. Hydrolysis of the N-acetyl group using hydrazine prior to nitrous acid digestion removes this block to deaminative cleavage. LC-MS analysis of the nitrous acid digests provided important clues to the structure of the contaminant. The majority of the contaminant sample was not digested without hydrazine de-acetylation (Fig. 13a). The most prominent digestion product from this process corresponded to the most prevalent heparin disaccharide (**1**), indicating the presence of heparin in the sample. The detection of small amounts of heparin in the contaminant sample was not surprising considering that the sample was isolated from a heparinase digest via membrane filtration that would not lead to a clear separation of all mixture components. In sharp contrast, the process of nitrous acid digestion after hydrazine treatment cleaved the contaminant molecules to form several products (Fig. 13b). These observations again confirmed that the amino groups in contaminant compound(s) were fully acetylated. Second, the measured molecular weights of the major fragments of nitrous acid digestion were in agreement with theoretical fragments of fully sulfated chondroitin or dermatan sulfate. These data, together with the data described above, pointed to fully sulfated chondroitin sulfate or dermatan sulfate as the contaminant. Definitive data for this postulate were derived from analysis of synthetic FSCS and oversulfated dermatan sulfate (OSDS, Maruyama et al. 1998). The HPLC-MS chromatograms of the nitrous acid digests of the contaminant and synthetic FSCS were very similar (Fig. 13b, c) except for a slight retention time shift that is justifiable because the two chromatograms were acquired on different days. OSDS (Fig. 13d), on the contrary, showed a very different elution pattern. These data confirmed the NMR determination that the contaminant was a fully sulfated chondroitin sulfate.

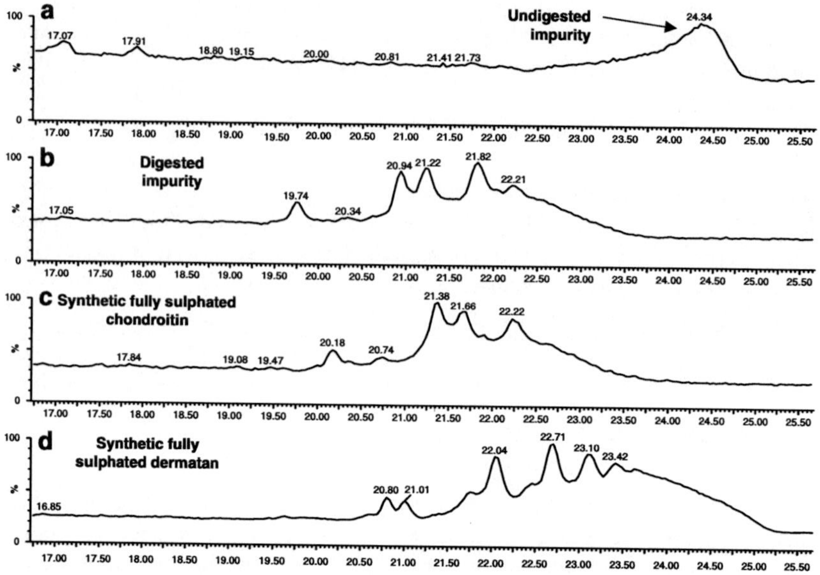


**Fig. 13** LC-MS chromatograms of nitrous acid digests of (**a**) the isolated impurity treated with hydrazine, (**b**) the isolated impurity without hydrazine treatment, (**c**) synthetic fully sulfated chondroitin sulfate, and (**d**) synthetic fully sulfated dermatan sulfate

## *3.9 Association of Contaminant with Chinese Crude Heparin*

During the structure determination activities, parallel efforts were underway to screen lots of API used in Baxter's finished products and to investigate the origin of the contamination. The most critical pieces of data toward determining the origin came from the NMR and CE analyses of several lots of crude heparin retains, obtained from the CZSPL manufacturing plant in China. Both CE and NMR analyses found the FSCS contaminant in one lot of crude material. These lots of crude heparin corresponded with recently manufactured APIs, and the batch records provided the link between lots of crude and lots of contaminated API. This provided a link between a batch of contaminated crude heparin from China and the contaminated Chinese-manufactured API. Because the link between contaminated API and contaminated lots of finished product had already been established, these data on the crude heparin and API provided the definitive proof that the source of the contamination observed in finished product was the Chinese crude heparin.

# 4 Biological/Medical Team Plan

The adverse events reported from the field were described as "allergic type" and were often characterized as hypotension, nausea and shortness of breath [see Blossom et al. (2008) for a full description of the clinical responses]. These reactions appeared to occur within minutes (from 5 to 15 min) of exposure to the heparin infusion, suggesting that the biological mediators were produced acutely. A potential list of possible acute "allergic type" mediators was generated and included complement anaphylatoxins, C3a and C5a, histamine released from blood basophils or mast cells, and bacteria contaminants such as peptidoglycan (PG). A suite of in vitro screening assays to cover as many of the potential mediator pathways as quickly possible were proposed (see Fig. 1), and the plan was to allow the results of these screening assays to help determine which in vivo assays would be appropriate.

# 5 Biological/Medical Team Results

## 5.1 *In Vitro Testing*

Analysis of contaminated heparin lots for PG demonstrated that no PG was present in these lots above the detection limits of the assay. Other in vitro screening tests conducted included the USP sheep plasma clot test, an anti-Factor IIa (aFIIA) test, and an anti-Factor Xa (aFXA) test. The USP clot test results showed that the contaminated heparin lots met the USP potency criterion of >140 U/mg. However, the average value of the test lots was 92% that of the control lots. Results from the aFIIA and aFXA screening showed that the test lots were significantly less potent than the controls for both assays. The aFIIA/aFXA ratio was 0.97 for the controls versus 0.92 for the test articles; however, this difference was not statistically different. Although these results provided the first differences between test and control lots of heparin observed from biological testing, they did not provide any insight into why the test lots were associated with "allergic type" responses. Testing then quickly focused on in vitro tests for complement activation and histamine release and in vivo testing of the contaminated lots in animal test systems.

### 5.1.1 In Vitro Complement Testing

Samples of recalled heparin lots (test lots) as well as control lots were tested in vitro for their propensity to produce an elevation in complement anaphylatoxins. These test and control lots were incubated for an hour at 37°C with human plasma samples (collected with heparin anticoagulant at a concentration of 1 U/ml) at a final concentration of 10 U/ml for the test and control articles. The reactions were stopped with EDTA and the plasma samples were analyzed for C3a levels. There was no difference observed between the samples incubated with test lots and

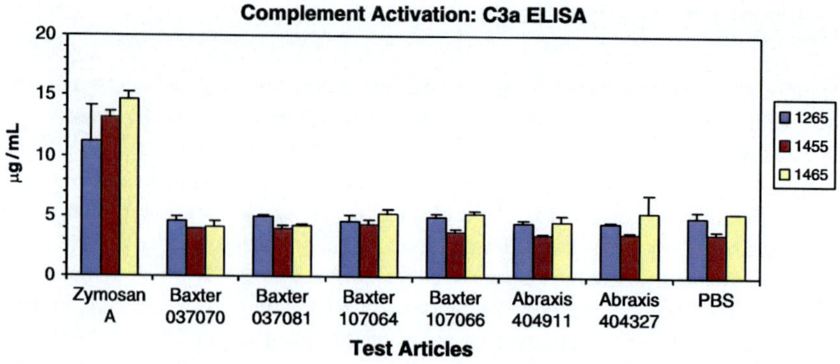


**Fig. 14** Effect of synthetic FSCS on complement activation in normal human plasma. In vitro complement activation in the presence of heparin. Various Baxter heparin final product lots (control: 037070 and 037081; test: 107064 and 107066) and two Abraxis heparin final product lots were incubated with whole blood to activate complement at 10 U/mL heparin. The amount of C3a generated after incubation was determined via ELISA. Zymosan A (50 µg/mL) was used as a positive control for complement activation, while PBS was used as a negative control. These data are the average responses from three healthy human donors

the samples incubated with control lots of heparin, and, in fact, there was no difference between any of the heparin-spiked plasma samples and a phosphate buffered saline (PBS) control plasma sample (Fig. 14). When the synthetic FSCS sample became available, these incubations were repeated with heparin-anticoagulated plasma. For all concentrations tested (ranging from 0.5 to 500 µg/ml), we saw no significant elevation in C3a or C5a levels above what was observed for the PBS control sample (Fig. 15). A positive control sample incubated with zymosan did show a large elevation in both C3a and C5a. These results are consistent with the conclusion that OSCS does not exhibit a significant propensity to activate the complement system.

### 5.1.2 In Vitro Histamine Release Testing

The lack of C5a production suggested that C5a-mediated histamine release was an unlikely consequence of OSCS exposure. However, to ensure that the OSCS-contaminated heparin was not inducing direct degranulation of blood basophils, whole blood samples (heparin-anticoagulated) were incubated with the test and control heparin lots at 37°C for 1 h. The blood samples were placed on ice and then centrifuged to obtain the plasma fractions. Each sample was analyzed for histamine content with a commercial enzyme immunoassay (EIA). While there was great variability in results between donors, there was no apparent difference between test and control heparin samples tested against a single donor. This analysis was repeated with isolated blood basophils that were incubated with heparin samples, and still no difference was observed between test and control heparin lots.

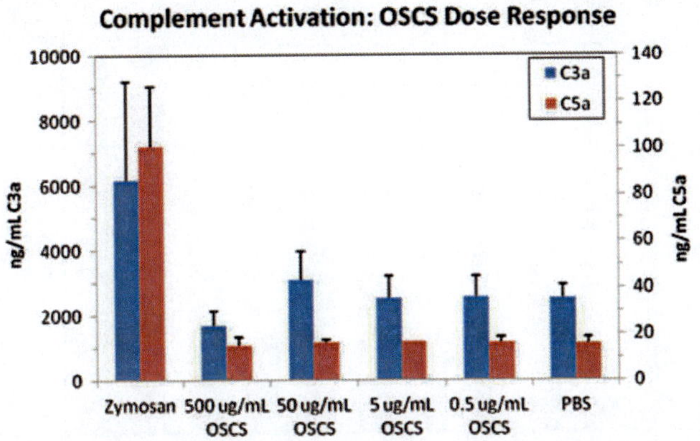


**Fig. 15** In vitro complement activation in the presence of OSCS. A range of 500–0.5 μg/mL of OSCS was incubated with whole blood and the amount of C3a and C5a were assessed via ELISA. Zymosan A (10 μg/mL) was used as a positive control for complement activation, while PBS was used as a negative control. These data shown are the average responses from three healthy human donors. Error is expressed as one standard deviation in all plots

Representative results, from a single donor, are shown in Fig. 16. Thus, the OSCS-contaminated heparin did not appear to elicit a direct histamine response from basophils. In separate experiments, we also investigated whether the OSCS would increase expression of CD11b on WBC (an activation marker), but we observed no such increased expression on cells exposed to OSCS.

## 5.2 *In Vivo Testing*

Concurrent with the in vitro testing, plans were developed to test the contaminated heparin lots in rats where physiological responses such as blood pressure could be monitored. When rats were infused with a control sample of uncontaminated heparin at 5,000 U/kg, they exhibited an immediate and transient hypotensive response that recovered within seconds (Fig. 17a) (McKee et al. 2010). This response was consistently observed in all finished lots of heparin regardless of the presence or absence of OSCS. When a test sample of heparin containing 28 mol% of OSCS was infused into this test species, the animals showed the same immediate and transient hypotensive response. However, this response was followed by a distinctly different secondary hypotensive response that had a delayed onset, longer duration or recovery time and exhibited baseline overshoot (Fig. 17b). The secondary hypotensive response was seen consistently in all animals tested with this solution (OSCS dose ~13 mg/kg), with a heparin product contaminated at 11 mol % (OSCS dose ~5 mg/kg), but not with a heparin lot contaminated at 3 mol% (OSCS dose ~1 mg/kg). When rats were pretreated with Bradyzide (a rodent-selective bradykinin $B_2$ receptor antagonist) and then treated with the heparin lot containing 28 mol% OSCS, the secondary hypotensive response was completely

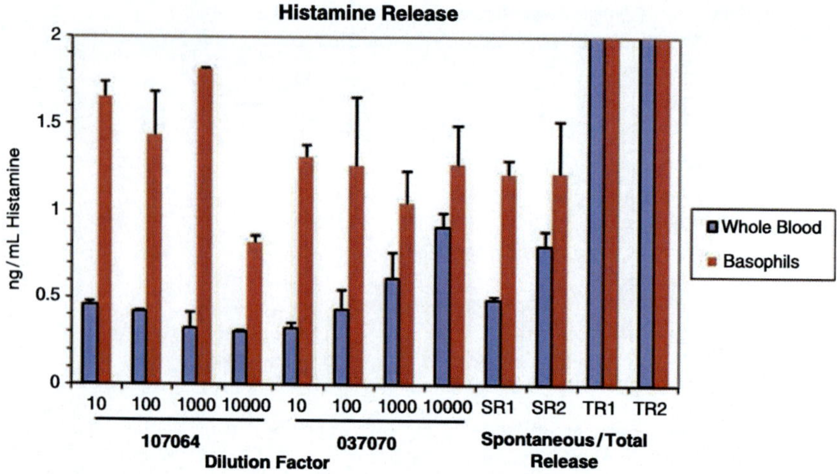


**Fig. 16** Histamine release from whole blood and purified basophils. Heparin test articles (contaminated lot 107064 and uncontaminated lot 037070) were diluted as indicated with PAGCM buffer (25 mM PIPES, 110 mM NaCl, 5 mM KCl, 0.003% HSA, 0.1% Glucose, 1 mM $CaCl_2$, 1 mM $MgCl_2$, pH 7.4) and incubated with either whole blood or isolated basophils from a single donor. PAGCM buffer was used to determine the spontaneous release of histamine (duplicate reactions are shown as SR1 and SR2). Total release of histamine (duplicate reactions are shown as TR1 and TR2) was accomplished by heating samples prepared with PAGCM buffer for 10 min at 95°C. Histamine levels were determined via EIA (Fitzgerald). These results were indicative of the responses observed for multiple donors

abrogated, suggesting bradykinin is the responsible mediator of the secondary hypotensive response (Fig. 17c). Bradyzide did not affect the initial, immediate, and transient hypotensive response that recovered within seconds as discussed above. Finally, when rats were infused with control (uncontaminated) heparin spiked with the synthetic FSCS derivative (FSCS dose ~8.8 mg/kg), all animals displayed the same secondary hypotensive response seen with the contaminated lot of heparin (Fig. 17d). In total, these data strongly implicate the OSCS contaminant in causing the hypotensive response, which appears to be mediated by bradykinin.

A subsequent set of in vivo experiments were performed in pigs to provide validation of the rat results in a second species, and to further explore the dose–response behavior of the synthetic FSCS derivative. Initial experiments were performed with the synthetic FSCS in 0.9% saline. When administered intravenously to pigs at a dose of 1 mg/kg, the FSCS derivative produced no physiologically significant changes in blood pressure (Fig. 18a). At doses of 2, 3, and 5 mg/kg, the FSCS derivative produced severe hypotension beginning shortly after administration. These results suggest a No Observable Effect Limit (NOEL) of ~1 mg/kg corresponding to a heparin dose of ~4,000 U/kg (or 4 mL/kg) contaminated with ~3% OSCS, which is similar to the NOEL observed with the rat. The timing of the onset and nadir in blood pressure, as well as the dose–response characteristics, was similar to the secondary hypotensive response observed in the rats.

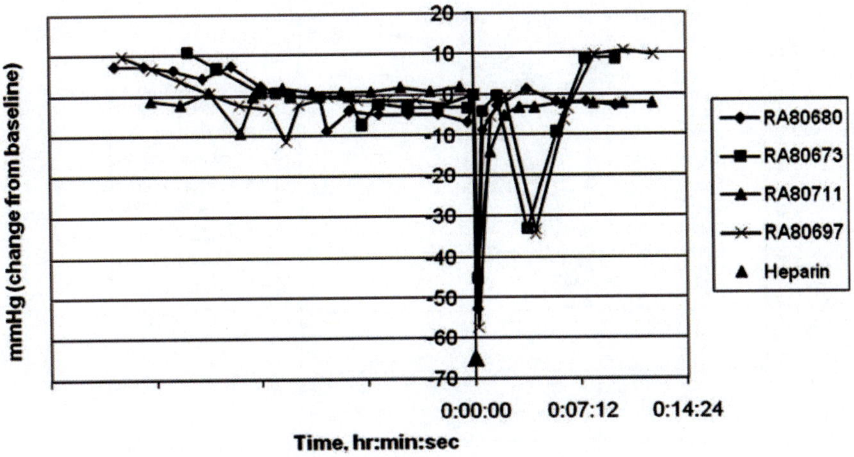


**Fig. 17** Changes in individual rat mean arterial pressure in response to heparin and OSCS. (**a**) The effect of 5,000 U/kg heparin containing no detectable OSCS on mean arterial pressure in an individual rat (RA80680, symbol *black diamond*). (**b**) The effect of 5,000 U/kg heparin containing 28% OSCS on mean arterial pressure in an individual rat (RA80673, symbol *black small square*). (**c**) The effect of 5,000 U/kg heparin containing 28% OSCS on mean arterial pressure in an individual rat (RA80711, symbol *black up-pointing triangle*) pretreated with 10 mg/kg Bradyzide, a rodent-selective $B_2$ bradykinin receptor antagonist (Bradyzide administered at −00:04:41). (**d**) The effect of 5,000 U/kg heparin containing no detectable OSCS spiked with synthetic FSCS (8.8 mg/kg) on mean arterial pressure in an individual rat (RA80697, symbol *sal tire*). In all panels, Heparin Sodium Injection was administered at time 0:00:00 (h:min:s) (black up-pointing triangle = time heparin administered). Five rats were dosed in each experiment and results from a representative animal are shown here for each dosing regimen

The effect of repeated exposure to synthetic FSCS formulated in 0.9% saline on mean arterial blood pressure in individual pigs is depicted in Fig. 18b. The first administration of 5 mg/kg of synthetic OSCS in animal PG8027 caused severe hypotension beginning shortly after administration as described above. A second administration of 5 mg/kg of synthetic FSCS did not result in any change in blood pressure indicating complete tachyphylaxis, suggesting that the activating system had been exhausted or the bradykinin receptor-signal transduction pathway was downregulated. Partial or complete exhaustion of the activating system is supported by the work of Waeber et al. (1988) who demonstrated that the intravenous administration of the active fragment of factor XII in conscious, normotensive rats induced a significant hypotension; whereas, rats depleted of circulating prekallikrein by the administration of dextran sulfate prior to the administration of the active fragment of factor XII did not exhibit any change in blood pressure (Waeber et al. 1988). In addition, the first administration of 1 mg/kg of synthetic OSCS in animal PG8032 did not result in any change in blood pressure but a second administration of 5 mg/kg of synthetic FSCS caused severe hypotension beginning shortly after administration.

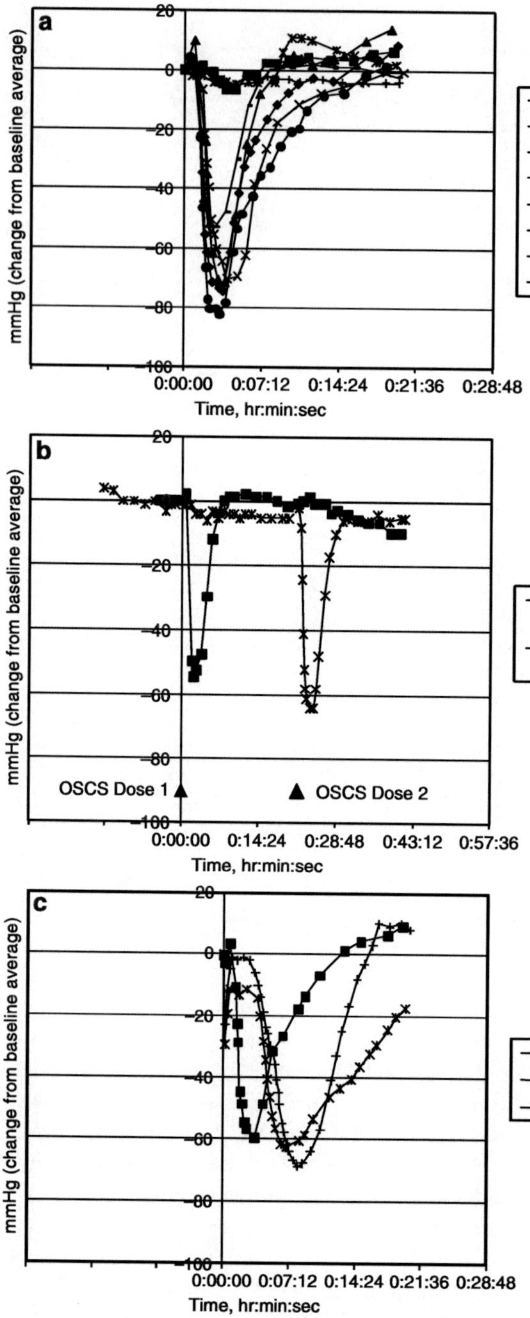


**Fig. 18** Changes in individual domestic pig mean arterial pressure in response to heparin and OSCS. (**a**) The effect of synthetic FSCS in saline on mean arterial pressure in individual domestic pigs (PG8027–PG8032, PG8036 and PG8037). FSCS was administered at time 0:00:00 (h:min:s).

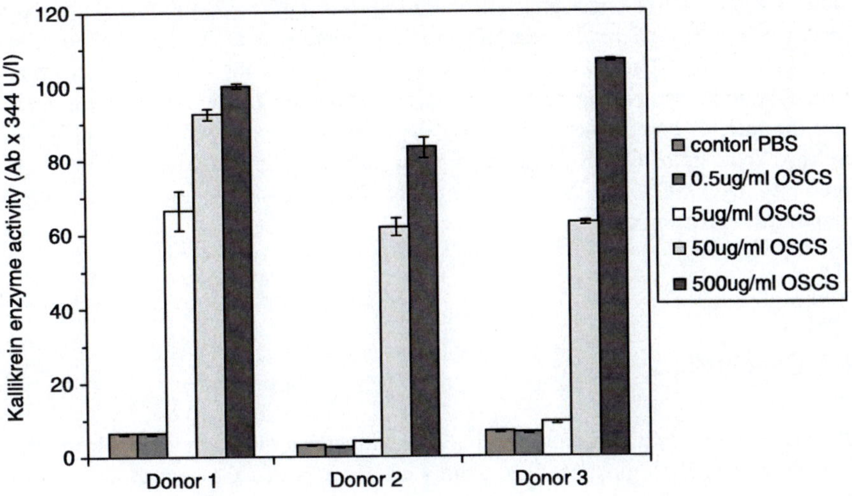


**Fig. 19** Effect of synthetic FSCS on contact activation in normal human plasma. Human plasma was incubated with various concentrations of FSCS as described in the text. The plasma was subsequently incubated with a chromogenic peptide to assess Kallikrein-like activity (shown as an increase in absorbance at 405 nm)

Finally, several animals were treated with the synthetic FSCS derivative spiked into the control heparin to evaluate the effect of heparin on the OSCS response. As shown in Fig. 18c, the co-administration of 5 mg/kg OSCS and 1,000 U/kg of heparin caused severe hypotension beginning shortly after administration. The temporal relationship with respect to the onset and nadir is consistent with that observed in pigs following the administration of 5 mg/kg of OSCS in saline (Fig. 18a). As the dose of heparin was increased to 3,000 U/kg, a primary hypotensive response became apparent (as was observed in the rat), and the higher dose of heparin appeared to delay the onset, and perhaps, the recovery time of the secondary hypotensive response.

These data are consistent with the hypothesis that OSCS activates the contact activation system in plasma (as also shown by Kishimoto et al. 2008), resulting in the production of bradykinin. Also consistent with the work of Kishimoto et al. (2008), we have observed the ability of OSCS to stimulate the amidolytic activity of plasma kallikrein in normal plasma. Incubation of synthetic OSCS in (heparin-anticoagulated) human plasma resulted in a dose–dependent activation of plasma kallikrein-like activity (Fig. 19), which could include FXIIa, beta-FXIIa, or

**Fig. 18** (continued) **(b)** The effect of repeat exposure to synthetic FSCS in saline on mean arterial pressure in individual domestic pigs (PG8027 and PG8032). First dose of FSCS was administered at time 0:00:00 (h:min:s). **(c)**The effect of heparin containing no detectable OSCS spiked with synthetic FSCS (5 mg/kg) on mean arterial pressure in individual domestic pigs (PG8033–PG8035). Heparin containing FSCS was administered at time 0:00:00 (h:min:s) kallikrein, in a range consistent with the in vivo response (50 μg/mL is approximately equivalent to 2 mg/kg assuming a 70 kg patient with a 7% blood volume and 45% hematocrit). The observations that OSCS activates the contact system that a bradykinin receptor antagonist can completely abrogate the hypotensive response, and the lack of complement activation or histamine production observed in the in vitro analysis of OSCS, strongly suggest that bradykinin is the sole mediator produced by OSCS and fully accounts for the observed hypotensive adverse responses seen clinically.

## 6 Conclusions

At the end of several months of intensive investigation, the results from the Analytical and Biological/Medical team's efforts provided the needed answers to four big scientific questions associated with the adverse events: What was different about the lots of heparin that were associated with the cluster of adverse events? What is the nature of the contaminant? Where did it come from? How does it create the observed hypotensive effects? Other questions remained: Exactly where in the crude heparin process was the contaminant introduced? Who perpetrated the adulteration? What was the motive? Although some of these latter questions may never be answered, there were clear scientific conclusions reached, and this event triggered, or at least accelerated, global recognition that more needs to be done to protect the supply chain in the pharmaceutical industry from accidental or purposeful introduction of contaminants. This event directly set the stage for global emergency adoption of NMR and CE screening methods for heparin API, which led to updates in the USP and EP monographs for heparin sodium, heparin calcium and injection solutions containing heparin salts. The revisions in the USP and EP monographs are striking both for the rapid manner in which they were implemented and for the extent of change. Prior to the 6/18/08 issuance of revised USP monographs for heparin sodium and heparin calcium, both methods contained a single identity test that was based on the identification of the counter ion and no direct test for the presence of the biomolecule other than potency. With the issuance of the Oct 1, 2009 revision of the Heparin Sodium USP monograph, three additional identity tests were added, including the use of modern NMR spectrometry, a strong anion exchange chromatography method, and a biological test (ratio of anti-factor Xa to anti-factor IIa activities). Additional tests for the absence of OSCS and limits for natural impurities, such as dermatan and chondroitin sulfates, nucleotidic residues, and proteins, were either added or strengthened. Similar changes in other pharmacopeia are anticipated. The overall regulatory and industrial responses to this unfortunate contamination issue will ultimately enhance the safety and control of the pharmaceutical supply chain, and have already provided new analytical methods for the more comprehensive screening of drug and incipient identities and impurity profiles (see, for example, Guerrini et al. 2009).

## References

Barzu T, Level M, Petitou M et al (1993) Preparation and anti-HIV activity of O-acylated heparin and dermatan sulfate derivatives with low anticoagulant effect. J Med Chem 36:3546–3555

Blossom DB, Kallen AJ, Patel PR et al (2008) Outbreak of adverse reactions associated with contaminated heparin . N Engl J Med 359:2674–2684

Food and Drug Administration (2008) Information on heparin sodium injection. http://www.fda.gov/cder/drug/infopage/heparin/default.htm#screening

Guerrini M, Besccati D, Schriver Z et al (2008) Oversulfated chondroitin sulfate is a contaminant in heparin associated with adverse clinical events. Nat Biotechnol 26:669–675

Guerrini M, Zhang Z, Shriver Z et al (2009) Orthogonal analytical approaches to detect potential contaminants in heparin. Proc Natl Acad Sci USA 106:16956–16961

Hu P, Fang L, Chess EK (2009) Source-induced fragmentation of heparin, heparin, and galactosaminoglycans and application. Anal Chem 81:2332–2343

Kishimoto TK, Viswanathan K, Ganguly T et al (2008) Contaminated heparin associated with adverse clinical events and activation of the contact system. N Engl J Med 358:2457–2467

Korir AK, Limtiaco JFK, Gutierrez SM et al (2008) Ultraperformance ion-pair liquid chromatography coupled to electrospray time-of-flight mass spectrometry for compositional profiling and quantification of heparin and heparan sulfate. Anal Chem 80:1297–1306

Maruyama T, Toida T, Imanari T et al (1998) Conformational changes and anticoagulant activity of chondroitin sulfate following its O-sulfonation. Carbohydr Res 306:35–43

McKee J, Bairstow S, Szabo C et al (2010) Structure elucidation and biological activity of the oversulfated chondroitin sulfate contaminant in Baxter heparin. J Clin Pharmacol [Epub ahead of print, 10 February 2010], doi:10.1177/0091270009355158

Patel RP, Narkowic C, Hutchinson JP et al (2008) A simple capillary electrophoresis method for the rapid separation and determination of intact low molecular weight and unfractionated heparins. J Pharm Biomed Anal 46:30–35

Pavão MSG, Mourão PAS, Mulloy B et al (1995) A unique dermatan sulfate-like glycosaminoglycan from ascidian. J Biol Chem 270:31027–31036

Petitou M, Coudert C, Level M et al (1992) Selectively O-acylated glycosaminoglycan derivatives. Carbohydr Res 236:107–119

Trehy ML, Reepmeyer JC, Kolinski RE et al (2009) Analysis of heparin sodium by SAX/HPLC for contaminants and impurities. J Pharm Biomed Anal 49:670–673

Viskov C, Boulet E, Hubert P et al (2009) Isolation and characterization of contaminants in recalled unfractionated heparin and low-molecular-weight heparin. Clin Appl Thromb Hemost 15:395–401

Waeber G, Schapira M, Waeber B et al (1988) Hypotensive effect of the active fragment derived from factor XII is mediated by an activation of the plasma kallikrein-kinin system. Circ Shock 26:75–382

# Low-Molecular-Weight Heparins: Differential Characterization/Physical Characterization

**Marco Guerrini and Antonella Bisio**

**Contents**



**Abstract** Low-molecular-weight heparins (LMWHs), derived from unfractionated heparin (UFH) through different depolymerization processes, have advantages with respect to the parent heparin in terms of pharmacokinetics, convenience of administration, and reduced side effects. Each LMWH can be considered as an independent drug with its own activity profile, placing significance on their biophysical characterization, which will also enable a better understanding of their structure–function relationship. Several chemical and physical methods, some involving sample modification, are now available and are reviewed.

**Keywords** Low-molecular-weight heparin • Structure • Affinity fractionation • Size fractionation • NMR • HSQC • Capillary electrophoresis • Depolymerization • Chromatography • Oligosaccharides • Monosaccharide composition

M. Guerrini (✉)
Istituto di Ricerche Chimiche e Biochimiche G. Ronzoni, Milan, Italy
e-mail: guerrini@ronzoni.it

R. Lever et al. (eds.), *Heparin - A Century of Progress*,
Handbook of Experimental Pharmacology 207,
DOI 10.1007/978-3-642-23056-1_7, 

## Abbreviations

| | |
|---|---|
| 1 6-an.M 2 | Amino-1, 6-anhydro-2-deoxy-β-D-mannopyranose |
| 1,6-an.A 2 | Amino-1,6-anhydro-2-deoxy-β-D-glucopyranose |
| $A^*$ | 2-deoxy-3-O-sulfo-2-sulfoamino-D-glucopyranose |
| $A_{6S}$ | 2-deoxy-6-O-sulfo-2-sulfoamino/acetylamino-D-glucopyranose |
| AM.ol | 2,5-anhydromannitol |
| $A_{NAc}$ | 2-deoxy-2-acetylamino-D-glucopyranose |
| $A_{NS}$ | 2-deoxy-2-sulfoamino-D-glucopyranose |
| Epox | Epoxide |
| G | D-glucuronic acid |
| $G_{2S}$ | 2-O-sulfo glucuronic acid |
| Gal | D-galactose |
| GalA | Galacturonic acid |
| $I_{2S}$ | 2-O-sulfo-L-iduronic acid |
| $M_{NS}$ | 2-deoxy-2-sulfamino-D-mannopyranose |
| α-red terminal | Reducing residue with α configuration |
| ΔU | 4-deoxy-α-L-threo-hex-4-enopyranosil uronic acid |
| $\Delta U_{2S}$ | 2-O-sulfo-4-deoxy-α-L-threo-hex-4-enopyranosil uronic acid |

## 1 Introduction

Low-molecular-weight heparins (LMWHs), derived from unfractionated heparin (UFH) through different depolymerization processes, possess several therapeutic advantages and reduced side effects compared to heparin, maintaining similar structural features in terms of monosaccharide composition, substitution pattern, and oligosaccharide sequence (Gray et al. 2008; Mousa 2005). Owing to their lower molecular size compared to their parent heparins, more predictable pharmacological action, sustained antithrombotic activity, better bioavailability, and longer half-life are the main advantages of LMWHs over UFH (Nader et al. 1999; Fareed et al. 2004).

Many of the anticoagulant, pharmacokinetic, and other biological differences between heparin and LMWHs can be explained by the relatively low affinity of the latter for several plasma proteins. UFH and LMWHs exert their antithrombotic and anticoagulant activities principally via two coagulation enzymes: factor-Xa and thrombin (factor IIa) through the activation of antithrombin (AT), a serpin known as a major physiological coagulation inhibitor (Hirsh and Raschke 2004). Owing to their reduced molecular size, LMWHs have lower AT-mediated anti-factor IIa activity than anti-factor-Xa activity. The binding of heparin chains to AT, which is an essential step for antithrombotic/anticoagulant activity, involves the interaction of the specific pentasaccharide sequence $Glc_{NAc/NS,6S}$–G–$Glc_{NS,3S,6S}$–$I_{2S}$–$Glc_{NS,6S}$ (AGA*IA). Whereas the inhibition of factor-Xa requires

AT activation by heparin binding through the AGA*IA sequence, to inhibit factor IIa heparin must bind both thrombin and AT, involving an oligosaccharide chain of at least 18 residues containing AGA*IA toward the nonreducing end (Lane et al. 1984). Most of the UFH chains are long enough to exert both inhibitory activities, but fewer than half of those of LMWHs are of sufficient length to bind both AT and thrombin. Consequently, the ratio of anti-factor-Xa/anti-factor IIa activity of UFH increases from 1:1 to 2:1 or even 4:1 for LMWHs (Gray et al. 2008).

Ideally, LMWH should combine the strength of the heparin interaction with AT through preservation of the active pentasaccharide structural integrity, with minimal interaction with thrombin. Structural modifications at the level of pentasaccharide may reduce or cancel the affinity to AT and the resulting anti-Xa activity (Petitou and van Boeckel 2004). The chemical structure of the highly complex collection of macromolecules constituting LMWHs is strongly influenced by the depolymerization used in their generation, mainly because of the different cleavage points along the heparin chain and the generation of different reducing and nonreducing residues. Moreover, the variation in the proportion of differently sulfated sequences resulting from different depolymerizing processes, might affect their binding with plasma proteins as well as their uptake by the endothelium, inducing a change in the pharmacokinetic profiles (Hirsh et al. 2001).

Recent studies have demonstrated that the extension of AGA*IA pentasaccharide sequence also has an influence on the AT-binding properties of heparin oligosaccharides and that this influence depends on the structure of the additional residues. Several chemical and physical methods are now available to better understand the structure–function relationship of different LMWHs. In this chapter, we discuss some of the techniques that are currently being used to investigate the structure of LMWHs and the differences in composition between the most common commercial LMWHs.

## 2 Methods for Preparing LMWHs

The first processes for LMWHs production were described in the early 1980s, while the first clinical trial for the prophylaxis of postsurgical DVT was published in 1986 (Planes et al. 1986). One of the first methods for LMWH production involved the depolymerization in the presence of peroxides of the acidic form of heparin followed by N-sulfation of glucosamine (Fussi 1980). Subsequently, many other strategies have been designed to obtain reproducibly uniformly high quality LMWHs, including size fractionation of heparin to select lower molecular weight material and partial cleavage of heparin chains through chemical or enzymatic processes. Four major strategies are used to prepare LMWHs: chemical deamination,

**Table 1** Commercial LMWHs: methods of preparation

| Trade name | Nonproprietary name | Manufacturer | Preparation method |
|---|---|---|---|
| Lovenox®, Clexane® | Enoxaparin | Sanofi Aventis | β-eliminative cleavage of the benzyl ester of heparin by alkaline treatment |
| Hibor® | Bemiparin | Rovi | Depolymerisation of Heparin benzethonium salt by $CTA^+$, $OH^-$ |
| Fragmin® | Dalteparin | Pfizer Inc. | Deaminative cleavage with nitrous acid |
| Fraxiparin® | Nadroparin | GlaxoSmithKline | |
| Seleparina® | | Italfarmaco | |
| Clivarin® | Reviparin | Knoll | |
| Sandoparin® | Certoparin | Novartis | Deaminative cleavage with isoamyl nitrite |
| Normiflo® | Ardeparin | Wyeth-Ayerst | Oxidative depolymerization with $H_2O_2$ |
| Fluxum® | Parnaparin | Alpha Wasserman | Oxidative depolymerization with $Cu^+/H_2O_2$ |
| Innohep® | Tinzaparin | Leo Pharma | β-eliminative cleavage by heparinase |

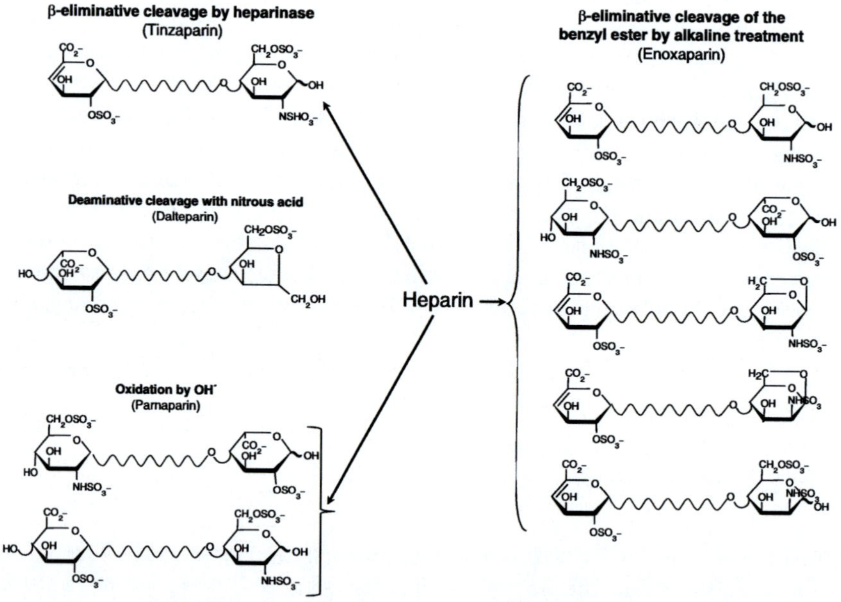


**Fig. 1** Scheme of depolymerization used to prepare the most common low-molecular-weight heparins. Major reducing and nonreducing residues are shown

chemical or enzymatic β-elimination and oxidative cleavage through radicals (Table 1 and Fig. 1) (Linhardt and Gunay 1999). According to the deaminative method, heparin is nitrosylated at the amino group of its N-sulfoglucosamine residues. The unstable N-nitroso-sulfonamide residues re-arrange to generate a carbocation at C2. Subsequent ring-contraction and hydrolysis of the adjacent glycosidic bond generates an anhydromannose residue that is stabilized by reduction with sodium borohydride to form terminal anhydromannitol (AM.ol) residues. The first patents describing use of nitrous acid to obtain LMWHs appeared more than 25 years ago (Barnett 1982) and more controlled deaminative cleavage conditions (temperature, pH, amount and type of nitrosation reagent) are currently applied to obtain an LMWH products with higher yields, higher activity, and desired molecular size (Lormeau et al. 1991). The chemical β-eliminative method is based on cleavage in alkaline medium of a quaternary ammonium salt of heparin, such as benzyltrimethylammonium, or any other hydroxide in alcoholic solution such as methanol, ethanol or isopropanol (Lopez 1991). As an alternative, cleavage in alkaline medium can be applied to the benzyl ester of heparin formed by treatment of the heparin benzethonium salt with benzyl chloride (Uzan 1998). Under these conditions, β-elimination generates an unsaturated uronate residue at the nonreducing end and an N-sulfo-glucosamine at the reducing end. Cleavage occurs at the iduronic acid residue without preference for 2-O-sulfated or nonsulfated residues. Similarly, in enzymatic depolymerization, the action of heparinase I generates an unsaturated uronate residue at the nonreducing position. However, this residue is mostly 2-O-sulfated due to the preference of the enzyme for the –$GlcN_{NS}$–$IdoA_{2S}$– sequence (Langer et al. 1983).

Heparin can also be oxidatively depolymerized using oxygen radicals generated by different methods such as hydrogen peroxide or ionizing γ-irradiation (Fussi 1982; De Ambrosi et al. 1991; Bisio et al. 2004). More specifically, in the case of parnaparin, hydrogen peroxide is decomposed in a water solution in the presence of catalytic amount of a transition metal of a low oxidation number (copper II). The strongly electrophilic hydroxyl radical generated by the process easily abstracts hydrogen from alcohols, ethers, and amides, inducing fragmentation of sugar residues and the consequent depolymerization of the chain. This cleavage generates oligomers with both even and odd number of residues (Linhardt and Gunay 1999; Vismara et al. 2007).

New LMWHs and ultra-low-molecular-weight heparin (ULMWH), at present in phase 2 multicenter trials, were recently described as novel rationally designed antithrombotics. One of them, an LMWH obtained by controlled cleavage of porcine intestinal mucosa heparin with modified forms of heparinase III with reduced enzymatic activity shows pronounced anti-Xa and anti-IIa activity, from two to three times higher than those of heparin on a mass basis (Shriver et al. 2007; Sundaram et al. 2003). The second product, a ULMWH, was obtained by selective β-eliminative cleavage of heparin benzyl esters through a phosphazene base. Selective depolymerization, preserving the AT-binding site, generates a ULMWH product with a mean molecular weight of 2.4 kDa with a high anti-Xa-activity and only residual anti-IIa-activity (Viskov et al. 2009).

## 3 Analytical Methods for Structural Analysis of LMWHs

The complete characterization of LMWHs should provide information on the size of the chains, the monosaccharidic composition in terms of glucuronic/iduronic, N-sulfated/N-acetylated glucosamine content and the sulfation pattern and the sequences of these residues along the chains. No single technique is capable of satisfying all these requirements and only a combination of orthogonal analytical methods can provide a through characterization. In this paragraph, we describe the advantages and disadvantages of the principal analytical procedures used to characterize heparins and LMWHs. In principle, all the biophysical methods for determining the di-oligosaccharide composition, which were developed for analyzing heparin/heparan sulfate chains, are also suitable for LMWHs. Such methods usually require chemical or enzymatic fragmentation of polymeric chains to smaller components and, depending on the technique selected for the analysis, preferably employ labeling of fragments with a fluorescent dye to improve sensitivity. The total oligosaccharide analysis of heparins and LMWHs has the potential of detecting subtle changes in their structure, which could influence their biological activity. A number of techniques, including polyacrylamide gel electrophoresis (PAGE), capillary electrophoresis (CE), low-pressure chromatography, high performance liquid chromatography (HPLC), and mass spectrometry (MS) are usually employed for the separation and identification of oligosaccharides. In addition, other analytical methods (e.g., NMR) or convenient modification of the former techniques, permit intact samples to be examined directly, or following appropriate labeling. In this chapter, special emphasis is given to the application of NMR methods for the characterization of LMWHs.

### 3.1 *Polyacrylamide Gel Electrophoresis*

Polyacrylamide gel electrophoresis (PAGE) has been widely used for the analysis of LMWHs, since discontinuous gradient PAGE together with specialized stacking gel buffers for high-resolution analysis of heparin oligosaccharides was introduced (Rice et al. 1987). The first comparative PAGE analysis of seven different LMWHs was described by Linhardt et al. (1990). The oligosaccharide maps of both intact and partially heparin lyase depolymerized LMWHs show (1) a remarkable level of variation in the content and distribution in polymer chains and (2) the unusual presence of major unidentified oligosaccharides in LMWHs obtained by nitrous acid and oxidative depolymerization procedures. The application of gradient PAGE for assessing molecular weight parameters (number-average molecular weight Mn, weight-average molecular weight Mw and polydispersity P) was demonstrated (Edens et al. 1992) and later applied by Malsch et al. (1996). Nevertheless, the most commonly used method for the measurement of molecular weight parameters is HPLC.

A novel application of a recently developed electrophoretic method based on focusing of LMWHs in polycationic polyacrylamide matrices has been reported (Zilberstein et al. 2009). Such matrices are made by incorporating a gradient of positively charged monomers (Immobilines) into the neutral polyacrylamide backbone. Separation of oligosaccharide components of polydisperse LMWHs occurs when polyanionic chains reach a steady-state position along the migration path and focus in an environment inducing charge neutralization. Such a focusing technique is able to discriminate complex LMWHs through mixed-mode fractionation in which both size and charge distribution along the polymer chains play an important role. The resolving power can be modulated using different gradients of Immobilines, e.g., in the intervals 0–5 or 0–10 mM in a linear or nonlinear gradient mode. Such a method offers a considerable improvement to oligosaccharide resolution of intact LMWH compared to gradient PAGE methods (Linhardt et al. 1990).

## 3.2 *Chromatography*

A number of chromatographic techniques have been explored for the analysis of LMWH or the separation of their oligosaccharide constituents. These include low-pressure size exclusion chromatography (SEC), high performance SEC (HP-SEC), strong anion exchange high performance liquid chromatography (SAX-HPLC), reverse phase ion-pairing chromatography, and affinity chromatography.

HP-SEC with different columns and eluents is the most commonly used method for the measurement of molecular weight and size distribution of LMWHs. Since these parameters are among the most important ones affecting LMWH biological activity, considerable efforts have been devoted to their accurate determination. Several procedures have been developed differing in the basis of chromatographic column calibration, the accuracy of such relative measurements closely depending on the quality of selected reference standard(s). The series of reference standards explored includes heparin fractions with narrow molecular weight distribution (Fareed et al. 1988; Malsch et al. 1996), a unique partially depolymerized heparin sample with broad polydispersity (Mulloy et al. 1997) and pullulan fractions (Guo et al. 2003). Knobloch and Shaklee developed an absolute chromatographic method for determining Mr distribution that employs multi-angle laser light scattering (MALLS) technology in conjunction with HP-SEC, thus overcoming the critical dependence on suitable calibration products (Knobloch and Shaklee 1997). This method improved the reliability of the response one that previously proposed by Komatsu et al. exploiting low angle laser light scattering (LALLS) detection (Komatsu et al. 1993). Recently, a new HP-SEC approach that exploits the simultaneous and combined action of three detectors, right angle and low angle laser light scattering (RALLS/LALLS), refractometer and viscometer, has been introduced (Bisio et al. 2009); also, this method does not require any chromatographic column calibration. The large number of studies published over the past two decades, during which the molecular weight and polydispersity values of various LMWHs have been evaluated, irrespective of the method used to assess such parameters, account for the wide diversity of size-related parameters reported for different products.

A number of improvements in different types of chromatography applied to the compositional analysis of heparin have been extended to the separation and characterization of heparin or HS-derived oligosaccharides. These include strong anion exchange chromatography SAX-HPLC (Turnbull 2001), HP-SEC combined with SAX-HPLC (Chuang et al. 2001), SAX-HPLC combined with PAGE (Vivès et al. 2001) and reversed-phase ion-pair HPLC (Thanawiroon and Linhardt 2003). In addition, Mourier and Viskov developed a particular application of SAX-HPLC, endowed with high selectivity for highly sulfated oligosaccharides, which exploits a stationary phase dynamically coated with cetyltrimethylammonium (CTA), so-called CTA-SAX chromatography (Mourier and Viskov 2004). However, none of these methods has been applied to LMWH preparations. More recently, low-pressure gel permeation chromatography on Bio-gel P10 was employed to separate three different LMWHs into their oligomeric components, obtaining distinct chromatographic profiles (Bisio et al. 2009). Results and additional discussion on this application will be presented in Sect. 4.

## 3.3 *Capillary Electrophoresis*

Being a very sensitive method with high resolving power, particularly suitable for strongly anionic compounds, Capillary electrophoresis (CE) has gained widespread acceptance in oligosaccharide analysis of heparin. Desai et al. reported the first application of CE for qualitative and quantitative di/oligosaccharide compositional analysis of six LMWHs, following exhaustive enzymatic digestion with a mixture of heparin lyase I, II and III (Desai et al. 1993). The optimum resolution of 17 defined di/oligosaccharides was achieved using 10 mM sodium borate and 50 mM sodium dodecyl sulfate at pH 8.8 and at a 20 kV constant voltage. The relative compositional analysis highlighted important differences between the various commercially available LMWHs. However, the number of unidentified oligosaccharide species and unresolved chromatographic peaks was so high as to preclude easy comparison. A number of protocol modifications, leading to much better sensitivity and resolving power of heparin/heparan sulfate disaccharides and also including the examination of larger oligosaccharides up to tetradecasaccharide, have been described subsequently and are reviewed elsewhere (Lamari et al. 2002; Mao et al. 2002; Volpi et al. 2008). A further significant improvement in CE detection sensitivity and selectivity has been the introduction of laser-induced fluorescence (LIF) along with pre-column derivatization of heparin oligosaccharides with fluorophores (Lamari et al. 2002; Kitagawa et al. 2002; Militsopoulou et al. 2002, 2003).

Few applications of CE to the analysis of intact LMWHs (i.e., notrequiring their fragmentation) have been reported. Toida and Linhardt, and Malsch et al. described the use of CE for the characterization of pharmaceutical preparations of heparin and LMWHs without the need for prior depolymerisation, by using acidic phosphate buffer in the presence or absence of copper ions, respectively (Toida and Linhardt 1996; Malsch et al. 1996). Both methods, however, provided poorly resolved CE profiles. Different profiles of LMWHs were obtained by a bare fused silica capillary under reverse polarity conditions in the presence of acidic copper sulfate buffer (Pervin et al. 1994; Ramasamy et al. 2003). Patel et al. developed a simple and rapid CE method that can differentiate diverse LMWH preparations and UFHs, with higher resolution than the previously reported CE methodologies, by using phosphate acidic buffer and 10-min run times (Patel et al. 2008). More recently, a further improvement in resolution by CE electropherograms of intact LMWHs was reported providing "fingerprints" of three different LMWHs (King and Desai 2008). Following pre-column derivatization with 2-aminoacridone (AMAC), the differential migration profiles of fluorescent-labeled LMWHs were achieved under reverse polarity conditions in the presence of appropriate concentrations of selected linear alkyl polyamines, including spermine, tetra- and penta-ethylenepentamine employed as resolving agents. The resolving power of polyamines, which assume a polycationic nature under the strongly acidic elution conditions, was accomplished through recognition of the heparin fine structure, which results in the modification of the overall charge density of the chains, altering the electrophoretic mobility and resulting in differential migration profiles. Accordingly, the electrophoretic resolution appears to be a function of the affinity of the polyamines for the LMWH chains.

### *3.4 Heparin Sensors*

Quantification of heparin and LMWH in biological media is a challenge in pharmacological studies. Generally, heparins are indirectly quantified according to their (in vitro) anticoagulant effect indicated in international units (IU). Due to its wide range of additional biological activities such as antithrombin-independent antithrombotic actions, anti-inflammatory, antimetastatic and/or antiangiogenic effects, there is a need for a direct quantitative assay, which detects the chemical substance heparin and not only its pharmacological effect.

The development of so-called heparin sensors, devices, or molecules producing a signal in the presence of heparin, has attracted attention of several groups (Ma et al. 1992; Briza et al. 2008).

Recently, a polymeric fluorescent sensors, polymer-H, was used to quantify heparins, LMWHs and other sulfated carbohydrates (Lühn et al. 2010). Polymer-H consists of three functional monomers based on a methacrylamide skeleton contributing both to fluorescence detection and interaction with sulphated carbohydrates. The low detection limit of ~ 0.09 μg/ml for LMWH (~0.01 aXa-IU/ml), of Polymer-H-based sensors showed a very promising method for sensitive detection of heparin. Moreover, a cross-reactive chemosensor for sulfated glycosaminoglycans was also described (Műller-Graff et al. 2010).

The method is based on the interaction between a dynamic mixture of Fe(II) complexes and sulfated polysaccharides followed by detection using UV-Vis spectroscopy. The assay allows to the differentiation of glycosaminoglycans mixtures with high precision.

## 3.5 Mass Spectrometry

Mass spectrometry (MS) represents a powerful technique for the structural elucidation of GAGs. The recent development of soft ionization methods, including electrospray ionization (ESI) and matrix-assisted laser desorption/ionization (MALDI), has made analysis of heparin/HS oligosaccharides, without loss of their sulfate groups possible. Heparin oligosaccharides have been analyzed by MALDI–MS for the first time, after complexation with a basic polypeptide (Juhasz and Biemann 1995). The method has been applied in combination with NMR spectroscopy, chemical and enzymatic depolymerisation to sequence heparin oligosaccharides up to 10 saccharide units (Shriver et al. 2000a, b). Mass spectrometric separation combined with different chromatographic techniques, including size exclusion, anion exchange, reversed-phase ion-pairing as well as capillary electrophoresis, represents one of the new promising approaches for sequencing complex mixtures of highly sulfated oligosaccharides. The analysis of an intact LMWH through online SEC/MS method that display, in a single run, more than 60 oligosaccharide components have been reported (Henriksen et al. 2004). By measuring the mass of each component, the number of constituent monosaccharides, sulfate and acetyl groups can be determined. Although SEC resolution is low, the combination with MS detection permits a helpful characterization of LMWH preparation. Detailed discussions on further advances in separation and characterization of heparin-derived oligosaccharides and particularly on separations techniques online with MS can be found in reviews (Korir and Larive 2009; Zaia 2009). However, as for chromatography, no specific studies have so far been devoted to comparative characterization of LMWH preparations.

## 3.6 NMR Spectroscopy

Although a full characterization of LMWHs cannot be achieved with any single method, nuclear magnetic resonance can be considered the technique of choice for structural analysis of LMWHs, which does not require any sample modification and provides the most direct fingerprint (Capila et al. 2005; Torri and Guerrini 2008), permitting a detailed quantitative compositional picture of their mono- and disaccharidic components (Guerrini et al. 2001, 2005; Desai and Linhardt 1995). The advantages of NMR in defining GAG structures have been evident since the first NMR spectra were published (Perlin et al. 1968, 1970). Since then, a large number of papers and reviews describing the structural features of heparin have been published (Mulloy and Johnson 1987; Yates et al. 1996; Capila et al. 2005). Although NMR spectroscopy has a lower sensitivity compared to the previously mentioned analytical tools, it has great potential for the quantitative determination of parameters that define sequences in intact glycosaminoglycan oligosaccharides. These parameters include monosaccharide composition, sulfation patterns and position and configuration of linkages between glucosamine and uronic acid residues.

NMR spectroscopy is particularly useful in the characterization of LMWHs. Even mono-dimensional NMR spectra permit the characteristic end-residues, typical of the different depolymerization processes (Figs. 5 and 6), to be distinguished at the glance (Linhardt and Gunay 1999; Casu and Torri 1999). However, although mono-dimensional NMR spectroscopy has been used to directly detect and quantify signals associated with major structural features of heparins (Guerrini et al. 2001; Casu et al. 1996), incomplete resolution of some signals in the more complex spectra of LMWHs often prevents their use for quantitative analysis. More recently, two-dimensional (2D) method involving the combined use of proton and carbon NMR spectroscopy was described to provide information on monosaccharidic composition sequences and sulfation patterns of heparin and HS. The quantitative evaluation of GAG features, achieved by proper selection of magnetically equivalent signals in the HSQC spectra, was also described (Guerrini et al. 2005).

The increase in resolution obtained by the 2D technique permits the quantitative evaluation of many signals that overlap in the corresponding mono-dimensional spectrum, a detailed analysis of the structural peculiarities of LMWHs and quantification, with an acceptable error, of minor monosaccharide components (Guerrini et al. 2007). The method is a powerful tool able to resolve and identify potential contaminants in heparins and LMWHs. Together with CE, it can provide detailed fingerprints of products making possible their use in analysis and of batch-to- batch comparisons of LMWH preparations (Guerrini et al. 2009). Currently, high magnetic field NMR spectrometers equipped with high sensitivity probes permit the accurate quantification of minor components present in amounts lower than 2%. NMR spectroscopy is also an extremely suitable technique with which to study the conformational properties of the active oligosaccharide components of LMWHs. Measurements of J-coupling, inter-proton Overhauser effects (NOEs) and relaxation parameters provide information on the conformation and dynamic properties of oligosaccharides in solution and how their conformational properties can change when bound to proteins (Mulloy and Forster 2000; Hricovini and Torri 1995; Hricovini et al. 2001).

A property-encoded nomenclature (PEN) computational platform for the structural analysis of heparin and HS has been developed. The PEN is a way of representing all possible structural diversity, assigning one letter code for each point of structural variation based on a hexadecimal notation system (Venkataraman et al. 1999). The combination of PEN analysis with CE and NMR methods considerably reduces the number of experimental constraints required for heparin/HS oligosaccharides sequence determination (Guerrini et al. 2002).

## 4 Structural Differences Between LMWHs

In principle, LMWHs should differ from their parent heparin only by their molecular weight, which is usually about one third of that of the original UFH. The internal structure of LMWHs should essentially be the same as that of the original UFH in terms of monosaccharide composition, substitution pattern, and oligosaccharide sequence. Even under the assumption that depolymerization processes did not modify the internal structure, procedures involved in the manufacturing of LMWHs cause some structural modifications of the monosaccharidic units at the site of cleavage and are characteristic of each depolymerization procedure (Fig. 1). Other differences, regarding the percentage of constituent saccharides and their sulfation pattern, can also be related to both the process and the structural features of the parent UFH used for the LMWH preparation (Bianchini et al. 2007).

The characteristic structural features of commercially available LMWHs have been described in different studies (Linhardt et al. 1990; Gray et al. 2008; Bisio et al. 2009). The most pronounced structural differences among LMWHs are the average molecular weight and oligosaccharide distributions (Table 2). Among the most common LMWHs, tinzaparin has the highest average Mw and exhibited the broadest polydispersity ($D = 1.40$); dalteparin has the lowest ($D = 1.22$), and is the most homogeneous; enoxaparin has the lowest average Mw and an intermediate $D$ value (1.30); parnaparin shows Mw similar to dalteparin but with a higher $D$ value (1.32). A comparison of the elution profiles obtained by HP-SEC/TDA of enoxaparin, tinzaparin, and dalteparin, together with a typical UFH displays their notable diversity (Fig. 2). The remarkable differences in molecular weight observed among LMWHs are associated with distinct oligomeric distributions, as determined by size exclusion chromatography on Bio-gel P10 (Fig. 3). Whereas enoxaparin is the richest in small oligomeric families from dp2 to dp12, dp8 is the smallest

**Table 2** Amount of linkage region, total degree of sulfation, anti-Xa, anti-Xa/anti IIa ratio, MW and polydispersity degree (in parenthesis) of LMWHs

| | Percentage of linkage region calculated by NMR HSQC method[a] | Total degree of sulfation (by NMR)[a] | Anti-Xa (U/mg) | Anti-Xa/anti-IIa activity | MW (Da) |
|---|---|---|---|---|---|
| UFH | 1.2[b] | 2.4 | 190[h] | 1 | 17,500 (1.14)[i] |
| Enoxaparin | 0.7 | 2.5 | 98[c]–104[e] | 3.9[d]–3.9[f] | 4,500[f]–5,300 (1.30)[g] |
| Tinzaparin | 3.1 | 2.4 | 79[c]–90[e] | 1.5[d]–1.6[f] | 6,500[f]–8,300 (1.40)[g] |
| Dalteparin | 0.6 | 2.6 | 130[c]–122[e] | 2.2[d]–2.5[f] | 6,000[f]–6,900 (1.22)[g] |
| Parnaparin | 0 | 2.5 | 82[e] | 4.1[d]–2.3[f] | 5,000[f]–6,500 (1.32)[e] |

[a]Guerrini et al. (2007), [b]Iacomini et al. (1999), [c]Cornelli and Fareed (1999), [d]Fareed et al. (2004), [e]Linhardt et al. (1990), [f]Gray et al. (2008), [g]Bisio et al. (2009), [h]Bisio et al. (2004), and [i]Bertini et al. (2005)

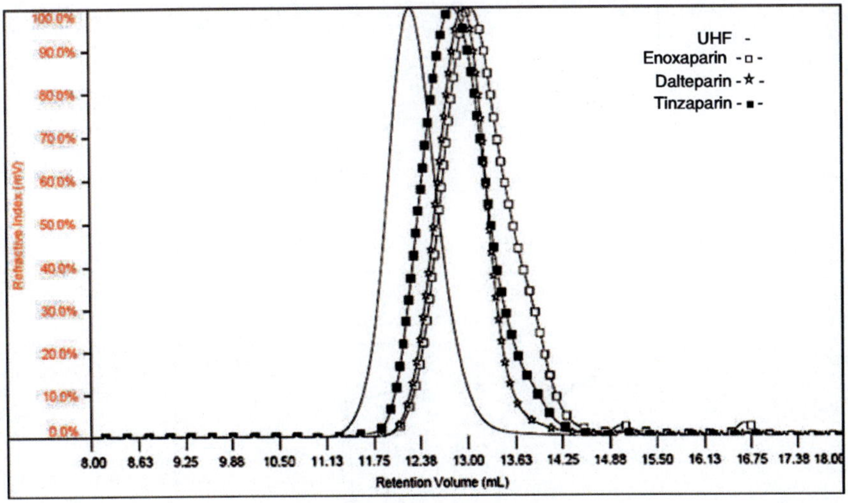


**Fig. 2** HP-SEC/TDA chromatograms (Refractive index response vs. Retention Volume) of enoxaparin, tinzaparin, and dalteparin in comparison with a typical UFH [redrawn from Bisio et al. (2009)]

oligomeric family found in dalteparin, which is also the richest in chains longer than dp12. Tinzaparin shows the highest polydispersity, all oligomeric families being represented. The lack of resolution observed in the gel permeation profile of parnaparin is related to the broad polydispersity of the sample, which is composed of equally represented odd and even numbered oligosaccharides (Vismara et al. 2010). The largest unresolved oligosaccharides (>dp16) are present in different proportions in each LMWH being particularly abundant in tinzaparin and dalteparin and less represented in both enoxaparin and parnaparin (Bisio et al. 2009).

A fingerprint pattern characteristic of individual LMWHs has been obtained by CE profiles of AMAC-labeled samples, tinzaparin, enoxaparin, and a Sigma preparation (King and Desai 2008). Their different electrophoretic profiles presented in Fig. 4, each one displaying a number of characteristic baseline resolved peaks, depends on the extent of interaction of each LMWH with the resolving agent. This approach permits easy identification of small and large compositional differences between products and is especially useful to assess product identity and batch-to-batch variability.

Mono-dimensional NMR analysis of LMWHs reveals the typical signature of the depolymerization process by which they have been produced (Linhardt and Gunay 1999; Casu and Torri 1999). Typical NMR signals of end-residues generated by depolymerization processes can be displayed in both $^{1}H$ and $^{13}C$ spectra (Figs. 5 and 6). Signals corresponding to C4, C2, and C5 of anhydromannitol residues (AM.ol) are clearly observable in the dalteparin carbon spectrum at 87.9, 85.9, and 82.3 ppm, respectively (Huckerby et al. 1985; Bienkowski and Conrad 1985).

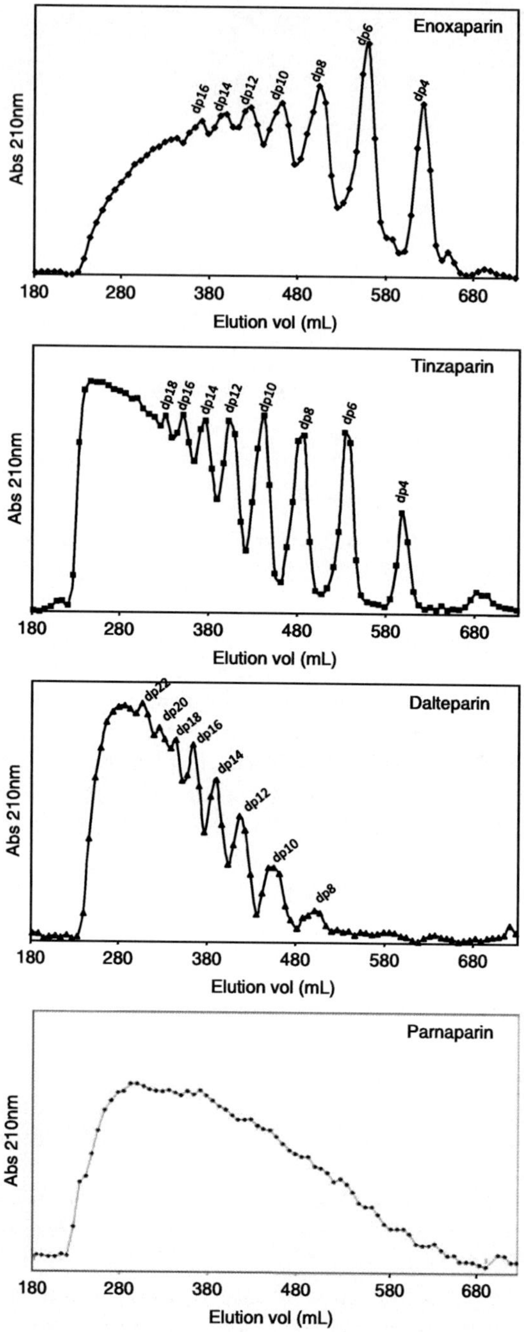


**Fig. 3** Gel permeation elution profiles on Bio-gel P10 of four low-molecular-weight heparins

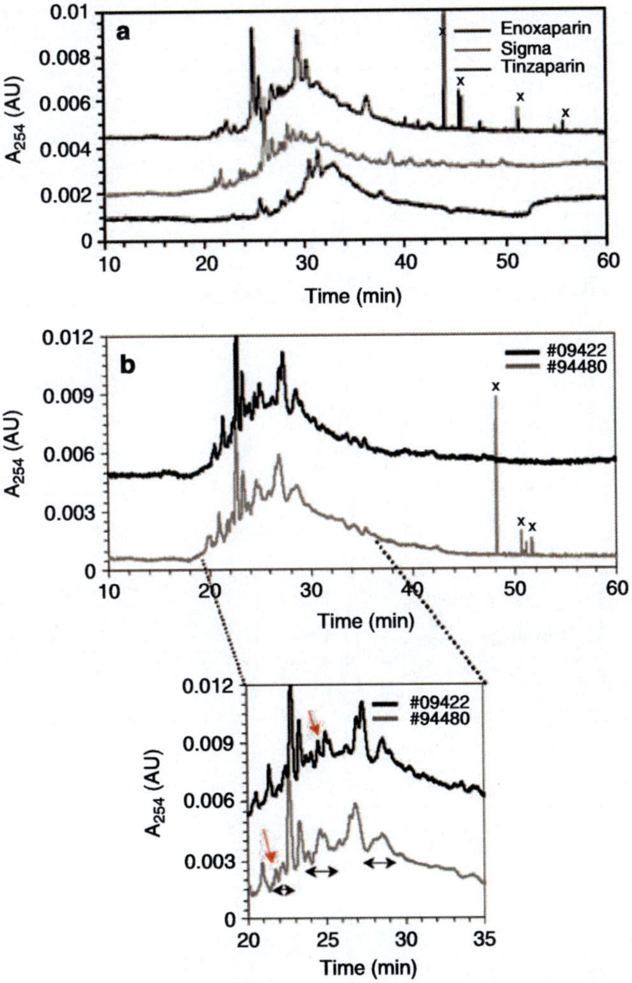


**Fig. 4** (**a**) Characteristic CE profiles of individual LMWHs. (**b**) Analysis of batch-to-batch variability of enoxaparin preparations. Note the difference in component pattern between the two lots in the 22, 25, and 27 min regions (marked in the *bottom figure*). *Arrows* at 22 and 25 min show new components present in the lot. In contrast, the component pattern is reversed for the region at 27 min. Peaks marked "x" are sudden disturbances due to bubble formation during the electrophoretic run. Reprinted from Analytical Biochemistry (King and Desai 2008), with permission from Elsevier

Proton and carbon signals at about 6.0 and 108 ppm, respectively, indicate the presence of a double bond in the uronic acid residue at the nonreducing end, characteristic of the unsaturated 2-O-sulfated uronate residue that is present in both enoxaparin and tinzaparin. One additional reducing anomeric signal at 95.6 ppm and other two signals in the C2 region of glucosamine at 58.6 and

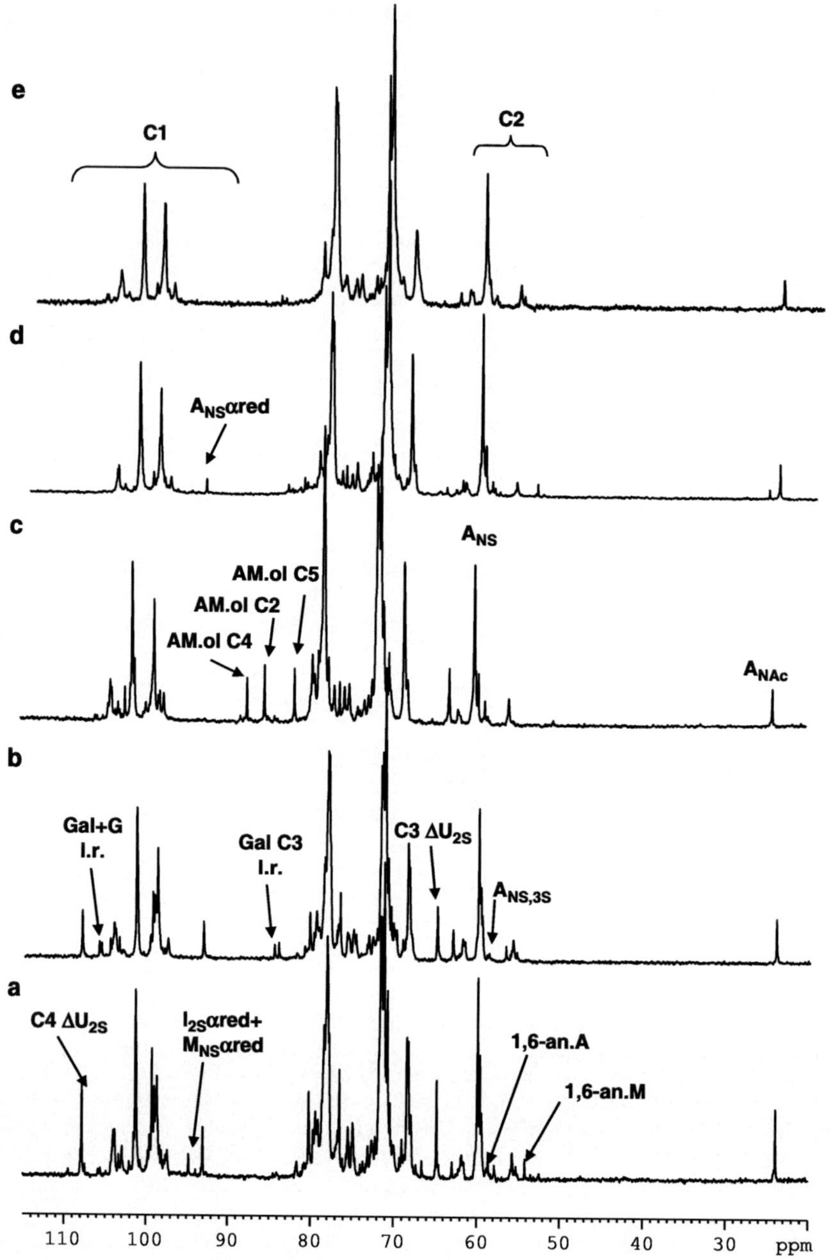


**Fig. 5** $^{13}$C-nuclear magnetic resonance spectra of **(a)** Enoxaparin, **(b)** Tinzaparin, **(c)** Dalteparin, **(d)** Parnaparin, and **(e)** UFH

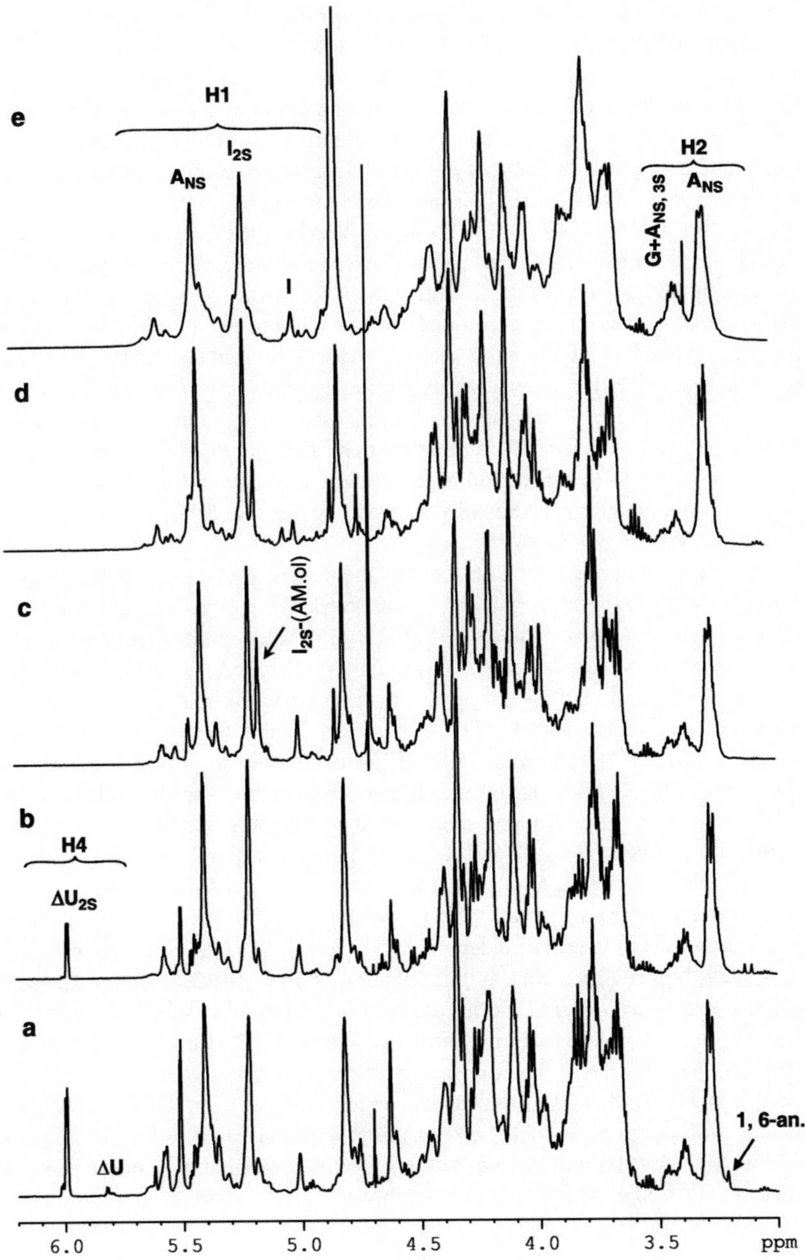


**Fig. 6** [1]H-nuclear magnetic resonance spectra of (**a**) Enoxaparin, (**b**) Tinzaparin, (**c**) Dalteparin, (**d**) Parnaparin, and (**e**) UFH

55.0 ppm are present in the enoxaparin spectrum. The first signal is due to the reducing N-sulfated, 6-O-sulfated mannosamine ($Man_{NS}$), whereas the other two belong to 2-sulfo-amino-1,6-anhydro-2-deoxy-β-D-glucopyranose (1,6-an.A) and 2-sulfo-amino-1,6-anhydro-2-deoxy-β-D-mannopyranose (1,6-an.M) residues, respectively. These two unique bicyclic structures at the reducing end originate from alkaline hydrolysis of the benzyl ester of heparin (Mourier and Viskov 2005a, b; Mascellani et al. 2007). In addition, signals corresponding to the "linkage region" (l.r.; i.e., the GlcA-β(1–3)-Gal-β(1–3)-Gal-β(1–4)-Xyl-(*Ser*)), a serine-linked tetrasaccharide characteristic of the first biosynthetic steps of the polysaccharide chain (Iacomini et al. 1999) are more abundant in the tinzaparin spectrum than in all other LMWHs. In particular, the signals of the CH group of serine (proton and carbon signals at 3.97 and 57.4 ppm, respectively) are present only in the tinzaparin HSQC spectrum (Fig. 8), confirming that the enzymatic treatment does not affect this sequence during the depolymerization process (Sugahara et al. 1995). On the contrary, chemical or physical treatments may also act on the linkage region sequence, resulting in the reduction of its content or modification of its structure, as observed in enoxaparin, dalteparin, and parnaparin $^{13}$C-NMR spectra (Fig. 5) (Mourier and Viskov 2005a, b).

The presence of epoxide was detected in the carbon spectra of enoxaparin and parnaparin and has already been observed in some preparations of nondepolymerized heparin (Guerrini et al. 2001). The epoxides can be generated by alkaline treatment of 2-O-sulfated iduronic acid ($IdoA_{2S}$) residues with the consequent loss of the sulfate group at position 2 and formation of an epoxide ring between carbons 2 and 3 (H2/C2 and H3/C3 at 3.74/54.2 ppm and 3.82/53.3 ppm, respectively) (Jaseja et al. 1989; Hricovini et al. 1995; Mourier and Viskov 2004). HSQC NMR spectra permitted resolution of signals hidden in the corresponding mono-dimensional spectra and highlighted further the structural peculiarity of LMWHs (Figs. 7 and 8). The signal corresponding to 2-O-sulfated glucuronic acid residue ($GlcA_{2S}$), which was found in very small amounts in natural GAG fragments (Yamada et al. 1995), but not in UFH spectra, was detected in enoxaparin HSQC spectra (Guerrini et al. 2007). This finding suggests that $GlcA_{2S}$ could be generated by C5-epimerization of IdoA residues under the basic conditions used during enoxaparin preparation. Moreover, HSQC spectra of enoxaparin and parnaparin exhibit typical signatures of galacturonic acid residues. The presence of epoxide in the oligosaccharidic chains may explain the formation of L-galacturonic acid (GalA) during the depolymerization process. In fact, the alkaline treatment could cleave the epoxide ring, with formation α–L-ido configuration or α-L-galacto configuration depending on the experimental conditions (Rej and Perlin 1990). Signals of GalA overlap those of IdoA, with exclusion of H5/C5 signals, which shift to 4.69/74.5 ppm (Fig. 7).

In addition to these structural peculiarities, different average contents of disaccharide components of different LMWHs, following depolymerization with heparin lyases, were observed. Linhardt and coauthors reported a compositional map of different LMWHs constructed by SAX-HPLC, which reveals remarkable variations in the relative percentage of different components measured against defined

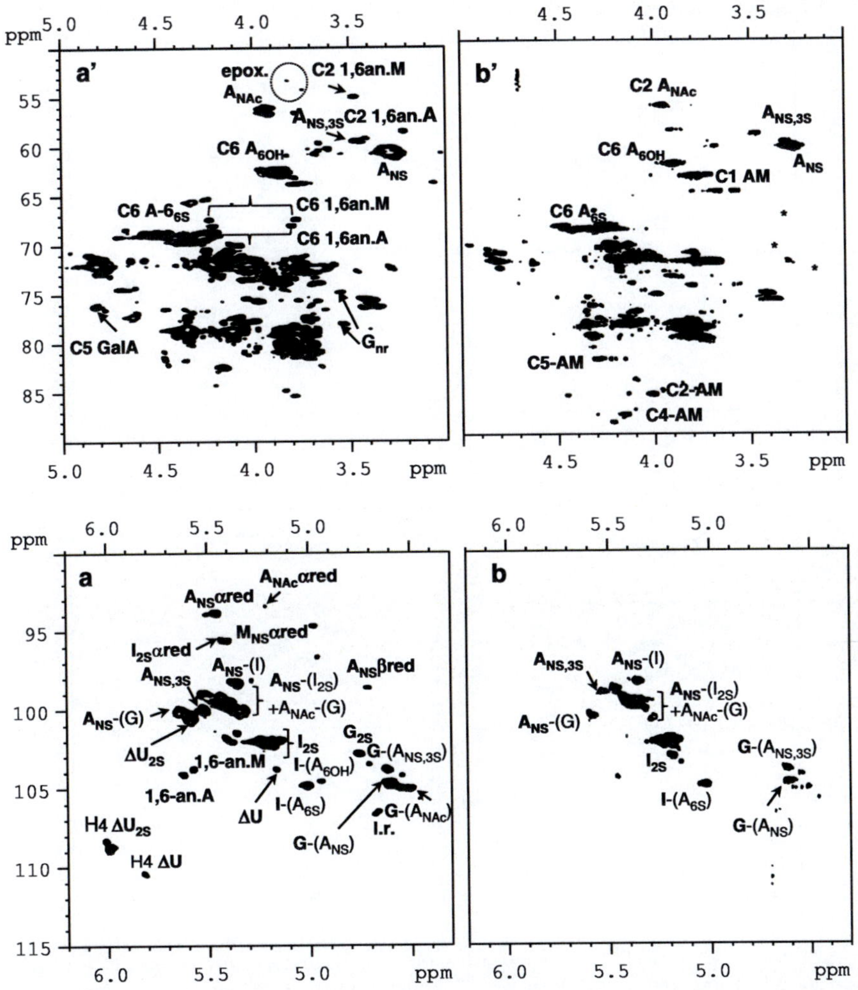

**Fig. 7** Anomeric region of two-dimensional heteronuclear single quantum coherence (2D-HSQC) spectra of (**a**) Enoxaparin, and (**b**) Dalteparin. Ring signals of 2D-HSQC spectra of (**a′**) Enoxaparin, and (**b′**) Dalteparin

oligosaccharide standards (Linhardt et al. 1990). For LMWHs prepared with an oxidative process using hydrogen peroxide, only a slightly reduced content of unsulfated uronic acids was observed (Desai et al. 1993). Nevertheless, for this LMWH preparation, the percentages of constituent disaccharides cannot be assessed precisely owing to the incomplete enzymatic cleavage of the product (Volpi et al. 1992). Possible structural modification introduced by the radical depolymerizing process inside the LMWH chains could be responsible for

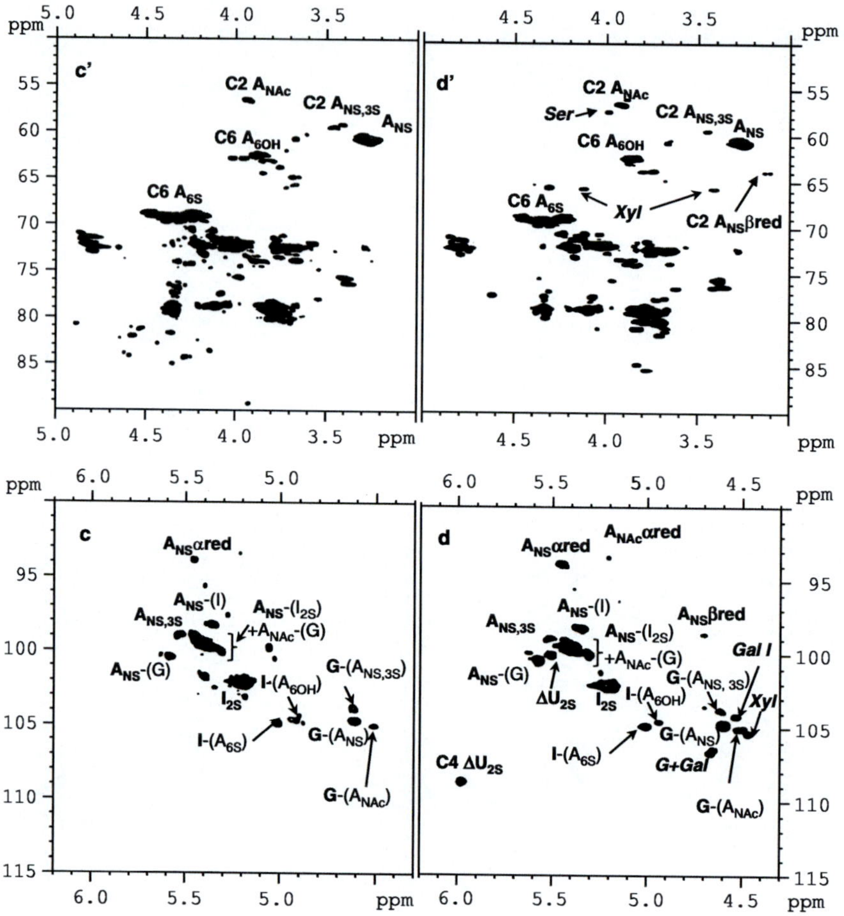

**Fig. 8** Anomeric region of two-dimensional heteronuclear single quantum coherence (2D-HSQC) spectra of (**c**) Parnaparin, and (**d**) Tinzaparin. Ring signals of 2D-HSQC spectra of (**c′**) Parnaparin, and (**d′**) Tinzaparin

incomplete heparinase action and, consequently, for the generation of larger oligosaccharides, which remain unidentified in analytical chromatography owing to the lack of proper model compounds. More recent SAX-HPLC analyses performed on heparinase digested UFH and enoxaparin preparations provided a similar and almost complete percentages of recovery of di- tetrasaccharides, since all the detected peaks were identified (Bianchini et al. 2007).

A qualitative fingerprint together with monosaccaride compositional analysis of heparin and LMWH samples can be obtained by NMR spectroscopy (Capila et al. 2005). The analysis of LMWHs by integration of HSQC NMR spectra showed variation in the heparin monosaccharide composition among commercial samples and compared to a typical UFH (Table 3) (Guerrini et al. 2007). The higher structural complexity of LMWHs compared to UFH, which is particularly pronounced in enoxaparin, is indicated by the presence of several signals arising from end residues. Whereas tinzaparin shows monosaccharide composition similar to UFH, reduction in the content of N-acetyl-glucosamine in enoxaparin, dalteparin, and parnaparin was observed. The latter LMWH also showed a parallel decrease of nonsulfated iduronate and glucuronate residues, suggesting the presence of more homogeneous oligomeric sequences, arising from the heparin NS-regions (Vismara et al. 2007). However, the total sulfation degree, calculated by adding all different sulfated monosaccharides, was almost the same (2.4/2.6) for all of the LMWHs examined, indicating that lower N-sulfation was compensated by somewhat higher O-sulfation (Table 2).

Although the observed structural differences clearly characterize each LMWH preparation, they cannot be straightforwardly correlated with their biological properties. The decrease of anti-Xa activity observed in all LMWHs compared to the parent heparin can be explained by partial modification of the AT-binding sequences arising from the different production processes (Fareed et al. 2004). On the other hand, the very strong reduction of thrombin inhibition of LMWHs with respect to the parent heparin is due to the reduction the number of chains endowed with the minimum length required for potentiation of thrombin inhibition (at least 13 saccharides proximal to the nonreducing side of the AGA*IA pentasaccharide (Table 2) (Lane et al. 1984). Although the distribution of the AGA*IA sequence along heparin chains is still not clearly defined, only a small part of fragments composed of 20–30 saccharides (corresponding to an Mn of 6,000–9,000 Da) contains the pentasaccharide in a position suitable for AT and thrombin bridging. Consequently, whereas the anti-IIa activity is correlated with the MW, differences of the anti-Xa activity are apparently not correlated with the average chain length of

**Table 3** Determination of variously substituted monosaccharide components (percentage) of commercial heparin, enoxaparin, tinzaparin, dalteparin, and laboratory prepared parnaparin, calculated by HSQC NMR spectra integration (Guerrini et al. 2005, 2007; Vismara et al. 2010)

| Glycosamine | $A_{NS}$-($I_{2S}$) | $A_{NS}$-(I) | $A_{NS}$-(G) | $A^*$ | $A_{NAc}$ | $A_{NAc\text{-}\alpha}$ red | $A_{NS\text{-}\alpha}$ red | $A_{NS\text{-}\beta}$ red | 1,6-an. A | 1,6-an. M | $M_{NS\text{-}\alpha}$ red | AM. ol | $A_{6S}$ |
|---|---|---|---|---|---|---|---|---|---|---|---|---|---|
| UFH[a] | 57.1 | 9.2 | 9.5 | 4.7 | 14.3 | 1.4 | 0.9 | 0 | 0 | 0 | 0 | 0 | 81.2 |
| Enoxaparin[a] | 44.4 | 8.8 | 14.5 | 4.4 | 10.7 | 0.5 | 8.2 | 1.0 | 2.1 | 2.3 | 3.2 | 0 | 84.2 |
| Tinzaparin | 55.7 | 6.9 | 8.2 | 2.0 | 14.6 | 0.4 | 10.5 | 1.6 | 0 | 0 | 0 | 0 | 81.9 |
| Dalteparin | 56.8 | 8.0 | 5.2 | 4.6 | 10.3 | 0 | 0 | 0 | 0 | 0 | 0 | 15.5 | 91.6 |
| Parnaparin | 68 | 8 | 7 | 4 | 10 | 0 | 3 | 0 | 0 | 0 | 0 | 0 | 88 |
| **Uronic acid** | $I_{2S}$ | I-($A_{6S}$) | I-($A_{6OH}$) | G-($A^*$) | G-($A_{NS}$) | G-($A_{NAc}$) | $G_{2S}$ | $\Delta U_{2S}$ | $\Delta U$ | $I_{2S\text{-}\alpha red}$ | $I_{2S\text{-}\beta red}$ | Gal A | Epox |
| UFH[a] | 70.7 | 7.1 | 2.5 | 2.7 | 8.3 | 6.6 | 0 | 0 | 0 | 0 | 0 | 0 | 0 |
| Enoxaparin[a] | 49.9 | 5.7 | 1.0 | 3.2 | 9.5 | 4.1 | 2.6 | 18.0 | 1.1 | 1.1 | 1.0 | 2.3 | 0.3 |
| Tinzaparin | 6.5 | 2.1 | 2.1 | 9.3 | 4.4 | 0.5 | 14,7 | 0 | 0 | 0 | 0 | 0 | 0 |
| Dalteparin | 75.4 | 9.5 | 0.5 | 4.1 | 6.8 | 3.6 | 0 | 0 | 0 | 0 | 0 | 0 | 0 |
| Parnaparin | 76 | 8 | 0 | 4 | 9 | 2 | 0 | 0 | 0 | 2 | 0 | 0.4 | 0.3 |

[a]Average values calculated by integration of HSQC spectra measured on three UFH samples

the LMWHs, depending exclusively on the presence of pentasaccharide containing oligosaccharides (Gray et al. 2008).

LMWHs have been shown to differ from each other in their affinity for AT and from UFH, presumably as a result of different structural changes inside their antithrombin binding site (AT-bs), due to the distinct methods employed during their preparation. To better characterize structural differences of AT, high-affinity fragments within different LMWHs, a preparative separation of high (HA) and no affinity (NA) fractions toward AT was recently described (Bisio et al. 2009). HA chains were consistently longer than NA chains, the first being in the range of 6,200–6,600 Da, without any appreciable amount of chains under 2,400 Da, irrespective of the MW of the starting LMWH.

A relevant aspect of this study was the comparison of the content of the N-sulfo-3,6-O-sulfated glucosamine (A*) and the glucuronic acid linked to A* (G–A*), both regarded as markers of the active pentasaccharide (Table 3). Owing to the fact that A* is contained in the pentasaccharidic sequence of the active site of heparin for AT, its content might be expected to correlate with anti-Xa activity. However, since signals associated with A* residues were also found in heparin fractions with no affinity for AT (Kusche et al. 1990), the correlation of A* with anti-Xa activity is not straightforward (Casu and Torri 1999). A more reliable correlation can be made between the amount of G–A*, since this disaccharide has been detected almost exclusively in active sequences (Bisio et al. 2009). Moreover, by treating heparin with heparinases I and II, a decasaccharide bearing the trisaccharide $A_{NAc,6S}$–G–$A_{NS,3S}$ (AGA*) at the reducing end and retaining significant anti-Xa activity was isolated (Shriver et al. 2000a, b). These data, in addition to demonstrating that the active pentasaccharide is susceptible to heparinase cleavage, are in agreement with the appreciable affinity toward AT exhibited by AGA* trisaccharide (Petitou et al. 1997). Furthermore, the G-(A*) content of enoxaparin, tinzaparin, and dalteparin (Table 3) correlates with the anti-Xa activity reported in Table 2. Such a correlation was not found for parnaparin, since the G-(A*) content reported in Table 3 refers to a laboratory preparation, whereas the anti-Xa activity shown in Table 2 was determined on a commercial preparation.

The most relevant difference emerging from comparison of the NMR spectra of the three LMWHs and their derived HA fractions is the content of the AT-bs per chain estimated through the G-(A*) content and the corresponding molecular weight Mn (Bisio et al. 2009). Interestingly, the content of G–A* per average chain of enoxaparin and dalteparin HA fractions is slightly higher than 1 (1.1 and 1.2, respectively), suggesting that some of the chains of these LMWHs, especially the longest chains, may contain two AT-bs, as already suggested for UFH (Kusche et al. 1990). In addition to the different content of heparin AT-bs oligosaccharides, the extensions of the active pentasaccharide sequence can also influence their AT-binding properties and regulate the consequent interaction with factor-Xa (Guerrini et al. 2008).

The interaction of AT with the pentasaccharide containing oligosaccharides was considered to be mainly due to the ionic interaction between the essential negatively charged groups of the pentasaccharide and complementary positive residues in the heparin-binding site of AT (van Boeckel et al. 1994; Petitou and van Boeckel 2004). In such a molecular assembly, an important role is also played by the peculiar conformational behavior of the L-iduronic acid (IdoA) moiety that can adopt three different conformations, $^1C_4$, $^4C_1$, and $^2S_0$ (Ferro et al. 1986) in solution. The unusual degree of flexibility of IdoA residues has been proposed to facilitate the most effective docking of anionic groups of heparin chains to the appropriate basic group of proteins (Mulloy and Linhardt 2001; Casu et al. 2004, Raman et al. 2003). In particular, both NMR and X-ray studies of AT-pentasaccharide complexes described so far reveal that the 2-O-sulfated IdoA residue assumes the $^2S_0$ conformation when bound to the protein, without being involved in any dipolar contact with AT amino acidic residues. The sulfate group of the IdoA residue is the main driving force in the shifting of the conformational equilibrium toward $^2S_0$, thereby facilitating contacts between the pentasaccharide moiety and the basic amino acid residues of the heparin-binding region of AT (Jin et al. 1997; Hricovini et al. 2001).

Many other studies, describing structural details of AT–heparin oligosaccharide complexes showed similar intra- and inter-residue geometries of the pentasaccharide moiety, independent of the ligand structure, supporting the idea that the AGA*IA pentasaccharide sequence binds AT in a specific manner (Li et al. 2004; Johnson et al. 2006). On the basis of these observations, the anti-Xa activity of heparin and LMWHs should be related mainly to the abundance of the intact pentasaccharide and not to its relative position in the heparin chains. Indeed, the anti-Xa activity measured for three LMWHs seems to be correlated with the amount of the G-(A*) sequence, marker of the pentasaccharide content (Table 3) (Fareed et al. 2004; Bisio et al. 2009). However, advances in analytical and separation methods permitted the isolation and sequencing of a large number of "active" oligosaccharides having AGA*IA positioned toward the reducing or nonreducing side of the chain (Loganathan et al. 1990). The active role of the residues prolonging the pentasaccharide sequence toward both its reducing and nonreducing side was recently demonstrated for octasaccharides isolated from enoxaparin (Guerrini et al. 2006). Affinity chromatography on immobilized AT of an octasaccharide mixture indicated a different order of elution, depending on the structure of the reducing and nonreducing pentasaccharide extension (Fig. 9) (Guerrini et al. 2008). NMR saturation transferred experiments showed that in all octasaccharides studied, the pentasaccharide sequence lay closer to the AT-binding site than its reducing or nonreducing extensions, confirming the specificity of the binding. However, some residues prolonging the pentasaccharide actively contribute to the binding in a manner dependent on their structure. In particular, the presence of D-glucuronic acid residue preceeding the AGA*IA sequence increases the affinity of the octasaccharide toward AT by one order of magnitude (i.e., this GlcA residue is probably generated by C5-epimerization of L-IdoA residue under basic conditions as described previously for 2-O sulfated GlcA). The higher nonionic contribution to AT affinity of this particular octasaccharide with respect to

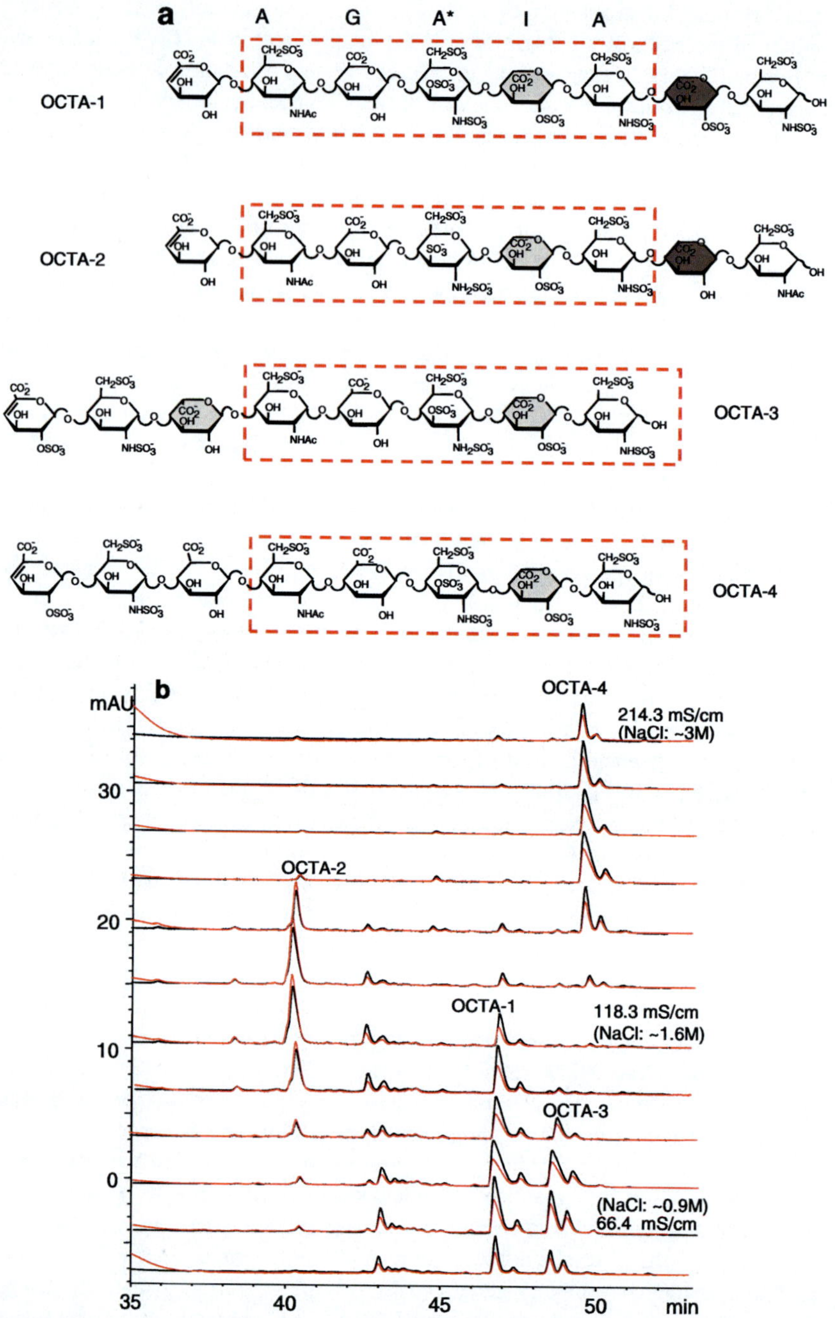


**Fig. 9** (**a**) Structure of four AT-binding octasaccharides containing the pentasaccharide sequence at the reducing (Octa 1 and Octa 2) and nonreducing end (Octa 3 and Octa 4). Iduronic acid residues assuming the $^{2}S_{0}$ or the $^{1}C_{4}$ conformation in the bound state are highlighted in *light gray* and *dark gray*, respectively. (**b**) AT affinity chromatography of reconstituted mixture of octasaccharides [redrawn from Guerrini et al. (2008)]

the synthetic pentasaccharide (fondaparinux) can be attributed to the stronger dipolar interaction between the hydroxyl group of GlcA and some amino acids of AT (Guerrini et al. 2008).

Particularly interesting is the role of the IdoA residues in the reducing and nonreducing extensions of the pentasaccharide. Besides partially contributing to the binding with AT, the conformation of this residue in the bound state seems to be independent of its 2-O-sulfation, or substitution of neighboring residues, in contrast with that observed in absence of protein (Ferro et al. 1990). Both sulfated and nonsulfated IdoA residues positioned at the reducing side of the pentasaccharide assume in the bound state the $^1C_4$ conformation, whereas the non sulfated IdoA residue located at the nonreducing side assumes the $^2S_0$ conformation, in spite of its conformational equilibrium being shifted toward almost pure $^1C_4$ form in the free state (Fig. 9). These findings suggest that these unexpected conformations of IdoA residues positioned near the active pentasaccharide could enhance the AT affinity, either by optimizing contacts between the pentasaccharide and AT or by promoting additional contacts involving the extension residues (Guerrini et al. 2008).

## 5 Conclusions

Over the past 10 years, considerable efforts have permitted the development of a series of methodologies for analyzing complex mixtures of heparin/HS oligosaccharides as well as LMWHs. Analysis of depolymerized heparins has provided the compositional map of their di/oligosaccharide components, which is important qualitative and quantitative structural information. The drawback of this approach is the dependence of results on the efficiency of enzymatic or chemical depolymerization processes. On the other hand, the analytical approach to intact LMWHs, which avoids any manipulation of samples, usually provides qualitative structural information, consisting of their electrophoretic or chromatographic profiles. The eligible technique for the structural analysis of LMWHs is represented by nuclear magnetic resonance. NMR, with a single approach and without requiring any sample manipulation, provides both qualitative and quantitative structural information. However, no single technique is adequate to identify all LMWH peculiarities, and only a combination of orthogonal analytical methods is suitable for their complete characterization.

All depolymerization methods generate low-molecular-weight products with a higher structural heterogeneity than the parent UFH. Owing to the different mechanisms of depolymerization, the resulting LMWHs differ not only at the level of the terminal reducing and nonreducing residues, but also in their internal oligosaccharide sequences. Even when the average monosaccharide and disaccharide composition does not significantly differ among different LMWHs, the oligosaccharidic sequences can vary depending on the different position of cleavage along the original heparin chains. Different cleavage positions can result in oligosaccharide chains containing the AT-bs in different positions, which presumably influences the strength of interaction with AT and related biological activities. Internal sequence and size of LMWH chains are also important for non-AT-mediated activities, since they can influence the interaction with several other plasma and endothelial proteins. The evaluation of the contribution of different oligosaccharide sequences to the binding with several proteins requires further systematic and comparative studies of size homogeneous oligomeric families of LMWHs. Such knowledge will also contribute to better understanding of different activities and pharmacokinetic profiles of LMWHs.

**Acknowledgments** We are grateful to Prof Benito Casu and Dr Giuseppe Cassinelli for critically reading this manuscript.

## References

Barnett WE (1982) Anticoagulant substances. US Patent #4351938

Bertini S, Bisio A, Torri G et al (2005) Molecular weight determination of heparin and dermatan sulfate by size exclusion chromatography with a triple detector array. Biomacromolecules 6:168–173

Bianchini P, Liverani L, Spelta F et al (2007) Variability of heparins and heterogeneity of low molecular weight heparins. Semin Thromb Hemost 33:496–502

Bienkowski MJ, Conrad HE (1985) Structure and characterization of the oligosaccharides formed with depolymerization of nitrous acid. J Biol Chem 260:356–365

Bisio A, Guglieri S, Frigerio M et al (2004) Controlled γ-ray irradiation of heparin generates oligosaccharides enriched in highly sulfated sequences. Carbohydr Polymer 455:101–112

Bisio A, Vecchietti D, Citterio L et al (2009) Structural features of low-molecular-weight heparins affecting their affinity to antithrombin. Thromb Haemost 102:865–873

Briza T, Kejik Z, Cisarova I et al (2008) Optical sensing of sulfate by polymethinium salt receptors: colorimetric sensor for heparin. Chem Commun 16:1901–1903

Capila I, Gunay NS, Shriver Z et al (2005) Methods for structural analysis of heparin and heparan sulfate. In: Garg HG, Linhardt RJ, Hales CA (eds) Chemistry and biology of heparin and heparan sulfate. Elsevier Ltd, Oxford

Casu B, Torri G (1999) Structural characterization of low molecular weight heparin. Semin Thromb Hemost 25(3):17–25

Casu B, Guerrini M, Naggi A et al (1996) Characterization of sulfation patterns of beef and pig mucosal heparins by nuclear magnetic resonance spectroscopy. Arzneimittelforschung 46:472–477

Casu B, Guerrini M, Torri G (2004) Structural and conformational aspects of the anticoagulant and antithrombotic activity of heparin and dermatan sulfate. Curr Pharm Des 10:939–949

Chuang WL, McAllister H, Rabenstein L (2001) Chromatographic methods for product-profile analysis and isolation of oligosaccharides produced by heparinase-catalyzed depolymerization of heparin. J Chromatogr A 932:65–74

Cornelli U, Fareed J (1999) Human pharmacokinetics of low molecular weight heparins. Semin Thromb Hemost 25(3):57–61

De Ambrosi L, Recchia W, Ferrari G (1991) Process for the controlled preparation of low molecular weight glycosaminoglycans. US Patent #4987222

Desai UR, Linhardt RJ (1995) Molecular weight of heparin using $^{13}$C nuclear magnetic resonance spectroscopy. J Pharm Sci 84:212–215
Desai UR, Wang H-M, Ampofo SA et al (1993) Oligosaccharide composition of heparin and low-molecular-weight heparins by capillary electrophoresis. Anal Biochem 213:120–127
Edens RE, al-Hakim A, Weiler JM et al (1992) Gradient polyacrylamide gel electrophoresis for determination of molecular weight of heparin preparations and low-molecular-weight heparin derivatives. J Pharm Sci 81:823–827
Fareed J, Walenga JM, Hoppensteadt D et al (1988) Comparative study on the in vitro and in vivo activities of seven low-molecular-weight heparins. Haemostasis 18:3–15
Fareed J, Ma Q, Florian M et al (2004) Differentiation of low molecular-weight heparins: impact on the future of the management of thrombosis. Semin Thromb Hemost 30(suppl 1):89–104
Ferro DR, Provasoli A, Ragazzi M et al (1986) Evidence for conformational equilibrium of the sulphated L-Iduronate residue in heparin and in synthetic heparin mono- and oligosaccharides: NMR and force-field studies. J Am Chem Soc 108:6773–6778
Ferro DR, Provasoli A, Ragazzi M et al (1990) Conformer populations of L-iduronic acid residues in glycosaminoglycan sequences. Carbohydr Res 195:157–167
Fussi F (1980) Process for obtaining low molecular weight heparins endowed with elevated pharmacological properties, and product so obtained. US Patent #4281108
Fussi F (1982) Process for obtaining low molecular weight heparins. European Patent GB2, 002, 406B
Gray E, Mulloy B, Barrowcliffe TW (2008) Heparin and low-molecular-weight heparin. Thromb Haemost 99:807–818
Guerrini M, Bisio A, Torri G (2001) Combined quantitative $^{1}$H and $^{13}$C-NMR spectroscopy for characterization of heparin preparations. Semin Thromb Hemost 27:473–482
Guerrini M, Raman R, Venkataraman G et al (2002) A novel computational approach to integrate NMR spectroscopy and capillary electrophoresis for structure assignment of heparin oligosaccharides. Glycobiology 12:713–719
Guerrini M, Naggi A, Guglieri S et al (2005) Complex glycosaminoglycans: profiling substitution patterns by two-dimensional nuclear magnetic resonance spectroscopy. Anal Biochem 337:35–47
Guerrini M, Guglieri S, Beccati D et al (2006) Conformational transitions induced in heparin octasaccharides by binding with antithrombin III. Biochem J 399:191–198
Guerrini M, Guglieri S, Naggi A et al (2007) Low molecular weight heparins: structural differentiation by bidimensional nuclear magnetic resonance spectroscopy. Semin Thromb Hemost 33:478–487
Guerrini M, Guglieri S, Casu B et al (2008) Antithrombin-binding octasaccharides and role of extensions of the active pentasaccharide sequence in the specificity and strength of interaction. J Biol Chem 283:26662–26675
Guerrini M, Zhang Z, Shriver Z et al (2009) Orthogonal analytical approaches to detect potential contaminants in heparin. Proc Natl Acad Sci USA 106:16956–16961
Guo X, Condra M, Kimura K et al (2003) Determination of molecular weight of heparin by size exclusion chromatography with universal calibration. Anal Biochem 312:33–39
Henriksen J, Ringborg LH, Roepstorrf P (2004) On-line size-exclusion chromatography/mass spectrometry of low molecular mass heparin. J Mass Spectrom 39:1305–1312
Hirsh H, Warkentin TE, Shaughnessy SG et al (2001) Heparin and low-molecular-weight heparin mechanisms of action, pharmacokinetics, dosing, monitoring, efficacy, and safety. Chest 119:64S–94S
Hirsh J, Raschke R (2004) Heparin and Low-Molecular-Weight Heparin. Chest 126:188S–203S
Hricovini M, Torri G (1995) Dynamics in aqueous solutions of the pentasaccharide corresponding to the binding site of heparin for antithrombin III studied by NMR relaxation measurements. Carbohydr Res 268:159–175
Hricovini M, Guerrini M, Torri G et al (1995) Conformational analysis of heparin epoxide in aqueous solution. An NMR relaxation study. Carbohydr Res 277:11–23

Hricovini M, Guerrini M, Bisio A et al (2001) Conformation of heparin pentasaccharide bound to antithrombin III. Biochem J 359:265–272

Huckerby TN, Sanderson PN, Nieduszynski A (1985) NMR Studies of the disulphated disaccharide obtained by degradation of bovine lung heparin with nitrous acid. Carbohydr Res 138:199–206

Iacomini M, Casu B, Guerrini M et al (1999) Linkage region sequences of heparins and heparan sulfates.Detection and quantification by NMR spectroscopy. Anal Biochem 274:50–58

Jaseja M, Rej RN, Sauriol F (1989) Novel region and stereoselective modifications of heparin in alkaline solution. Nuclear magnetic resonance spectroscopic evidence. Can J Chem 67: 1449–1456

Jin L, Abrahams JP, Skinner R et al (1997) The anticoagulant activation of antithrombin by heparin. Proc Natl Acad Sci USA 94:14683–14688

Johnson DJ, Li W, Adams TE et al (2006) Antithrombin-S195A factor Xa-heparin structure reveals the allosteric mechanism of antithrombin activation. EMBO J 25:2029–2037

Juhasz P, Biemann K (1995) Utility of non-covalent complexes in the matrix assisted laser desorption ionization mass spectrometry of heparin-derived oligosaccharides. Carbohydr Res 270:131–147

King JT, Desai UR (2008) A capillary electrophoretic method for fingerprinting low molecular weight heparins. Anal Biochem 380:229–234

Kitagawa H, Kinoshita A, Sugahara K et al (2002) Microanalysis of glycosaminoglycan-derived disaccharides labelled with the fluorophore 2-aminoacridone by capillary electrophoresis and high-performance liquid chromatography. Anal Biochem 232:114–121

Knobloch JE, Shaklee PN (1997) Absolute molecular weight distribution of low molecular weight heparins by size-exclusion chromatography with multiangle laser light scattering detection. Anal Biochem 245:231–241

Komatsu H, Yoshii K, Ishimitsu S et al (1993) Molecular mass determination of low molecular mass heparins. Application of wide collection angle measurements of light scattering using a high-performance gel permeation chromatography system equipped with a low angle laser light-scattering photometer. J Chromatogr 644:17–24

Korir A, Larive CK (2009) Advances in the separation, sensitive detection, and characterization of heparin and heparan sulfate. Anal Bioanal Chem 393:155–169

Kusche M, Torri G, Casu B et al (1990) Biosynthesis of heparin. Availability of 3-O-sulfation sites. J Biol Chem 265:7292–7300

Lamari FN, Militsopoulou M, Mitropoulou TN et al (2002) Analysis of glycosaminoglycan-derived disaccharides in biologic samples by capillary electrophoresis and protocol for sequencing glycosaminoglycans. Biomed Chromatogr 16:95–102

Lane DA, Denton J, Flynn AM et al (1984) Anticoagulant activities of heparin oligosaccharides and their neutralization by platelet factor 4. Biochem J 218:725–732

Langer RS, Linhardt RJ, Cooney LC et al (1983) Heparinase derived anticoagulants. US Patent #4396762

Li W, Johnson DJ, Esmon CT et al (2004) Structure of the antithrombin–thrombin–heparin ternary complex reveals the antithrombotic mechanism of heparin. Nat Struct Mol Biol 11:857–862

Linhardt RJ, Gunay NS (1999) Production and chemical properties of low molecular weight heparins. Semin Thromb Hemost 25(3):5–16

Linhardt RJ, Loganathan D, Al-Hakim A et al (1990) Oligosaccharide mapping of low molecular weight heparins: structure and activity differences. J Med Chem 33:1639–1645

Loganathan D, Wang HM, Mallis LM et al (1990) Structural variation in the antithrombin III binding site region and its occurrence in heparin from different sources. Biochemistry 29:4362–4368

Lopez LL (1991) Depolymerization method of heparin. US Patent #4981955

Lormeau JC, Petitou M, Choay J (1991) Method for obtaining biologically active mucopolysaccharides of high purity, by controlled depolymerization of heparin. US Patent #5019649

Lühn S, Schrader T, Sun W, Alban S (2010) Development and evaluation of a fluorescence microplate assay for quantification of heparins and other sulfated carbohydrates. J Pharm Biomed Anal 52:1–8

Ma SC, Yang VC, Meyerhoff ME (1992) Heparin-responsive electrochemical sensor: a preliminary study. Anal Chem 64:694–697

Malsch R, Harenberg J, Piazolo L et al (1996) Chromatographic and electrophoretic application for the analysis of heparin and dermatan sulfate. J Chromatogr B 685:223–231

Mao W, Thanawiroon C, Linhardt RJ (2002) Capillary electrophoresis for the analysis of glycosaminoglycans and glycosaminoglycan-derived oligosaccharides. Biomed Chromatogr 16:77–94

Mascellani G, Guerrini M, Torri G et al (2007) Characterization of di- and monosulfated, unsaturated heparin disaccharides with terminal N-sulfated 1,6-anhydrohydro-β-D-glucosamine or N-sulfated 1,6-an β-D-mannosamine residues. Carbohydr Res 342:835–842

Militsopoulou M, Lamari FN, Hjerpe A et al (2002) Determination of twelve heparin- and heparan sulfate-derived disaccharides as 2-aminoacridone derivatives by capillary zone electrophoresis using ultraviolet and laser-induced fluorescence detection. Electrophoresis 23:1104–1109

Militsopoulou M, Lecomte C, Bayle C et al (2003) Laser-induced fluorescence as a powerful detection tool for capillary electrophoretic analysis of heparin/heparan sulfate disaccharides. Biomed Chromatogr 17:39–41

Mourier PAJ, Viskov C (2004) Chromatographic analysis and sequencing approach of heparin oligosaccharides using cetyltrimethylammonium dynamically coated stationary phases. Anal Biochem 332:299–313

Mourier P, Viskov C (2005) Method for determining specific group constituting heparins or low molecular weight heparin. US Patent 2005/0119477 A1; Chem Abstr 142:89363

Mourier P, Viskov C (2005) Process for oxidizing unfractionated heparins and detecting presence of glycoserine in heparin and heparin products. European patent WO/2005/090411

Mousa SA (2005) Heparin and low molecular weight heparin in thrombosis and inflammation: emerging link. In: Garg HG, Linhardt RJ, Hales CA (eds) Chemistry and biology of heparin and heparan sulfate. Elsevier Ltd, Oxford

Műller-Graff PK, Szelke H, Kay S, Kramer R (2010) Pattern-based sensing of sulfated glycosaminoglycans with a dynamic mixture of iron complexes. Org Biomol Chem 8:2327–2331

Mulloy B, Forster MJ (2000) Conformation and dynamics of heparin and heparan sulfate. Glycobiology 10:1147–1156

Mulloy B, Johnson EA (1987) Assignment of the $^{1}$H NMR spectra of heparin and heparan sulfate. Carbohydr Res 170:151–165

Mulloy B, Linhardt RJ (2001) Order out of complexity-protein structures that interact with heparin. Curr Opin Struct Biol 11:623–628

Mulloy B, Gee C, Wheeler SF et al (1997) Molecular weight measurement of low molecular weight heparins by gel permeation chromatography. Thromb Haemost 77:668–674

Nader HB, Walenga JM, Berkowitz SD et al (1999) Preclinical differentiation of low molecular weight heparins. Semin Thromb Hemost 25(suppl 3):63–72

Patel RP, Narcowicz C, Hutchinson JP et al (2008) A simple capillary electrophoresis method for the rapid separation and determination of intact low molecular weight and unfractionated heparins. J Pharm Biomed Anal 46:30–35

Perlin AS, Mazurek M, Jaques LB et al (1968) Proton magnetic resonance spectral study of heparin: L-iduronic acid residues in commercial heparin. Carbohydr Res 7:369–379

Perlin AS, Casu B, Sanderson GR et al (1970) 220 MHz spectra of heparin, chondroitins, and other mucopolysaccharides. Can J Chem 220(14):2260–2268

Pervin A, al-Hakim A, Linhardt RJ (1994) Separation of glycosaminoglycan-derived oligosaccharides by capillary electrophoresis by using reverse polarity. Anal Biochem 221:182–188

Petitou M, van Boeckel CAA (2004) A synthetic Antithrombin III binding pentasaccharide is now a drug! What comes next? Angew Chem Int Ed Engl 43:3118–3133

Petitou M, Barzu T, Herault J et al (1997) A unique trisaccharide sequence in heparin mediates the early step of antithrombin III activation. Glycobiology 7:323–327
Planes A, Vochelle N, Ferru J et al (1986) Enoxaparin, low molecular weight heparin: its use in the prevention of deep vein thrombosis following total hip replacement. Haemostasis 16:152–158
Raman R, Venkataraman G, Ernst S et al (2003) Structural specificity of heparin binding in the fibroblast growth factor family of proteins. Proc Natl Acad Sci USA 100:2357–2362
Ramasamy I, Kennedy J, Tan K (2003) Capillary electrophoresis for characterization of low molecular weight heparins. Lab Hematol 9:64–66
Rej RN, Perlin AS (1990) Base-catalyzed conversion of the α-Liduronic acid 2-sulfate unit of heparin into a unit of α-Lgalacturonic acid, and related reactions. Carbohydr Res 200:437–447
Rice KG, Rottink MK, Linhardt RJ (1987) Fractionation of heparin-derived oligosaccharides by gradient polyacrylamide-gel electrophoresis. Biochem J 244:515–522
Shriver Z, Raman R, Venkataraman G et al (2000a) Sequencing of 3-O-sulphate containing heparin decasaccharides with a partial antithrombin III binding site. Proc Natl Acad Sci USA 97:10359–10364
Shriver Z, Sundaram M, Venkataraman G et al (2000b) Cleavage of the antithrombin III binding site in heparin by heparinases and its implication in the generation of low molecular weight heparin. Proc Natl Acad Sci USA 97:10365–10370
Shriver Z, Sundaram M, Venkataraman G et al (2007) Low Molecular weight heparin composition and uses thereof. US Patent #2007/0287683
Sugahara K, Tsuda H, Yoshida K et al (1995) Structure determination of the octa- and decasaccharide sequences isolated from the carbohydrate-protein linkage region of porcine intestinal heparin. J Biol Chem 270:22914–22923
Sundaram M, Qi Y, Shriver Z et al (2003) Rational design of low-molecular weight heparins with improved in vivo activity. Proc Natl Acad Sci USA 100:651–656
Thanawiroon C, Linhardt RJ (2003) Separation of a complex mixture of heparin-derived oligosaccharides using reverse-phase high-performance liquid chromatography. J Chromatogr A 1014:215–223
Toida T, Linhardt RJ (1996) Detection of glycosaminoglycans as a copper (II) complex in capillary electrophoresis. Electrophoresis 17:341–346
Torri G, Guerrini M (2008) Quantitative 2D NMR analysis of glycosaminoglycans. In: Holzgrabe U, Wawer I, Diehl B (eds) NMR Spectroscopy in pharmaceutical analysis. Integra, India
Turnbull JE (2001) Analytical and preparative strong anion-exchange HPLC of heparan sulfate and heparin saccharides. Methods Mol Biol 171:141–147
Uzan A (1998) Sulfated polysaccharides obtained from heparin, preparation process, pharmaceutical composition and use thereof. US Patent #5849721
van Boeckel CAA, Grootenhuis PDJ, Visser A (1994) A mechanism for heparin-induced potentiation of antithrombin III. Nat Struct Biol 1:423–425
Venkataraman G, Shriver Z, Raman R et al (1999) Sequencing complex polysaccharides. Science 286:537–542
Viskov C, Just M, Laux V et al (2009) Description of the chemical and pharmacological characteristics of a new hemisynthetic ultra-low-molecular-weight heparin, AVE5026. J Thromb Haemost 7(7):1143–1151
Vismara E, Pierini M, Guglieri S et al (2007) Structural modification induced in heparin by a fenton-type depolymerization process. Semin Thromb Hemost 33:466–477
Vismara E, Pierini M, Mascellani G et al (2010) Low molecular weight heparin from $Cu^{2+}$ and $Fe^{2+}$ Fenton's type depolymerisation processes. Thromb Haemost 103(3):613–622
Vivès RR, Goodger S, Pye DA (2001) Combined strong-anion exchange HPLC and PAGE approach for the purification of heparan sulfate oligosaccharides. Biochem J 354:141–147
Volpi N, Mascellani G, Bianchini P (1992) Low molecular weight heparins (5 kDa) and oligoheparins (2 kDa) produced by gel permeation enrichment or radical process: comparison of structures and physiochemical and biological properties. Anal Biochem 200:100–107

Volpi N, Maccari F, Linhardt RJ (2008) Capillary electrophoresis of complex natural polysaccharides. Electrophoresis 29:3095–3106

Yamada S, Murakami T, Tsuda H et al (1995) Isolation of the porcine heparin tetrasaccharides with glucuronate 2-O-sulfate. J Biol Chem 270:8696–8705

Yates EA, Santini F, Guerrini M et al (1996) $^{1}H$ and $^{13}C$ NMR spectral assignment of the major sequences of twelve systematically modified heparin derivatives. Carbohydr Res 294:15–27

Zaia J (2009) On-line separations combined with MS for analysis of glycosaminoglycans. Mass Spectrom Rev 28:254–272

Zilberstein G, Shlar I, Korol L et al (2009) Focusing of low-molecular-mass heparins in polycationic polyacrylamide matrices. Anal Chem 81:6966–6971

# Heparin and Heparan Sulfate: Analyzing Structure and Microheterogeneity

**Zachary Shriver, Ishan Capila, Ganesh Venkataraman, and Ram Sasisekharan**

## Contents



**Abstract** The structural microheterogeneity of heparin and heparan sulfate is one of the major reasons for the multifunctionality exhibited by this class of molecules. In a physiological context, these molecules primarily exert their effects extracellularly by mediating key processes of cellular cross-talk and signaling leading to the modulation of a number of different biological activities including development, cell proliferation, and inflammation. This structural diversity is biosynthetically imprinted in a nontemplate-driven manner and may also be dynamically remodeled as cellular function changes. Understanding the structural information encoded in these molecules forms the basis for attempting to understand the complex biology they mediate. This chapter provides an overview of the origin of the structural

Z. Shriver • R. Sasisekharan (✉)
77 Massachusetts Avenue, Building E25-Room 519, Cambridge, MA 02139, USA

Koch Institute for Integrative Cancer Research, Massachusetts Institute of Technology, Cambridge, MA 02139, USA

Department of Biological Engineering, Massachusetts Institute of Technology, Cambridge, MA 02139, USA
e-mail: rams@mit.edu

I. Capila • G. Venkataraman
Momenta Pharmaceuticals Inc., 675 West Kendall Street, Cambridge, MA 02142, USA

R. Lever et al. (eds.), *Heparin - A Century of Progress*,
Handbook of Experimental Pharmacology 207,
DOI 10.1007/978-3-642-23056-1_8, 

microheterogeneity observed in heparin and heparan sulfate, and the orthogonal analytical methodologies that are required to help decipher this information.

**Keywords** Heparan sulfate • Heparin • Microheterogeneity • Biosynthesis • Structural analysis • Conformation • Mass spectrometry • Nuclear magnetic resonance • Structure–function relationships.

# 1 Introduction

Heparin and heparan sulfate are complex, linear, acidic polysaccharides belonging to the glycosaminoglycan (GAG) family. In higher organisms, they can be found primarily on the cell surface or in the extracellular matrix, attached to a protein core. The structural diversity of heparin and heparan sulfate (HS) lies at the core of the varied range of physiological processes these molecules tend to modulate. Heparin is a well-known anticoagulant drug and is extensively used in medical practice (Wardrop and Keeling 2008). Heparin is isolated from animal organs, predominantly porcine intestinal mucosa, and goes through an extensive process of purification before it can be used for pharmaceutical purposes (Linhardt and Gunay 1999; Liu et al. 2009). The molecular basis for the anticoagulant function of heparin was elucidated in the early 1980s when a distinct pentasaccharide sequence within heparin chains was identified as being crucial for binding and activating antithrombin, leading to accelerated inhibition of the coagulation cascade (Lindahl et al. 1979; Rosenberg and Lam 1979). Identification of this unique structure–function relationship further strengthened the hypothesis that the variable sulfation within heparin and heparan sulfate encodes information that forms the basis for regulating other physiological activities as well.

Over the last 20 years, heparin and heparan sulfate have been shown to interact with a large number of important proteins thereby regulating a range of biological activities including cell proliferation, inflammation, angiogenesis, viral infectivity and development. Due to the structural diversity exhibited by these molecules, it is believed that possibly unique (in some cases) or most likely an ensemble of structural motifs might be responsible for different interactions. Therefore, it has become increasingly important to interpret the structural information represented in these complex molecules in order to enable a better understanding of their structure–function relationships.

The structural microheterogeneity is predominantly biosynthetically imprinted in a nontemplate-driven manner, as chains of heparin or heparan sulfate are elongated in the Golgi (Salmivirta et al. 1996). Furthermore, the purification and manufacture of heparin also leads to the introduction of additional chemical heterogeneity. Structural heterogeneity, coupled with a lack of ex vivo tools to amplify specific structures, makes the structural analysis of heparin/heparan sulfate extremely challenging. It is becoming increasingly clear that efficient, accurate structural analysis requires orthogonal analytical approaches to help decipher the information encoded in the saccharide sequences. As such, this chapter focuses on understanding the structural microheterogeneity of heparin and heparan sulfate; how it arises, and methods that have been developed to analyze this structural diversity.

## 2 Biosynthesis of Heparin: Encoding Microheterogeneity

Heparin and heparan sulfate are built up of linear chains of repeating disaccharide units consisting of a glucosamine and uronic acid. The initial disaccharide unit that constitutes the growing chain during biosynthesis has a D-glucuronic acid β1 → 4 linked to a D-N-acetylglucosamine. These units are linked to each other via an α1 → 4 linkage. The subsequent modifications proceed in a sequential manner, beginning with the N-deacetylation and N-sulfonation of glucosamine residues within the chains. This is followed by epimerization of the glucuronic acid to iduronic acid and O-sulfonation at either the C-2 of the uronic acid or the C-6 of the glucosamine. The final modification step in this pathway is the O-sulfonation at the C-3 of the glucosamine. Each of these biosynthetic reactions is dependent on the previous modification to some extent, as products of one step can often act as substrates for subsequent steps. Another aspect of this pathway that has an important bearing on saccharide structural diversity is that each of these biosynthetic modification steps likely does not proceed to completion. Thus, the resulting chain can be differentially modified in various regions, accounting for a significant component of the structural heterogeneity observed in these molecules (Salmivirta et al. 1996; Sugahara and Kitagawa 2002). The structure of the basic disaccharide sequence in heparin and heparan sulfate along with the sites of variable sulfation is shown in Fig. 1. The average heparin disaccharide contains ~2.7 sulfate groups, whereas heparan sulfate contains ≥1 sulfate group per disaccharide (Toida et al. 1997).

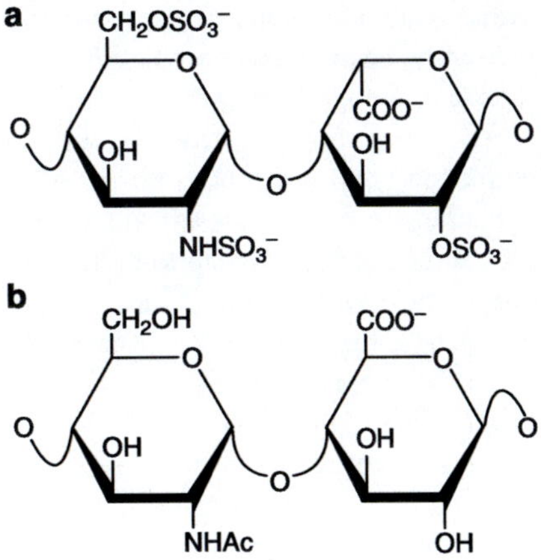


**Fig. 1** The major repeating disaccharide unit in (**a**) heparin and (**b**) heparan sulfate. Structural heterogeneity arises due to the variable presence of either acetyl or sulfo groups at the N – position of the glucosamine, sulfation at the 2-O – position on the uronic acid or 6-O – and 3-O – positions on the glucosamine, and epimerization at the C-5 of the uronic acid

Also, while L-iduronic acid predominates in heparin, the D-glucuronic acid epimer represents the majority of the uronic acid present in heparan sulfate (Lindahl et al. 1998). Heparin is often referred to as the more completely modified version of heparan sulfate and also possesses the highest negative charge density of any known biological macromolecule. While heparan sulfate contains all of the structural variations found in heparin, the frequency of occurrence of the minor sequence variations is greater in HS; thus, the extent of structural heterogeneity observed in heparan sulfate is usually greater than that observed in heparin. However, both heparin and heparan sulfate chains are polydisperse, with a broad molecular weight distribution. Heparan sulfate chains are generally longer than heparin chains, and have an average molecular weight of ~30 kDa as compared to ~15 kDa for heparin. This structural variability at multiple levels makes heparin and heparan sulfate very challenging molecules to characterize.

In addition to the structural heterogeneity, heparin and heparan sulfate also exhibit conformational flexibility due to the presence of iduronic acid in their linear sequence (Ferro et al. 1990; Sanderson et al. 1987). This provides an additional dimension through which chains of heparin and heparan sulfate can alter the spatial orientation of their sulfate groups to allow for productive binding to various proteins. Co-crystal structures of a heparin hexasaccharide with basic fibroblast growth factor demonstrate this phenomenon (Canales et al. 2005; Faham et al. 1996). One of the sulfated iduronic acid residues in the saccharide is observed to contact the protein in a $^{1}C_{4}$ conformation. This conformation is typically not favored by the residue in aqueous solution.

# 3 Purification and Isolation of Heparin

Heparin is one of the most widely used carbohydrate drugs, and has been used as a pharmaceutical product for several decades. Clinically, heparin is used as a prophylactic agent to prevent the formation of thrombi, as well as for their initial treatment. Heparin is a natural product that is isolated from animal tissues. Currently, heparin is sourced almost exclusively from a single source, viz., porcine intestine, though other animal sources of heparin have been reported (Casu et al. 1995, 1996). Heparin from bovine sources, especially bovine lung, had been employed previously, but current use is not widespread due to concerns of BSE contamination. There are efforts ongoing to create heparin through defined chemical sulfonation of the appropriate saccharide backbone, coupled with in vitro enzymatic modification of chains through the use of recombinantly expressed enzymes of the biosynthetic pathway (Kuberan et al. 2003; Liu et al. 2010).

As behooves the fact that it is derived biosynthetically, the purification of heparin is a multistep involved process. Several basic steps are used in the initial purification of heparin from porcine intestinal mucosa to form crude heparin. Preparation of raw or crude heparin is typically performed outside of cGMP (Good Manufacturing Practices) and involves the physical separation of mucosa from the gut lining, solubilization through the addition of proteases, multiple precipitation and resolubilization steps, followed by a final ethanolic precipitation step (Liu et al. 2009). At this stage, crude heparin is typically a mixture of many components. Besides residual biomaterials, such as proteins, nucleic acids, and lipids, crude heparin contains a mixture of glycosaminoglycan complex polysaccharides, such as hyaluronic acid, chondroitin sulfate, as well as the two predominant components – heparin and dermatan sulfate.

Crude heparin is then transported into a cGMP facility for additional purification, testing and release. Depending on the manufacturer, different purification strategies are employed; broadly these involve additional precipitations, ion capture and exchange, and chemical steps to decolorize and depyrogenate heparin. Upon completion of final polishing steps, heparin is a fairly well-defined material, predominantly composed of glycosaminoglycan chains with some degree of polydispersity, but an average molecular weight that consistently is within the range of 12–20 kDa. It also has a fairly defined composition, primarily consisting of repeating units of the trisulfated disaccharide, i.e., a 2-O-sulfonated iduronic acid 1 → 4 linked to a 6-O, N-sulfonated glucosamine (Fig. 1).

Differential precipitation steps used in the purification of heparin are partly the reason for unfractionated heparin being one of the most highly sulfated glycosaminoglycans. Studies looking at waste material, so-called "side stream" heparin, indicate that undersulfated and/or lower molecular weight material is removed from the final product, yielding heparin chains with an average sulfate-to-carboxylate ratio, a measure of the degree of sulfation per average disaccharide, of 2.4-2.6 (Liu et al. 2009). Furthermore, this purification strategy likely limits the amount of undersulfated materials, including porcine-based heparan sulfate, which can be present within a sample. Porcine heparan sulfate, which may be created biosynthetically and characterized by a sulfate-to-carboxylate ratio of ~0.3 (Toida et al. 1997), is probably removed early on in the purification process, if even present at appreciable levels. Indeed, examination of crude heparin preparations indicates few, if any, of the signs of the presence of heparan sulfate. Additionally, considering that, at later polishing stages, precipitation results in the removal of heparin material with a higher degree of sulfation than HS, it is highly unlikely that appreciable levels of "true" heparan sulfate are present within heparin preparations. Conversely, dermatan sulfate, a higher sulfated glycosaminoglycan that contains a high degree of iduronic acid, has been shown to be an impurity in heparin preparations.

Apart from the inherent structural variability among the heparin chains, both in terms of sulfation and chain length, there are structural variations that are introduced as a function of the purification process, especially chemical treatment to decolor, deodorize, and depyrogenate heparin samples. Because a variety of chemical agents can be employed, including peroxide treatment under alkaline conditions, permanganate, hypochlorite, and ozone, the type and extent of modifications that can occur are variable and can be indicative of the manufacturing process employed.

Under basic conditions, it is clear that some amount of 2-O sulfonated iduronic acid can be converted to an intermediate 2, 3-epoxide, followed by conversion

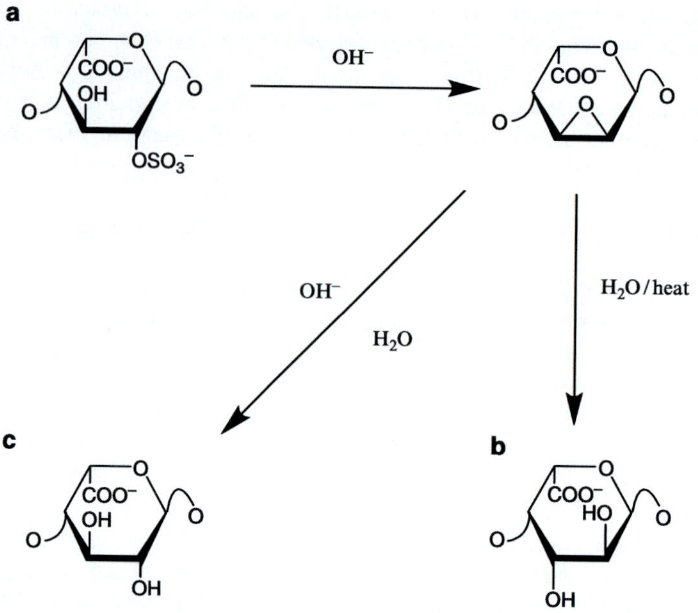


**Fig. 2** Sensitivity of the 2-O-sulfated iduronic acid in heparin to manufacturing conditions. Conversion of 2-O-sulfated iduronic acid in heparin to (**a**) 2,3 epoxide upon treatment with base. Subsequently, the 2,3 epoxide can convert to either (**b**) a galacturonic acid species on heat treatment, or (**c**) an iduronic acid (at higher pH)

to galacturonic acid (Fig. 2). We and others have found that this can constitute up to ~2–3% of the total disaccharide content of heparin (Rej et al. 1989). Additionally, under basic conditions, 6-O sulfonated glucosamine can undergo cyclization to form a 1,6 anhydro ring. Finally, peeling reactions via Millard chemistry can also occur, resulting in loss of a monosaccharide and the production of "odd-numbered" chains.

In addition to these reactions, which can occur to an appreciable extent, there are other minor modifications that can occur. Several groups have identified the fact that addition of oxidizing agents, such as peracetic acid, can result in transient O-acetylation. Potassium permanganate-based oxidation can extend to modification of N-acetylglucosamine moieties at the reducing ends of the chain resulting in formation of an N-acetylglucosaminic acid (Fig. 3). Presence of such a species has recently been shown to produce a characteristic signal in the $^{1}$H-NMR spectrum at 2.10 ppm, and is detected in heparin from several manufacturers (Beccati et al. 2010).

Taken together, it is clear that the production and purification of heparin starts with a diverse set of polysaccharide chains, with some degree of sequence variability. This variability is both reduced through purification of the final heparin product, which results in a narrowing of the polydispersity and a homogenization of the sulfation pattern of individual chains, and is increased, through introduction of purification process-specific alterations.

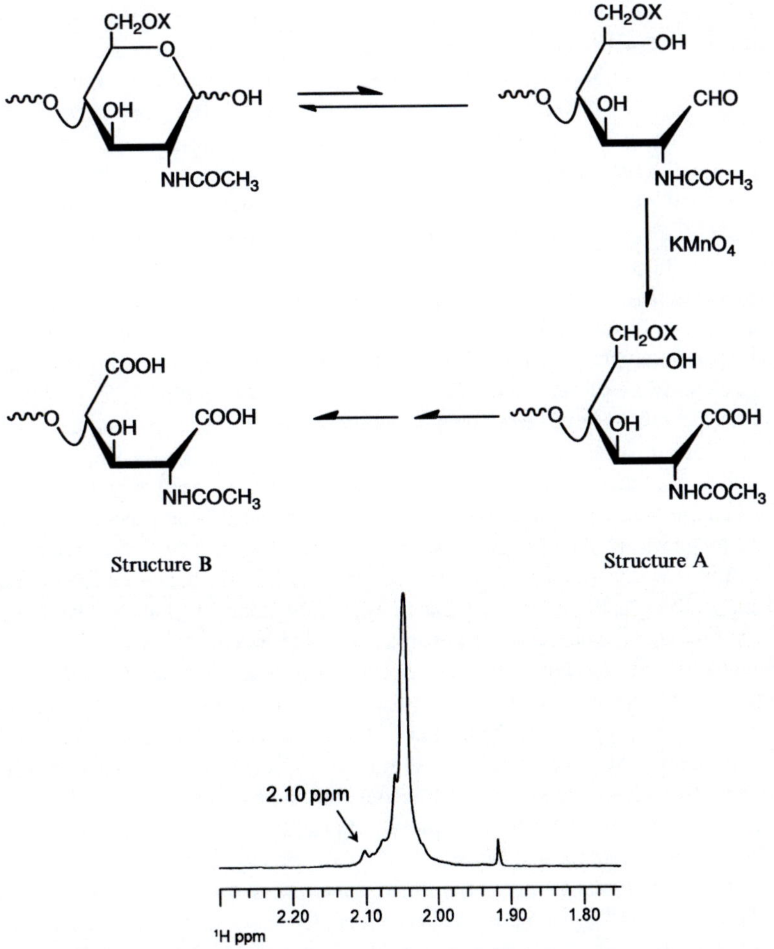


**Fig. 3** Potassium permanganate oxidation of the reducing end N-acetylglucosamine to N-acetylglucosaminic acid. The formation of Structure A causes a shift in the N-acetyl protons to 2.10 ppm, as shown in the 1 H-NMR spectrum. If X=H in Structure A, then the formation of Structure B may also be possible due to further oxidation

## 4 Structural Analysis of Heparin and Heparan Sulfate

Due to differential sulfation, domain structure (i.e., nonuniform distribution of disaccharides), and variable polysaccharide length, the structure and sequence of heparan sulfate (HS) is highly variable. This structural variability arises mostly from the fact that HS is synthesized in a nontemplate manner, through the concerted action of biosynthetic enzymes, including the O-sulfotransferases, the N-deactylase/N-sulfotransferase, and the C5-epimerase (Salmivirta et al. 1996).

However, this variability, while significant, is controlled, through the tissue/cell-specific expression of enzyme isoforms. This has been confirmed experimentally by examining HS composition across different tissue preparations (Guimond et al. 2009; Shi and Zaia 2009) where there are primarily six major disaccharides (Table 1), with the most abundant being a nonsulfated uronic acid linked to an N-acetylglucosamine (I/G $\rightarrow H_{NAc}$). In this context, heparin can be thought of as a "specialized" HS, one that is more highly sulfated, and hence contains less variability. Especially in the context of HS, the challenge of detailed structural analysis is exacerbated by the fact that, because synthesis is nontemplate-driven, strategies employed for other template-driven biopolymers, such as DNA or proteins, are not amenable to sequencing complex polysaccharides. Thus, new glycan-specific techniques must either be developed de novo or existing techniques (e.g., proteomic-based techniques) must be substantially adapted.

Numerous efforts have been made to facilitate the sequencing of heparin and HS chains, primarily to determine structure/function relationships for important protein-binding saccharides. In recent years, these efforts have been enhanced through the development of sophisticated analytical tools, including mass spectrometry and nuclear magnetic resonance (Kuberan et al. 2002; Yates et al. 1996; Saad et al. 2005). Additionally, heparin-degrading enzymes, primarily the bacterially derived heparinases, but also the mammalian enzyme heparanase, have proven to be important tools for sequencing of heparin and HS (Bisio et al. 2007).

Broadly, efforts to provide structural information for heparin and HS include both top-down approaches, where key structural information is obtained on intact chains, and bottom-up approaches, where either partial or complete digestion of the chains is carried out, followed by identification /quantification of the resulting fragments. In terms of top-down approaches, monodimensional ($^{1}H$ and $^{13}C$) and multidimensional NMR have been employed. Proton and $^{13}C$ NMR have been used to differentiate animal source and key structural motifs of heparin and HS (Guerrini et al. 2001, 2005). Additionally, for shorter heparin fragments, up to tetradecasaccharides, multidimensional NMR can be used to obtain complete sequence information. In the case of complex mixtures, including intact heparin and low-molecular-weight heparin, multidimensional NMR can be used to identify mono- and disaccharide constituents as well as unusual building blocks produced in low-molecular-weight heparins as a function of the process (Fig. 4) (Guerrini et al. 2007). Notably, multidimensional NMR was the key experimental technique to identify the structure of the heparin contaminant, oversulfated chondroitin sulfate (Guerrini et al. 2008).

In addition to recent advances in NMR analysis of heparin/HS, mass spectrometry has been used to examine the fine structure of heparin and HS. While information from intact analysis of heparin chains is limited, in-line placement of high-resolution gel permeation chromatography has been successfully used to determine composition of low-molecular-weight heparin chains up to octadecasaccharide (Henriksen et al. 2004). Additionally, such an approach has recently been used to determine detailed structural information for the chondroitin sulfate proteoglycan, bikunin (Chi et al. 2008).

**Table 1** Six disaccharides that constitute the major building blocks of heparan sulfate chains

| Disaccharide structure | Structural notation |
|---|---|
| 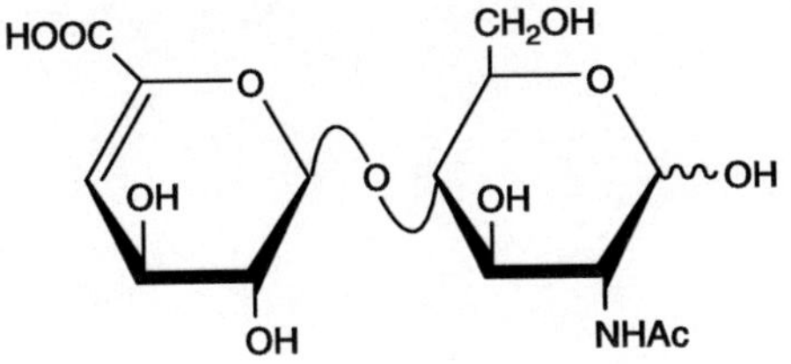 | $\Delta U\text{-}H_{NAc}$ |
| 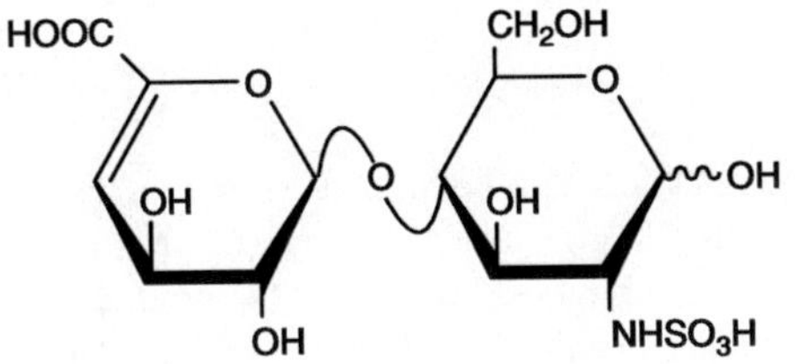 | $\Delta U\text{-}H_{NS}$ |
| 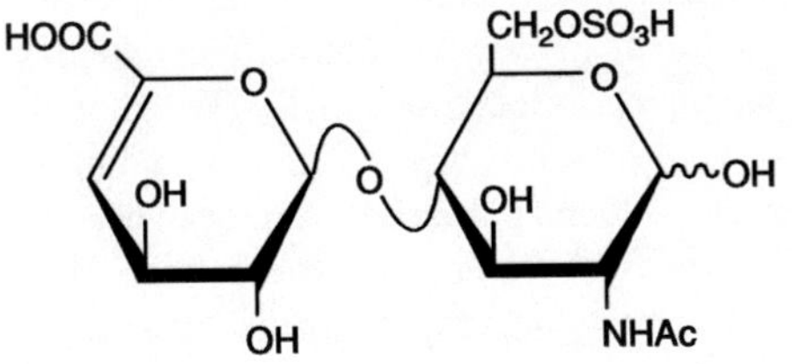 | $\Delta U\text{-}H_{NAc6S}$ |
| 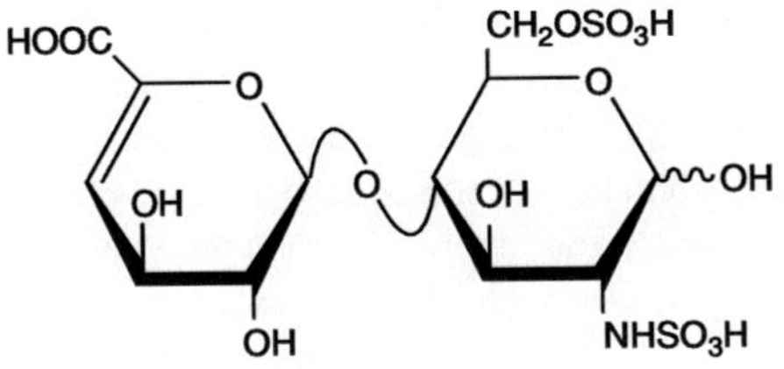 | $\Delta U\text{-}H_{NS6S}$ |
| 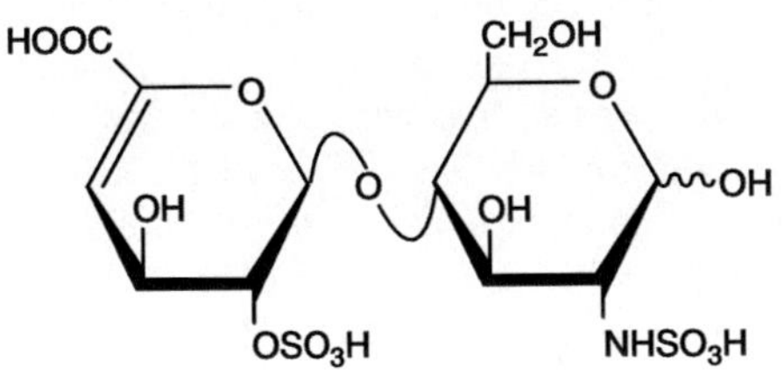 | $\Delta U_{2S}\text{-}H_{NS}$ |

(continued)

**Table 1** (continued)

| Disaccharide structure | Structural notation |
|---|---|
| 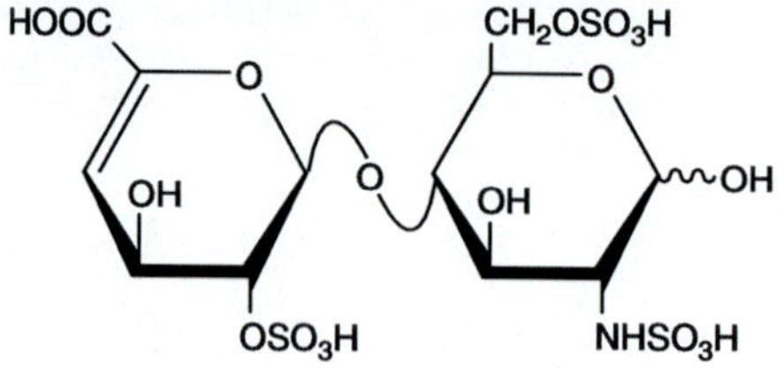 | $\Delta U_{2S}$-$H_{NS6S}$ |

Coupled with the above advances, progress has also been made in strategies to fragment HS chains. Historically, application of high energy sufficient for fragmentation (and even ionization) has typically resulted in desulfation and the production of low information content fragments. These limitations have been overcome through several strategies, including formation of noncovalent complexes with cationic buffers or basic peptides (Juhasz and Biemann 1994; Rhomberg et al. 1998), application of gentle ionization conditions, as well as alternative strategies for fragmentation, including electron detachment dissociation (EDD) (Wolff et al. 2007). These advances have led to strategies where informative ions, including B and Y ions (Fig. 5) can be formed and detected. Such approaches have been used to sequence heparin-derived saccharides. However, there are limitations with these approaches; while sequence information has been obtained for sequences up to hexasaccharides, sequencing longer fragments still suffers from loss of structural information due to substantial sulfate loss during fragmentation.

Bottom-up approaches, involving either partial or complete degradation of the heparin/HS chains and determination of the monosaccharide constituents of the fragments has been largely perfected. Quantification is typically completed through separating the resulting fragments by ion exchange or ion-pairing reverse phase HPLC or via capillary electrophoresis. Identification is typically completed either through co-injection with reference standards or through MS and $MS^n$ analysis by online coupling, especially for ion-pairing HPLC where labile pairing agents, such as dibutylamine, have simplified detection (Kuberan et al. 2002). An alternative strategy that has been employed is direct infusion of a digested mixture and detection and quantification by MS (Saad and Leary 2003, 2005).

In terms of digestion strategies, chemical means such as nitrous acid have been successfully used to determine disaccharide composition and have the added advantage that such strategies maintain the epimeric state of the uronic acid. However, the fact that no chromophore is generated and that detection typically must be completed through labeling (most often using radioactive sodium borohydride) has limited the adoption of this technique. An alternative strategy, employing bacterially derived heparinases, has been more widely adopted. Here, addition of

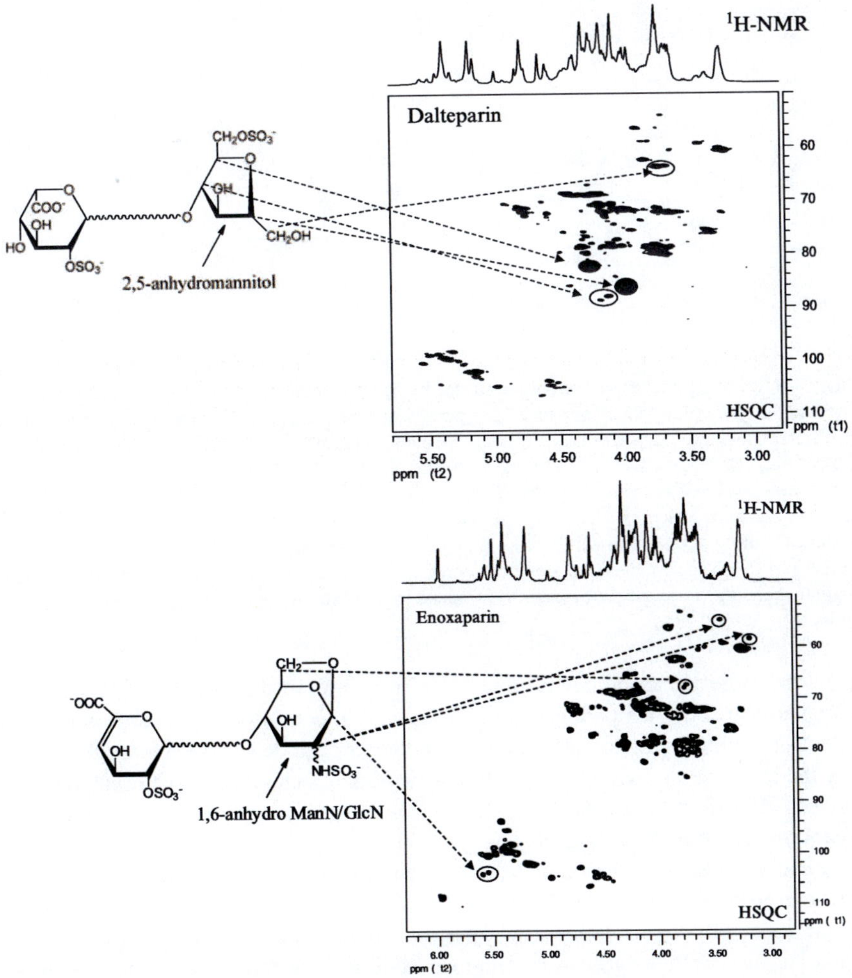


**Fig. 4** HSQC spectra of low-molecular-weight heparins. 2D-NMR spectra of dalteparin and enoxaparin exhibit unique cross peaks arising from signature structures. These structures result from the specific chemistry used to depolymerize unfractionated heparin into lower molecular weight chains. The 2,5 anhydromannitol in dalteparin, and the 1,6-anhydro structure in enoxaparin, along with their associated signature cross peaks are shown

one or more heparin-degrading enzymes, each with distinct, but overlapping substrate specificity, results in cleavage of heparin or HS (Ernst et al. 1995). Both complete digestion down to di- and tetrasaccharides, and partial digestion have been completed (Linhardt et al. 1988, 1990; Karamanos et al. 1997). Since these enzymes are lyases, digestion results in the formation of a Δ4,5 double bond on the uronic acid. Such species absorb UV light strongly, providing a tag for detection. Furthermore, structural information, specifically the presence or absence of an

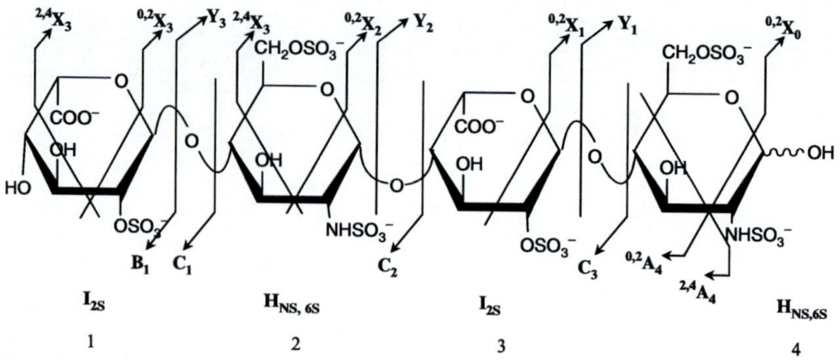


**Fig. 5** Sequencing of heparin oligosaccharides by MS fragmentation. Several researchers have focused on developing MS-based sequencing procedures for heparin oligosaccharides. Shown here is different potential fragments for a representative tetrasaccharide: $I_{2S}H_{NS,6S}I_{2S}H_{NS,6S}$. In general, "cross-ring" cleavages, such as those that form fragments such as ${}^{0,2}X_{0,1,2,3}$ and/or ${}^{2,4}X_{0,1,2,3}$ are more informative of sequence. However, fragments resulting from glycosidic bond cleavage are generally more prevalent. Additionally, fragments are often detected with some loss of sulfate (and water), resulting in sometimes ambiguous assignments; for example, the inability to distinguish B and Z ions from C and Y ions (the most common fragments). Finally, from published reports, more fragments and/or cross-ring fragmentations result from hexosamine (H) units than uronic (I/G) units

N-acetyl moiety, can be determined. Also of benefit is that the digestion can be completed at room temperature at pH of around 7, conditions conducive with the retention of labile sulfates. Alternatively, detection can be achieved through labeling the reducing end via reductive amination. Such approaches have dramatically increased sensitivity, especially in conjunction with laser-induced fluorescence, enabling detection of femtomoles of material.

Finally, given their high charge density and isomeric configurations (as well as the presence of α and β isoforms), often saccharide components will co-elute/co-migrate with one another. This greatly complicates both identification and quantification. While MS can be employed online to differentiate isomeric states, an alternative, easily implemented strategy is the use of additional bacterial or mammalian degradation enzymes, including the Δ4,5 glycuronidase (Myette et al. 2002), and the 2-O and 6-O sulfatases (Myette et al. 2003, 2009). Concerted or stepwise addition of these enzymes has been used to resolve di- and oligosaccharides, enabling accurate quantification and determination of mass balance.

Given the structural complexity of heparin and HS, as well as the increasingly sophisticated insight that we are obtaining into biological processes, it is likely that integration of several independent approaches, including both top-down and bottom-up approaches will prove important. Thus, while some methodologies, such as those outlined above, have provided superior information content, dataset integration enables more complete and rapid assessment of sequence and structure–function than any individual method used in isolation. As such development of frameworks to enable integration of datasets from multiple techniques will continue to prove important for the sequencing of heparin and HS (Venkataraman et al. 1999). As one example of such an integrative approach, results from NMR-based analysis of heparin fragments has been complemented and extended by CE-based disaccharide analysis after digestion with a cocktail of enzymes. In this manner, use of both datasets in an iterative and integrative manner enables one to "walk" through the saccharide sequence (Guerrini et al. 2002).

While the above approach has been successfully applied to isolated saccharides, or simple mixtures, for more complex mixtures, such as what has been isolated from cell surface HS, use of multiple approaches, or constraints, will likely be necessary; however, the approach is still valid. Depending on the complexity of the mixture, use of multiple, orthogonal constraints will likely be necessary, including both top-down and bottom-up approaches. Indeed, recently such logic was employed by the U.S. FDA in the approval of the first generic low-molecular-weight heparin.

## 5 Addressing Structure–Function Relationships for Heparan Sulfate

Heparan sulfate is known to interact with a wide variety of proteins and modulate their activity (Capila and Linhardt 2002; Lindahl 2007). A partial list of proteins, the role of HS, and structural motifs (i.e., length, sulfation pattern, etc.) that they recognize is listed in Table 2. The first (and still best) example of a defined protein-

**Table 2** Physiological and pathophysiological role of heparin and heparan sulfate

| Physiology/ pathology | Binding proteins | Characteristics of heparin/HS binding |
|---|---|---|
| Anticoagulation | Antithrombin | Unique 5-mer sequence; requires 3-O-sulfonation |
| | Thrombin | Trisulfated disaccharide repeat |
| Cell proliferation and metastasis | FGF-2 | 4-mer to 6-mer; IdoA2S-GlcNS |
| | FGF-1 | 4-mer to 6-mer; IdoA2S-GlcNS6S |
| | Selectins | $\geq$4-mer; requires highly sulfated sequences, with presence of free amine residues |
| Inflammation | IL-8 | 18-mer to 20-mer; requires alternating domains of high sulfation-low sulfation-high sulfation |
| | SDF1$\alpha$ | 12-mer to 14-mer; requires highly sulfated domains |
| Viral infectivity | HSV gD | 8-mer; requires a unique 3-O-sulfated, glucosamine at the reducing end |
| | HIV gp120 | 10-mer; highly sulfated sequences |
| | Dengue virus envelope protein | 10-mer; highly sulfated sequences |
| Development | HB-GAM | 16-mer to 18-mer; requires highly sulfated domains |

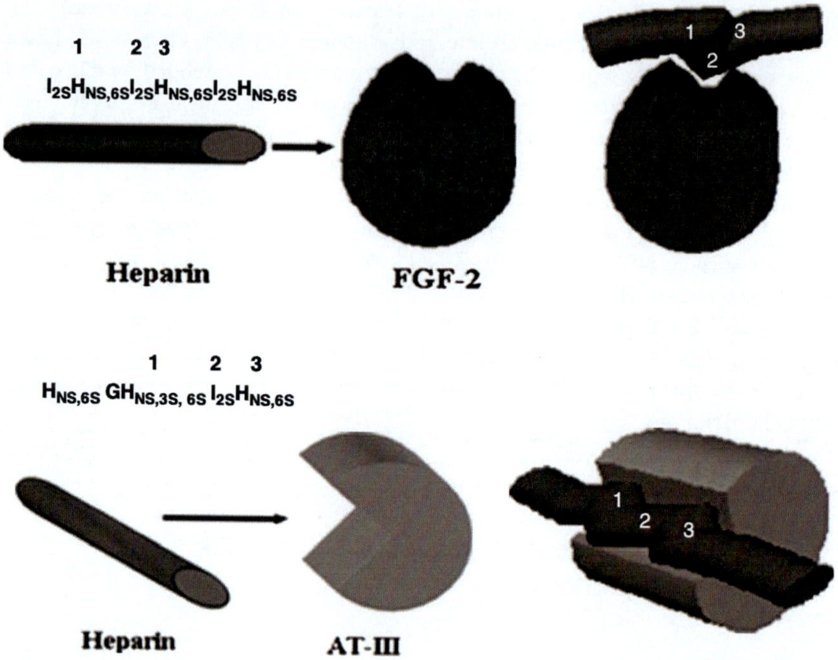


**Fig. 6** Specificity of binding to heparin is governed by sequence and topology. Two examples of this "kink" modulating specificity include the FGF system where binding to a trisaccharide spanning kink [$H_{NS,6S}I_{2S}H_{NS,6S}$] is enhanced by an internal iduronate residue adopting a particular conformation ($^{1}C_{4}$). Similarly, with the antithrombin system, a separate "kink", namely a trisaccharide of the sequence $H_{NS,3S,6S}I_{2S}H_{NS,6S}$ influences "overwinding" of the helical axis, mediated in part through the $^{2}S_{0}$ conformation of iduronate

binding motif within heparin was the discovery of a discrete pentasaccharide sequence within heparin required for binding antithrombin. This pentasaccharide sequence is rare, occurring in only about one-third of heparin chains, and with even less frequency in heparan sulfate. Its most distinguishing feature is the presence of an unusual 3-O-sulfo group on an internal GlcNpS6S residue, which is absolutely essential for its high affinity binding to antithrombin. Notably, 3-O-sulfonation likely occurs late in the biosynthesis of heparin and heparan sulfate, and there are multiple 3-O-sulfotransferase isoforms having different substrate specificities. Given its relative rarity as well as the cloning and characterization of the biosynthetic enzymes, 3-O sulfonation has been used as a marker to identify sequences that are important in several biological systems, including viral entry and bridging of growth factor-receptor complexes, among others (Liu et al. 2002; Shukla et al. 1999).

In addressing structure–function relationships for heparin/HS, reference to the antithrombin-heparin system provided a set of biochemical and analytical tools to identify high affinity sequences in heparin. With this system, the presence of a 3-O sulfate resulted in a ~100-fold increase in affinity, almost representing an "on/off switch". Unlike this situation, many heparin/HS–protein interactions are often gradated, where sequence matters, but strength of binding to various structures depends on more than primary sequence. Addressing the structure–function relationships in these instances not only requires sequencing of HS from biologically relevant sources, but it also requires an integration of information from complementary approaches that permit a move from the biological to the structural space in an iterative and transitive manner. Approaching the problem in this way enables a fuller appreciation of all aspects of high-affinity interactions between soluble factors, including the key role that conformation plays in the high-affinity interaction between HS/heparin and growth factors. Introduction of conformational considerations has led to the introduction of the "kink" concept, where binding of growth factors to particular sequences induces a local distortion in the secondary structure of the polysaccharide chain, aligning sulfate groups for strong interaction with the protein, particularly basic residues (Fig. 6) (Raman et al. 2003). Thus, beyond primary sequence, conformational considerations, as well as the positioning of protein binding sites within the chain (thus modulating protein oligomerization) are essential to fully understanding structure–function relationships for this class of molecules.

## 6 Conclusions

Heparin, used clinically in both the treatment and prophylaxis of thrombosis, is one of the most widely used drugs. Given that the manufacture of heparin has been widely optimized, and given that a large number of proteins of physiological and pathophysiological importance interact with heparin and HS, raises the potential for the use of heparin/HS in a range of potential therapeutic applications, ranging from cancer to neurological disorders to inflammatory diseases. The major limitations in utilizing heparin/HS-based oligosaccharides for a wide range of disorders are (1) derivation of definitive structure–function relationships; and (2) the source and manufacturing of heparin has been optimized for high anticoagulant potency, which in essence becomes a side-effect, for nonanticoagulant applications. For the first consideration, the development of analytical technologies for the sequencing of heparin/HS as well as structural tools to understand high affinity binding in the context of protein structure have enabled and will continue to enable development of robust structure-activity understanding. For the second consideration, the introduction of low-molecular-weight heparins and the preparation of heparin oligosaccharides and synthetic analogs devoid of anticoagulant activity may open up a wide variety of new potential therapeutic applications in the treatment of cancer, and viral and bacterial infection, among others.

**Acknowledgement** The authors thank Dr. Daniela Beccati for providing Fig. 4.

## References


Beccati D, Roy S, Yu F, Gunay NS, Capila I, Lech M, Linhardt RJ, Venkataraman G (2010) Identification of a novel structure in heparin generated by potassium permanganate oxidation. Carbohydr Polymer 82:699–705

Bisio A, Mantegazza A, Urso E, Naggi A, Torri G, Viskov C, Casu B (2007) High-performance liquid chromatographic/mass spectrometric studies on the susceptibility of heparin species to cleavage by heparanase. Semin Thromb Hemost 33:488–495

Canales A, Angulo J, Ojeda R, Bruix M, Fayos R, Lozano R, Gimenez-Gallego G, Martin-Lomas M, Nieto PM, Jimenez-Barbero J (2005) Conformational flexibility of a synthetic glycosylaminoglycan bound to a fibroblast growth factor. FGF-1 recognizes both the (1)C(4) and (2)S(O) conformations of a bioactive heparin-like hexasaccharide. J Am Chem Soc 127: 5778–5779

Capila I, Linhardt RJ (2002) Heparin-protein interactions. Angew Chem Int Ed Engl 41:391–412

Casu B, Guerrini M, Naggi A, Torri G, De-Ambrosi L, Boveri G, Gonella S, Ronzoni G (1995) Differentiation of beef and pig mucosal heparins by NMR spectroscopy. Thromb Haemost 74: 1205

Casu B, Guerrini M, Naggi A, Torri G, De-Ambrosi L, Boveri G, Gonella S, Cedro A, Ferro L, Lanzarotti E, Paterno M, Attolini M, Valle MG (1996) Characterization of sulfation patterns of beef and pig mucosal heparins by nuclear magnetic resonance spectroscopy. Arzneimittelforschung 46:472–477

Chi L, Wolff JJ, Laremore TN, Restaino OF, Xie J, Schiraldi C, Toida T, Amster IJ, Linhardt RJ (2008) Structural analysis of bikunin glycosaminoglycan. J Am Chem Soc 130:2617–2625

Ernst S, Langer R, Cooney CL, Sasisekharan R (1995) Enzymatic degradation of glycosaminoglycans. Crit Rev Biochem Mol Biol 30:387–444

Faham S, Hileman RE, Fromm JR, Linhardt RJ, Rees DC (1996) Heparin structure and interactions with basic fibroblast growth factor. Science 271:1116–1120

Ferro DR, Provasoli A, Ragazzi M, Casu B, Torri G, Bossennec V, Perly B, Sinay P, Petitou M, Choay J (1990) Conformer populations of L-iduronic acid residues in glycosaminoglycan sequences. Carbohydr Res 195:157–167

Guerrini M, Bisio A, Torri G (2001) Combined quantitative (1)H and (13)C nuclear magnetic resonance spectroscopy for characterization of heparin preparations. Semin Thromb Hemost 27: 473–482

Guerrini M, Raman R, Venkataraman G, Torri G, Sasisekharan R, Casu B (2002) A novel computational approach to integrate NMR spectroscopy and capillary electrophoresis for structure assignment of heparin and heparan sulfate oligosaccharides. Glycobiology 12:713–719

Guerrini M, Naggi A, Guglieri S, Santarsiero R, Torri G (2005) Complex glycosaminoglycans: profiling substitution patterns by two-dimensional nuclear magnetic resonance spectroscopy. Anal Biochem 337:35–47

Guerrini M, Guglieri S, Naggi A, Sasisekharan R, Torri G (2007) Low molecular weight heparins: structural differentiation by bidimensional nuclear magnetic resonance spectroscopy. Semin Thromb Hemost 33:478–487

Guerrini M, Beccati D, Shriver Z, Naggi A, Viswanathan K, Bisio A, Capila I, Lansing JC, Guglieri S, Fraser B, Al-Hakim A, Gunay NS, Zhang Z, Robinson L, Buhse L, Nasr M, Woodcock J, Langer R, Venkataraman G, Linhardt RJ, Casu B, Torri G, Sasisekharan R (2008) Oversulfated chondroitin sulfate is a contaminant in heparin associated with adverse clinical events. Nat Biotechnol 26:669–675

Guimond SE, Puvirajesinghe TM, Skidmore MA, Kalus I, Dierks T, Yates EA, Turnbull JE (2009) Rapid purification and high sensitivity analysis of heparan sulfate from cells and tissues: toward glycomics profiling. J Biol Chem 284:25714–25722

Henriksen J, Ringborg LH, Roepstorff P (2004) On-line size-exclusion chromatography/mass spectrometry of low molecular mass heparin. J Mass Spectrom 39:1305–1312

Juhasz P, Biemann K (1994) Mass spectrometric molecular-weight determination of highly acidic compounds of biological significance via their complexes with basic polypeptides. Proc Natl Acad Sci USA 91:4333–4337

Karamanos NK, Vanky P, Tzanakakis GN, Tsegenidis T, Hjerpe A (1997) Ion-pair high-performance liquid chromatography for determining disaccharide composition in heparin and heparan sulphate. J Chromatogr A 765:169–179

Kuberan B, Lech M, Zhang L, Wu ZL, Beeler DL, Rosenberg RD (2002) Analysis of heparan sulfate oligosaccharides with ion pair-reverse phase capillary high performance liquid chromatography-microelectrospray ionization time-of-flight mass spectrometry. J Am Chem Soc 124:8707–8718

Kuberan B, Beeler DL, Lech M, Wu ZL, Rosenberg RD (2003) Chemoenzymatic synthesis of classical and non-classical anticoagulant heparan sulfate polysaccharides. J Biol Chem 278: 52613–52621

Lindahl U (2007) Heparan sulfate-protein interactions – a concept for drug design? Thromb Haemost 98:109–115

Lindahl U, Backstrom G, Hook M, Thunberg L, Fransson LA, Linker A (1979) Structure of the antithrombin-binding site in heparin. Proc Natl Acad Sci USA 76:3198–3202

Lindahl U, Kusche-Gullberg M, Kjellen L (1998) Regulated diversity of heparan sulfate. J Biol Chem 273:24979–24982

Linhardt RJ, Gunay NS (1999) Production and chemical processing of low molecular weight heparins. Semin Thromb Hemost 25(Suppl 3):5–16

Linhardt RJ, Rice KG, Kim YS, Lohse DL, Wang HM, Loganathan D (1988) Mapping and quantification of the major oligosaccharide components of heparin. Biochem J 254:781–787

Linhardt RJ, Loganathan D, al-Hakim A, Wang HM, Walenga JM, Hoppensteadt D, Fareed J (1990) Oligosaccharide mapping of low molecular weight heparins: structure and activity differences. J Med Chem 33:1639–1645

Liu J, Shriver Z, Pope RM, Thorp SC, Duncan MB, Copeland RJ, Raska CS, Yoshida K, Eisenberg RJ, Cohen G, Linhardt RJ, Sasisekharan R (2002) Characterization of a heparan sulfate octasaccharide that binds to herpes simplex virus type 1 glycoprotein D. J Biol Chem 277: 33456–33467

Liu H, Zhang Z, Linhardt RJ (2009) Lessons learned from the contamination of heparin. Nat Prod Rep 26:313–321

Liu R, Xu Y, Chen M, Weiwer M, Zhou X, Bridges AS, DeAngelis PL, Zhang Q, Linhardt RJ, Liu J (2010) Chemoenzymatic design of heparan sulfate oligosaccharides. J Biol Chem 285: 34240–34249

Myette JR, Shriver Z, Kiziltepe T, McLean MW, Venkataraman G, Sasisekharan R (2002) Molecular cloning of the heparin/heparan sulfate delta 4,5 unsaturated glycuronidase from Flavobacterium heparinum, its recombinant expression in *Escherichia coli*, and biochemical determination of its unique substrate specificity. Biochemistry 41:7424–7434

Myette JR, Shriver Z, Claycamp C, McLean MW, Venkataraman G, Sasisekharan R (2003) The heparin/heparan sulfate 2-O-sulfatase from Flavobacterium heparinum. Molecular cloning, recombinant expression, and biochemical characterization. J Biol Chem 278:12157–12166

Myette JR, Soundararajan V, Shriver Z, Raman R, Sasisekharan R (2009) Heparin/heparan sulfate 6-O-sulfatase from Flavobacterium heparinum: integrated structural and biochemical investigation of enzyme active site and substrate specificity. J Biol Chem 284:35177–35188

Raman R, Venkataraman G, Ernst S, Sasisekharan V, Sasisekharan R (2003) Structural specificity of heparin binding in the fibroblast growth factor family of proteins. Proc Natl Acad Sci USA 100: 2357–2362

Rej R, Jaseja M, Perlin AS (1989) Importance for blood anticoagulant activity of a 2-sulfate group on L-iduronic acid residues in heparin. Thromb Haemost 61:540

Rhomberg AJ, Ernst S, Sasisekharan R, Biemann K (1998) Mass spectrometric and capillary electrophoretic investigation of the enzymatic degradation of heparin-like glycosaminoglycans. Proc Natl Acad Sci USA 95:4176–4181

Rosenberg RD, Lam L (1979) Correlation between structure and function of heparin. Proc Natl Acad Sci USA 76:1218–1222
Saad OM, Leary JA (2003) Compositional analysis and quantification of heparin and heparan sulfate by electrospray ionization ion trap mass spectrometry. Anal Chem 75:2985–2995
Saad OM, Leary JA (2005) Heparin sequencing using enzymatic digestion and ESI-MSn with HOST: a heparin/HS oligosaccharide sequencing tool. Anal Chem 77:5902–5911
Saad OM, Ebel H, Uchimura K, Rosen SD, Bertozzi CR, Leary JA (2005) Compositional profiling of heparin/heparan sulfate using mass spectrometry: assay for specificity of a novel extracellular human endosulfatase. Glycobiology 15:818–826
Salmivirta M, Lidholt K, Lindahl U (1996) Heparan sulfate: a piece of information. FASEB J 10: 1270–1279
Sanderson PN, Huckerby TN, Nieduszynski IA (1987) Conformational equilibria of alpha-L-iduronate residues in disaccharides derived from heparin. Biochem J 243:175–181
Shi X, Zaia J (2009) Organ-specific heparan sulfate structural phenotypes. J Biol Chem 284: 11806–11814
Shukla D, Liu J, Blaiklock P, Shworak NW, Bai X, Esko JD, Cohen GH, Eisenberg RJ, Rosenberg RD, Spear PG (1999) A novel role for 3-O-sulfated heparan sulfate in herpes simplex virus 1 entry. Cell 99:13–22
Sugahara K, Kitagawa H (2002) Heparin and heparan sulfate biosynthesis. IUBMB Life 54: 163–175
Toida T, Yoshida H, Toyoda H, Koshiishi I, Imanari T, Hileman RE, Fromm JR, Linhardt RJ (1997) Structural differences and the presence of unsubstituted amino groups in heparan sulphates from different tissues and species. Biochem J 322(Pt 2):499–506
Venkataraman G, Shriver Z, Raman R, Sasisekharan R (1999) Sequencing complex polysaccharides. Science 286:537–542
Wardrop D, Keeling D (2008) The story of the discovery of heparin and warfarin. Br J Haematol 141:757–763
Wolff JJ, Amster IJ, Chi L, Linhardt RJ (2007) Electron detachment dissociation of glycosaminoglycan tetrasaccharides. J Am Soc Mass Spectrom 18:234–244
Yates EA, Santini F, Guerrini M, Naggi A, Torri G, Casu B (1996) $^{1}H$ and $^{13}C$ NMR spectral assignments of the major sequences of twelve systematically modified heparin derivatives. Carbohydr Res 294:15–27

# Part III
# Clinical Use of Heparin and LMWH

# Heparin in the Prophylaxis and Treatment of Venous Thromboembolism and Other Thrombotic Diseases

Paolo Gresele, Chiara Busti, and Gloria Paganelli

## Contents



P. Gresele (✉)
Division of Internal and Cardiovascular Medicine, Department of Internal Medicine, University of Perugia, Via E. dal Pozzo, 06126 Perugia, Italy
e-mail: grespa@unipg.it

R. Lever et al. (eds.), *Heparin - A Century of Progress*,
Handbook of Experimental Pharmacology 207,
DOI 10.1007/978-3-642-23056-1_9, 

**Abstract** In this chapter, we discuss the key-role of heparin in the prophylaxis and treatment of venous thromboembolism (VTE) and other thrombotic disorders. Heparin exerts its antithrombotic effects by facilitating the ability of antithrombin (AT), a plasma serum protease inhibitor, to inhibit thrombin (factor IIa) and factor Xa. Different heparin formulations can be used for the prophylaxis of thrombosis and treatment, going from unfractionated heparin (UFH), different low-molecular-weight heparin (LMWH) preparations, to the recently introduced synthetic pentasaccharide fondaparinux. All heparin formulations can be administrated only by the parenteral route, including the intravenous (iv) and the subcutaneous (sc) route. We will overview the clinical evidence for the use of different heparin formulations in the prophylaxis and treatment of venous thromboembolism, of superficial vein thrombosis and of acute coronary syndromes (ACS). Special issues, like the use of heparins in pregnancy or in children, will also be discussed. Although heparin is an almost one century-old drug it remains a cornerstone of antithrombotic treatment.

**Keywords** Acute Coronary Syndrome (ACS) • Fondaparinux • Heparin induced thrombocytopenia (HIT) • Low-Molecular-Weight heparin (LMWH) • Pulmonary Embolism (PE) • Superficial Vein Thrombosis (VVT) • Unfractionated Heparin (UFH) • Venous Thromboembolism (VTE)

## 1 Introduction

In this chapter we discuss the key-role of heparin in the prophylaxis and treatment of venous thromboembolism (VTE) and other thrombotic disorders. Heparin exerts its antithrombotic effects by facilitating the ability of antithrombin (AT), a plasma serine protease inhibitor, to inhibit thrombin (factor IIa) and factor Xa. Different heparin formulation can be used for the prophylaxis and treatment of thrombosis, including unfractionated heparin (UFH), a mixture of sulfated glycosaminoglycans with a wide range of molecular weights extracted from lung and intestinal tissue of cows and swine, different low-molecular-weight heparin (LMWH) preparations, produced through chemical or enzymatic treatment of UFH to decrease the size of the polysaccharide chains yielding formulations with molecular weight distribution ranging between 4,000 and 5,000 Da and provided with a prevalent anti-Xa activity over the anti-IIa activity, and the recently introduced synthetic pentasaccharide exerting a selective anti-Xa activity. All heparin formulations can be administrated only by parenteral route, including the intravenous (iv) and the subcutaneous (sc) ways of administrations. We will overview the clinical evidence for the use of different heparin formulations in the prophylaxis and treatment of venous thromboembolism of superficial vein thrombosis and of acute coronary syndromes (ACS).

## 2 Heparin in the Prophylaxis of Venous Thromboembolism

The risk of VTE is increased in patients hospitalized for an acute medical illness or trauma or surgery. Routine antithrombotic prophylaxis with heparin reduces morbidity, mortality, and costs in patients at risk of deep vein thrombosis (DVT) and pulmonary embolism (PE), as highlighted by several national and international guidelines (American College of Chest Physicians 2008). The incidence of DVT in medical patients not undergoing anticoagulant prophylaxis is 10–15%. With an appropriate prophylaxis, the absolute risk reduction is 57% for fatal PE and 53% for DVT (Geerts et al. 2001).

### *2.1 Risk Stratification*

The incidence of VTE in hospitalized patients is high, ranging from 10 to 90%, depending on the clinical condition (Table 1).

In order to identify patients requiring VTE prophylaxis, one can follow two different approaches (Prandoni and Samama 2008): (1) Consider the predisposing factors associated with the risk of VTE in the individual patient, and then prescribe thromboprophylaxis according to the estimated risk; (2) Define the level of risk depending on general criteria, such as type of surgery (minor, major), age (<40 years, 40–60 years, and >60 years), and the presence of additional risk factors (such as cancer or previous VTE) (Fig. 1).

However, these methods are not routinely used in clinical practice. According to the American College of Chest Physicians (ACCP) 2008 guidelines (American College of Chest Physicians 2008), a more useful and easy method in clinical practice is the identification of specific risk-groups who may benefit of routine VTE thromboprophylaxis, e.g., patients undergoing major general surgery or major orthopedic surgery (Table 2). Prophylaxis of VTE can be obtained by pharmacological or physical means. Pharmacological prophylaxis includes:

- Unfractionated heparin (UFH)
- Low-molecular-weight heparin (LMWH)

**Table 1** Incidence of DVT in hospitalized patients

| Patients | DVT prevalence (%) |
|---|---|
| Medical patients | 10–20 |
| General surgery | 15–40 |
| Major gynecological surgery | 15–40 |
| Major urological surgery | 15–40 |
| Neurosurgery | 15–40 |
| Stroke | 20-50 |
| Hip or knee arthroplasty | 40–60 |
| Major trauma | 40–90 |
| Critical care patients | 10–90 |

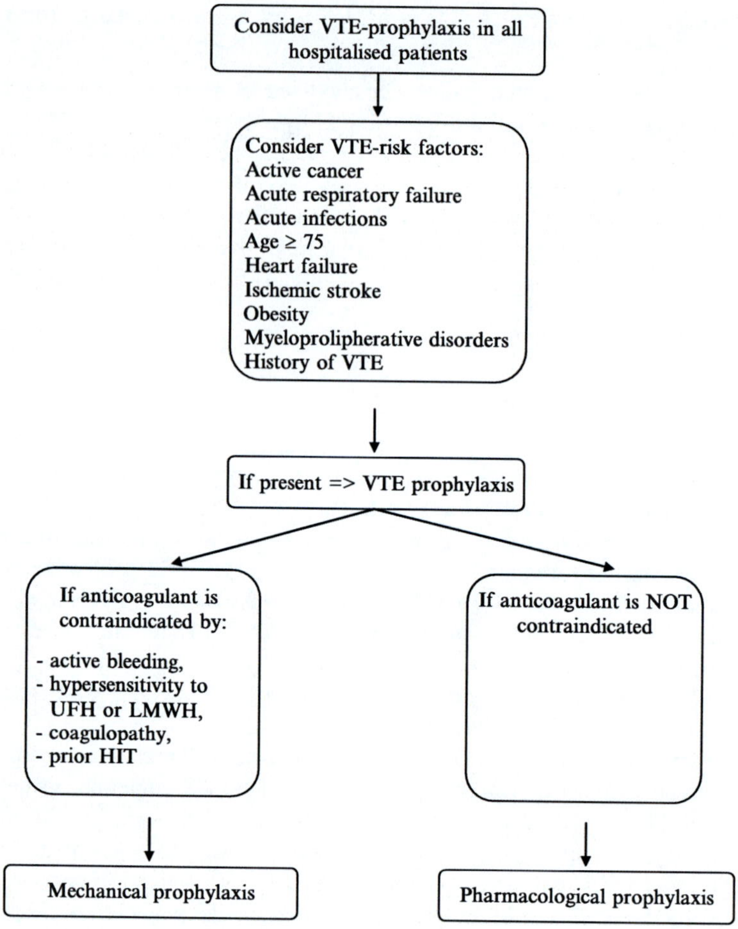


**Fig. 1** Algorithm for VTE prophylaxis in the hospitalized medical patient

- Inhibitors of factor Xa (e.g., fondaparinux)
- Oral vitamin K antagonists.

Antiplatelet agents, such as aspirin, are not indicated for the prevention of venous thromboembolism (American College of Chest Physicians 2008).

Prophylaxis by physical means includes the use of graduated elastic compression stockings (GECS), intermittent pneumatic compression (IPC) devices, and mechanical foot pumps, and aims to increase venous outflow and/or to reduce stasis within the leg veins.

Physical tools are the preferred option for patients at high risk for bleeding. However, the protection given by these devices is inferior if compared with pharmacological prophylaxis, and they are contraindicated for patients with peripheral arterial disease or with some dermatological diseases.

**Table 2** Incidence of DVT and thromboprophylaxis options in different clinical settings depending on the level of risk

| Levels of risk | DVT risk without prophylaxis (%) | Thromboprophylaxis options |
|---|---|---|
| Low risk<br>– Minor surgery in mobile patients<br>– Medical patients fully mobile | <10 | No specific thromboprophylaxis |
| Moderate risk<br>– Gynecological, urological surgery<br>– Medical patients bed rest | 10–40 | sc LMWH, LDUH (2 or 3 times a day), fondaparinux |
| High risk<br>– Hip or Knee arthroplasty<br>– Hip fracture surgery | 40–80 | sc LMWH, fondaparinux or oral VKA (INR 2–3) |

**Table 3** Risk factors associated with VTE

- Age
- Obesity
- Varicose veins
- Previous VTE
- Congenital or acquired thrombophilia: low coagulation inhibitors (antithrombin, protein C or S); activated protein C resistance (factor V Leiden); high coagulation factors (I, II, VIII, IX, XI); antiphospholipid antibody syndrome; high homocysteine
- Surgery
- Trauma
- Malignancy
- Respiratory failure
- Heart failure and myocardial infarction
- Hematological disorders (paroxysmal nocturnal hemoglobinuria, polycythaemia, essential thrombocythemia, myeloma)
- Severe infection
- Pregnancy (mostly third trimester and puerperium)
- Oral contraceptives
- Immobility
- Hospitalization (medical illness, trauma, surgery)

Patients undergoing surgery should be individually assessed for risk of VTE and several risk factors have been identified to be associated with VTE (Table 3). In addition, it should be considered that different types of surgery are associated with a different risk of VTE: major orthopedic surgery and neurosurgery are associated with an elevated incidence of VTE compared with vascular or general surgery.

## 2.2 *General Surgery*

The incidence of asymptomatic DVT in patients undergoing general surgery without prophylaxis varies between 15% and 30%, and that of fatal PE between 0.2%

and 0.9% (Nicolaides et al. 1973). Over the last years, several improvements have been introduced in the management of the patient undergoing surgery that reduce the risk of VTE, including improvement in general perioperative care, more precocious mobilization of the patient, and greater use of local anesthesia (associated with a lower incidence of VTE) and of thromboprophylaxis.

However, there are also new factors emerging which tend to increase the risk of VTE in patients undergoing general surgery, including the more frequent employment of surgical procedures in older and sicker patients, the use of preoperative chemotherapy and shorter lengths of stay in the hospital (with consequent shorter duration of thromboprophylaxis).

LMWH or UFH reduce the risk of DVT and symptomatic PE with the same efficacy, but bleeding complications are more frequent with doses of LMWH higher than 3,400 IU compared with UFH (Koch et al. 1997).

Usually, the first dose of LMWH or UFH is given 2–3 h before surgery. If the patient is at increased risk of bleeding, prophylaxis can be initiated after surgery. Prophylactic doses are: 5,000 IU b.i.d, or in particular cases t.i.d., for UFH, 4,000 IU/day for enoxaparin or 5,000 IU/day for dalteparin, and 2.5 mg/day for fondaparinux, the latter starting 6–8 h after surgery.

Current guidelines recommend early mobilization after surgery but no thromboprophylaxis for low-risk patients without additional thomboembolic risk factors undergoing minor procedures, while thromboprophylaxis with LMWH, UFH, or fondaparinux is recommended for not neoplastic patients undergoing major surgery (American College of Chest Physicians 2008).

In patients undergoing major surgery for cancer, LMWH, or low-dose (LDUH) three times daily, or fondaparinux are recommended. Finally, in high-risk patients with multiple risk factors for VTE, a combination of pharmacological and mechanical thromboprophylaxis is recommended.

Thromboprophylaxis should be continued for 2–3 weeks after surgery because, although the risk of DVT is highest in the first week after general surgery, VTE complications may occur even later (Lausen et al. 1998). A randomized, blinded clinical trial compared prophylaxis with the selective factor Xa inhibitor fondaparinux (2.5 mg sc qd, started 6 h postoperatively) with the LMWH dalteparin (5,000 IU sc qd, starting with 2,500 IU given 2 h before surgery and 2,500 IU 12 h postoperatively) among almost 3,000 patients undergoing major abdominal surgery. There were no significant differences between the two groups in the rates of VTE (4.6% vs. 6.1%, respectively), major bleeding (3.4% vs. 2.4%), or death (1.0% vs. 1.4%). Postoperative fondaparinux was at least as effective as perioperative dalteparin in patients undergoing high-risk abdominal surgery (Agnelli et al. 2005).

## 2.3 *Orthopedic Surgery*

The incidence of venographic DVT (7–14 days) after major orthopedic surgery in patients not undergoing prophylaxis is about 40–60% (Table 4) (Cordell-Smith

**Table 4** Incidence of VTE in orthopedic surgery

| Type of surgery | DVT (total) | DVT (proximal) | PE (total) | PE (fatal) |
|---|---|---|---|---|
| Hip arthroplasty | 42–57% | 18–36% | 0.9–28.0% | 0.1–2.0% |
| Knee arthroplasty | 41–85% | 5–22% | 1.5–1.0% | 0.1–1.7% |
| Hip fracture surgery | 46–60% | 23–30% | 3.0–11.0% | 0.3–7.5% |

et al. 2004). DVT is frequently asymptomatic and occurs mainly in the operated limb; however, DVT of the contralateral limb occurs in 20% of patients undergoing total hip replacement and in 14% of those undergoing knee arthroplasty. Most of the symptomatic VTE occurs after hospital discharge, and the risk continues to be high for up to 2 months after surgery (Leclerc et al. 1998).

Therefore, adequate thromboprophylaxis is strongly recommended for all patients undergoing major orthopedic surgery of lower limbs.

### 2.3.1 Hip Surgery

Both asymptomatic and symptomatic VTE is very common after elective hip arthroplasty and thromboprophylaxis can be accomplished by mechanical and/or pharmacological means. Mechanical thromboprophylaxis includes the use of graduated elastic compression stockings (GECS), intermittent pneumatic compression devices (IPC), or mechanical foot pumps. Risk reduction of VTE attained with these devices is of 20–70%. Pharmacological prophylaxis can be carried out with LMWH, which is safer and more effective than UFH. LMWH is easier to use in clinical practice because it does not require monitoring, it carries a lower risk of adverse reactions (heparin-induced thrombocytopenia-HIT) and, different from UFH that requires twice-daily administration, it can be administered once a day. Current guidelines recommend to start LMWH 12 h before orthopedic surgery or 12–24 h after surgery, or half of the usual dose 4–6 h after surgery and then increasing to the usual dose starting on the following day. Enoxaparin can be administered at the dose of 4,000 IU sc once a day starting 10–12 h before surgery. Dalteparin can be administered at the dose of 5,000 IU sc starting 8–12 h before surgery and then continuing with 5,000 IU sc once a day from 12 to 24 h after surgery (American College of Chest Physicians 2008).

Fondaparinux is also recommended for the thromboprophylaxis of VTE, at the dose of 2.5 mg once a day started 6–24 h after surgery. However, the longer half-life of fondaparinux as compared with LMWH and its main renal excretion render this anticoagulant more risky for patients with renal failure (creatinine clearance of <30 ml/h). Finally, international guidelines recommend also adjusted-dose VKA started preoperatively or the evening of the surgical day (INR range, 2.0–3.0).

Hip fracture surgery (HFS) is associated with an increased risk of VTE. Around 50% and 1.4–7.5% of patients undergoing HFS without thromboprophylaxis will develop a DVT or a fatal PE, respectively. A time interval of longer than 48 h between hip fracture and surgery increases the risk of VTE in these patients. Therefore, prophylaxis should be started immediately after hospitalization using heparin or fondaparinux. Current guidelines recommend to use routinely thromboprophylaxis with fondaparinux, or LMWH, or adjusted-dose VKA (INR range: 2.0–3.0), or LDUH, at the doses above indicated for hip arthroplasty. Aspirin or other antiplatelet agents are not recommended alone for the prophylaxis of VTE in hip surgery patients (American College of Chest Physicians 2008).

### 2.3.2 Knee Surgery

Patients undergoing elective knee arthroplasty are at high risk of VTE (40–60% of VTE without prophylaxis in the 7–14 days after surgery). For these patients thromboprophylaxis is strongly recommended routinely. Mechanical devices can also be used, but always in association with pharmacological prophylaxis. Neither aspirin nor UFH is indicated alone for thromboprophylaxis. LMWH (4,000 IU sc once a day starting 10–12 h after surgery) or fondaparinux (2.5 mg once a day starting 6–8 h after surgery) is the best choice for prophylaxis in this type of surgery. Thromboprophylaxis is usually continued until hospital discharge; however, several clinical trials show that the risk of VTE is increased for up to 3 months after surgery. Hospitalization lasts often 5–6 days, an insufficient time for a correct protection from VTE. Therefore extending prophylaxis for 35–40 days after discharge is advised as it reduces significantly the incidence of VTE.

Diagnostic or therapeutic knee arthroscopy is also associated with an increased risk of VTE, a risk, however, considerably lower than that of knee arthroplasty; for this reason, when there are no additional thromboembolic risk factors, thromboprophylaxis is not routinely recommended and these patients should only be encouraged to early mobilization.

## *2.4 Other Types of Surgery*

For patients undergoing minor surgery or laparoscopic procedures and without additional risk factors for VTE, current guidelines recommend against the use of routine thromboprophylaxis. For patients undergoing other types of major surgery, prophylaxis is recommended and it should be started just before the surgery and continued until the patient is not ambulating.

### 2.4.1 Spinal Surgery

VTE is relatively rare in patients undergoing spinal surgery. There are only few prospective data about VTE prevention in these patients (Brambilla et al. 2004). Current guidelines suggest that in patients undergoing spinal surgery, and without other risk factors of VTE, such as advanced age, previous VTE, an anterior surgical approach, malignancy, a prolonged procedure and reduced preoperative or postoperative mobility, prophylaxis should not be administered, and early mobilization should be advised.

Instead, for patients with additional VTE risk factors, thromboprophylaxis is indicated, using postoperative LDUH, postoperative LMWH or, if heparin is contraindicated, perioperative IPC.

Finally, for patients undergoing spinal surgery and who have multiple risk factors for VTE, guidelines suggest that a pharmacological method (LDUH or LMWH) should be combined with a mechanical method (i.e., GCS and/or IPC) (American College of Chest Physicians 2008).

### 2.4.2 Other Surgery

For patients undergoing vascular, gynecological, urological, or laparoscopic surgery, the indications concerning thromboprophylaxis differ depending on the type of surgery (minor or major) and the presence of VTE risk factors. For patients undergoing minor surgery or laparoscopic procedures and without additional risk factors for VTE, current guidelines recommend against the use of routine thromboprophylaxis other than early ambulation. For patients undergoing major open surgery or laparoscopic surgery but in the presence of VTE risk factors, prophylaxis is recommended with LMWH (enoxaparin 4,000 IU/day) or LDUH (5,000 IU b.i.d.) or fondaparinux (2.5 mg/day), and it should be started just before surgery and continued until the patient is fully ambulating. For patients undergoing extensive oncological surgery and for patients with additional VTE risk factors, guidelines recommend routine thromboprophylaxis with LMWH or LDUH three times daily or IPC, starting just before surgery and continuing it until the patient is ambulating. For patients at high risk of VTE (major oncological surgery, previous VTE), guidelines suggest to consider the extension of thromboprophylaxis for up to 28 days (American College of Chest Physicians 2008).

## *2.5 Cancer Patients*

Cancer is one of the most important risk factors for VTE. Thromboembolic complications represent the second cause of death in cancer patients and VTE recurrence rate is high after the end of anticoagulation but is consistent also during anticoagulation. VTE risk is particularly high in patients with some malignancies (malignant brain tumors, adenocarcinomas of the lung, ovary, pancreas, colon, stomach, prostate, kidney, and hematological malignancies), in those who undergo surgery or chemotherapy, hormonal therapy or that carry a central venous catheter (CVC).

Current guidelines recommend thromboprophylaxis for all cancer patients bedridden with an acute medical illness. Routine heparin prophylaxis is not recommended instead for cancer patients receiving chemotherapy or hormonal therapy or for those with indwelling CVC to prevent catheter-related thrombosis (American College of Chest Physicians 2008).

## 2.6 Medical Patients

Patients hospitalized for medical reasons have an eight-fold increased risk of VTE, with an absolute incidence of VTE estimated to 15%. These patients represent a heterogeneous group; therefore, it is important to define risk factors for VTE in order to identify those who need prophylaxis.

Medical patients at high risk of VTE are those hospitalized for stroke, COPD exacerbation, myocardial infarction, cardiac failure (NYHA III–IV), sepsis and other acute illnesses with immobilization.

Some clinical trials have shown that heparin thromboprophylaxis reduces the incidence of VTE in medical patients. The MEDENOX trial (Prophylaxis in MEDical patients with ENOXaparin) enrolled 1,102 medical patients older than 40 years, with a short immobilization period (less than 3 days) and affected by acute cardiac failure or acute respiratory disease or sepsis or inflammatory bowel disease. Enrolled patients (except those with acute cardiac or respiratory failure) had to have at least another risk factor for VTE, including an age over 75, previous VTE, obesity, cancer, hormonal therapy, and chronic cardiac or respiratory disease. Enrolled patients were randomized to enoxaparin 4,000 IU/d, enoxaparin 2,000 IU/d or placebo for 6–14 days and the primary endpoint was VTE within the first 14 days. This clinical trial demonstrated a statistically significant 63% reduction of VTE in patients treated with enoxaparin 4,000 IU. Bleeding complications were comparable in the 3 groups (Samama et al. 1999).

The ARTEMIS trial (ARixtra for ThromboEmbolism prevention in a Medical Indications Study) compared fondaparinux 2.5 mg/d with placebo both administered for 6–14 days for VTE prophylaxis in medical patients. Primary endpoint was the incidence of asymptomatic or symptomatic VTE. The trial showed a statistically significant risk reduction of VTE in the fondaparinux group (Cohen et al. 2006).

An optimal duration of thromboprophylaxis for medical patients is extremely important as VTE can occur also after the conventional 10 days of heparin administration. The EXCLAIM trial (Extended Clinical prophylaxis in Acutely Ill Medical patients) enrolled patients with an acute medical illness and reduced mobilization, initiating treatment with the LMWH enoxaparin 4,000 IU for 10 days and then randomizing the patients to two arms, one continuing enoxaparin at the same dosage and the other treated with placebo, both for up to 28 days. Patients treated for 28 days with enoxaparin had a reduction of the incidence of VTE of 44% compared with those receiving only 10 days of thromboprophylaxis, with a small increase in bleeding complications (Hull et al. 2006). Therefore, postdischarge prophylaxis with LMWH extended for 28 days should be considered for hospitalized medical patients with reduced mobility who are older than 75 or have a history of cancer or of previous VTE.

## 3 Heparin in the Treatment of Venous Thromboembolism

In this section, we discuss the role of heparin in the treatment of patients with acute venous thromboembolism (VTE), lower limb deep venous thrombosis (DVT), upper extremity DVT (UEDVT), pulmonary embolism (PE) and superficial vein thrombosis (SVT). Venous thrombosis of other anatomical sites, such as cerebral or abdominal venous thrombosis, the incidence of which is much lower than that of DVT, is treated with essentially the same modalities as used for DVT.

### *3.1 Treatment of Deep Vein Thrombosis of the Legs*

Anticoagulation is the main therapy for acute DVT of the legs and patients should be treated with anticoagulants as soon as the diagnosis is confirmed by objective testing. If the clinical suspicion is high and there is a delay before confirmatory diagnostic testing can be performed, treatment should be started before such testing.

#### 3.1.1 Heparin in the Acute Phase of DVT

The objectives of anticoagulant therapy in the initial treatment of this condition are to prevent thrombus extension and pulmonary embolism and to reduce the risk of early and late recurrences of VTE.

In patients with DVT, it has been shown that an initial course of heparin in addition to vitamin K antagonists (VKA) is more effective in preventing recurrent VTE as compared to starting treatment with VKA alone (Brandjes et al. 1992).

Four options are available for the initial treatment of DVT:

1. Intravenous (iv) unfractionated heparin (UFH), with monitoring
2. sc UFH, with monitoring
3. Weight-based low-molecular-weight heparin (LMWH), administered subcutaneously (sc), without laboratory monitoring
4. sc fondaparinux, without monitoring.

Two randomized, controlled clinical trials (RCTs) compared a short term (5–7 days) with a longer term (10–14 days) heparin regimen before starting VKA, showing that the first option is as effective as the second if followed by adequate long-term oral anticoagulant therapy (Gallus et al. 1986; Hull et al. 1990). In addition, shortening the duration of initial heparin therapy reduces the risk of heparin-induced thrombocytopenia (HIT).

Guidelines recommend an initial treatment with heparin for at least 5 days, starting both heparin and VKA at the time of diagnosis, and then to discontinue heparin after 5 days, provided the international normalized ratio (INR), i.e., the test used to monitor VKA anticoagulation, is >2.0 for at least 24 h.

### Intravenous Unfractionated Heparin

UFH is highly effective but is associated with the risk of hemorrhage caused by excessive dosing, or with the risk of therapeutic failure caused by subtherapeutic dosing.

To reduce the risk of bleeding, UFH is administered by continuous iv infusion while the intermittent iv boluses modality, that was associated with excessive hemorrhage, is no longer recommended.

To reduce the risks associated with the use of UFH, therapy requires frequent monitoring of the activated partial thromboplastin time (aPTT) and the adjustment of the UFH dose in response to monitoring.

To improve efficacy and safety, guidelines recommend that the starting dose of iv UFH for the treatment of DVT be one of the following:

1. A bolus dose of 5,000 IU, followed by a continuous infusion of at least 30,000 IU for the first 24 h
2. A weight-adjusted regimen of an 80 IU/kg bolus, followed by 18 IU/kg/h

Aging is a risk factor for major bleeding; therefore, total dose requirements are decreased in the elderly. At 3–6 h after the start of infusion (3 h in absence of the bolus), aPTT is measured and the infusion dose is adjusted using a standardized nomogram (Fig. 2).

The therapeutic range is based on studies that showed that optimal prevention of thrombus extension is obtained by a heparin dose that prolongs the aPTT to 1.5–2.5 basal. This value corresponds to therapeutic heparin plasma levels of 0.3–0.7 IU/mL of anti-Xa activity as assessed by the amidolytic assay. The reference aPTT value to determine the ratio should be the baseline value of the patient, but often the control value of the local laboratory is used. aPTT should be repeated every 3–6 h until the value is stable and every 6–12 h thereafter. Checking aPTT earlier could lead to erroneous dosing calculation because a steady-state kinetic cannot be assumed. Dosage changes must be made prudently (by 100–200 IU/h).

### Subcutaneous Unfractionated Heparin

For the initial treatment of DVT, sc UFH heparin twice daily can be used in alternative to the iv continuous infusion. A meta-analysis of eight clinical trials (Raschke et al. 1993) showed that this regimen is as effective and safe as iv infusion. When heparin is given by sc injection the starting of the anticoagulant effect is delayed by 1 h, and peak plasma levels occur after 3 h. Therefore, the

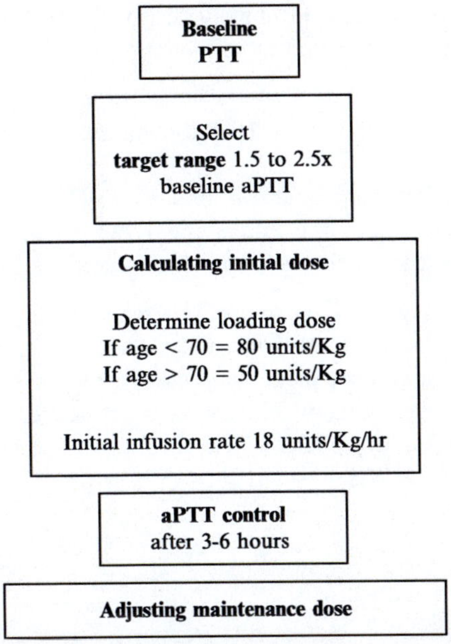


| aPTT ratio | Loading Dose | Maintenance dose |
|---|---|---|
| Less than 1.2 times control | 50 IU/kg | Increase by 4 IU/kg/hr |
| 1.2-1.3 times control | 25 IU/kg | Increase by 3 IU/kg/hr |
| 1.3-1.4 times control | None | Increase by 2 IU/kg/hr |
| 1.5-2.5 times control | None | No change |
| 2.6-2.9 times control | None | Decrease by 2 IU/kg/hr |
| 3-4 times control | Hold for 1 hour | Decrease by 3 IU/kg/hr |
| Greater than 4 times control | Hold for 2 hours | Decrease by 4 IU/kg/hr |

**Fig. 2** Intravenous therapeutic heparin dosing flow chart

patient usually receives an initial iv bolus of 5,000 IU followed by 250 IU/kg bid, with dose adjustments to achieve an aPTT ratio of 1.5 to 2.5 normal 6 h after the morning dose.

## Low-Molecular-Weight Heparin

LMWHs have been associated with fewer thrombotic complications, less major bleeding and fewer deaths as compared with sc or iv UFH in many clinical trials and meta-analysis (Raschke et al. 1993).

LMWHs have favorable pharmacokinetic features, a more predictable effect and greater bioavailability. All the studies comparing LMWHs and UFH showed that LMWHs:

- Can be administered sc once or twice daily without monitoring, in the majority of patients
- Are safe and effective in the treatment of acute DVT
- Are safe and effective both in inpatient and in outpatient administration
- There are no significant differences among different preparations of LMWHs.

In patients with acute DVT, all guidelines recommend initial treatment with LMWH administered sc once or twice daily, on an outpatient basis whenever possible, or in a hospitalized patient if necessary, rather than treatment with iv UFH. No routine laboratory monitoring, with the measurement of anti-factor Xa activity, is recommended. In patients with acute DVT and severe renal failure, UFH is recommended over LMWH.

Subcutaneous Fondaparinux

Fondaparinux is a synthetic pentasaccharide that exerts a selective antithrombin-mediated inhibition of factor Xa, therefore producing a dose-dependent inhibition of thrombin generation. It has 100% bioavailability after sc injection and an elimination half-life of 17 h and can therefore be given once daily. It is eliminated mainly by the renal route; therefore, its use is contraindicated if the creatinine clearance is lower than 30 mL/min. It is insensitive to inactivation by platelet-released PF4. Moreover, because it does not induce the formation of heparin–PF4 complexes, HIT is unlikely to occur with fondaparinux.

A recent trial compared once-daily sc fondaparinux (7.5 mg if body weight 50–100 kg; 5.0 mg if <50 kg; 10 mg if >100 kg) with twice-daily sc LMWH (enoxaparin, 1 mg/kg of body weight) in the treatment of acute DVT, using a blinded design: there was no difference in terms of recurrent VTE, major bleeding, or death at three months (Buller et al. 2004).

### 3.1.2 Heparin in the Long-Term Treatment of DVT

Patients that cannot use VKA for the long-term treatment of DVT can be treated with sc LMWHs. Several randomized clinical trials, using different regimens and different LMWH preparations, showed no substantial differences between the use of LMWHs and VKA in the long-term treatment of DVT, in terms of prevention of recurrent VTE and incidence of major bleeding (Kher and Samama 2005). Recent clinical trials studied the role of LMWHs in active cancer patients: the recurrence of VTE after 3 or 6 months of therapy was lower in the LMWHs group compared with the VKA group (Meyer et al. 2002).

**Table 5** Recommendations for the treatment of DVT and PE

| |
|---|
| *Four options for the initial treatment of DVT*: |
| 1. *iv unfractionated heparin* (UFH), with monitoring |
| 2. *sc UFH*, with monitoring |
| 3. Weight-based *low-molecular-weight heparin* (LMWH), administered subcutaneously (SC), without monitoring |
| 4. *sc fondaparinux*, without monitoring |
| *Duration of initial heparin treatment*: at least 5 days; start both heparin and VKA at the time of diagnosis; discontinue heparin after 5 days provided the INR is $\geq$2.0 for at least 24 h |
| *Start acute DVT treatment with LMWHs sc* twice daily, on an outpatient basis if possible, or as an inpatient if necessary, rather than iv UFH. No routine monitoring with anti-factor Xa level measurements is recommended. In patients with acute DVT and severe renal failure, UFH over LMWH is recommended |
| *Using iv UFH*, the starting dose for the treatment of DVT is either of the following: |
| 1. *A bolus dose of 5,000 IU*, followed by a continuous infusion of at least *30,000 IU for the first 24 h* |
| 2. A weight-adjusted regimen of an *80 IU/kg bolus*, followed by *18 IU/kg/h* |
| At 3–6 h after the start of infusion (3 h in absence of the bolus), aPTT is measured and the infusion rate is adjusted using a standard nomogram |
| *Using sc UFH*, the patient usually receives an initial iv bolus of 5,000 IU followed by 250 IU/kg bid, with dose adjustment to achieve an aPTT ratio prolongation of 1.5–2.5 |
| *Fondaparinux*: fondaparinux (7.5 mg if 50–100 kg; 5.0 mg if <50 kg; 10 mg if >100 kg) sc once a day |
| *Recommendations for the treatment of PE* |
| Treatment of PE is similar to treatment of DVT |
| *LMWHs* is as effective as *UFH* and has a tendency to give less VTE recurrence, less major bleeding and similar mortality |
| Weight-adjusted *fondaparinux* is effective and safe in the treatment of acute PE |

### 3.1.3 Heparin in the Acute Phase of Pulmonary Embolism

Treatment of PE is similar to treatment of DVT because the two conditions are manifestations of the same disease process. A large number of studies and meta-analysis about the treatment of the acute phase of PE showed that LMWHs are as effective as UFH, with similar mortality but a tendency to less recurrence and major bleeding (Quinlan et al. 2004). The Matisse PE study showed that also fixed dose, weight-adjusted fondaparinux is effective and safe in the treatment of acute PE (Buller et al. 2003). A summary of the recommendations for heparin treatment of DVT and PE is given in Table 5.

## *3.2 Treatment of Superficial Vein Thrombosis*

Superficial vein thrombosis (SVT) occurs in lower extremity superficial veins, frequently in patients with chronic venous insufficiency where it involves a varicose vein (more frequently the saphena magna), or in upper extremity superficial

**Table 6** Incidence of SVT depending on the lower limbs vein

| SVT in lower limbs veins | Incidence |
|---|---|
| Great saphena | 60–80% |
| Short Saphena | 10–20% |
| Other veins | 10–20% |

veins, associated with the use of peripheral iv catheters for drug infusion (Table 6) (Tagalakis et al. 2002). Incidence of SVT has been estimated to be 1 per 1,000, higher than that of DVT (Barrellier 1993).

Predisposing risk factors for SVT are very similar to those for VTE: postoperative periods, pregnancy, active malignancies, auto-immune disease, use of oral contraceptives, and a history of previous VTE. SVT has been considered a benign disease even if it is associated with the occurrence of DVT (6–44% of cases) or of PE (20–33% asymptomatic; 2–13% symptomatic) (Barrellier 1993).

There are many controversies about the management of SVT and only few clinical trials are available. Possible treatments include heparin, NSAID, elastic compression stockings, or surgery.

In the Stenox study, a RCT, 436 patients with SVT were randomized to placebo, to nonsteroidal anti-inflammatory drugs (NSAIDs), or to 2 doses of LMWH (a prophylactic or a therapeutic dose). All patients wore elastic compression stockings. The placebo group had a higher incidence of recurrent SVT than the other three groups. VTE incidence resulted similar with the two doses of LMWH and higher with NSAIDs; however, no statistical difference between the groups was found (The Superficial Thrombophlebitis Treated by Enoxaparin Study Group 2003).

To identify the optimal dose of LMWH in patients with SVT, the VESALIO clinical trial has compared a prophylactic dose of LMWH with a therapeutic dose of LMWH for 10 days and then half the dose for other 20 days in patients with acute SVT of the saphena magna. No statistically significant differences in VTE incidence or in SVT extension have been shown between the two treatment groups. Therefore, it seems that therapeutic doses of LMWH do not protect more effectively than prophylactic doses (Prandoni et al. 2005).

Current guidelines distinguish between the treatment of spontaneous SVT and infusion-related SVT. For spontaneous SVT, prophylactic or intermediate doses of LMWH or UFH for 4 weeks (or VKA for 4 weeks) are indicated. For peripheral vein infusion-related thrombophlebitis oral diclofenac or another NSAID, topical diclofenac gel or heparin gel, until resolution of symptoms or for up to 2 weeks, are recommended (American College of Chest Physicians 2008). Regarding surgical therapy, a small randomized trial of 60 patients with an extended thrombosis of the saphenous vein compared treatment with LMWH with surgical saphenous vein ligation. Patients in the LMWH group did not experience episodes of VTE but had a 10% incidence of recurrent SVT. Patients treated surgically suffered 2 PE (6.7%) and 1 recurrent SVT (3.3%). Therefore, LMWH seems to be the preferred treatment, even because extremely cheaper than surgery (Lozano and Almazan 2003).

# 4 Heparin in the Treatment of Acute Coronary Syndromes

Coronary thrombosis is crucial in the pathogenesis of unstable angina, acute myocardial infarction (MI), and sudden cardiac death (Hirsh et al. 2001). Heparin is used in the initial management of patients with unstable angina or acute myocardial infarction and during and after coronary angioplasty or stent placement to inhibit thrombin generation and/or activity, thus reducing thrombus-related events.

In most patients, heparin contributes to limit thrombosis in ACS, but it is no longer used as the sole antithrombotic drug in this setting. Today, heparin is always used in combination with aspirin in patients with acute myocardial ischemia eligible to receive thrombolytic therapy, in those treated with platelet GP IIb/IIIa antagonists for unstable angina and in those undergoing high-risk coronary angioplasty (Oler et al. 1996a; PRISM PLUS Study Investigators 1998).

## *4.1 Non-ST-Elevation MI*

Heparin has been evaluated in a number of randomized, double-blind, placebo-controlled clinical trials for the short-term treatment of unstable angina or non-Q-wave MI (NSTEMI).

Several anticoagulant options are available for the treatment of the acute phase of NSTE-ACS:

- Unfractionated heparin (UFH) by iv infusion
- Low-molecular-weight heparin (LMWH) by sc injection
- Fondaparinux by sc injection
- Direct thrombin inhibitors (DTIs) by iv infusion.

We will shortly describe only the first three options, which concern heparin.

### 4.1.1 Unfractionated Heparin

A pooled analysis of six trials testing short-term UFH vs. placebo or untreated controls showed a significant 33% risk reduction of death and MI with UFH. In trials comparing the combination of UFH plus aspirin vs. aspirin alone, a trend towards a benefit was observed in favor of the UFH–aspirin combination, at the cost of a moderate increase in the risk of bleeding. A meta-analysis of 6 small randomized trials ($n = 1{,}353$ subjects) reported a risk reduction of 33% (95% confidence interval [CI] −2% to 56%) in cardiovascular death and MI with the combination of UFH and aspirin, which was of borderline significance (Fig. 3) (Oler et al. 1996b).

The available evidence supports a weight-adjusted dosing regimen with UFH. An initial bolus of 60–70 IU/kg (maximum 5,000 IU) followed by an infusion of

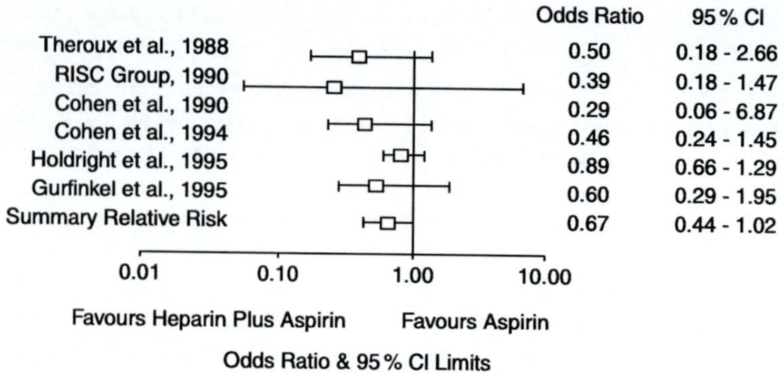


**Fig. 3** Meta-analysis of heparin plus aspirin versus aspirin alone in unstable angina: relative risk of MI or death during hospitalization (reproduced from ref (Oler et al. 1996b))

12–15 IU/kg/h (maximum 1,000 IU/h), titrated to a target aPTT of 50–75 s, may be optimal.

### 4.1.2 Low-Molecular-Weight Heparin

Several trials have assessed the relative efficacy and safety of various LMWH preparations in comparison with UFH in NSTE-ACS patients. Only enoxaparin was shown to be superior to UFH in reducing the 30-day composite endpoint of death or MI. The same study showed no significant differences in the requirement of blood transfusions or in the incidence of major bleeding at 7 days (Petersen et al. 2004). Importantly, also in the subgroup of patients undergoing PCI, there were no differences in ischemic events (including abrupt coronary closure) between the two treatment groups. SYNERGY was the largest trial to test enoxaparin against UFH in the context of a clinically updated approach, i.e., with a high rate of primary invasive procedures (PCI/revascularization, stent implantation) and concomitant active antiplatelet therapy with aspirin, clopidogrel, and glycoprotein IIb/IIIa (GP IIb/IIIa) receptor inhibitors. No significant difference was observed in terms of death and MI at 30 days for enoxaparin vs. UFH. More bleeding occurred with enoxaparin, with a statistically significant increase in TIMI major bleeding (9.1 vs. 7.6%, $p < 0.008$) but a non-significant excess in GUSTO severe bleeding (2.7 vs. 2.2%, $p < 0.08$), and requirement of blood transfusions (Ferguson et al. 2004).

The LMWH dosages used in NSTE-ACS are body weight adjusted and are essentially the same as those used in the treatment of VTE. The optimal level of anti-factor Xa activity has not been determined for patients with ACS receiving LMWH. The available information derived from nonrandomized clinical studies in patients undergoing PCI suggests that an anti-Xa activity $\leq$0.5 IU/mL is associated with a low incidence of ischemic/thrombotic and hemorrhagic events. LMWHs are commonly administered subcutaneously every 12 h in NSTE-ACS to avoid the risk of transiently inadequate anti-Xa levels during treatment. With the doses currently used in clinical practice, monitoring of anti-Xa activity is not required, except in special populations of patients, such as those with renal failure or obesity.

Prolongation of treatment with LMWH did not show superior protection against recurrence of ischemic events while more bleeding occurred; therefore, discontinuation of LMWH is recommended at hospital discharge (Fragmin During Instability in Coronary Artery Disease (FRISC) Study Group 1996). The risk of bleeding with LMWH is dose related and is increased with higher age, female gender, lower body weight, reduced renal function, and interventional procedures.

LMWH should be stopped at least 12 h before surgery, and if angiography is performed within 8 h from LMWH administration, peri-procedural addition of UFH is not required.

### 4.1.3 Factor Xa Inhibition with Fondaparinux

The safety and efficacy of fondaparinux in patients with unstable angina and non-ST-segment myocardial infarction was evaluated in the OASIS 5 trial, a large, randomized, double-blind, multicenter study designed to compare fondaparinux with enoxaparin. The rates of the combined ischemic endpoint (death, MI, or recurrent intervention), as well as the individual endpoints at Day 9, were identical in both arms of the study while major and minor bleeding complications were less frequent in patients treated with fondaparinux (Petersen et al. 2004).

In ACS, fondaparinux at 2.5 mg/day fixed dose is recommended. No dose adjustment and no monitoring of anti-Xa activity is required.

In two small, phase II studies, fondaparinux also showed promising results as a substitute for enoxaparin or UFH in NSTE-ACS and PCI. Catheter thrombus formation during PCI was observed in both groups, though at a significantly higher rate with fondaparinux than with enoxaparin. On the basis of the OASIS-5 trial (Yusuf et al. 2006a), if fondaparinux is chosen as anticoagulant therapy, it should be maintained for up to 5 days or until hospital discharge, and it should not be used as the sole anticoagulant during PCI.

A summary of the recommendations for heparin treatment in patients with NSTE-ACS is given in Table 7.

## *4.2 ST-Elevation ACS*

The first goal in patients presenting with a STEMI within 12 h after clinical onset, or with persistent pain, ST-elevation or new left bundle-branch block, is to obtain early reperfusion either by mechanical or by pharmacological means.

The goals of antiplatelet and antithrombotic drugs are to establish and maintain patency of the culprit coronary artery and to reduce the systemic tendency to thrombosis (mural thrombosis, cardioembolic stroke, pulmonary embolism).

**Table 7** Recommendations for the treatment of NSTE-ACS

| |
|---|
| *Anticoagulant options for the treatment of the acute phase of NSTE-ACS* |
| – Unfractionated heparin (UFH) as iv infusion |
| – Low-molecular-weight heparin (LMWH) as sc injection |
| – Fondaparinux as sc injection |
| – Direct thrombin inhibitors (DTIs) as iv infusion |
| *Unfractionated heparin (UFH)* |
| A weight-adjusted dosing regimen with UFH is recommended |
| – Initial bolus of 60–70 IU/kg (maximum 5,000 IU) |
| – Initial infusion of 12–15 IU/kg/h (maximum 1,000 IU/h) titrated to a target aPTT of 50–75 s |
| *Low-molecular-weight heparin (LMWHs)* |
| – Only enoxaparin was shown to be superior to UFH in reducing the 30-day composite endpoint of death or MI |
| – The LMWH dosages used in NSTE-ACS are body weight adjusted and are identical to those used in the treatment of VTE |
| – LMWHs are commonly administered sc every 12 h |
| – Monitoring of anti-Xa activity is not necessary, except in special populations of patients, such as those with renal failure and obesity |
| – Discontinuation of LMWH is recommended at hospital discharge |
| – LMWH should be stopped at least 12 h before surgery |
| *Fondaparinux* |
| – A 2.5 mg fixed dose is recommended |
| – No dose adjustment and no monitoring of anti-Xa activity is required |
| – It should be maintained for up to 5 days or until hospital discharge |
| – It cannot be used as the sole anticoagulant during PCI procedures |

### 4.2.1 Patients Undergoing Reperfusion Therapy

#### Unfractionated Heparin

UFH is the standard anticoagulant therapy used in conjunction with thrombolytic therapy or mechanical reperfusion. UFH is given as an initial iv bolus of 100 IU/Kg body weight (60 IU/kg of body weight if GPIIb/IIIa antagonists are also used) followed by an initial infusion of 12 IU/kg per hour (maximum 1,000 IU/h) adjusted to maintain an aPTT level of 1.5–2 times the control value. Evidence suggests that addition of UFH to reperfusion therapy helps to maintain coronary patency and to reduce the infarct area at the cost of a small risk of bleeding.

After reperfusion, UFH should be: discontinued after 48 h in low-risk patients; continued subcutaneously in patients at high risk of systemic embolization; continued iv in patients at high risk for coronary reocclusion. No differences have been shown between UFH when administered by the iv or sc route with close monitoring of aPTT.

#### Low-Molecular-Weight Heparin

Limited evidence supports the use of LMWHs after mechanical reperfusion. In patients undergoing primary PCI, UFH should be used periprocedurally and LMWH initiated 1 hour after sheath removal.

In patients treated with fibrinolysis, enoxaparin was shown to give a net clinical benefit (a composite endpoint that combines safety and efficacy) over UFH (Antman et al. 2006). Enoxaparin can be used in all patients aged less than 75 and without renal impairment, starting with a 30 mg i.v. bolus followed, after 15 min, by a sc dose of 1 mg/Kg every 12 h, until hospital discharge or for a maximum of 8 days. In patients other than 75 or with creatinine level >2.5 mg/dL, no iv bolus should be given. In patients with creatinine clearance <30 ml/h, it is reasonable to start with 0.75 mg/kg and to continue with the same dose every 24 h.

#### Fondaparinux

In the PCI setting, fondaparinux was associated with a non-significantly higher incidence of death and reinfarction at 30 days compared to heparin. These findings, together with the evidence of catheter-related thrombosis, do not support the use of fondaparinux as the sole anticoagulant in patients undergoing PCI (The OASIS-6 Trial Group 2006).

In patients undergoing thrombolytic therapy, fondaparinux is a good alternative to heparin. In the absence of renal impairment (creatinine level <3 mg/dl), fondaparinux should be administered as an initial 2.5 mg iv bolus, followed by 2.5 mg sc once daily until 8 days or hospital discharge (Yusuf et al. 2006b).

### 4.2.2 Patients Treated Without Reperfusion Therapy

In patients in whom reperfusion therapy is not possible, guidelines recommend administration of aspirin, clopidogrel, and antithrombotic agents as soon as possible.

Enoxaparin, UFH, and fondaparinux are all valid therapeutic options in this setting using the same doses as those used in combination with fibrinolytics.

#### Long-Term Anticoagulation Therapy

Patients with an anterior myocardial infarction, severe LV dysfunction, heart failure, a history of embolism, echocardiographic evidence of mural thrombosis, or atrial fibrillation, should receive full-dose antithrombotic therapy (LMWHs or UFH) for the period of hospitalization, followed by at least 3 months of warfarin therapy.

A summary of the recommendations for heparin treatment in patients with STE-ACS is given in Table 8.

## 5 Heparin in Pregnancy

During pregnancy, heparin can be safely used for both prevention and treatment of VTE.

**Table 8** Recommendations for the treatment of STE-ACS

| |
|---|
| *With primary PCI* |
| – UFH: iv bolus of 100 IU/Kg weight (60 IU/Kg weight if GPIIb/IIIa antagonists are used) |
| – To be continued for 24–48 h at an initial infusion of 12 IU/Kg per hour (maximum 1,000 IU/h) adjusted to maintain an aPTT level 1.5–2 times the control value |
| – The use of LMWHs periprocedurally is not recommended. If chosen, it should be started one hour after sheath removal |
| – The use of fondaparinux is not recommended |
| *With fibrinolysis* |
| – UFH: iv bolus of 100 IU/Kg weight (60 IU/Kg weight if GPIIb/IIIa antagonists are used) |
| – To be continued for 24–48 h at an initial infusion of 12 IU/Kg/h (maximum 1,000 IU/h) adjusted to maintain an aPTT level 1.5–2 times the control value |
| – Enoxaparin: in patients <75 years or with creatinine level <2.5 mg/dl: 30 mg bolus followed, after 15 min, by a subcutaneous dose of 1 mg/Kg every 12 h until hospital discharge or for 8 days; in patients >75 years or with creatinine level >2.5 mg/dl: no iv bolus; start with 0.75 mg/kg; in patients with creatinine clearance <30 ml/h doses are administered at 24 h intervals. |
| – Fondaparinux could be a good alternative to heparin: as 2.5 mg iv bolus followed by 2.5 mg s. c. once daily until 8 days or hospital discharge |
| *Patients treated without reperfusion therapy* |
| Enoxaparin, UFH, and fondaparinux are all valid therapeutic options in this setting with the same doses as used with fibrinolytics |

Given that UFH and LMWHs do not cross the placenta, there is no risk of fetal teratogenicity or bleeding, although bleeding at the uteroplacental junction may occur.

Breastfeeding during heparin treatment is also allowed given the low capacity of heparin to be excreted into breast milk, due to its negative charge and its high molecular weight, and to its low capacity to be absorbed from the infant's gut.

## *5.1 Unfractionated Heparin*

UFH can be administered by continuous iv infusion with laboratory-guided dose adjustment to achieve a target therapeutic aPTT, or subcutaneously twice-daily in doses sufficient to achieve a therapeutic aPTT 6 h after injection. During pregnancy, the aPTT response to heparin is blunted; therefore, higher doses are necessary to achieve the therapeutic aPTT target. The rate of bleeding in pregnant women treated for deep vein thrombosis reported in the literature (2%) suggests that higher heparin levels do not imply an adjunctive risk of bleeding in pregnant women (Ginsberg et al. 1989).

Subcutaneous UFH in pregnancy can be associated with an anticoagulant effect (prolongation of the aPTT) persisting for up to 28 h after the last injection, with a mechanism that remains unclear, and this can render its use risky prior to delivery (Anderson et al. 1991).

The frequency of HIT in pregnant and postpartum women treated with UFH is not established. Generally speaking, there is a consensus that the risk of HIT in pregnant patients undergoing treatment with UFH is low. HIT should be differentiated from other causes of thrombocytopenia in pregnancy, including incidental thrombocytopenia of pregnancy and HELLP (hemolysis, elevated liver enzymes, and low platelets) syndrome. If HIT occurs, then danaparoid sodium is recommended because it is an effective antithrombotic agent that does not cross the placenta.

## 5.2 *Low-Molecular-Weight Heparins*

Although there are only few data from controlled clinical trials in pregnant women, LMWHs are commonly used for the prophylaxis and treatment of thromboembolism in this setting. This change in clinical practice derives mainly from retrospective analyses that reported a low rate of bleeding and a low incidence of HIT and osteoporosis with LMWH as compared with UFH. Concerning breast feeding, in a series of women receiving 2,500 IU of LMWH after cesarean section, there was evidence of excretion of only small amounts of LMWH into milk (Table 9) (Richter et al. 2001). Moreover, given the extremely low oral bioavailability of heparin, even in case it is excreted in breast milk, LMWH is very unlikely to have any clinically relevant effect in the brest-fed infant.

## 5.3 *Thromboprophylaxis and Cesarean Delivery*

Cesarean section alone does not justify the use of thromboprophylaxis. Current guidelines recommend an individual assessment of the risk of thrombosis in all women undergoing cesarean section to determine the need for thromboprophylaxis.

The risk of symptomatic VTE attributable to cesarean section itself appears similar to that seen in low-risk surgical patients, being about 0.4% for proximal

**Table 9** Excretion of low-molecular-weight heparin in human milk

| Study/Year | Intervention | Patients analyzed, no./total | Length of follow-up after delivery | Infant hemorrhage, no./total | Presence in breast milk, no./total | Effect in infant blood, no./total |
|---|---|---|---|---|---|---|
| Richter et al. (2001) | 2,500 IU sc dalteparin during breastfeeding | Breast-fed infants: 15/15 | Up to 8 days | Not relevant | Anti-Xa LMWH level: 11/15 (range 0.007–0.028 IU/mL)[a] | 0/15 |

[a]Therapeutic anti-Xa LMWH level: 0.5–1.5 IU/mL, 4–6 h after injection

Table 10 VTE risk factors that favor the use of thromboprophylaxis in cesarean delivery

| VTE risk factors |
|---|
| – Increased age |
| – Prior VTE |
| – Obesity |
| – Thrombophilia |
| – Lower limb paralysis |
| – Immobilization |
| – Extended surgery such as hysterectomy |
| – Preeclampsia |
| – Comorbid medical conditions, such as heart failure |

DVT and about 0.2% for symptomatic PE (Geerts et al. 2004): thus, routine thromboprophylaxis is not recommended but early mobilization should be encouraged.

The concomitant presence of other risk factors, especially if combined (Table 10), places the patient in a moderate-to-high risk VTE class with the consequent indication for thromboprophylaxis.

Given the lack of data from controlled clinical trials in this population, recommendations are extrapolated from other patient populations or based on experts opinions. Current guidelines suggest:

- The use of either mechanical or pharmacological prophylaxis, during hospital stay, for women considered at increased risk of VTE after cesarean section due to the presence of at least one risk factor.
- Pharmacological prophylaxis combined with the use of graduated compression stockings for women with multiple additional risk factors for VTE.
- Extended prophylaxis (up to 4–6 weeks after delivery) following discharge from hospital for selected high-risk women in whom important risk factors persist following delivery.

## 5.4 *Treatment of VTE During Pregnancy*

For the treatment of VTE during pregnancy LMWH is the safest drug for both the mother and the fetus and is the preferred choice for its better bioavailability, longer plasma half-life, no requirement of aPTT monitoring and lower incidence of osteoporosis and thrombocytopenia compared to UFH. The dosage is weight adjusted, and during pregnancy it has to be increased as compared to that used in nonpregnant patients because the volume of distribution of LMWH changes and glomerular filtration rate increases from the second trimester of pregnancy. There are controversies concerning dose adjustment in proportion to the change in weight during pregnancy and anti-Xa monitoring. Guidelines recommend therapy with LMWH throughout pregnancy and the puerperium (increased VTE risk during this period) for pregnant women with acute VTE. There are no specific trials to guide the duration of postpartum anticoagulation for women diagnosed with VTE during pregnancy. In general, at least 6 months of anticoagulant therapy with treatment continued until at least 6 weeks postpartum is a reasonable duration (American College of Chest Physicians 2008).

# 6 Heparin in Children

## 6.1 *Use of Heparin in Pediatric Patients*

Many factors contribute to the difference existing between heparin use in adults and children. These include the fact that the distribution, binding, and clearance of antithrombotic drugs are age dependent; limited vascular access in children reduces the possibility to effectively deliver antithrombotic therapy; often, the only vascular access available is used for drug delivery, and so accurate monitoring of blood anticoagulation is difficult; specific pediatric formulations of antithrombotic drugs are not available (for example, LMWHs are available in predosed syringes based on adult weights).

## 6.2 *Unfractionated Heparin*

UFH is commonly used in pediatric patients (Newall et al. 2003). Many factors may alter the activity of UFH in children, such as reduced antithrombin levels, low capacity to generate thrombin in children, age-related differences in anti-FXa activities and anti-FIIa activities, age-related differences in plasma protein binding of UFH, age-related differences in the amount of tissue factor pathway inhibitor released for the same amount of UFH.

### 6.2.1 Therapeutic Ranges

The aPTT therapeutic ranges are universally calculated using adult plasma, but extrapolating the aPTT range from adults to pediatric patients may not be valid.

There are also significant differences in the results of anti-FXa assays in children depending on the assay method used. Nonetheless, in the absence of further information and specific pharmacodynamic studies, extrapolation of the adult therapeutic range to pediatric patients remains necessary.

### 6.2.2 Doses

Based on the scanty evidence obtained from studies in children (Andrew et al. 1994), it is suggested to use an iv bolus of 70–100 IU/kg to achieve therapeutic aPTT values in 4–6 h.

Maintenance UFH doses are age dependent:

- Infants (up to 2 months corrected for gestational age) have the highest requirements, on average 28 IU/Kg/h
- Children <1 year of age have lower requirements, on average 20 IU/kg/h
- Older children: doses similar to the weight-adjusted requirements in adults, i.e., 18 IU/kg/h.

Indeed, UFH clearance is faster in newborns than in older children.

## 6.3 *Monitoring*

Optimal monitoring of UFH is problematic for the reasons reported above. While there is no published data to support the practice, many clinicians use anti-FXa assays (anti-FXa level between 0.35 and 0.7 IU/mL) in children less than 1 year old and aPTT values (with adults ranges) in older children. An example of protocol for UFH administration and monitoring in children is reported in Table 11 (Andrew et al. 1994).

### 6.3.1 Adverse Effects

The rate of bleeding in children treated with UFH is undetermined. One cohort study reported a rate of bleeding of 1.5% in children treated for DVT or PE, but many children were treated with subtherapeutic doses. Other studies conducted in ICUs reported much higher incidences.

Osteoporosis is a potentially serious but uncommon side effect. There are only three case reports of UFH-induced osteoporosis in children and all patients received also steroids. The effect of heparin on the bone of growing children is not known. Therefore, long-term treatment with heparin in children should be avoided.

**Table 11** Protocol for UFH administration and monitoring in children

| aPTTs | Bolus IU/kg | Hold for min | Rate change % | aPTT control h |
|---|---|---|---|---|
| <50 | 50 | 0 | +10 | 4 h |
| 50–59 | 0 | 0 | +10 | 4 h |
| 60–85 | 0 | 0 | 0 | Next day |
| 86–95 | 0 | 0 | −10 | 4 h |
| 96–120 | 0 | 30 | −10 | 4 h |
| >120 | 0 | 60 | −15 | 4 h |

Loading dose: 75–100 IU/kg in 10 min

Initial maintaining dose: 28 IU/kg/h for infants <1 year old, and 20 IU/kg/h for children >1 year old

Adjust therapy to maintain an aPTT between 55 and 85 s as follows

aPTT control after 4 h from the loading dose, after 4 h from any dose change and once daily when aPTT values are in the therapeutic range

Cell blood count once daily for platelet monitoring

HIT has been reported in children hospitalized in pediatric ICU (Schmugge et al. 2002). A high clinical suspicion is necessary to diagnose HIT in children because pediatric ICU patients have many reasons for thrombocytopenia. A review of case reports showed that clinical symptoms include not only isolated thrombocytopenia but also venous, arterial, and intracardiac thromboembolism, sometimes catheter related. Pulmonary embolism and major bleeding were rarely described. Confirmatory functional or antigenic testing for HIT antibodies is warranted. An unfavorable outcome (death/limb amputation) occurred in 42.1% of children with HIT without therapy, and in 18% of children treated with danaparoid, lepirudin, or argatroban (Risch et al. 2004).

## *6.4 LMWHs*

LMWHs are now preferred to UFH in pediatric patients for their minimal requirement for monitoring, possibility to be administered subcutaneously and the lower risk of HIT and osteoporosis. The predictability of the anticoagulant effect of weight-adjusted doses of LMWH in children is uncertain because of the altered plasma protein binding of heparin. Therapeutic ranges are extrapolated from trials in adults and include an anti-FXa level, determined 4–6 h following sc injection, of between 0.5 and 1 IU/ml for the therapeutic range and between 0.1 and 0.3 IU/ml for the prophylactic range.

Doses required to achieve a prophylactic or therapeutic anti-FXa level with enoxaparin, dalteparin, tinzaparin, and reviparin in pediatric patients are reported in Table 12.

### 6.4.1 Adverse Effects

There are no clear data on the rates of bleeding in children treated with LMWHs. Bleeding can occur at the site of sc injections; therefore, LMWH should be used with caution in newborns with little subcutaneous tissue. There are no data on HIT, osteoporosis, and hypersensitive reactions in children treated with LMWH.

# 7 Conclusions

Heparin has now almost one century of life, having been originally identified approximately in 1920 within an aqueous extract of liver that exhibited anticoagulant activity in vitro, from which it derives its name, but is still a cornerstone of treatment of many thrombotic diseases. Although the original mixture of sulphated glycosaminoglycans forming UFH is less used, several new formulations, going from different LMWH preparations obtained from fractionation of UFH to the

**Table 12** Doses of LMWHs in pediatric patients

| Heparin | Doses | Anti-FXa monitoring frequency[a] |
|---|---|---|
| *Reviparin* (weight dependent doses) | IU/kg q12 h | |
| • Treatment: weight <5 kg | 150 | |
| • Treatment: weight <5 kg | 100 | |
| • Prophylaxis: weight <5 kg | 50 | |
| • Prophylaxis: weight <5 kg | 30 | |
| *Enoxaparin* (age dependent doses) | mg/kg q12h | |
| • Treatment: age <2 months | 1.5 | After second or third dose and weekly after therapeutic dose achieved |
| • Treatment: age >2 months <18 year | 1 | After second or third dose and weekly after therapeutic dose achieved |
| • Prophylaxis: age <2 months | 0.75 | |
| • Prophylaxis: age >2 months | 0.5 | |
| *Dalteparin* (all ages) | IU/kg q24h (mean ± SD) | |
| • Treatment | 129 ± 43 | |
| • Prophylaxis | 92 ± 52 | |
| *Tinzaparin* (age dependent) | IU/kg q24h | |
| • Treatment: 0–2 months | 275 | |
| • Treatment: 2–12 months | 250 | |
| • Treatment: 1–5 year | 240 | |
| • Treatment: 5–10 year | 200 | |
| • Treatment: 10–16 year | 175 | |

[a]In case of renal failure consider increased monitoring

completely synthetic pentasaccharide fondaparinux, are progressively replacing UFH with advantages in terms of safety and ease of use. It is likely that heparin and its derivatives will still continue to represent a fundamental weapon in the fight against thrombosis also for the years to come.

**Acknowledgment** The skilful help of Dr. Sara Orsini with editorial handling is gratefully acknowledged. This study was supported by grants from Fondazione Cassa di Risparmio di Perugia (project n. 2009-020-0097), MIUR (2007, Prot. 20073KBBHC_003 ) and the Italian Ministry of Health (RFPS-2006-8-334062) to P. G.

## References

Agnelli G, Bergqvist D, Cohen AT et al (2005) Randomized clinical trial of postoperative fondaparinux versus perioperative dalteparin for prevention of venous thromboembolism in high-risk abdominal surgery. Br J Surg 92:1212–1220

American College of Chest Physicians (2008) Antithrombotic and thrombolytic therapy: the American College of Chest Physicians evidenced-based clinical practice guidelines 8th edition. Chest 133:67s–968s

Anderson DR, Ginsberg JS, Burrows R, Brill-Edwards P (1991) Subcutaneous heparin therapy during pregnancy: a need for concern at the time of delivery. Thromb Haemost 65:248–250
Andrew M, Marzinotto V, Massicotte P et al (1994) Heparin therapy in pediatric patients: a prospective cohort study. Pediatr Res 35:78–83
Antman EM, Morrow DA, McCabe CH et al (2006) Enoxaparin versus unfractionated heparin with fibrinolysis for ST-elevation myocardial infarction. N Engl J Med 354:1477–1488
Barrellier MT (1993) Superficial venous thromboses of the legs. Phlebologie 46:633–639
Brambilla S, Ruosi C, La Maida GA et al (2004) Prevention of venous thromboembolism in spinal surgery. Eur Spine J 13:1–8
Brandjes DPM, Heijboer H, Buller HR et al (1992) Acenocoumarol and heparin compared with acenocoumarol alone in the initial treatment of proximal-vein thrombosis. N Engl J Med 327:1485–1489
Buller HR, Davidson BL, Decousus H et al (2003) Subcutaneous fondaparinux versus intravenous unfractionated heparin in the initial treatment of pulmonary embolism. N Engl J Med 349:1695–1702
Buller HR, Davidson BL, Decousus H et al (2004) Fondaparinux or enoxaparin for the initial treatment of symptomatic deep venous thrombosis: a randomized trial. Ann Intern Med 140:867–873
Cohen AT, Davidson BL, Gallus AS et al (2006) Efficacy and safety of fondaparinux for the prevention of venous thromboembolism in older acute medical patients: randomised placebo controlled trial. BMJ 332:325–329
Cordell-Smith JA, Williams SC, Harper WM et al (2004) Lower limb arthroplasty complicated by deep venous thrombosis: prevalence and subjective outcome. J Bone Joint Surg Br 86:99–101
Ferguson JJ, Califf RM, Antman EM et al (2004) Enoxaparin vs. unfractionated heparin in high-risk patients with non-ST-segment elevation acute coronary syndromes managed with an intended early invasive strategy: primary results of the SYNERGY randomized trial. JAMA 292:45–54
Fragmin During Instability in Coronary Artery Disease (FRISC) Study Group (1996) Low-molecular-weight heparin during instability in coronary artery disease. Lancet 347:561–568
Gallus AS, Jackaman J, Tillett J et al (1986) Safety and efficacy of warfarin started early after submassive venous thrombosis or pulmonary embolism. Lancet 2:1293–1296
Geerts WH, Heit JA, Clagett GP et al (2001) Prevention of venous thromboembolism. Chest 119:132–175
Geerts WH, Pineo GF, Heit JA, Bergqvist D, Lassen MR, Colwell CW, Ray JG (2004) Prevention of venous thromboembolism. Chest 126:338S–400S
Ginsberg JS, Kowalchuk G, Hirsh J, Brill-Edwards P, Burrows R (1989) Heparin therapy during pregnancy: risks to the fetus and mother. Arch Intern Med 149:2233–2236
Hirsh J, Anand SS, Halperin JL, Fuster V; American Heart Association. (2001) Guide to Anticoagulant Therapy: Heparin. A Statement for Healthcare Professionals From the American Heart Association. Circulation 103:2994–3018
Hull RD, Raskob GE, Rosenbloom D et al (1990) Heparin for 5 days as compared with 10 days in the initial treatment of proximal venous thrombosis. N Engl J Med 322:1260–1264
Hull RD, Schellong SM, Tapson VF et al (2006) Extended-duration thromboprophylaxis in acutely ill medical patients with recent reduced mobility: methodology for the EXCLAIM study. J Thromb Thrombolysis 22:31–38
Kher A, Samama MM (2005) Primary and secondary prophylaxis of venous thromboembolism with low-molecular-weight heparins: prolonged thromboprophylaxis, an alternative to vitamin K antagonists. J Thromb Haemost 3:473–481
Koch A, Bouges S, Ziegler S et al (1997) Low molecular weight heparin and unfractionated heparin in thrombosis prophylaxis after major surgical intervention: update of previous meta-analyses. Br J Surg 84:750–759
Lausen I, Jensen R, Jorgensen LN et al (1998) Incidence and prevention of deep venous thrombosis occurring late after general surgery: randomised controlled study of prolonged thromboprophylaxis. Eur J Surg 164:657–663

Leclerc JR, Gent M, Hirsh J et al (1998) The incidence of symptomatic venous thromboembolism during and after prophylaxis with enoxaparin: a multi-institutional cohort study of patients who underwent hip or knee arthroplasty. Arch Intern Med 158:873–878

Lozano FS, Almazan A (2003) Low-molecular-weight heparin versus saphenofemoral disconnection for the treatment of above-knee greater saphenous thrombophlebitis: a prospective study. Vasc Endovascular Surg 37:415–420

Meyer G, Marjanovic Z, Valcke J et al (2002) Comparison of low-molecular-weight heparin and warfarin for the secondary prevention of venous thromboembolism in patients with cancer: a randomized controlled study. Arch Intern Med 162:1729–1735

Newall F, Barnes C, Ignjatovic V et al (2003) Heparin-induced thrombocytopenia in children. J Paediatr Child Health 39:289–292

Nicolaides A, Irving D, Pretzell M et al (1973) The risk of deep-vein thrombosis in surgical patients. Br J Surg 60:312

Oler A, Whooley MA, Oler J et al (1996a) Adding heparin to aspirin reduces the incidence of myocardial infarction and death in patients with unstable angina: a meta-analysis. JAMA 276:811–815

Oler A, Whooley MA, Oler J, Grady D (1996b) Adding heparin to aspirin reduces the incidence of myocardial infarction and death in patients with unstable angina. JAMA 276:811–815

Petersen JL, Mahaffey KW, Hasselblad V et al (2004) Efficacy and bleeding complications among patients randomized to enoxaparin or unfractionated heparin for antithrombin therapy in non-ST-segment elevation acute coronary syndromes: a systematic overview. JAMA 292:89–96

Prandoni P, Samama MM (2008) Risk stratification and venous thromboprophylaxis in hospitalized medical and cancer patients. Br J Haematol 141:587–597

Prandoni P, Tormene D, Pesavento R, Vesalio Investigators Group (2005) High vs. low doses of low-molecular-weight heparin for the treatment of superficial vein thrombosis of the legs: a double-blind, randomized trial. J Thromb Haemost 3:1152–1157

PRISM PLUS Study Investigators (1998) Inhibition of the platelet glycoprotein IIb/IIIa receptor with tirofiban in unstable angina and non–Q-wave myocardial infarction. N Engl J Med 338:1488–1497

Quinlan DJ, McQuillan A, Eikelboom JW (2004) Low-molecular weight heparin compared with intravenous unfractionated heparin for treatment of pulmonary embolism: a metaanalysis of randomized, controlled trials. Ann Intern Med 140:175–183

Raschke RA, Reilly BM, Guidry JR et al (1993) Heparin dosing review. Ann Intern Med 119:104–109

Richter C, Sitzmann J, Lang P, Weitzel H, Huch A, Huch R (2001) Excretion of low molecular-weight heparin in human milk. Br J Clin Pharmacol 52:708–710

Risch L, Fischer JE, Herklotz R, Huber AR (2004) Heparin-induced thrombocytopenia in paediatrics: clinical characteristics, therapy and outcomes. Intensive Care Med 30:1615–1624

Samama MM, Cohen AT, Darmon JY et al (1999) A comparison of enoxaparin with placebo for the prevention of venous thromboembolism in acutely ill medical patients. Prophylaxis in Medical Patients with Enoxaparin Study Group. N Engl J Med 341:793–800

Schmugge M, Risch L, Huber AR et al (2002) Heparin-induced thrombocytopenia-associated thrombosis in pediatric intensive care patients. Pediatrics 109:E10

Tagalakis V, Kahn SR, Libman M, Blostein M (2002) The epidemiology of peripheral vein infusion thrombophlebitis: a critical review. Am J Med 113:146–151

The OASIS-6 Trial Group (2006) Effects of fondaparinux on mortality and reinfarction in patients with acute ST-segment elevation myocardial infarction: the OASIS-6 randomized trial. JAMA 295:1519–1530

The Superficial Thrombophlebitis Treated by Enoxaparin Study Group (2003) A pilot randomized double-blind comparison of a low-molecular-weight heparin, a nonsteroidal anti-inflammatory agent, and placebo in the treatment of superficial vein thrombosis. Arch Intern Med 163:1657–1663

Yusuf S, Mehta SR, Chrolavicius S et al (2006a) Efficacy and safety of fondaparinux compared to enoxaparin in 20,078 patients with acute coronary syndromes without ST segment elevation. The OASIS (Organization to Assess Strategies in Acute Ischemic Syndromes)-5 Investigators. N Engl J Med 354:1464–1476

Yusuf S, Mehta SR, Chrolavicius S et al (2006b) Effects of fondaparinux on mortality and reinfarction in patients with acute ST-elevation myocardial infarction; the OASIS-6 randomized trial. JAMA 295:1519–1530

# Adverse Effects of Heparin

**S. Alban**

## Contents



S. Alban (✉)
Pharmazeutisches Institut, Abteilung Pharmazeutische Biologie, Christian-Albrechts-Universität zu Kiel, Kiel Gutenbergstr. 76, 24118 Kiel, Germany
e-mail: salban@pharmazie.uni-kiel.de

R. Lever et al. (eds.), *Heparin - A Century of Progress*,
Handbook of Experimental Pharmacology 207,
DOI 10.1007/978-3-642-23056-1_10, 

**Abstract** All the adverse effects of heparins are related to their wide variety of biological activities, with bleeding being the most important safety issue, resulting directly from the potency of heparin as an anticoagulant. However, it is hard to define the bleeding risk, since it depends on numerous parameters including the indication, dosage, method, and duration of heparin application, the clinical study design and definition of bleeding as well as patient characteristics and determinants of bleeding such as type of surgery and co-medication. Nonbleeding complications of heparins are caused by binding of heparin molecules to proteins other than antithrombin and to cells, which is generally more pronounced with unfractionated heparin than with low-molecular-weight heparins. Accordingly, heparin-induced thrombocytopenia, the most severe nonbleeding adverse reaction, occurs about 10 times less with low-molecular-weight heparins than with unfractionated heparin. Frequent and therefore important adverse reactions of heparins are skin lesions resulting from delayed-type hypersensitivity reactions. All the other undesirable effects are discussed as well, but they are mostly clinically irrelevant.

**Keywords** Heparin • Bleeding • Heparin-induced thrombocytopenia (HIT) • Osteoporosis

## Abbreviations

| | |
|---|---|
| ACS | Acute coronary syndrome |
| APTT | Activated partial thromboplastin time |
| AT | Antithrombin |
| aXa | anti-Factor Xa |
| BMD | Bone mineral density |
| BW | Body weight |
| D-ITP | Drug-induced thrombocytopenic purpura |
| DTHR | Delayed-type hypersensitivity reactions |
| DTI | Direct thrombin inhibitor |
| DVT | Deep vein thrombosis |
| EIA | Enzyme immunoassay |
| EMA | European Medicines Agency |
| GAG | Glycosaminoglycan |
| GPI | Glycoprotein IIb/IIIa inhibitors |

| | |
|---|---|
| GUSTO | Global Utilization of Streptokinase and Tissue plasminogen activator for Occluded coronary arteries |
| HIPA | Heparin-induced platelet activation assay |
| HIT | Heparin-induced thrombocytopenia |
| ISTH | International Society on Thrombosis and Hemostasis |
| LMWH | Low-molecular-weight heparin |
| MHRA | Medicines and Healthcare Products Regulatory Agency |
| NSAID | Nonsteroidal anti-inflammatory drug |
| OR | Odds ratio |
| PCI | Percutaneous coronary intervention |
| PE | Pulmonary embolism |
| PF4 | Platelet factor 4 |
| RANK | Receptor activator for nuclear factor-κB |
| RANKL | Receptor activator for nuclear factor-κB ligand |
| RCT | Randomized controlled trial |
| RIETE | Registry of Patients with Venous Thromboembolism |
| SRA | Serotonin release assay |
| TE | Thromboembolism |
| TIMI | Thrombolysis In Myocardial Infarction |
| UFH | Unfractionated heparin |
| VKA | Vitamin K antagonist |
| VTE | Venous thromboembolism |

## 1 Introduction

All the adverse effects of heparins are related to their wide variety of biological activities, with bleeding being the most important safety issue, resulting directly from the potency of heparin as an anticoagulant. This anticoagulant activity is mainly caused by heparin molecules with high affinity to antithrombin (AT), which amount to only 30–50% of unfractionated heparin (UFH) and less than 20% of low-molecular-weight heparins (LMWHs) (Alban 2008a).

Nonbleeding complications as well as other beneficial activities are caused by binding of heparin molecules to proteins other than AT and cells, whereby the structural requirements for many of such interactions are unknown (Mulloy 2005). Generally, the binding tendency of the negatively charged heparin molecules increases with their chain length. As is known with UFH, competing binding partners may even interfere with the high-affinity binding to AT and thus reduce the anticoagulant effect (Hirsh and Raschke 2004). These so-called “nonspecific bindings” are much less pronounced with LMWHs, which results not only in improved effect-based pharmacokinetics, but also may explain their reduced risk of nonbleeding complications.

Nonbleeding complications have to be regarded in terms of their incidence and severity (Eikelboom 2007). The prime example for a major adverse effect of UFH is immune heparin-induced thrombocytopenia (HIT). HIT is an uncommon but potentially life-threatening complication so that it has to be considered in every patient treated with UFH. Osteoporosis should be observed in all pregnant or elderly patients who require long-term heparin anticoagulation (e.g., 1–2 weeks or longer), although it might be reversible. In recent years, the most frequent unwanted side effects of heparins were reported to be cutaneous delayed-type hypersensitivity reactions (Schindewolf et al. 2010b). Albeit they are usually not life-threatening, they need attention, since they impair the compliance of the patients and may lead to discontinuation of important anticoagulant therapy. Minor elevations of potassium levels or liver transaminases occur often as well, but they are of little clinical relevance in most patients. In these cases, routine monitoring of potassium levels is not necessary.

# 2 Bleeding

As with other antithrombotics, bleeding is the major complication of heparins (Schulman et al. 2008). By inhibiting blood coagulation, antithrombotics generally cause a balancing act between prevention of thrombus formation and inhibition of physiological coagulation, whenever blood vessels are damaged. The incidence of bleeding under heparin therapy is hard to define, as it depends on numerous parameters including the indication, dosage, method, and duration of heparin application, the clinical study design and definition of bleeding, patient characteristics and determinants of bleeding such as type of surgery and co-medication.

## *2.1 Types and Classification of Bleeding Complications*

### 2.1.1 Types of Bleeding Complications

Hemorrhagic complications due to heparins range from life-threatening, i.e., fatal, intracranial, or retroperitoneal bleeding, or bleeding requiring surgical intervention, transfusion or hospitalization, to hematoma at the injection site. Epidural or spinal hematomas, which may occur with the associated use of heparins and spinal/epidural anesthesia or spinal puncture, are rare, but severe complications, since they can result in long-term or permanent paralysis.

In general, bleeding can occur at any site during therapy. An unexplained fall in hematocrit or blood pressure should lead to a search for a bleeding site. Whereas major bleeds are rare adverse events (<0.1%), hematomas are commonly (1–10%) observed. As the latter may impair the compliance, especially in patients under long-term treatment, they should not be considered clinically irrelevant.

### 2.1.2 Risk of Bleeding in Clinical Studies

In clinical studies, the bleeding risk varied considerably even with the same heparin in one indication. For example, frequency of major hemorrhage in total hip replacement surgery amounted to 0.0–1.5% for dalteparin, 0.0–2.0% for enoxaparin, and 0.9–2.8 for tinzaparin (Lopez 2001). In clinical trials in general surgery, bleeding incidence ranged from 0.0% to 20.0% for dalteparin, and 0.9 to 4.1% for enoxaparin (Lopez 2001). One reason, among others, is differences in the definition and judgment of bleeding complications.

The European Medicines Agency (EMA) addresses the problem of bleeding assessment in current guidelines on clinical investigation of antithrombotic drugs (EMEA/CHMP 2006, 2007). Here, the use of validated and clinically relevant classification of bleeding and the adjudication of bleeding events by central independent and blinded committees of experts is suggested. Bleeding should be classified into major or minor according to internationally accepted standards. The guidelines give examples of major bleeding (Table 1), which represents the main safety criteria. In addition, the recording of other bleeding-related parameters is recommended (Table 1), as the current criteria for major bleeding often underestimate the risk of clinically important bleeding.

Moreover, the bleeding risk associated with heparins, as known from trials performed up to now, is assumed to be lower than in clinical reality, since patients

**Table 1** Examples of major bleeding and other useful bleeding-related parameters (EMEA/CHMP 2007)

*Major bleeding (main safety criteria)*

- Fatal bleeding
- Clinically overt bleeding associated with a decrease in the hemoglobin level of >20 g/l compared with the pre-randomization level
- Clinically overt bleeding leading to transfusion of ≥2 units of whole blood or packed cells
- Critical bleeding (intracerebral, intraocular, intraspinal, pericardial, or retroperitoneal)
- Bleeding warranting treatment cessation
- Bleeding located at the surgical site and leading to reoperation or to any unusual medical intervention or procedure for relief (e.g., draining or puncture of an haematoma at the surgical site, transfer to an intensive care unit or emergency room)

*Other bleeding-related parameters*

- Hemoglobin plasma level, hematocrit and red cell count changes during the treatment period, creatinine, serum protein level
- Measured blood loss (perioperative, postoperative) quantified by an objective method (weight of swabs and operative drapes, volumes in the suction bottles after surgery, and drain collectors on admission to the postanesthesia care unit and thereafter for the two postoperative days)
- Calculated blood loss (perioperative, postoperative) using the following formula: Calculated bleeding (expressed in ml of red blood cells (RBC), hematocrit (Ht) 100%) = estimated blood volume (EBV) × (preoperative Ht – day 2 Ht) + 150 ml per RBC or cell salvage unit, assuming an EBV of 70 ml/kg (men), 65 ml/kg (women, obese men) and 60 ml/kg (obese women), respectively
- Incidence of patients receiving transfusion of packed red cells and transfused quantities during the treatment period (homologous and autologous transfusions need to be distinguished)

with bleeding risk had been almost systematically excluded (EMEA/CHMP 2007). In particular, patients with renal insufficiency, which is very frequent and related to patients' age and surgery itself, had been underrepresented in clinical trials.

### 2.1.3 Definition of Bleeding Complications

Despite the recommendations of regulatory authorities (Table 1), recent clinical studies sometimes still result in different bleeding risks associated with heparins. An example is the phase III trials in major orthopedic surgery of the thrombin inhibitor dabigatran etexilate and the factor Xa inhibitor rivaroxaban, both in comparison with enoxaparin (Zikria and Ansell 2009). Similar to corresponding trials on fondaparinux (Turpie et al. 2002), the incidence of major bleeding in the enoxaparin group ranged from 1.3% to 1.7% in the studies on dabigatran etexilate, whereas it amounted to only 0.1–0.3% in the RECORD trials on rivaroxaban.

This discrepancy highlights the importance of an internationally standardized definition and assessment of bleeding complications. Such standards are formulated by the International Society on Thrombosis and Haemostasis (ISTH) Guidelines for medical patients, but there are currently none for surgery patients. In trials and registries on acute coronary syndrome (ACS), a variety of bleeding definitions is used (Rao et al. 2009). They rely either predominantly on laboratory data elements (e.g., definition by the "Thrombolysis in Myocardial Infarction" (TIMI) group) or solely on clinical data (e.g., definition by the "Global Utilization of Streptokinase and Tissue plasminogen activator for Occluded coronary arteries" (GUSTO) investigators) or on a mixture of them as applied in newer trials such as CURE, OASIS-5, REPLACE-2, and ACUITY. Due to the lack of consistency, it is difficult to compare results on the bleeding risk across clinical trials.

## 2.2 *Risk Factors for Bleeding*

The propensity to bleed is certainly increased by heparins, but the actual bleeding risk of a patient is determined by many other factors. It depends on the patient's baseline characteristics as well as on the clinical situation and the drug exposure. In contrast to the individual disposition of the patient, the exposure-related risk factors can partly be modified (Manoukian et al. 2007).

### 2.2.1 Disposition-Related Risk Factors

In all patients treated with antithrombotic therapy, increased age and renal insufficiency are strong independent baseline predictors of major bleeding (Bounameaux and Perrier 2008; Eikelboom et al. 2009; Kinnaird et al. 2003; Macie et al. 2004; Manoukian et al. 2007; Schulman et al. 2008).

There are further important risk factors, but different ones depending on the patient group. In VTE prophylaxis in surgery or VTE prophylaxis in medical patients patients receiving VTE prophylaxis, lower body weight, male gender, absence of history of VTE represent additional strong predictors as revealed from analysis of 8 large randomized controlled trials (RCTs) ($n = 13{,}085$) (Eikelboom et al. 2009). According to the computerized registry of patients with venous thromboembolism RIETE registry‘ bleeding score, major risk factors for bleeding in VTE patients are age >75 years, recent bleeding, anemia, cancer, creatinine levels >1.2 mg/dL, and pulmonary embolism (PE) (Ruiz-Gimenez et al. 2008; Trujillo-Santos et al. 2008). In patients with ACS, anemia is a major predictor of bleeding as well (Manoukian et al. 2007). Moreover, female gender, black race, low body weight, ST-segment deviation ≥1 mm and elevated cardiac biomarkers were found to be associated with an increased bleeding risk of ACS patients.

A disadvantage of risk factor analyses based on clinical trials is that they ignore the patients who are mostly excluded from clinical trials due to a very high propensity for bleeding or severe complications caused by bleeding (Schulman et al. 2008). Such conditions are defined as contraindications or special warnings and precaution for use of heparins as well as other antithrombotics (Table 2) (MHRA 2011).

**Table 2** Disposition-related risk factors for bleeding and conditions with increased potential for critical bleeding

*Disposition-related risk factors for bleeding*

- Clinically relevant bleeding (≤30 days ago)
  - Gastrointestinal bleeding
  - Macroscopically visible urogenital bleeding
  - Other abnormal bleeding
- Clinically relevant impaired hemostasis (acute or historical)
  - Hemorrhagic diathesis
  - Coagulation factor deficiencies
  - Thrombocytopenia
- Severe liver- and pancreas diseases
- Chronic renal insufficiency
- Increased age

*Conditions with increased potential for critical bleeding*

- For bleeding in the central nervous system (CNS):
  - Intracranial or spinal injury or surgery (≤6 weeks ago)
  - Hemorrhagic or ischemic stroke (acute or ≤6 months ago)
  - Other intracranial bleeding (acute or ≤6 months ago)
  - Intracranial disease (acute or historical) (neoplasia, arteriovenous malformation, aneurysm)
- For ocular and aural bleeding:
  - Injury or surgery (≤6 weeks ago)
  - Vascular retinopathy
  - Vitreous hemorrhage
  - Other intraocular bleeding
- For bleeding in other organs
  - History of peptic ulcer
  - Uncontrolled severe arterial hypertension
  - Bacterial endocarditis
  - Abortus imminens

### 2.2.2 Exposure-Related Risk Factors

Among the exposure-related risk factors, both surgery and treatment of ACS are associated with a high risk of bleeding (Eikelboom et al. 2009; Rao et al. 2009). Especially challenging are patients receiving anticoagulant or antiplatelet therapy due to indications such as arterial or venous thromboembolism (TE), atrial fibrillation, mechanical heart valve or high risk for TE, who require a surgical or other invasive procedure, since they have a high risk for both bleeding and TE (Douketis et al. 2008).

*VTE prophylaxis in surgery*. Compared to VTE prophylaxis in medical patients, surgical patients were found to have an odds ratio (OR) for major bleeding of 3.53 (1.54–8.13), whereby abdominal surgery has been associated with a significantly higher risk than orthopedic surgery (Eikelboom et al. 2009). Surgical and other invasive procedures associated with a high bleeding risk include: coronary artery bypass or heart valve replacement surgery; intracranial or spinal surgery, aortic aneurysm repair, peripheral artery bypass, and other major vascular surgery; major orthopedic surgery, reconstructive plastic surgery, major cancer surgery; and prostate and bladder surgery (Douketis et al. 2008).

As found in clinical trials comparing rivaroxaban or dabigatran etexilate with enoxaparin in othopedic surgery, about 90% of the major bleeding events occurred at the surgical site and about 50% of the major bleeding events started before treatment and >80% within the first 48 h of the procedure (Eriksson et al. 2006, 2007).

*Spinal/epidural anesthesia and spinal puncture*. A special topic in the context of surgery is the risk of spinal or epidural hematomas. Neuraxial blockade generally has several advantages including a reduced bleeding risk compared with general anesthesia or systemic analgesia (Geerts et al. 2008). With the concomitant use of heparin or other antithrombotics, it has, however, to be considered as a risk factor for bleeding, since spinal or epidural hematomas represent rare, but potentially devastating complications by resulting in long-term or permanent paralysis (Geerts et al. 2008). The risk of these hematomas is higher with needle or catheter insertion and catheter removal in the presence of high levels of anticoagulants and with the combined use of drugs affecting hemostasis.

*Acute coronary syndrome*. In cardiology, bleeding complications seem to occur even more often than in surgery. In clinical trials, up to 30% of patients with ACS or undergoing percutaneous coronary intervention (PCI) experience bleeding complications, and even higher rates have been reported in contemporary practice (Manoukian et al. 2007). On the one hand, procedural characteristics, and on the other hand, the use of antithrombotic drugs, have an impact on the bleeding risk. It was shown to increase with longer, repeated and unsuccessful PCI, with the size of the vascular sheath and with the use of an intra-aortic balloon pump.

During the last two decades, the treatment with multiple antithrombotic drugs, i.e., acetyl salicylic acid (aspirin), clopidogrel, heparins, and glycoprotein IIb/IIIa inhibitors (GPI), has continually reduced ischemic events, but has also increased bleeding (Bassand et al. 2007; Van de Werf et al. 2008). This is, however, not only due to the synergistic effects of these drugs, but also due to inadequate dosing including excess doses (Alexander et al. 2005; Macie et al. 2004). According to data

**Table 3** Interactions between heparins and other drugs potentially increasing the risk of bleeding

| Antithrombotic drugs for ACS treatment | Other antithrombotic drugs | Drugs with other indications |
|---|---|---|
| Acetyl salicylic acid | Vitamin K antagonists (e.g., warfarin, phenprocoumon) | NSAIDs (e.g., indometacin, ketorolac, propionic acid derivatives such as ibuprofen) |
| P2Y12 antagonists (clopidogrel, prasugrel, ticagrelor) | Direct thrombin inhibitors (lepirudin, argatroban, dabigatran etexilate) | Glucocorticoids |
| Glycoprotein IIb/IIIa inhibitors (abciximab, eptifibatide, tirofiban) | Direct factor Xa inhibitors (rivaroxaban, apixaban) | Sulfinpyrazone |
| Other anticoagulants (fondaparinux, bivalirudin) | Other glyco-anticogualants (danaparoid, pentosan polysulfate) | Cytostatics |
| Thrombolytics (streptokinase, alteplase, reteplase, tenecteplase) | Antithrombin | Dextrans (plasma substitutes) |
| | Ticlopidine | Trapidil |
| | Dipyridamole | Piracetam |
| | | Drotrecogin alfa |

from the CRUSADE registry, initial UFH bolus and infusion dosing were higher than the recommended weight-adjusted ranges in 35% of the cases and these excess doses were associated with more bleeding (Melloni et al. 2008). Especially elderly patients are subject to the risk of overdosage, when the dosages of enoxaparin, some other LMWHs as well as eptifibatide or tirofiban are not carefully adapted to their renal function. Moreover, ACS patients are often on additional medication impairing hemostasis such as nonsteroidal anti-inflammatory drugs (NSAIDs) (Table 3) (Alexander et al. 2005; Macie et al. 2004).

*Drug interactions.* In general, pharmacodynamic drug interactions between heparins and other drugs affecting hemostasis may notably contribute to bleedings appearing in clinical practice (Table 3). Patients treated with heparin are frequently comorbid and have clear indications for such drugs (e.g., NSAIDs in patients undergoing orthopedic VTE prophylaxis in surgery). In some cases, the interactions may not be realized or the physician might even be not informed about all the drugs taken by the patient.

## 2.3 *Heparin-Related Determinants of Bleeding*

### 2.3.1 Dosing and Intensity of Anticoagulation

The risk of bleeding should be related to the dose of heparin or the intensity of anticoagulation measured in the plasma. Accordingly, in the OASIS-2 trial, every

10-s increase in the APTT induced by UFH was associated with a 7% increase of major bleeding (Anand et al. 2003). Further, in high-risk patients with ACS (NSTEMI), where the initial UFH bolus and infusion doses were frequently higher than recommended BW-adjusted ranges, excess bolus (OR 1.03, 95% CI 1.00–1.06) and infusion rate (OR 1.16, 95% CI 1.05–1.28) dosing was associated with more bleeding and was common among elderly and females (Melloni et al. 2008).

Yet, many other studies did not demonstrate a relationship between the UFH dose or its Monitoring and the bleeding rate (Schulman et al. 2008). This might be due to the widely varying bioavailability of UFH as well as due to the insufficiency of APTT usually used for dose adjustment (Rapaport et al. 2004).

In contrast to UFH, subcutaneously (s.c.) applied LMWHs lead to dose-dependent anti-factor Xa (aXa) plasma levels. But the aXa level correlates only with the incidence of hemorrhage at aXa levels >0.8 U/mL (Leizorovicz et al. 1993; Nieuwenhuis et al. 1991). Correspondingly, a prospective trial did not reveal improved safety by adjusting the dose of dalteparin by aXa monitoring (Alhenc-Gelas et al. 1994). The low number of dose-ranging trials with LMWHs suggests a significant dose-dependency of the bleeding risk only at therapeutic (Thery et al. 1992; TIMI-11A-Trial-Investigators 1997), but not conclusively at prophylactic dosage (Heit et al. 1997; Kakkar et al. 1986; Samama et al. 1999).

This relative robustness is consistent with the finding that prophylaxis in orthopedic trauma patients with fixed-dose nadroparin was as safe as body weight (BW)-adjusted doses of nadroparin (Haentjens 1996). For therapeutic dosage, such a direct comparison is missing. Consequently, LMWH doses are traditionally BW-adjusted in VTE therapy, although the distribution volume of heparins does not vary to a greater extent in relation to body weight. Only certoparin is applied as a twice daily fixed dose of 8000 aXa-IU, since there was no sign for increased bleeding at low body weight in a corresponding trial (Riess et al. 2003).

### 2.3.2 Method of Administration

In contrast to LMWHs being only approved for s.c. injection, UFH is applied by either intravenous (i.v.) infusion or s.c. injection. Whereas the rate of major bleeding was found to be higher with intermittent than with continuous i.v. infusion, continuous i.v. UFH and s.c. UFH were associated with a similar amount of bleeding (Schulman et al. 2008).

### 2.3.3 Timing of the First LMWH Dose

Depending on the local practice, the first dose of LMWH prophylaxis is given either pre- or postoperatively. In orthopedic VTE prophylaxis in surgery, the aggregate clinical research evidence indicates that initiation ≤2 h preoperatively or <6 h postoperatively increases major bleeding without improved efficacy compared to initiation at 6 h postoperatively (Hull et al. 2001; Raskob and Hirsh 2003). Furthermore, a tinzaparin dosage of 50 aXa-U/kg started 2 h before surgery turned out to be less safe and less protective than a tinzaparin dosage of 75 aXa-U/kg started 12 h before surgery (Lassen et al. 2000). A recent study shows that 2,500 IU dalteparin started 6 h after surgery significantly reduced blood loss and the need for transfusions compared to 5,000 IU dalteparin injected 12 h before surgery, but was equally efficacious (Borgen et al. 2010).

### 2.3.4 Once Versus Twice Daily LMWH

The half-life of LMWH justifies the once or twice daily administration for acute VTE therapy. A meta-analysis of five clinical trials revealed that once daily LMWH is as effective (OR 0.82, 0.49–1.39) and safe (OR 0.77, 0.40–1.45) as twice daily LMWH (van Dongen et al. 2005). However, there is inadequate data to be able to exclude the possibility of a higher frequency of fatal bleeding with once-daily therapy.

### 2.3.5 Accumulation in Renal Insufficiency

In contrast to UFH, LMWHs are partly excreted via urine in their active form and therefore accumulate in patients with impaired renal function. In patients with a creatinine clearance (ClCr) of $\leq$30 mL/min, major bleeding was significantly increased when a standard therapeutic dose of enoxaparin was used [8.3% vs. 2.4%; odds ratio, 3.88 (CI, 1.78–8.45)] (Lim et al. 2006). The bleeding risk under prophylactic dosage is lower, but increases with the duration of LMWH application in patients with renal insufficiency.

However, the various LMWHs differ in their tendency to accumulate (Alban 2008b), so that dose reduction of LMWHs cannot generally be recommended (Gouin-Thibault et al. 2010). Moreover, it has to be considered that patients with renal insufficiency basically have an increased bleeding risk. For example, in the MATISSE PE trial, the bleeding rates under UFH amounted to 6.9% (CrCl $>$ 80 mL/min), 3.1% (CrCl 50–80 mL/min), 11.1% (CrCl 30–50 mL/min), and 10.7% (CrCl $<$ 30 mL/min) in dependence on renal function.

## *2.4 Bleeding Risk in Clinical Trials*

### 2.4.1 Heparins Versus Placebo

For the application in VTE prophylaxis, abundant data have demonstrated little or no increase in the rates of clinically important bleeding by low-dose UFH and LMWHs, which represents a desirable benefit-to-risk ratio with respect to complications of TE (Geerts et al. 2008). This benefit is not limited to prophylaxis in VTE prophylaxis in surgery, but applies to patients with lower-leg immobilization also, where the rate of major bleeding was found to be 0.3% (Testroote et al. 2008).

In general VTE prophylaxis in medical patients, who have often additional risk factors for bleeding (see above), UFH caused more major and minor bleeding (0.5% and 3.7%) than placebo (0.2% and 2.0%, respectively). Compared to UFH, LMWH reduced the risk of major bleeding to 0.3%. The VTE risk reduction of ~50% by heparins is thus paid by an increased rate of minor (RR 1.74; 95% CI 1.26–2.41; $P = 0.0008$) and major bleeding (RR 2.18; 95% CI 1.28–3.72; $P = 0.004$) (Alikhan and Cohen 2009).

In acute VTE therapy, the risk of bleeding associated with heparins is <3% in recent trials (Schulman et al. 2008). A meta-analysis on the use of heparins in ACS showed an increased incidence of minor bleeds (RR = 6.80, 95% CI 1.23–37.49, NNH = 17), but not major bleeds, compared to placebo (Magee et al. 2008).

### 2.4.2 LMWHs Versus UFH

According to an early meta-analysis on VTE prophylaxis in orthopedic VTE prophylaxis in surgery, LMWHs induce significantly less minor bleeding than UFH (RR 0.76 (95% CI: 0.64–0.92) and are superior to UFH in the prevention of DVT (RR 0.76 (95% CI: 0.60–0.91) (Palmer et al. 1997).

In general VTE prophylaxis in surgery, LMWHs at doses <3,400 aXa-U were found to be safer and as effective as UFH, while higher doses yielded increased hemorrhagic risk, including that of major hemorrhage, but slightly superior efficacy (Mismetti et al. 2001). Similarly, a meta-analysis of 16 studies on VTE prophylaxis after abdominal surgery revealed reduced bleeding complications at equal efficacy with LMWHs compared with UFH. However, at higher doses of LMWH, no increased bleeding risk was seen despite superior efficacy (Bergqvist 2004).

The first large study comparing an LMWH with UFH in general VTE prophylaxis in medical patients showed a trend for less major (0.4% vs. 0.6%) and minor (2.8% vs. 4.0%) bleeding under LMWH (Riess et al. 2010).

There is thus an indication of reduced bleeding with LWMHs compared with UFH in VTE prophylaxis trials. According to the most recent meta-analysis of studies on acute VTE therapy, major hemorrhages occurred in 1.1% of participants treated with LMWH compared with 1.9% treated with UFH (OR 0.58; 95% CI 0.40–0.83) (Erkens and Prins 2010). In addition, LMWHs significantly reduced the incidence of thrombotic complications and the overall mortality at follow-up.

According to a meta-analysis of 12 trials comparing enoxaparin with UFH in patients with ACS (NSTEMI or STEMI), major bleeding was increased with the LMWH (4.3 vs. 3.4%, OR 1.25, $P = 0.019$), but this increase was offset by a reduction in death or MI resulting in a net clinical benefit in favor of enoxaparin among the STEMI population (Murphy et al. 2007). In contrast, a meta-analysis of 13 trials comparing LMWHs with UFH during PCI showed that the use of intravenous LMWH during PCI is associated with a significant reduction in major bleeding events compared with UFH (OR, 0.57; 95% CI, 0.40–0.82; $P = 0.002$) and a trend toward a reduction in minor bleeding (OR, 0.75; 95% CI, 0.47–1.20; $P = 0.24$), without compromising outcomes on hard ischemic end points (Dumaine et al. 2007).

Altogether, LMWH seems to be somewhat safer than UFH regarding the incidence of bleeding.

### 2.4.3 Heparins Versus Other Anticoagulants

Besides heparins, the most widely used anticoagulants are vitamin K antagonists (VKA), which are however primarily applied for long-term primary and secondary prophylaxis and thus in other indications than heparins. Studies on long-term treatment of DVT with LMWH instead of VKA showed trends toward both less major bleeding (OR, 0.45; 95% CI, 0.18–1.11) and less recurrent VTE (OR, 0.66; 95% CI, 0.41–1.07) with 3 months of LMWH compared with VKA (Kearon et al. 2008). But prolonged VTE therapy with LMWHs is currently only recommended in patients with cancer (Kearon et al. 2008). According to an analysis of five trials comparing LMWH and VKA in cancer patients, the pooled RR of major bleeding was 0.98 (95% CI: 0.49–1.93, $P = 0.95$) at a ~50% reduction of VTE (Louzada et al. 2009).

Depending on the indication, approved alternatives to LMWHs are fondaparinux, bivalirudin, and the novel oral anticoagulants dabigatran etexilate, rivaroxaban, and apixaban. So far there is no one agent that, across all indications, is safer regarding major hemorrhage than heparins. Comparisons between heparins and non-heparins may yield different relative risks depending on the selected intensity of treatment, concomitant or antecedent anticoagulant, antiplatelet or thrombolytic drugs, characteristics of the patient population, and also the condition that is being treated (Schulman et al. 2008). For instance, fondaparinux was safer than enoxaparin in unstable angina or NSTEMI, but enoxaparin caused less surgical site related bleeding after orthopedic VTE prophylaxis in surgery. Another example was bivalirudin, which was safer than UFH in unstable angina, NSTEMI or PCI but appeared to have higher bleeding risk than UFH in STEMI.

## 2.5 *Impact of Bleeding on Outcome*

During recent years, the appraisal of bleeding complications in anticoagulated patients has changed. Although bleeding is a treatable complication and reversible in most cases, there is emerging evidence that its clinical impact is considerable and, perhaps, greater than previously appreciated (Douketis et al. 2008). Bleeding has proven to be a strong predictor of mortality in patients with ACS and stroke. The OASIS-5 trial demonstrated that >90% of the excess deaths in patients receiving enoxaparin compared with fondaparinux occurred in patients who experienced major bleeding (Budaj et al. 2009). Moreover, major bleeding was shown to be associated with an increase in mortality in the VTE prophylaxis in surgery population as well (Eikelboom et al. 2009). At 30 days, the risk of death was 7-fold higher among patients with a major bleeding event.

There might be several reasons for the increased mortality in patients who develop major bleeding (Eikelboom et al. 2009). Apart from fatal bleeding, discontinuation of antithrombotic therapy has the potential to increase the risk for cardiovascular events or to provoke the progression of asymptomatic DVT resulting in fatal PE. Furthermore, red cell transfusion given in the case of major bleeding could lead to adverse outcomes including increased mortality. Finally, the relationship between major bleeding and death is likely to be confounded by patient characteristics, because male subjects, older patients with lower body weight, and those with renal impairment were at increased risk of both major bleeding and death.

Regarding this, bleeding remains the most important complication of heparins as well as other antithrombotic agents. In clinical practice, the risk of hemorrhage has to be weighed against the reduction of thromboembolic events. Without doubt, heparins can trigger hemorrhage, however, available data highlight that many other factors crucially affect the occurrence of bleeding in the individual patient.

## 3 Heparin-Induced Thrombocytopenia

Heparin-induced thrombocytopenia type II (HIT) is an immunological adverse effect of heparins and represents a potentially devastating complication. HIT typically occurs in the second week of heparin therapy. Antibody-mediated platelet activation and consequent thrombin generation result in a fundamental paradox: despite thrombocytopenia induced by an anticoagulant, the major clinical effect in HIT is a substantially enhanced risk for venous and/or arterial thrombosis (Greinacher 2009). Depending on the patient population affected, it amounts to 30–75% and is thus 20–40 times higher than that of patients without HIT (Warkentin et al. 2008a). Among patients who are recognized as having isolated HIT (subsequently confirmed serologically) and who are managed by simple discontinuation of heparin, or substitution of heparin by warfarin, the risk of symptomatic thrombosis ranges from 25 to 50%, including an overall risk of fatal thrombosis of about 5% (Warkentin et al. 2008a).

Consequently, prompt diagnosis, cessation of heparin and initiation of alternative anticoagulation are important to prevent further complications in HIT. Since the two leading clinical symptoms, i.e., thrombocytopenia and/or thrombosis, are not specific for HIT, the diagnosis of HIT is challenging. Therefore, it has to be based on both clinical picture and detection of pathogenic antibodies; that is, HIT is a clinicopathological syndrome (Warkentin et al. 1998).

## 3.1 *Differentiation*

*HIT type I*. The immunological HIT has to be distinguished from the nonimmune heparin-induced thrombocytopenia type I (HIT type I), which occurs in 10–30% of the patients at the beginning of therapy with UFH, less often with LWMH (Shantsila et al. 2009). HIT type I is characterized by a transient platelet count fall to about 100,000–150,000/μL, which is assumed to be due to direct platelet activation by heparins. Usually, HIT type I goes without complications and does not require cessation of heparin.

*Immune thrombocytopenia*. Although the pathogenesis of HIT is associated with antibodies, it differs from all other types of immune thrombocytopenia. The platelet count fall is not caused by increased phagocytosis of antibody-opsonized platelets, but is caused by intravascular platelet activation and subsequent aggregation (Chong et al. 1994).

## 3.2 *Pathogenesis*

The pathogenesis of HIT represents a connection between the immune system and hemostasis. It comprises the following steps:

1. Antigen formation, mostly by binding of heparin to platelet factor 4 (PF4) and induction of antibodies
2. Formation of multimolecular PF4/heparin/IgG complexes.
3. Platelet activation by binding of the multimolecular complexes to platelet FcγRIIa receptors
4. Release of PF4 from activated platelets and generation of platelet microparticles
5. Activation of monocytes and endothelial cells by the multimolecular complexes
6. Activation of coagulation resulting in massive thrombin generation

### 3.2.1 HIT Antigens

Most frequently, HIT antigens are found on PF4, a tetrameric member of the CXC chemokine subfamily, bound to heparin or other sulfated polysaccharides (Amiral et al. 1992). The epitopes are conformationally altered sites on PF4 resulting from its binding to heparin (Suh et al. 1998) and/or because of close approximation of PF4 tetramers by heparin charge neutralization (Greinacher et al. 2006). They are present on multimolecular clusters of PF4 and heparin that require an optimal stoichiometric ratio of PF4: heparin of 1:1–2:1 (Rauova et al. 2005). In addition, PF4 can bind to endothelial cells and platelets and also produce the antigen independently of the presence of heparin (Areapally et al. 2007; Rauova et al. 2006). Antigens formed by other heparin-binding proteins (interleukin 8, neutrophil activating protein-2) play a minor role in HIT.

The immunogenicity of the PF4/heparin complexes is influenced by their relative size, amount and stability and is thus dependent on the structural characteristics of the heparin or sulfated glycan, respectively. Ultralarge complexes of PF4 and UFH with molecular masses ($M_r$) > 670,000 are central to the pathogenesis of HIT (Rauova et al. 2005). LMWHs are less likely to trigger both antibodies and HIT, compared with UFH. The pentasaccharide anticoagulant, fondaparinux, does not form the antigens with PF4 recognized by HIT antibodies (Amiral et al. 1997), although the occurrence of HIT antibodies in orthopedic VTE prophylaxis in surgery patients under fondaparinux was similar to that in those receiving treatment with LMWH (Warkentin et al. 2005a).

### 3.2.2 HIT Immune Complexes

HIT antibodies bind to PF4 via their F(ab) domains. As serial PF4 molecules become aligned, several antibodies bind, leading to formation of large immune complexes (Greinacher et al. 2006). Among the IgG, IgM, and IgA antibodies recognizing the antigens, the IgG isotype antibodies are crucial for the development of clinical HIT (Warkentin et al. 2009). They can bind to platelet FcγIIa receptors resulting in platelet activation, whereas IgA and IgM antibodies have no Fc-parts enabling the binding to these receptors.

### 3.2.3 Platelet Activation and Thrombin Generation

Cross-linking of the platelet FcγIIa receptors by the immune complexes causes platelet activation. The activated platelets trigger a cascade of events that ultimately lead to activation of coagulation pathways resulting in thrombin generation and potentially development of TE (Qian et al. 2010; Tardy-Poncet et al. 2009).

Activated platelets release their α-granule proteins, including PF4, enabling formation of more multimolecular PF4-heparin complexes and setting up a vicious cycle of platelet activation. Moreover, activated platelets bind to fibrinogen, recruit other platelets, and begin to form platelet aggregates.

During shape change, procoagulant platelet-derived microparticles are released, providing a phospholipid surface for amplifying thrombin generation (Warkentin and Sheppard 1999).

The released PF4 also binds to endothelial cell heparan sulfate, forming local antigen complexes recognized by HIT antibodies (Blank et al. 2002). Finally, tissue factor expression on activated endothelial cells and monocytes further enhances thrombin generation (Arepally and Mayer 2001; Pouplard et al. 2001).

## 3.3 Clinical Presentation

### 3.3.1 Typical HIT

HIT is usually associated with a fall in platelet count of >50% (from the highest value after day 4 of heparin treatment) and often new thrombosis, typically occurring 5–14 days after start of prophylactic or therapeutic doses of heparin (Arepally and Ortel 2006). Thromboembolic complications predominantly affect the venous system (Greinacher 2009). Other complications include skin lesions including skin necrosis, limb artery thrombosis and acute limb ischemia, adrenal hemorrhagic necrosis (secondary to adrenal vein thrombosis), and post-intravenous heparin bolus anaphylactoid reactions (Warkentin 2006, 2007; Warkentin et al. 2005b). The more unusual a new thrombosis during heparin treatment seems, the more HIT should be considered.

This typical clinical presentation of HIT is consistent with the temporal features of antibody-mediated pathogenesis, which is initiated when heparin is given under circumstances that favor immunization (Greinacher et al. 2009; Warkentin 2007; Warkentin et al. 2009). One such setting is VTE prophylaxis in surgery (release of PF4 from activated platelets; perioperative inflammation). HIT antibodies first become detectable 4 days (median) after intra- or postoperative heparin administration. Two days later (median, day 6), the platelet count begins to decrease, and after 2 further days (median, day 8), a >50% platelet count fall is reached (Warkentin et al. 2009). On average, HIT-associated thrombosis occurs close to the 50% platelet count decrease (median, day 10–11), but it can even precede it by 1–2 days (~30% of thrombi) in a subset of patients (Greinacher et al. 2005). Antibody levels typically peak between days 10 and 14 and then were found to decline even with continued heparin administration (Greinacher et al. 2009; Warkentin et al. 2009).

### 3.3.2 Platelet Counts

The mean platelet count in patients with HIT is 60,000/μL and ranges from 15,000 to 150,000/μl in 90% of patients (Warkentin 2007). Patients with platelet counts <15,000/μl rarely have HIT. In these patients, other reasons for the decrease in platelet count are much more probable, e.g., other drug-induced immune thrombocytopenias, e.g., GPIIb/IIIa inhibitor induced thrombocytopenia, or post-transfusion purpura. However, even in the minority of HIT patients with very low platelet counts, thrombosis, rather than bleeding, dominates.

On the other hand, there are patients with HIT-associated thrombosis having a platelet count nadir of >150,000/μL (e.g., patients with myeloproliferative disorders), which makes the term "thrombocytopenia" somewhat misleading. Comprehensive studies indicated that a "platelet count fall >50% from the highest value after day 4 of heparin treatment" is a much better characteristic of HIT.

### 3.3.3 Platelet Count Monitoring

The time course of HIT has implications for the mandatory monitoring of platelet counts during heparin therapy. The platelet count should always be determined before start of heparin administration. Monitoring of platelet counts at days 5, 7 and 9 is then sufficient to recognize HIT in the vast majority of patients (Hinz et al. 2009).

However, it has to be considered that especially in surgical patients it is not appropriate to compare the platelet count value at the time of suspicion of HIT with the baseline value, because surgical patients typically show a reactive increase in platelet counts with an overshooting of platelet counts in the second week after major surgery (Greinacher et al. 2010a). Comparison of the actual platelet count with the baseline value will underestimate the magnitude of the platelet count fall.

Moreover, certain patients, like those undergoing cardiac surgery, can have platelet counts <100,000/μl post-surgery for several days. Here, persistence of the low platelet counts beyond day 5–10 has to be rated as high probability of HIT.

### 3.3.4 Rapid-Onset HIT

A platelet count fall within the first 4 days of heparin treatment is usually not HIT, unless the patient has been preimmunized and already has circulating anti-PF4/heparin antibodies (Lubenow et al. 2002; Warkentin and Kelton 2001). In these preimmunized patients, the platelet count typically decreases within the first hours after start of heparin.

Heparin-dependent antibodies tend to rapidly decrease in titers and become undetectable in >90% of patients within 100 days (Kelton and Warkentin 2008). In fact, in most patients, the heparin-dependent, platelet-activating antibodies disappear even sooner, usually within one month. Thus, rapid-onset HIT should be expected primarily in patients who did receive heparin within the preceding 30 days.

In patients with recent heparin exposure, the platelet count should be determined before start of heparin and 12–24 h thereafter to recognize rapid-onset HIT (Warkentin et al. 2008a).

### 3.3.5 Delayed-Onset HIT and Spontaneous HIT

In a few cases, HIT begins several days after discontinuation of heparin or persists for several weeks even though heparin administration has been stopped. Sera from patients with "delayed-onset HIT" activate platelets strongly in vitro in the absence of added heparin. The most likely explanation is that these patients develop autoantibodies, which recognize PF4 bound to platelet surface GAGs (Alban and Greinacher 2007; Rauova et al. 2006).

There is increasing awareness of a prothrombotic syndrome resembling HIT in patients despite the absence of preceding heparin therapy ("spontaneous HIT") (Suzuki et al. 1997; Warkentin et al. 2008b). Circumstances other than heparin use can trigger HIT-like antibodies such as preceding infections, inflammatory events or concomitant antiphospholipid antibodies. This is thought to lead to a spontaneous disorder that closely mimics HIT, further supporting the autoimmune disorder nature of this adverse drug reaction (Alberio 2008; Cines et al. 2007; Greinacher et al. 2009).

## *3.4 Frequency*

Due to the widespread exposure of patients to heparins, the absolute number of patients affected by HIT is the highest of all immune-mediated, drug-induced blood cell disorders.

### 3.4.1 Large Variations of HIT Frequency

The frequency of HIT varies widely, ranging from a negligible risk (e.g., LMWH given during pregnancy) to a high risk of 5–10% (e.g., females receiving UFH prophylaxis for 2 weeks after orthopedic surgery or during therapeutic-dose UFH post-implantation of a ventricular assist device) (Warkentin 2010).

The risk of HIT is higher in VTE prophylaxis in surgery than in VTE prophylaxis in medical patients. Analysis of the database of the National Hospital Discharge Survey including >10 million patients revealed an incidence of only ~0.5% HIT among patients receiving heparin prophylaxis in medical indications or VTE therapy (Stein et al. 2009). Similarly, analysis of three prospective studies on critically ill patients found an incidence of HIT of 0.4%, although significant thrombocytopenia occurred in 9.5% of the patients (Crowther et al. 2010).

Other non-drug risk factors that appear to influence the risk of HIT include gender, age, body mass index, trauma severity, and duration and timing of heparin prophylaxis: The incidence of HIT is higher in females than in males (female vs. male, OR 1,5-2), but occurs rarely in pregnant women and women following delivery (Warkentin 2010). HIT is generally rare among patients aged <40 years. According to a recent analysis, postoperative patients with higher body mass index are more likely to form HIT antibodies during fixed-dose prophylaxis with LMWH (Warkentin et al. 2010). Another study showed that the greater the degree of trauma and thus the release of PF4, the more likely the patient went on to develop HIT antibodies and clinical HIT (Lubenow et al. 2010). This may also explain why the HIT risk associated with heparins for VTE prophylaxis is much higher in orthopedic surgery patients than in medical patients (Greinacher 2006). Furthermore, the risk of HIT is more duration-related than dose-related, and higher with UFH when used for an extended duration. Regarding the timing, starting prophylaxis post-surgery was found to result in a higher frequency of antibody formation than starting pre-surgery in patients undergoing knee or hip replacement VTE prophylaxis in surgery, but the opposite occurred with hip fracture surgery, that is, antibody formation was more frequent when prophylaxis was given pre-surgery (Warkentin et al. 2010).

These observations are in line with the experimental findings on the important role of surface PF4 antigenic complexes in the etiology of HIT (Rauova et al. 2006). In addition, they indicate that the concentrations of both PF4 and heparins and especially their ratio influence the immunization risk, which fits to the stoichiometry concept, that is, formation of immunogenic complex requires certain optimal ratios of PF4 and heparin (Greinacher et al. 2008).

### 3.4.2 HIT Due to UFH vs. LMWH

According to a meta-analysis of prospective clinical trials (mostly orthopedic surgery studies), the incidence of clinical HIT in orthopedic surgery was about 10 times lower under LMWH than under UFH (OR, 0.1; 95% CI, 0.03–0.33; $P < 0.001$) (Martel et al. 2005). The absolute risk for clinical HIT with LMWH was 0.2%, and with UFH the risk was 2.6%. However, the occurrence of HIT antibodies in orthopedic surgery patients receiving heparin prophylaxis is much higher and differs less between UFH (~30%, IgG ~15%) and LWMH (~15%, IgG ~7%) (Warkentin et al. 2005b, c).

Compared with the considerably higher HIT risk associated with UFH in VTE prophylaxis, VTE therapy resulted in smaller differences of the incidence between UFH and LMWH (Stein et al. 2009).

## *3.5 Diagnosis*

Since the fall in platelet count in the presence of HIT antibodies is the classical feature of HIT, routine platelet count monitoring during heparin administration, as well as the detection of antibodies, are the two central points of HIT diagnosis. Meanwhile, there are distinct recommendations for platelet monitoring depending on the clinical situation and the use of either UFH and LMWH (Warkentin et al. 2008a).

Notably, screening for subclinical HIT antibodies should not be performed (Warkentin et al. 2008a). This suggestion is based on the "iceberg" model of HIT, which means that the visible component of the iceberg represents clinically evident features of HIT, whereas its mass (below the waterline) corresponds to the entire spectrum of HIT antibodies with only a certain part being potentially pathogenic (Kelton and Warkentin 2008).

For that reason, the clinical likelihood of HIT should be assessed before laboratory testing for HIT antibodies. Assignment of pretest probability is particularly worthwhile in clinical decision making in the case of a reduction in platelet count in the absence of other sequelae of HIT, such as thromboembolic disease (Warkentin 2003).

In addition, patients suspected to have HIT should be investigated for lower-limb DVT by duplex ultrasonography (Warkentin 2006).

### 3.5.1 The 4Ts Score

As HIT is a complex disorder where frequently the diagnosis must be made against a background of equally complex and distracting medical conditions, it is useful to have a systematic approach consisting of both clinical and laboratory parameters to either rule in or rule out the diagnosis (Warkentin and Heddle 2003). For this, the so-called 4T's score, which summarizes the typical features of HIT, was developed (Warkentin and Heddle 2003) and validated in several studies (Bryant et al. 2008; Denys and Devreese 2009; Denys et al. 2008; Lo et al. 2006; Pouplard et al. 2007) (Table 4). It gives points (0, 1, or 2) according to four clinical criteria:

1. *T*hrombocytopenia
2. *T*iming of thrombocytopenia
3. *T*hrombosis (or other clinical sequelae),
4. o*T*her reasons for thrombocytopenia.

The determined score indicates the clinical probability of HIT, which is low (0–3 points), intermediate (4–6 points), or high (7–8 points).

For patients with a score <4, the probability of having HIT is <5% and testing for HIT antibodies is not recommended. A laboratory test for HIT antibodies should rather be performed in patients with an intermediate score of 4–6, while for patients with a high clinical probability (score 7–8) a direct switch of treatment to an alternative anticoagulant is recommended and laboratory testing should be performed in retrospect.

Increasing knowledge of the complex HIT syndrome has successively resulted in refinement of the scoring system increasing its utility in the estimation of the probability of HIT. These changes include (Warkentin and Linkins 2010):

1. Platelet declines occurring within 3 days of surgery are scored a maximum of 1 point, even if the platelet count fall exceeds 50%.
2. For assessing rapid-onset HIT (i.e., platelet count fall within 1 day of reinitiating heparin), an exposure to heparin within the past 5–30 days counts as 2 points, whereas an exposure within the past 31–100 days counts only as 1 point.
3. Adrenal hemorrhage counts as 2 points (because of its strong association with adrenal vein thrombosis secondary to HIT).
4. Clarity has been added to the sorts of alternative non-HIT diagnoses ("oTher") that would lead to scores of 1 or 0 points, including information on helping to judge the probability of a (non-HIT) drug-induced immune thrombocytopenic purpura (D-ITP) being present.
5. Concerning the "Timing", the beginning of the platelet count fall alleged to be HIT should be used, not the point at which the platelet count decline reaches some arbitrary threshold indicating thrombocytopenia (e.g., 50% decline).

**Table 4** The "4Ts Score" in the latest version (Warkentin and Linkins 2010)

| 4Ts | 2 points | 1 point | 0 points |
|---|---|---|---|
| *Thrombocytopenia*[a] | • Platelet count fall >50% AND platelet nadir ≥20 AND no surgery within preceding 3 days | • Platelet count fall 30–50%<br>• Platelet nadir 10–19/nL<br>• Platelet count fall > 50% and surgery within preceding 3 days | • Platelet count fall <30%<br>• Platelet nadir <10/nL |
| *Timing of platelet count fall or thrombosis*[b] | • Clear platelet fall days 5–10<br>• Platelet fall ≤1 day AND prior heparin exposure within 30 days | • Consistent with platelet fall days 5–10 but not clear (e.g. missing counts)<br>• Platelet fall ≤1 day AND prior heparin exposure within 31-100 days<br>• Platelet fall >10 days | • Platelet count fall <4 days AND no heparin exposure within 100 days |
| *Thrombosis sequelae or other* | • Confirmed new thrombosis<br>• Skin necrosis at injection site<br>• Acute systemic reaction to i.v. UFH<br>• Adrenal hemorrhage | • Suspected thrombosis (not proven)<br>• Non-necrotizing (erythematous) skin lesions at UFH injection site<br>• Progressive or recurrent thrombosis | • Thrombosis not suspected |
| *OTher causes for thrombocytopenia* | • No alternative explanation | Possible other cause is present:<br>• Sepsis without proven microbial source<br>• Thrombocytopenia associated with initiation of ventilator<br>• Other | Probable other cause is present:<br>• Within 72 h of surgery<br>• Non-necrotizing (erythematous) skin lesions at LMWH injection site (DTH)<br>• Confirmed bacteremia/fungemia<br>• Chemotherapy or radiation within past 20 days<br>• Disseminated intravascular coagulation (DIC) due to non-HIT cause<br>• Posttransfusion purpura (PTP)<br>• Thrombotic thrombocytopenic purpura (TTP)<br>• Platelet count <20/nL AND given a drug causing D-ITP |

*DTH* delayed-type hypersensitivity, *DIC* disseminated intravascular coagulation, *D-ITP* drug-induced immune thrombocytopenia (D-ITP)

[a]% decrease from the highest platelet count within the decline

[b]Day 0 = first day of most recent heparin exposure. In some circumstances, "Timing" should be judged on the onset of "other sequelae"

Based on recent results on heparin-induced skin reactions being one of the other sequelae of HIT (Schindewolf et al. 2010a), two further modifications were proposed (Warkentin and Linkins 2010): In some circumstances, it may be appropriate to restrict the scoring of "Timing" not on thrombocytopenia, but to consider other clinical events such as skin lesions at heparin injection sites. Furthermore, non-necrotizing skin lesions at LMWH injections sites should be judged as "other cause," whereas necrotizing skin lesions are rated as "other clinical sequelae" of HIT.

### 3.5.2 Laboratory Testing

Besides application of the scoring system, testing for HIT antibodies is the second most important measurement to exclude or to confirm the diagnosis of HIT in patients with clinically suspected HIT. Two groups of serological assays are available. The first group are the antigen tests. These are enzyme immunoassays (EIA), particle gel immunoassays or particle immunofiltration assays, which recognize binding of antibodies to PF4/polyanion complexes. The advantage of antigen assays is their high negative predictive value. The second group are the platelet activation assays, which detect heparin-dependent activation of platelets by patient serum. In contrast to antigen assays, these functional HIT assays reflect in vitro the pathogenic cascade of HIT one step further downstream than the antigen tests. Thus, they have a higher positive predictive value for HIT. In fact, only about 50% of patients testing positive in the antigen assays also test positive in the functional assays (Greinacher et al. 2007).

*PF4/heparin antigen assays*. Antigen assays detect the antibodies based upon binding to PF4/heparin complexes or PF4/polyvinyl sulfonate complexes bound to a solid phase. There are at least three commercially available EIAs: PF4 Enhanced X-HAT 45 (Genetics Testing Institute), Asserachrom PF4 (Diagnostica Stago), Zymunt-HIA (Hyphen BioMed). They detect anti-PF4/heparin antibodies of the IgG/IgA/IgM class, or IgG only, depending on the secondary antibodies used. Although these assays differ considerably in their structure, including the source of PF4, they give only slightly different results and show an overall acceptable concordance, especially with high titer HIT antibodies (Greinacher et al. 2010a).

The high sensitivity of the EIAs implies a high negative predictive value. However, a positive test does not necessarily indicate clinical HIT. Given that IgA and IgM antibodies are unlikely to cause clinical HIT (see Sect. 3.2.1), the specificity of EIAs is increased by detecting only IgG isoforms (Greinacher et al. 2007). Another strategy to improve the positive predictive value of EIAs is to consider the measured optical density (OD) (expressed as normalized OD), as the OD has been shown to correlate with the likelihood of antibodies to be clinically relevant (Greinacher et al. 2010b). If an IgG EIA is weakly positive (OD $< 1.0$), the antibodies are most likely non-platelet-activating. A confirmatory step using high concentrations of heparin should be performed; if reactivity is not inhibited, HIT is very unlikely and heparin can be maintained (Greinacher 2009). An IgG EIA

(OD > 1.0) indicates an increased risk for platelet-activating antibodies. These sera should ideally be assessed by a washed platelet activation assay, having a higher positive predictive value.

*Point-of-care antigen assays*. Currently, two assays are offered for the rapid detection of HIT antibodies. The particle gel immunoassay (ID-PaGIA heparin/PF4 antibody test, DiaMed) uses PF4/heparin complexes bound to red, high-density polystyrene beads, which are agglutinated by anti-PF4/heparin antibodies (IgG, IgA, and IgM). It is at best semiquantitative and has been shown to produce false negative results in a few patients with high probability of HIT (Pouplard et al. 2007), so that in case of high clinical suspicion, a negative result should prompt testing in another test system. However, its biggest advantage is its fast turn-around time of less than 15 min. In contrast, the second point-of-care assay, the particle immunofiltration assay, needs to be improved and is currently not recommended (Greinacher et al. 2010a).

*Platelet activation assays*. Functional assays based on whole blood or platelet-rich plasma of healthy donors are widely established in clinical labs, but have a sensitivity of only ~30% compared with washed platelet assays and sometimes give false positive results in critically ill patients (Greinacher et al. 2010a). If possible, a washed platelet assay should therefore be performed, that is either the serotonin release assay (SRA), which detects the release of radioactive serotonin from labeled donor platelets as a platelet-activating endpoint, or the heparin-induced platelet activation assay (HIPA). Both assays are performed on microtiter plates and allow for several quality controls. Demonstration of platelet-activating antibodies makes HIT very likely. A negative functional assay makes HIT unlikely and heparin can be maintained or restarted, respectively.

## 3.6 *Therapy*

If there is high clinical suspicion for HIT, stopping heparin alone is insufficient. To prevent new thrombosis, non-heparin anticoagulant therapy is required. Particularly during the intense period of HIT-associated hypercoagulability, which is seen ~7–14 days after the immunizing heparin exposure, effective anticoagulation is needed to prevent microvascular thrombosis from contributing to tissue ischemic necrosis that often occurs in the setting of preceding macrovascular thrombosis.

### 3.6.1 Alternative Anticoagulants

Currently, three drugs are approved for anticoagulation in HIT (Warkentin et al. 2008)

- Lepirudin (recombinant hirudin, direct thrombin inhibitor (DTI))
- Argatroban (synthetic small-molecule DTI)
- Danaparoid (glycosaminoglycan mixture).

**Table 5** Six treatment principles for management of confirmed or strongly suspected HIT

| | |
|---|---|
| Two do's | • Stop heparin, including heparin "flushes"<br>• Start alternative, non-heparin and non-VKA anticoagulant, usually in therapeutic doses |
| Two don'ts | • Do not initiate VKA therapy before substantial platelet count recovery is apparent (give intravenous vitamin K, if VKA has already been given when HIT is recognized[a])<br>• Do not administer platelet transfusions |
| Two diagnostics | • Test for HIT antibodies<br>• Investigate for lower-limb DVT (duplex ultrasonography) as the most common complication of HIT |

[a]There are two reasons to give intravenous vitamin K when HIT is recognized after VKA has been given. First, it may reduce the risk of VKA-induced microthrombosis (venous limb gangrene and skin necrosis syndromes); second, it reduces the risk of underdosing of direct thrombin inhibitor therapy because of prolongation of the APTT by VKA.

Rational therapies for HIT are also

- Bivalirudin (synthetic hirudin-analogue peptide, DTI), and
- Fondaparinux (synthetic pentasaccharide, selective Factor Xa inhibitor fondaparinux).

These alternative anticoagulants differ considerably in their chemical characteristics, their pharmacodynamics and -kinetics and various aspects of their clinical application (Table 6) (Alban 2008b). Their application in HIT patients including data on their efficacy and safety are described in detail elsewhere (Hursting and Soffer 2009; Warkentin 2008, 2010; Warkentin et al. 2008a).

*Long- versus short-acting anticoagulants*. The five drugs can be classified into two groups: the short-acting, selective DTIs (lepirudin, argatroban, bivalirudin), and the long-acting sulfated glycans danaparoid and fondaparinux. There are differences with specific implications for HIT between the two groups. According to Warkentin, long-acting inhibitors of factor Xa, danaparoid and fondaparinux, have several advantages over DTIs for management of HIT (Warkentin 2010):

- Their long half-life is considered an advantage by avoiding potential for rebound hypercoagulability, which is not excluded with the short-acting reversible DTIs argatroban and bivalirudin.
- Danaparoid and fondaparinux do not interfere with the protein C pathway, whereas the DTIs may inhibit the thrombin-mediated protein C activation and so impair an important anticoagulant mechanism.
- Although danaparoid and fondaparinux do not have to be continuously monitored for dose adjustment, specific assays are available to permit accurate determination of drug levels. In contrast, the dosage of DTIs is usually monitored by APTT, which is influenced by many parameters. A major problem is inappropriate dose reduction of DTI, when low prothrombin levels (e.g., because of VKA therapy) result in falsely prolonged APTTs during DTI treatment (Lindhoff-Last et al. 2000).
- Another advantage of the two long-acting anticoagulants is convenience of their s.c. fixed-dose regimens (for danaparoid after the acute phase), allowing for

**Table 6** Comparison of anticoagulants available for treatment of HIT patients

| | Danaparoid | Fondaparinux | Lepirudin | Bivalirudin | Argatroban |
|---|---|---|---|---|---|
| *HIT Indications* | | | | | |
| EMA | Prophylaxis, and therapy | Not approved for HIT[a] | Therapy[b] | Not approved for HIT | Prophylaxis, and therapy |
| FDA | Not approved for HIT | Not approved for HIT[a] | Therapy[b] | PCI | Prophylaxis, therapy, and PCI[c] |
| *Characteristics of the drug substance* | | | | | |
| Chemistry | Partially degraded GAGs mixture isolated from porcine mucosa | Synthetic, chemical defined sulfated pentasaccharide | Recombinant protein | Synthetic peptide | Synthetic arginine derivative |
| Type of anticoagulant | Multivalent activities including AT-dependent aXa activity | Selective AT-dependent Factor Xa inhibitor | Bivalent direct thrombin inhibitor | Bivalent direct thrombin inhibitor | Univalent direct thrombin inhibitor |
| Quantification | aXa-units/mg | Gravimetric (mg) | Gravimetric (mg) | Gravimetric (mg) | Gravimetric (mg) |
| *Pharmacokintetics* | | | | | |
| T(max) | 4–5 h (s.c.) (aXa activity)[d] | 2 h (s.c.) | Immediately | Immediately | Immediately |
| T(steady state) | 4–5 days (aXa activity)[4] | 3–4 days | ~4 h | n.d. | 1–2 h |
| Terminal half-Life | ~25 h (aXa activity)[4] | 17–21 h | ~ 80 min | ~25 min | ~52 min |
| Metabolism | Unknown | Not metabolized | Metabolized in kidneys | Proteolysis | Metabolized in liver |
| Exkcretion | Partly unchanged in urine | Unchanged in urine | 45% in urine (35% unchanged)[e] | 20% unchanged in urine | 65% in feces, 22% in urine |
| Influence on clearance | Renal function | Renal function | Renal function[f], antibodies | Renal function | Liver function |
| *Posology, method of administration* | | | | | |
| Dosage regime | Various i.v./s.c. regimes, partly aXa activity-adjusted[g] | Fixed-dose regime | BW-adjusted and aPTT-adjusted | BW-adjusted | BW-adjusted and aPTT-adjusted |
| Usual dosage | Variable[g] | *[Once daily s.c. 2.5 mg (propylaxis), 7.5 mg (therapy)]*[a] | Usually no bolus, ≤0.1 mg/kg/h | 0.75 mg/kg i.v. bolus, 1.75 mg/kg/h | No bolus, ≤2 μg/kg/min |

| | | | | | |
|---|---|---|---|---|---|
| Monitoring | aXa activity in therapy and in special patients | No, but possible by aXa activity | APTT (target ratio 1.5–2.5) | ACT (365 ± 100 s) only initially | APTT (target ratio 1.5–3.0) |
| Dose reduction | Renal impairment, if necessary | *(CrCl 20–50 ml/min: 1.5 mg CrCl <20 ml/min: contraindicated)*[a] | CrCl 15–60 ml/min: 15–50% CrCl <15 ml/min: contraindicated | CrCl 30–60 ml/min: 15–50% CrCl <30 ml/min: contraindicated | Hepatic failure: 0.5 µg/kg/min Severe: contraindicated |
| To attend | ~5% serological and ~3% clinical cross-reactivity with HIT antibodies | Occurence of HIT-antibodies, but no cross-reactivity | Narrow therapeutic window Antibody formation (40%) Anaphylaxis (0.01–0.1%) | | Falsely positive increase of INR at concomitant use with VKA |

[a]According to the approved use, fondaparinux can be applied in patients with previous HIT. Since the use in acute HIT is based on case reports and small clinical trials, appropriate dosing is unkown.

[b]Indication: Anticoagulation in adult patients with HIT and thromboembolic disease mandating parenteral antithrombotic therapy

[c]Indication: Anticoagulation in patients with or at risk for HIT undergoing PCI

[d]Only effect kinetics (aXa activity) known

[e]% means percentage of the applied dose

[f]The bleeding risk of lepirudin resulting from accumulation in patients with renal insufficiency is higher than that of the other alternative anticoagulants

[g]For dosage of danaparoid, see "Summary of Product Characteristics" and 8th ACPP Practice Guidelines on HIT (Warkentin et al. 2008a)

prolonged application. In this way, the risk of early transition to VKA may be avoided. Contrary to DTIs (especially argatroban), they do not interfere with the INR determination complicating the transition to VKA
- Finally, there are certain drug-specific favorable characteristics, i.e., the inhibitory effect of danaparoid on platelet activation by HIT antibodies (see below) and the extensive experience with fondaparinux in a wide range of approved indications, respectively.

*Danaparoid versus fondaparinux.* Despite their similarities (see above), there are pronounced differences between danaparoid and fondaparinux, which illustrate two strategies for glycan-based non-heparin anticoagulants applicable in HIT (Alban and Greinacher 2007). Fondaparinux is a synthetic, chemically defined, highly sulfated pentasaccharide, which selectively binds to AT with high affinity and in this way inhibits Factor Xa, whereas danaparoid consists of a complex mixture of partially degraded (molecular mass 4,000–7,000), low-sulfated GAGs isolated from porcine mucosa. Notably, only ~4% of the danaparoid molecules have AT-mediated aXa activity, which is used for dosing and monitoring in default of other options (Alban 2008a). Moreover, fondaparinux does not exhibit any cross-reactivity with HIT antibodies and so does not form pathogenic immune complexes (Chong and Chong 2010; Greinacher et al. 2008; Lobo et al. 2008). In contrast, danaparoid shows certain cross-reactivity but, at therapeutic concentrations, it was found to have the intriguing property to inhibit HIT antibodies-induced platelet activation in a large percentage of HIT sera (Krauel et al. 2008). Specifically, danaparoid was found to displace PF4/heparin complexes from the platelets, to reduce the PF4/heparin complex size and to inhibit HIT antibody binding to PF4/heparin complexes.

*Bleeding complications.* All alternative anticoagulants confer significant risk for major bleeding (0.8–1.25% per treatment day) and no specific antidotes are available (Greinacher et al. 2010a). In only 5% of patients clinically suspected to have HIT, the diagnosis is confirmed (even fewer in ICU patients). Therefore, in patients with low/moderate clinical probability for HIT, some experts use prophylactic-dose alternative anticoagulation, pending results of laboratory testing (Greinacher et al. 2010a), others propose fondaparinux in prophylactic dosage (2.5 mg s.c. once daily (Warkentin 2010).

### 3.6.2 VKA are Contraindicated During Acute HIT

A crucial tenet is avoiding VKA during acute HIT, which includes postponing long-term VKA therapy until the platelet count has substantially and stably recovered (i.e., usually to at least 150,000/μL). Acute HIT is a major risk factor for so-called coumarin necrosis, which can manifest either as venous limb gangrene or as classic skin necrosis and at worst can require limb amputation (Warkentin et al. 1997). The risk of coumarin necrosis in HIT is approximately 5% to 10%, a frequency much greater than the background risk (overall, <0.01%).

The pathogenesis is microthrombosis due to early depletion of the endogenous anticoagulant protein C by VKA in the setting of increased thrombin generation from HIT.

If a diagnosis of HIT is made after VKA has been given, intravenous application of vitamin K is recommended (Warkentin et al. 2008a). On the one hand, vitamin K may reduce the risk of VKA-induced microthrombosis; on the other hand, it reduces the risk of underdosing of DTI therapy because of prolongation of the APTT and thus simulating DTIs by VKA (Warkentin 2010).

### 3.6.3 Heparin Exposure for Patients with Previous HIT

For patients with a previous history of HIT who have an important indication for heparin (e.g., cardiac or vascular surgery), it is recommended to use heparin, provided that HIT antibodies are no longer detectable, or (in emergencies) that sufficient time has passed so that clinically significant levels of HIT antibodies are not likely to be present (Warkentin et al. 2008a). This advice is based on the experience that a previous history of HIT does not appear to be a risk factor for a higher frequency of antibody formation during subsequent exposure, and in any event such antibodies would not regenerate in high levels for at least 5 days, at which time a non-heparin anticoagulant can be used (Warkentin 2010).

## 4 Heparin-Induced Skin Lesions

Heparins can induce skin lesions, which are immunologically mediated adverse effects such as HIT. Traditionally, three types of cutaneous reactions were distinguished (Bick and Frenkel 1999; Hirsh et al. 1998):

1. Urticarial lesions (immediate hypersensitivity reactions)
2. Erythematous papules and plaques (delayed-type hypersensitivity reactions)
3. Skin necrosis (most serious dermal reactions)

Typically, skin lesions develop >5 days following s.c. administration and occur at the sites of heparin injection, although some reports have implicated skin necrosis at sites remote from heparin injection (Hirsh et al. 1998).

Meanwhile, further mechanisms have been recognized as cause of skin lesions (e.g., pustulosis (Komericki et al. 2007)) with HIT as the most important one from a clinical point of view (Jappe 2006; Warkentin 2006; Warkentin et al. 2005b). But the vast majority of skin lesions are caused by delayed-type hypersensitivity reactions and their incidence is considerably higher than initially thought (Schindewolf et al. 2009). Since heparin-induced skin lesions can represent potentially life-threatening adverse events, accurate diagnosis and identification of alternative anticoagulants is mandatory.

## 4.1 Types of Skin Lesions

### 4.1.1 Delayed-Type (Type IV) Hypersensitivity Reactions

*Frequency*. Cutaneous delayed-type (type IV) hypersensitivity reactions (DTHR) after s.c. UFH have been reported from as early as in 1952 by Plancherel. After a long period without new cases, a continuously increasing number of reports on DTHR were published since the 1980s (Bircher et al. 2006; Jappe 2006; Ludwig et al. 2006). In a prospective study of 320 enrolled patients receiving s.c. UFH or LMWHs, 24 patients (7.5%, 95% CI 4.7%–10.6%) had heparin-induced skin lesions and DTHR were identified as cause in all 24 patients (Schindewolf et al. 2009). Consequently, heparin-induced skin lesions and particularly DTHR are common adverse reactions of heparins.

*Time course*. The DTHR to heparin occur after a sensitization or induction period of at least 7–10 days, but often after weeks and sometimes even after 5 months (Bircher et al. 1995, 2006; Jappe 2006). The latency period in previously sensitized individuals lasts typically from 1 to several days. Due to the widespread use of heparins, it should generally be considered that patients receiving heparins can be already sensitized. This is reflected by the analysis of Ludwig et al. (2006), where ~60% of the DTHR to heparin were noticed within one week of initiation of s.c. anticoagulant treatment. If treated accordingly, lesions usually resolve within a week.

*Characteristics*. Based on the analysis of 223 patients with heparin-induced DTHR described in the literature, patients present with itching, sometimes infiltrated, and blistering erythemas at the injection sites of heparins (Ludwig et al. 2006). The size of individual lesions may vary, ranging from very small papules to rather large erythematous plaques with papulovesicles and bullae. Seldom, generalized eczematous reactions may result.

Rarely reported types of heparin-induced DTHR include generalized maculopapular, sometimes bullous and febrile exanthemas (DRESS), Lyell syndrome, and flexural exanthemas (so-called Baboon Syndrome or SDRIFE) (Bircher et al. 2006).

*Differential diagnosis*. The main differential diagnoses include local hematoma, infections, such as erysipelas, contact dermatitis (e.g., from disinfectants), and particularly HIT-associated skin necrosis (see Sect. 4.1.4).

*Risk factors*. Evaluation of cases described in the literature revealed female gender (>90% females), obesity, prolonged heparin exposure (e.g., during pregnancy) as patient-related risk factors for the development of heparin-induced DTHR (Bircher et al. 2006; Ludwig et al. 2006). In the prospective study by Schindewolf et al. (Schindewolf and Ludwig 2010; Schindewolf et al. 2009), these risk factors for heparin-induced skin lesions were verified: body mass index >25 (OR 4.6, 95% CI 1.7–15.3), duration of heparin therapy >9 days (OR 5.9, 95% CI 1.9–26.3), and female sex (OR 3.0, 95% CI 1.1–8.8). The gender imbalance seems to be even greater than that for HIT (OR 1.5–2) (Warkentin et al. 2006). Hormonal factors, longer persistence of heparins in subcutaneous adipose tissue or a relationship to the lipase activity of heparins have been proposed to explain this striking gender difference (Hohenstein et al. 2004).

In addition, drug-related risk factors have to be considered. Induction of DTHR was found to increase with the chain length of the applied GAG, i.e., UFH vs. LMWH > danaparoid > fondaparinux (Ludwig et al. 2005; Schindewolf et al. 2010b). Concerning the role of the molecular mass within the group of LMWHs, other structural characteristics seem to be at least as important (Grims et al. 2007; Ludwig et al. 2008; Schindewolf et al. 2009). Recently, a prospective evaluation of 231 patients treated with fondaparinux, confirmed the low allergenic potential of the pentasaccharide (Schindewolf et al. 2010b). The incidence of cutaneous DTHR (0.4%; 95% CI, 0.01–2.4%) was almost 20 times lower compared to that with commonly used heparins.

### 4.1.2 Immediate (Type I, Allergic) Hypersensitivity Reactions

Urticarial, often pruritic, skin lesions as a cutaneous symptom of immediate hypersensitivity reactions at the site of injection occur very rarely (Harr et al. 2006). This is probably due to the continuously improved pharmaceutical quality of heparins. In former times, such reactions have been attributed to contaminants such as proteins within the heparin preparation (Hirsh et al. 1998). However, pharmaceutical additives such as preservatives in multidose dosage forms, sodium sulfite and sodium metabisulfite, can provoke allergic reactions as well. In contrast to DTHR, immediate hypersensitivity reactions are treatable with H1 receptor antagonists and show a positive result in skin prick tests.

### 4.1.3 Skin Necrosis

Skin necrosis is a rare, but the most serious form of dermal reaction induced by both UFH and LMWHs. Early recognition is vital as continued treatment with heparin may precipitate life-threatening complications in other organ systems.

Middle-aged women with a history of thrombotic disease are typically affected (Drew et al. 1999). In contrast to DTHR, both s.c. and i.v. administration have led to the condition. Painful areas of erythema and inflammation of varying size appear up to 13 days after commencing heparin and progress to become necrotic (Drew et al. 1999). The lesions occur locally or distant from the injection site, i.e., on the extremities, abdominal wall, nose, or on the dorsum of the hand (Bick and Frenkel 1999). Clinically, it is not distinguishable from coumarin-induced necrosis.

The etiology of heparin-induced skin necrosis is unclear, but various mechanisms may be involved including allergic reactions, local trauma, localized Arthus type hypersensitivity reactions, and allergic vasculitis (Drew et al. 1999; Handschin et al. 2005a; Jappe 2006). Most frequently, however, skin necrosis seems to occur as part of the HIT syndrome (see Sect. 4.1.4). HIT antibodies were detected in 9 of 11 patients, who experienced skin necrosis induced by LMWH (Handschin et al. 2005a). Moreover, necrosis resulting from occlusion of dermal vessels due to the prothrombotic HIT-associated immune complexes seems plausible. Since many of the cases were published at times when the awareness of HIT was low, the question arises whether there is really any non-HIT explanation for skin necrosis at heparin injection sites.

### 4.1.4 HIT-Associated Skin Lesions

HIT is caused by platelet-activating, heparin-dependent IgG antibodies and predisposes the patient to thrombosis. However, some patients with HIT antibodies can experience other unusual clinical events including skin lesions at heparin injection sites irrespective of whether thrombocytopenia occurs (Warkentin 1996, 2006, 2007; Warkentin et al. 2005b).

In a nested case–control study, four of 20 patients with HIT IgG, but none of 100 control patients, developed inflammatory ($n = 3$) or necrotic skin lesions ($n = 1$), respectively, at heparin injection sites on postoperative day 7 or later (Warkentin et al. 2005b). The patient with necrosis additionally developed thrombocytopenia and thrombosis.

Therefore, skin reactions in the presence of HIT antibodies signal an acute need to discontinue heparin therapy and select an appropriate alternative agent (Warkentin 1996; Warkentin et al. 2005b). Given the rare occurrence and the severity of skin necrosis as well as the very frequent occurrence of HIT antibodies in these patients (Handschin et al. 2005a), early switch of anticoagulation is certainly warranted.

Although the initial clinical presentation of HIT-associated skin lesions is similar to that of DTHR (see Sect. 4.1.1), immediate discontinuation of heparin in the case of non-necrotizing, erythematous skin lesions at LMWH injection sites, however, seems questionable (Schindewolf et al. 2010a). In a recent cohort study evaluating 87 consecutive patients with heparin-induced skin lesions (85 LMWH), all the lesions were caused by DTHR, none was due to HIT. Even the cutaneous reaction in one patient with concomitant HIT could be classified histologically as DTHR. As a result, non-necrotizing skin lesion induced by LMHW does not seem to be strongly associated with HIT and might rather be due to DTHR than due to microvascular thrombosis (Schindewolf et al. 2010a). As a consequence, a correspondingly refined pretest probability 4Ts score for HIT has been proposed (Warkentin and Linkins 2010) (see Sect. 3.5.1).

## 4.2 *Diagnosis*

Due to the high probability of developing cutaneous DTHR to heparins, routine inspection of heparin injection sites, especially for patients with the identified risk factors, is recommended (Schindewolf et al. 2010a).

*Differentiation between DTHR and HIT-associated skin lesions.* Despite DTHR being the main cause for heparin-induced skin lesions, HIT as cause of a cutaneous reaction needs to be ruled out (Ludwig et al. 2006). This includes clinical evaluation of pretest probability of HIT (4T's score) and laboratory testing (see Sect. 3.5). To differentiate between the two different conditions is of utmost importance, as treatment options (see below) for DTHR are of potential great risk for patients with HIT-associated skin lesions.

*Skin biopsy.* Usually, DTHR toward heparins are diagnosed based on the history and clinical presentation of the patient. However, in some situations, especially those where lesions are present as erythemas only and not localized at the injection sites, a skin biopsy may be helpful to distinguish between DTHR and HIT-associated skin reactions (e.g., erythemas and/or necrosis).

Histological examination of a DTHR reveals typical features of a contact dermatitis with epidermal spongiosis, dermal edema, and a lymphocytic infiltrate (predominantly CD4-positive T cells) accompanied by numerous eosinophils in the papillary dermis (Bircher et al. 2006; Ludwig et al. 2006). In contrast, HIT-associated skin lesions present dermal vascular occlusion secondary to microvascular platelet/fibrin plugs, or "white clots". In developed lesions, the overlying epidermis shows ischemic necrosis.

Notably, the histological diagnosis of DTHR was shown not to rule out the diagnosis of HIT, as the DTHR may represent a coinciding complication of the heparin therapy rather than part of the HIT syndrome (Schindewolf and Ludwig 2010; Warkentin and Linkins 2010).

*Allergy testing.* Among the in vivo tests, the intracutaneous test is the most sensitive one for the diagnosis, whereas prick tests and patch tests (epicutaneous) have low sensitivity and specificity (Bircher et al. 1995; Jappe 2006; Ludwig et al. 2005, 2006). After application of the undiluted anticoagulant, test sites should be monitored for development of urticaria 20 min after exposure and for eczematous reactions 24, 48, and 96 h after exposure. Tolerance toward negatively tested agents (see Sect. 4.3) is best proven with an s.c. or i.v. provocation test. Thus, skin and provocation tests are reliable tools to identify safe anticoagulants in individuals.

Most in vitro tests are not universally available and have a low sensitivity (Scherer et al. 2008). For instance, lymphocyte transformation assays as usually performed have been shown to be unsuitable for the diagnosis of DTHR to heparins (Bircher et al. 2006; Jappe 2006; Lopez et al. 2009; Ludwig et al. 2005)

It has been recommended to use allergy testing only for the identification of a tolerated anticoagulant and to make the diagnosis based on the history, clinical presentation and histology (Schindewolf and Ludwig 2010). In patients with skin necrosis or suspicion of HIT, skin tests are contraindicated anyway, because even minute doses of any heparin may exacerbate thrombocytopenia (Bircher et al. 2006; Scherer et al. 2008).

### 4.3 Therapy and Cross-Reactivity

In case of skin lesions during treatment with heparins, treatment should be stopped. In patients with a high pretest probability of HIT, treatment with a non-heparin anticoagulant (approved are argatroban, lepirudin, danaparoid) has to be started directly to prevent potentially fatal thrombotic events (see Sect. 3.6). This is in contrast to coumarin-induced skin necrosis, where therapy may be continued or restarted at a lower dose (Trautmann and Seitz 2010).

If HIT is excluded, either i.v. administration of UFH (or of the same LMWH as off-label use) (Gaigl et al. 2005; Trautmann and Seitz 2010) or s.c. fondaparinux can be used for further anticoagulation (Bircher et al. 2006; Ludwig et al. 2006; Schindewolf et al. 2010a, b; Schindewolf and Ludwig 2010; Trautmann and Seitz 2009). After change of the anticoagulant and application of topical glucocorticosteroids if necessary, lesions usually resolve within a week.

The tolerance of intravenously applied heparins, as opposed to s.c., may be because of the lack of a particular protein to form an allergen, different antigen-presenting cells or a compartment effect because of T-cell homing (Bircher et al. 2006).

In a recent prospective examination on 231 patients treated with fondaparinux, the incidence of allergic skin lesions amounted to only 0.4% (95% CI, 0.01–2.4%) and thus was almost 20 times lower compared with LMWHs (Schindewolf et al. 2010b). Additionally, no cross-allergies were observed in patients with DTHR to heparins. For patients who had developed heparin allergy in the past, or with corresponding risk factors, fondaparinux seems to represent the anticoagulant of choice. Albeit not approved for therapy of HIT, current data suggest that it may even be applicable in patients with HIT antibodies (Savi et al. 2005; Warkentin 2010).

Since LMWHs show extensive cross-reactivity, one preparation should not be replaced by another in patients with skin reactions (Ludwig et al. 2005; Trautmann and Seitz 2009). Prolonged exposure to various heparins may even increase the cross-reactivity (Schindewolf and Ludwig 2010). Also, danaparoid cannot be recommended as an alternative anticoagulant, since this heparinoid exhibits up to >40% cross-reactivity (Koch et al. 2000; Ludwig et al. 2005; Seitz et al. 2008; Szolar-Platzer et al. 2000). The cross-reactivity of the heparinoid pentosan polysulfate may be lower (Schindewolf and Ludwig 2010); however, its high degree of sulfation suggests a potential to induce HIT and the evidence of its safety and efficacy is relatvely low.

## 5 Allergic, Anaphylactic, and Anaphylactoid Reactions

Besides heparin-induced skin lesions, systemic allergic and anaphylactic immune reactions to heparins have been described more than 50 years ago (Bernstein 1956; Chernoff 1950; Turcotte et al. 1965). Today, immediate-type reactions to heparins, probably involving IgE-mediated mechanisms, seem to be very rare. One reason might be the increased pharmaceutical quality of heparins.

Moreover, reported acute reactions immediately after UFH application may often have been rapid-onset HIT (Warkentin and Greinacher 2009) (see Sect. 3.3.4). Intravenous bolus UFH administered to patients with circulating HIT antibodies, usually as a result of recent heparin therapy, can produce anaphylactoid reactions (i.e., not anaphylactic), probably as a consequence of in vivo activation of platelets and, possibly, leukocytes. Affected patients often exhibit fever/chills, hypertension and/or acute respiratory compromise ("pseudo-pulmonary embolism").

In contrast, heparin-induced anaphylaxis is caused by activation of the contact system, with formation of vasoactive kinins (bradykinin, des-arg 9-bradykinin) (Warkentin and Greinacher 2009). This syndrome has been linked in an epidemic form to administration of OSCS-contaminated heparin (see Chess et al. 2011); the reactions feature prominent hypotension and laryngeal edema. Measures to improve the quality control and to prevent further falsifications of heparins may hopefully prevent the occurrence of such severe adverse reactions in future.

Patients with an increased risk for both anaphylactic and anaphylactoid (i.e., HIT) reactions are those undergoing hemodialysis. However, heparin-mediated reactions have to be distinguished from infrequent anaphylactic or anaphylactoid reactions during hemodialysis, which can be due to ethylene oxide hypersensitivity or toxicity, reactions induced by dialyser membrane, toxic effects of formaldehyde or glutaraldehyde or idiopathic responses (Jappe 2006).

## 6 Osteoporosis

With a 2.2–5% incidence of heparin-induced osteoporotic fracture, osteoporosis is the most common serious side effect of long-term use of UFH (Le Templier and Rodger 2008; Lefkou et al. 2009; Nelson-Piercy 1997; Rajgopal et al. 2008). Short-term, high-dose UFH therapy as used to treat acute TE is not associated with osteoporosis (Le Templier and Rodger 2008). Available clinical data suggest that the risk of LWMHs is lower, but should be considered in sensitive patients who are pregnant or elderly and in children.

Osteoporosis is a systemic skeletal disease characterized by low bone mass and micro-architectural deterioration resulting in increased bone fragility and a consequent increase in fracture risk (Christiansen 1995). Fractures represent the most serious and clinically relevant complication of osteoporosis, whereas osteoporosis itself is usually asymptomatic. Therefore, the diagnosis of osteoporosis is based on the assessment of bone mineral density (BMD), which is best determined by dual-energy X-ray absorptiometry (DXA). Osteoporosis is diagnosed when bone mineral density (BMD) is $\geq$2.5 SD below the mean BMD of healthy young adults (Miller 2006).

Osteoporosis is primarily a disease of the aged population. Besides, there are many forms of secondary osteoporosis caused by factors other than menopause and aging, such as long-term heparin use, pregnancy, and lactation (Gennari et al. 1998). In pregnancy, osteoporosis therefore represents the principal risk associated with long-term use of UFH (Le Templier and Rodger 2008).

## 6.1 UFH and Osteoporosis

The association between UFH therapy and osteoporosis dates back to 1965 when it was reported (Griffith et al. 1965) that 6 of 10 patients who had received daily UFH doses of 15,000–30,000 IU for at least 8 months sustained spontaneous Osteoportoic fracture.

Since then, numerous case reports as well as some observational and several controlled trials have been published regarding this phenomenon [reviewed by Bick and Frenkel (1999) and Hirsh et al. (2001)].

Limitations of the studies are small size, distinct study designs and end points, and the fact that the patient populations were mainly pregnant women, who have an increased osteoporosis risk anyway (Rajgopal et al. 2008). This makes it difficult to extrapolate the incidence of UFH-induced osteoporosis.

Despite considerable heterogeneity, the overall results led to the belief that up to one third of all patients receiving long-term UFH will experience a significant decrease in Bone mineral density (Hirsh et al. 2001; Rajgopal et al. 2008). These patients frequently developed fractures, but there was not always a correlation with the BMD and also not between BMD and dose or duration of UFH therapy, respectively. The incidence ranged from 2.2% vertebral fractures among 184 pregnant women receiving 13,000–40,000 IU/d UFH for 25 weeks (Dahlman 1993) up to 15% spinal fractures among 40 patients receiving 10,000 IU/d UFH for 3–6 months (Monreal et al. 1994). In the first study, only women with severe back pain were tested for fracture, whereas patients in the second study were significantly older and were routinely screened for spinal fracture.

According to a study on 23 pregnant women treated with UFH for 5–8 months, a significant reduction in BMD was still detectable at 3 years postpartum (Pettila et al. 2002). This suggests that patients who have experienced long-term UFH treatment may be at an increased risk of osteoporosis later in life.

## 6.2 LMWH and Osteoporosis

As meanwhile LMWHs have largely replaced UFH in long-term prophylaxis and therapy, it is important to know the associated risk of osteoporosis. In the literature, only 11 cases of LMWH-induced osteoporotic fractures or symptomatic osteoporosis have been reported until December 2008 (Lefkou et al. 2009). The used LMWHs were dalteparin, enoxaparin, fraxiparin, and tinzaparin, and 8 of the 11 cases occurred during pregnancy. This may reflect a generally low risk of LMWHs to induce osteoporosis and fragility fractures.

So far, the only situation in which LMWH-induced osteoporosis has been evaluated is pregnancy (Lefkou et al. 2009). In the largest systematic data review of the use of LMWH during pregnancy, which included 64 studies reporting 2,777 pregnancies, there was only one case of osteoporotic fracture (Greer and Nelson-Piercy 2005). Lefkou et al. (2009) currently reviewed the studies investigating the LWMH-associated risk of osteoporosis. The eleven small studies, each including 16–89 patients and mostly performed on pregnant women, showed a certain trend of Bone mineral density reduction. However, in two trials including control groups not receiving anticoagulation, no significant differences in BMD changes were detected between the LMWH and control groups (Carlin et al. 2004; Rodger et al. 2007). Thus, the bone loss associated with the use of long-term LMWH is possibly not significantly different from physiological losses during pregnancy and lactation.

Concerning the risk of LMWH compared with UFH, two studies found a significantly lower incidence of fractures (Monreal et al. 1994; Pettila et al. 1999) and reduction in BMD (Pettila et al. 2002), respectively, in the LMWH group, whereas another study revealed a 2–2.5% incidence of a >10% bone loss regardless of the type of heparin (Casele et al. 2006).

In summary, the absence of evidence of osteoporosis associated with LMWH is not evidence of the absence of the risk (Greer and Hunt 2005). Therefore, large clinical trials investigating pre- and post-treatment BMD and comparing different dosages of LMWH are needed to make safe conclusions (Lefkou et al. 2009; Tooher et al. 2010).

## *6.3 Anticoagulation in Pregnancy*

In view of current knowledge, pregnant women requiring long-term anticoagulation should be reassured regarding the risk of osteoporosis associated with LMWHs (Le Templier and Rodger 2008). A safe and effective alternative to LMWHs in these patients may be fondaparinux (Eikelboom 2007; Hawkins and Evans 2005). In contrast to UFH and LMWHs, fondaparinux does not appear to have a negative effect on bone metabolism. It was found neither to inhibit osteoblast proliferation, mitochondrial activity and protein synthesis nor to affect the matrix collagen type II content and calcification (Handschin et al. 2005b; Matziolis et al. 2003).

## *6.4 UFH in Animal Experiments*

As reviewed by Rajgopal et al. (2008), animal experiments have supported the clinical finding that long-term UFH treatment can cause bone loss.

Treatment of Sprague Dawley rats with UFH (0.25–1.0 aXa U/g/d) for a period of 28 days was shown to decrease cancellous bone volume in both a dose- and time-dependent manner, with most of the bone loss occurring in the first 8 days of treatment (Muir et al. 1996). The bone loss was caused by both decreasing bone formation and increasing bone resorption. Specifically, rats treated with 1 U/g/d of UFH demonstrated a 37% decrease in the number of osteoblasts, and a 43% increase in the number of osteoclasts. In addition, the amount of unmineralized collagen (osteoid) lining the cancellous bone surface was reduced by 81%. These findings were supported by dose-dependent changes of biochemical markers, that is decrease in alkaline phosphatase, a marker of bone formation, and increase in urinary type I collagen-derived cross-linked pyridinoline, a marker of bone resorption.

Further studies in rats showed that the bone loss detected after UFH treatment for 28 days was unchanged, when measured four weeks later (Shaughnessy et al. 1999). UFH was found to accumulate in the bone during the course of its administration and to be retained there for an extended period of time. These results indicate that effects of UFH on bone are not readily reversible.

## *6.5 LMWH in Animal Experiments*

In line with the clinical findings, a convincing number of animal studies comparing various LMWHs with UFH demonstrated that LMWHs are considerably less deleterious to bone than UFH, but still cause some bone loss (Folwarczna et al. 2004, 2005; Rajgopal et al. 2008).

For example in Sprague Dawley rats, the LMWH tinzaparin (0.5–1.0 aXa-IU/g/d) produced a much lower dose-dependent decrease in cancellous bone volume than UFH (0.5 and 1.0 U/g/d) (Muir et al. 1997). While both LMWHs and UFH decreased osteoblast number and activity and so the bone formation, the number and activity of osteoclasts, i.e., bone resorption, were increased solely by UFH (Folwarczna et al. 2005). Corresponding to this, UFH stimulated the release of calcium from fetal rat calvaria, as an index of bone resorption, at clinically used concentrations, whereas various LMWH preparations caused an equivalent effect only at >50-fold higher concentrations (Shaughnessy et al. 1995).

## *6.6 Mechanisms*

How heparins impair bone formation and increase bone resorption is not yet clear.

It is supposed that heparin binds to bone matrix and interacts with a variety of cell types present in the bone microenvironment, including cells of the osteoblast lineage (Hirsh et al. 2001). Such interactions may alter mesenchymal stem cell differentiation reducing the number of mature osteoblasts (Muir et al. 1996), reduce collagen synthesis by osteoblasts (Shaughnessy et al. 1995), and release specific growth factors and/or cytokines capable of inducing the formation of osteoclasts from pluripotent mononuclear precursors in the bone marrow. Hence, there are probably several molecular mechanisms contributing to the finally observed effects.

A direct inhibition of osteoblast function by heparins was found in an osteoblast cell culture model (Handschin et al. 2005b, 2006). The LMWH dalteparin was shown to inhibit not only the osteoblast proliferation and protein synthesis, but also the expression of two important regulators of osteoblast differentiation, osteocalcin and Cbfa-1 (Handschin et al. 2005b, 2006).

Newer data suggest that UFH interacts with the OPG/RANK/RANKL pathway (Irie et al. 2007; Vik et al. 2007). This signaling pathway plays a key role in the regulation of the activity of osteoblasts and osteoclasts and is meanwhile considered a promising target in the development of agents to prevent osteoporosis (Trouvin and Goeb 2010).

RANKL (receptor activator for nuclear factor-κB ligand = osteoprotegerin ligand (OPG-L) = osteoclast differentiation factor (ODF)) is produced by osteoblast lineage cells and represents one of the two final effectors of osteoclast formation. Binding of RANKL to RANK, its receptor on osteoclast lineage cells, results in rapid differentiation of osteoclast precursors in bone marrow to mature osteoclasts and in increased functional activity and reduced apoptosis of mature osteoclasts (Hofbauer et al. 2000). The biological activity of RANKL is neutralized by binding to OPG (osteoprotegerin = osteoclastogenesis inhibitory factor (OCIF)), that also is secreted by osteoblast lineage cells. Targeted ablation of OPG was shown to lead to severe osteoporosis, whereas targeted ablation of RANKL or overexpression of OPG resulted in osteopetrosis.

In an in vitro osteoclast culture assay, various GAGs were shown to enhance the activity of osteoclasts as measured by the formation of resorption pits on dentin slices (Irie et al. 2007). This effect proved to be due to binding of GAGs to OPG, preventing its inhibitory activity on the osteoclastic bone resorption. Notably, the effect was dependent on the GAG structure and was observed with UFH, but not with LMWH.

These experimental data agree with the observation that UFH administration to human volunteers caused a strong increase in plasma OPG (Vik et al. 2007). UFH vs. LMWH seems to mobilize OPG from the bone microenvironment, where it exerts its biological activity, into the circulation. UFH caused a more pronounced vascular mobilization of OPG than LMWH, indicating that UFH has a higher affinity for OPG than LMWH.

## 7 Elevation of Liver Enzymes

A common side effect of heparins is significant but transient increases of liver enzymes, especially serum transaminase levels.

### 7.1 Transaminases and Hepatotoxicity

Abnormal levels of alanine aminotransferase (ALT) and aspartate aminotransferase (AST) are believed to represent sensitive indicators of drug-induced hepatocellular injury, although elevations in ALT and AST can occur from conditions other than liver injury (Arora and Goldhaber 2006; Pratt and Kaplan 2000; Russo and Watkins 2004; Senior 2003). Usually, levels of ALT >3 times the upper limit of normal (ULN) are used as a possible signal for drug-induced hepatotoxicity (Pratt and Kaplan 2000). Strong elevation in transaminase levels in conjunction with a rise in bilirubin level >2 ULN or >3 mg/dL, respectively, is a more ominous marker for drug-induced hepatocellular injury affecting the global liver function.

### 7.2 Transaminases and Heparins

Older studies report frequencies of elevated ALT and AST of more than 30% and 20%, respectively, after administration of high UFH doses, whereby the incidence in males was higher than that in females (Dukes et al. 1984; Monreal et al. 1989). However, both pre-existing increased enzyme levels and other reasons for elevation have to be taken into account (Russo and Watkins 2004) so that an incidence of up to 15% seems more realistic (Arora and Goldhaber 2006; Guevara et al. 1993).

Increased ALT/AST levels occur also with short-term low-dose UFH vs. LMWH prophylaxis (van der Wiel et al. 1993) and LMWH (Arora and Goldhaber 2006; Carlson et al. 2001; Hui et al. 2001). According to the study of Monreal et al. (1989), there might be a dose dependence and a lower incidence with LMWHs. The latter is in line with the observation that the pentasaccharide fondaparinux uncommonly (0.1–1%) leads to increased hepatic enzymes (EMA 2010).

The enzyme concentrations usually increase within the first days of heparin administration and return to normal after discontinuation, sometimes even during heparin therapy (Dukes et al. 1984).

Transaminase elevations by heparins are generally asymptomatic and are not associated with adverse sequelae. In contrast to VKA (Arora and Goldhaber 2006), there are only sporadic case reports of clinically significant hepatitis associated with heparins (Hui et al. 2001).

*Mechanisms*. The specific mechanism of transaminase elevation after heparin has not been identified. Speculations include direct toxicity, hepatocyte membrane modification, and immune-mediated hypersensitivity reaction (Carlson et al. 2001). Direct toxicity seems rather unlikely considering the protective actions of heparins on liver found in animal experiments (Harada et al. 2006; Kukner et al. 2010; Taha et al. 2009). Another hypothesis might be an association with the metabolism of heparins, since liver sinusoidal endothelial cells are described as the principal site for elimination of heparin from the circulation (Oie et al. 2008).

# 8 Further Side Effects

## *8.1 Alopecia*

For decades, alopecia has been known as an occasional adverse effect of long long-term use of UFH (Hirschboeck et al. 1954). However, cases of alopecia after therapy with various LMWHs are reported only infrequently (Apsner et al. 2001; Barnes et al. 2000; Sarris et al. 2003; Wang and Po 2006). After discontinuation of heparin, hair growth usually returns to normal within some weeks. Although hair loss is not a serious adverse reaction, medically, it has to be considered a serious cosmetic problem for the patient such that it might affect compliance.

Heparin-induced alopecia results from a premature transformation of growing hairs into the resting phase (telogen effluvium), which leads to hair loss 6 weeks to 3 months after heparin exposure, whereby the hairs shed are normal in appearance (Barnes et al. 2000). It thus differs from the rapid loss of dystrophic hairs caused by chemotherapy due to an abrupt cessation of the normal growth phase (anagen effluvium).

*Mechanisms*. The mechanism responsible for alopecia caused by heparins is still unknown. Hair growth studies in anagen-induced mice revealed that heparin exerts dose- and hair cycle-dependent, differential effects on skin epithelial cell functions (Paus 1991). Intraperitoneal, but not topical, application of heparin suppressed the development of anagen follicles as assessed by morphometry. In an organ culture assay, the epidermis of heparin-treated mice showed a significant reduction of epithelial bulb, but not epidermal cell proliferation, and an increase in the synthesis of arginine-rich proteins (ARP).

Other conceivable AT-independent mechanisms of heparin are its inhibitory effect on heparanase and/or direct interactions with heparin-binding growth factors, as heparanase was identified as an important regulator of hair growth (Zcharia et al. 2005). Experimental data suggest that heparanase supports hair growth by facilitating the migration of follicular stem cell progeny and releasing extracellular matrix-resident, heparan sulfate-bound growth factors.

*Alopecia induced by other anticoagulants.* Since other highly sulfated polysaccharides often exhibit biological effects similar to heparin, it is not surprising that alopecia is also a known side effect of pentosan polysulfate and dextran sulfate (Greten et al. 1978; Tudhope et al. 1958). The concept of interference by sulfated polysaccharides with endogenous GAGs involved in hair growth is supported by the observation that switch from anticoagulation with dalteparin in hemodialysis patients to regional anticoagulation with citrate led to cessation of excessive hair loss after 6 weeks to 3 months (Apsner et al. 2001). However, the fact that alopecia is induced also by VKA (Umlas and Harken 1988) points to a possible direct association between alopecia and anticoagulation.

## 8.2 *Hypoaldosteronism, Hyperkalemia, and Metabolic Acidosis*

A rare adverse effect of heparins is hypoaldosteronism, associated with hyperkalemia and metabolic acidosis.

Hypoaldosteronism is probably due to the inhibitory effect of heparins on the step where corticosterone is converted to 18-hydroxycorticosterone (Hirsh et al. 1998). This is usually of no clinical importance, but there are a few case reports of heparin-induced hypoaldosteronism leading to severe hyperkalemia and other metabolic derangements and even death (Hirsh et al. 1998; Monreal et al. 1989). The hyperkalemia and hypoaldosteronism reverted to normal in two of these patients when the heparin therapy was discontinued.

In a prospective study including 171 patients, serum potassium concentrations higher than the upper limit of normal were measured in 4–8% of the patients independently of the dose and type of heparin (UFH vs. dalteparin) (Monreal et al. 1989). However, there was no case of severe hyperkalemia.

It is therefore not necessary to monitor potassium levels routinely in patients treated with UFH or LMWH, but it would seem prudent to do so in patients with renal insufficiency, diabetes mellitus, or those taking other drugs that can cause hyperkalemia (Eikelboom 2007).

## 8.3 *Priapism*

Heparin has been associated with priapism in a number of case reports (Bick and Frenkel 1999). It is uncertain whether it was the heparin or the underlying thrombotic condition, potentially as part of a HIT event, that caused the priapism.

## 8.4 *Impaired Fracture Healing*

Although VTE prophylaxis is indispensable in surgical patients, heparins might have some counterproductive effects by their antiproliferative and antiangiogenic activities. However, apart from the fact that hematoma formation at surgical sites may delay wound healing, there is no reliable evidence that heparins significantly impede surgical healing in general (Eikelboom 2007).

In contrast, analysis of literature revealed that heparins, VKA as well as aspirin retard fracture healing (Lindner et al. 2008). Experimental data suggest that this may be related to effects of anticoagulants on normal bone metabolism. However, regarding the evidence-based benefit to patients with bone fractures from anticoagulation, such adverse reactions seem to be acceptable and not clinically relevant.

## References

Alban S (2008a) Natural and synthetic glycosaminoglycans. Molecular characteristics as the basis of distinct drug profiles. Hamostaseologie 28:51–61

Alban S (2008b) Pharmacological strategies for inhibition of thrombin activity. Curr Pharm Des 14:1152–1175

Alban S, Greinacher A (2007) Role of sulfated polysaccharides in the pathogenesis of heparin-induced thrombocytopenia. In: Warkentin TGA (ed) Heparin-induced thrombocytopenia, 4th edn. Informa Healthcare, New York, pp 167–186

Alberio L (2008) Heparin-induced thrombocytopenia: some working hypotheses on pathogenesis, diagnostic strategies and treatment. Curr Opin Hematol 15:456–464

Alexander KP, Chen AY, Roe MT, Newby LK, Gibson CM, Allen-LaPointe NM, Pollack C, Gibler WB, Ohman EM, Peterson ED (2005) Excess dosing of antiplatelet and antithrombin agents in the treatment of non-ST-segment elevation acute coronary syndromes. JAMA 294:3108–3116

Alhenc-Gelas M, Jestin-Le Guernic C, Vitoux JF, Kher A, Aiach M, Fiessinger JN (1994) Adjusted versus fixed doses of the low-molecular-weight heparin fragmin in the treatment of deep vein thrombosis. Fragmin-Study Group. Thromb Haemost 71:698–702

Alikhan R, Cohen AT (2009) Heparin for the prevention of venous thromboembolism in general medical patients (excluding stroke and myocardial infarction). Cochrane Database Syst Rev: CD003747

Amiral J, Bridey F, Dreyfus M, Vissoc AM, Fressinaud E, Wolf M, Meyer D (1992) Platelet factor 4 complexed to heparin is the target for antibodies generated in heparin-induced thrombocytopenia. Thromb Haemost 68:95–96

Amiral J, Lormeau JC, Marfaing-Koka A, Vissac AM, Wolf M, Boyer-Neumann C, Tardy B, Herbert JM, Meyer D (1997) Absence of cross-reactivity of SR90107A/ORG31540 pentasaccharide with antibodies to heparin-PF4 complexes developed in heparin-induced thrombocytopenia. Blood Coagul Fibrinolysis 8:114–117

Anand SS, Yusuf S, Pogue J, Ginsberg JS, Hirsh J (2003) Relationship of activated partial thromboplastin time to coronary events and bleeding in patients with acute coronary syndromes who receive heparin. Circulation 107:2884–2888

Apsner R, Horl WH, Sunder-Plassmann G (2001) Dalteparin-induced alopecia in hemodialysis patients: reversal by regional citrate anticoagulation. Blood 97:2914–2915

Areapally GM, Poncz M, Cines DB (2007) Immune vascular injury in heparin-induced thrombocytopenia. In: Warkentin TGA (ed) heparin-induced thrombocytopenia, 4th edn. Informa Healthcare, New York, pp 209–226

Arepally GM, Mayer IM (2001) Antibodies from patients with heparin-induced thrombocytopenia stimulate monocytic cells to express tissue factor and secrete interleukin-8. Blood 98:1252–1254

Arepally GM, Ortel TL (2006) Clinical practice. Heparin-induced thrombocytopenia. N Engl J Med 355:809–817

Arora N, Goldhaber SZ (2006) Anticoagulants and transaminase elevation. Circulation 113: e698–e702

Barnes C, Deidun D, Hynes K, Monagle P (2000) Alopecia and dalteparin: a previously unreported association. Blood 96:1618–1619

Bassand JP, Hamm CW, Ardissino D, Boersma E, Budaj A, Fernandez-Aviles F, Fox KA, Hasdai D, Ohman EM, Wallentin L, Wijns W (2007) Guidelines for the diagnosis and treatment of non-ST-segment elevation acute coronary syndromes. Eur Heart J 28:1598–1660

Bergqvist D (2004) Low molecular weight heparin for the prevention of venous thromboembolism after abdominal surgery. Br J Surg 91:965–974

Bernstein IL (1956) Anaphylaxis to heparin sodium; report of a case, with immunologic studies. J Am Med Assoc 161:1379–1381

Bick RL, Frenkel EP (1999) Clinical aspects of heparin-induced thrombocytopenia and thrombosis and other side effects of heparin therapy. Clin Appl Thromb Hemost 5(Suppl 1):S7–S15

Bircher AJ, Itin PH, Tsakiris DA, Surber C (1995) Delayed hypersensitivity to one low-molecular-weight heparin with tolerance of other low-molecular-weight heparins. Br J Dermatol 132:461–463

Bircher AJ, Harr T, Hohenstein L, Tsakiris DA (2006) Hypersensitivity reactions to anticoagulant drugs: diagnosis and management options. Allergy 61:1432–1440

Blank M, Shoenfeld Y, Tavor S, Praprotnik S, Boffa MC, Weksler B, Walenga MJ, Amiral J, Eldor A (2002) Anti-platelet factor 4/heparin antibodies from patients with heparin-induced thrombocytopenia provoke direct activation of microvascular endothelial cells. Int Immunol 14:121–129

Borgen PO, Dahl OE, Reikeras O (2010) Preoperative versus postoperative initiation of dalteparin thromboprophylaxis in THA. Hip Int 20:301–307

Bounameaux H, Perrier A (2008) Duration of anticoagulation therapy for venous thromboembolism. Hematology Am Soc Hematol Educ Program: 252–258

Bryant A, Low J, Austin S, Joseph JE (2008) Timely diagnosis and management of heparin-induced thrombocytopenia in a frequent request, low incidence single centre using clinical 4T's score and particle gel immunoassay. Br J Haematol 143:721–726

Budaj A, Eikelboom JW, Mehta SR, Afzal R, Chrolavicius S, Bassand JP, Fox KA, Wallentin L, Peters RJ, Granger CB, Joyner CD, Yusuf S (2009) Improving clinical outcomes by reducing bleeding in patients with non-ST-elevation acute coronary syndromes. Eur Heart J 30:655–661

Carlin AJ, Farquharson RG, Quenby SM, Topping J, Fraser WD (2004) Prospective observational study of bone mineral density during pregnancy: low molecular weight heparin versus control. Hum Reprod 19:1211–1214

Carlson MK, Gleason PP, Sen S (2001) Elevation of hepatic transaminases after enoxaparin use: case report and review of unfractionated and low-molecular-weight heparin-induced hepatotoxicity. Pharmacotherapy 21:108–113

Casele H, Haney EI, James A, Rosene-Montella K, Carson M Bone density changes in women who receive thromboprophylaxis in pregnancy. Am J Obstet Gynecol;195:1109–1113

Chernoff AI (1950) Anaphylactic reaction following injection of heparin. N Engl J Med 242:315–319

Chess ED et al (2011) Case study: contamination of heparin with oversulfated chondroitin sulfate. In: Lever R, Mulloy B, Page CP (eds) Heparin – a century of progress. Springer, Heidelberg 99–125

Chong BH, Chong JJ (2010) Heparin-induced thrombocytopenia associated with fondaparinux. Clin Adv Hematol Oncol 8:63–65

Chong BH, Murray B, Berndt MC, Dunlop LC, Brighton T, Chesterman CN (1994) Plasma P-selectin is increased in thrombotic consumptive platelet disorders. Blood 83:1535–1541

Christiansen C (1995) Osteoporosis: diagnosis and management today and tomorrow. Bone 17:513S–516S

Cines DB, Rauova L, Arepally G, Reilly MP, McKenzie SE, Sachais BS, Poncz M (2007) Heparin-induced thrombocytopenia: an autoimmune disorder regulated through dynamic autoantigen assembly/disassembly. J Clin Apher 22:31–36

Crowther MA, Cook DJ, Albert M, Williamson D, Meade M, Granton J, Skrobik Y, Langevin S, Mehta S, Hebert P, Guyatt GH, Geerts W, Rabbat C, Douketis J, Zytaruk N, Sheppard J, Greinacher A, Warkentin TE (2010) The 4Ts scoring system for heparin-induced thrombocytopenia in medical-surgical intensive care unit patients. J Crit Care 25:287–293

Dahlman TC (1993) Osteoporotic fractures and the recurrence of thromboembolism during pregnancy and the puerperium in 184 women undergoing thromboprophylaxis with heparin. Am J Obstet Gynecol 168:1265–1270

Denys B, Devreese K (2009) A clinical-laboratory approach contributing to a rapid and reliable diagnosis of heparin-induced thrombocytopenia: an update. Thromb Res 124:642–643

Denys B, Stove V, Philippe J, Devreese K (2008) A clinical-laboratory approach contributing to a rapid and reliable diagnosis of heparin-induced thrombocytopenia. Thromb Res 123:137–145

Douketis JD, Berger PB, Dunn AS, Jaffer AK, Spyropoulos AC, Becker RC, Ansell J (2008) The perioperative management of antithrombotic therapy: American College of Chest Physicians Evidence-Based Clinical Practice Guidelines (8th Edition). Chest 133:299S–339S

Drew PJ, Smith MJ, Milling MA (1999) Heparin-induced skin necrosis and low molecular weight heparins. Ann R Coll Surg Engl 81:266–269

Dukes GE Jr, Sanders SW, Russo J Jr, Swenson E, Burnakis TG, Saffle JR, Warden GD (1984) Transaminase elevations in patients receiving bovine or porcine heparin. Ann Intern Med 100:646–650

Dumaine R, Borentain M, Bertel O, Bode C, Gallo R, White HD, Collet JP, Steinhubl SR, Montalescot G (2007) Intravenous low-molecular-weight heparins compared with unfractionated heparin in percutaneous coronary intervention: quantitative review of randomized trials. Arch Intern Med 167:2423–2430

Eikelboom J (2007) Nonbleeding complications of unfractionated and low-molecular weight heparins. Clin Adv Hematol Oncol 5:871–873

Eikelboom JW, Quinlan DJ, O'Donnell M (2009) Major bleeding, mortality, and efficacy of fondaparinux in venous thromboembolism prevention trials. Circulation 120:2006–2011

EMA (2010) Arixtra, Fondaparinux – EMEA/H/C/000403-II/0045. EPAR – Product information 31/08/2010

EMEA/CHMP (2006) Guideline on clinical investigation of medicinal products for prophylaxis of venous thromboembolic risk in non-surgical patients. CPMP/EWP/6235/04

EMEA/CHMP (2007) Guideline on clinical investigation of medicinal products for prophylaxis of high intra- and post-operative venous thromboembolic risk. CPMP/EWP/707/98 Rev.1 corr

Eriksson BI, Borris L, Dahl OE, Haas S, Huisman MV, Kakkar AK, Misselwitz F, Kalebo P (2006) Oral, direct Factor Xa inhibition with BAY 59-7939 for the prevention of venous thromboembolism after total hip replacement. J Thromb Haemost 4:121–128

Eriksson BI, Dahl OE, Rosencher N, Kurth AA, van Dijk CN, Frostick SP, Kalebo P, Christiansen AV, Hantel S, Hettiarachchi R, Schnee J, Buller HR (2007) Oral dabigatran etexilate vs. subcutaneous enoxaparin for the prevention of venous thromboembolism after total knee replacement: the RE-MODEL randomized trial. J Thromb Haemost 5:2178–2185

Erkens PM, Prins MH (2010) Fixed dose subcutaneous low molecular weight heparins versus adjusted dose unfractionated heparin for venous thromboembolism. Cochrane Database Syst Rev: CD001100

Folwarczna J, Janiec W, Sliwinski L (2004) Effects of heparin and low-molecular-weight heparins on bone mechanical properties in rats. Thromb Haemost 92:940–946

Folwarczna J, Sliwinski L, Janiec W, Pikul M (2005) Effects of standard heparin and low-molecular-weight heparins on the formation of murine osteoclasts in vitro. PR Pharmacol Rep 57:635–645

Gaigl Z, Pfeuffer P, Raith P, Brocker EB, Trautmann A (2005) Tolerance to intravenous heparin in patients with delayed-type hypersensitivity to heparins: a prospective study. Br J Haematol 128:389–392

Geerts WH, Bergqvist D, Pineo GF, Heit JA, Samama CM, Lassen MR, Colwell CW (2008) Prevention of venous thromboembolism: American College of Chest Physicians Evidence-Based Clinical Practice Guidelines (8th Edition). Chest 133:381S–453S

Gennari C, Martini G, Nuti R (1998) Secondary osteoporosis. Aging (Milano) 10:214–224

Gouin-Thibault I, Siguret V, Pautas E (2010) A need for evidence-based clinical practice guidelines for the use of heparins in the elderly. Clin Interv Aging 5:119–121

Greer I, Hunt BJ (2005) Low molecular weight heparin in pregnancy: current issues. Br J Haematol 128:593–601

Greer IA, Nelson-Piercy C (2005) Low-molecular-weight heparins for thromboprophylaxis and treatment of venous thromboembolism in pregnancy: a systematic review of safety and efficacy. Blood 106:401–407

Greinacher A (2006) Heparin-induced thrombocytopenia: frequency and pathogenesis. Pathophysiol Haemost Thromb 35:37–45

Greinacher A (2009) Heparin-induced thrombocytopenia. J Thromb Haemost 7(Suppl 1):9–12

Greinacher A, Farner B, Kroll H, Kohlmann T, Warkentin TE, Eichler P (2005) Clinical features of heparin-induced thrombocytopenia including risk factors for thrombosis. A retrospective analysis of 408 patients. Thromb Haemost 94:132–135

Greinacher A, Gopinadhan M, Gunther JU, Omer-Adam MA, Strobel U, Warkentin TE, Papastavrou G, Weitschies W, Helm CA (2006) Close approximation of two platelet factor 4 tetramers by charge neutralization forms the antigens recognized by HIT antibodies. Arterioscler Thromb Vasc Biol 26:2386–2393

Greinacher A, Juhl D, Strobel U, Wessel A, Lubenow N, Selleng K, Eichler P, Warkentin TE (2007) Heparin-induced thrombocytopenia: a prospective study on the incidence, platelet-activating capacity and clinical significance of antiplatelet factor 4/heparin antibodies of the IgG, IgM, and IgA classes. J Thromb Haemost 5:1666–1673

Greinacher A, Alban S, Omer-Adam MA, Weitschies W, Warkentin TE (2008) Heparin-induced thrombocytopenia: a stoichiometry-based model to explain the differing immunogenicities of unfractionated heparin, low-molecular-weight heparin, and fondaparinux in different clinical settings. Thromb Res 122:211–220

Greinacher A, Kohlmann T, Strobel U, Sheppard JA, Warkentin TE (2009) The temporal profile of the anti-PF4/heparin immune response. Blood 113:4970–4976

Greinacher A, Althaus K, Krauel K, Selleng S (2010a) Heparin-induced thrombocytopenia. Hamostaseologie 30(17–8):20–28

Greinacher A, Ittermann T, Bagemuhl J, Althaus K, Furll B, Selleng S, Lubenow N, Schellong S, Sheppard JI, Warkentin TE (2010b) Heparin-induced thrombocytopenia: towards standardization of platelet factor 4/heparin antigen tests. J Thromb Haemost 8:2025–2031

Greten H, Schettler G, Mordasini R, Zipperle G, Klar E, Laible V (1978) Mechanism of action of synthetic heparinoids: results in patients with hypertriglyceridaemia (author's transl). Dtsch Med Wochenschr 103:204–211

Griffith GC, Nichols G Jr, Asher JD, Flanagan B (1965) Heparin osteoporosis. JAMA 193:91–94

Grims RH, Weger W, Reiter H, Arbab E, Kranke B, Aberer W (2007) Delayed-type hypersensitivity to low molecular weight heparins and heparinoids: cross-reactivity does not depend on molecular weight. Br J Dermatol 157:514–517

Guevara A, Labarca J, Gonzalez-Martin G (1993) Heparin-induced transaminase elevations: a prospective study. Int J Clin Pharmacol Ther Toxicol 31:137–141

Haentjens P (1996) Thromboembolic prophylaxis in orthopaedic trauma patients: a comparison between a fixed dose and an individually adjusted dose of a low molecular weight heparin (nadroparin calcium). Injury 27:385–390

Handschin AE, Trentz O, Kock HJ, Wanner GA (2005a) Low molecular weight heparin-induced skin necrosis – a systematic review. Langenbecks Arch Surg 390:249–254

Handschin AE, Trentz OA, Hoerstrup SP, Kock HJ, Wanner GA, Trentz O (2005b) Effect of low molecular weight heparin (dalteparin) and fondaparinux (Arixtra) on human osteoblasts in vitro. Br J Surg 92:177–183

Handschin AE, Egermann M, Trentz O, Wanner GA, Kock HJ, Zund G, Trentz OA (2006) Cbfa-1 (Runx-2) and osteocalcin expression by human osteoblasts in heparin osteoporosis in vitro. Clin Appl Thromb Hemost 12:465–472

Harada N, Okajima K, Uchiba M (2006) Dalteparin, a low molecular weight heparin, attenuates inflammatory responses and reduces ischemia-reperfusion-induced liver injury in rats. Crit Care Med 34:1883–1891

Harr T, Scherer K, Tsakiris DA, Bircher AJ (2006) Immediate type hypersensitivity to low molecular weight heparins and tolerance of unfractioned heparin and fondaparinux. Allergy 61:787–788

Hawkins D, Evans J (2005) Minimising the risk of heparin-induced osteoporosis during pregnancy. Expert Opin Drug Saf 4:583–590

Heit JA, Berkowitz SD, Bona R, Cabanas V, Corson JD, Elliott CG, Lyons R (1997) Efficacy and safety of low molecular weight heparin (ardeparin sodium) compared to warfarin for the prevention of venous thromboembolism after total knee replacement surgery: a double-blind, dose-ranging study. Ardeparin Arthroplasty Study Group. Thromb Haemost 77:32–38

Hinz P, Lubenow N, Wessel A, Greinacher A, Ekkernkamp A (2009) [Thrombosis prophylaxis in trauma surgery units in Germany: a survey]. Unfallchirurg 112(12):1029–1033

Hirschboeck JS, Madison FW, Pisciotta AV (1954) Alopecia and other toxic effects of heparin and synthetic heparinoids. Am J Med Sci 227:279–282

Hirsh J, Raschke R (2004) Heparin and low-molecular-weight heparin: the Seventh ACCP Conference on Antithrombotic and Thrombolytic Therapy. Chest 126:188S–203S

Hirsh J, Warkentin TE, Raschke R, Granger C, Ohman EM, Dalen JE (1998) Heparin and low-molecular-weight heparin: mechanisms of action, pharmacokinetics, dosing considerations, monitoring, efficacy, and safety. Chest 114:489S–510S

Hirsh J, Warkentin TE, Shaughnessy SG, Anand SS, Halperin JL, Raschke R, Granger C, Ohman EM, Dalen JE (2001) Heparin and low-molecular-weight heparin: mechanisms of action, pharmacokinetics, dosing, monitoring, efficacy, and safety. Chest 119:64S–94S

Hofbauer LC, Khosla S, Dunstan CR, Lacey DL, Boyle WJ, Riggs BL (2000) The roles of osteoprotegerin and osteoprotegerin ligand in the paracrine regulation of bone resorption. J Bone Miner Res 15:2–12

Hohenstein E, Tsakiris D, Bircher AJ (2004) Delayed-type hypersensitivity to the ultra-low-molecular-weight heparin fondaparinux. Contact Dermatitis 51:149–151

Hui CK, Yuen MF, Ng IO, Tsang KW, Fong GC, Lai CL (2001) Low molecular weight heparin-induced liver toxicity. J Clin Pharmacol 41:691–694

Hull RD, Pineo GF, MacIsaac S (2001) Low-molecular-weight heparin prophylaxis: preoperative versus postoperative initiation in patients undergoing elective hip surgery. Thromb Res 101: V155–V162

Hursting MJ, Soffer J (2009) Reducing harm associated with anticoagulation: practical considerations of argatroban therapy in heparin-induced thrombocytopenia. Drug Saf 32:203–218

Irie A, Takami M, Kubo H, Sekino-Suzuki N, Kasahara K, Sanai Y (2007) Heparin enhances osteoclastic bone resorption by inhibiting osteoprotegerin activity. Bone 41:165–174

Jappe U (2006) Allergy to heparins and anticoagulants with a similar pharmacological profile: an update. Blood Coagul Fibrinolysis 17:605–613

Kakkar VV, Kakkar S, Sanderson RM, Peers CE (1986) Efficacy and safety of two regimens of low molecular weight heparin fragment (Fragmin) in preventing postoperative venous thrombolism. Haemostasis 16(Suppl 2):19–24

Kearon C, Kahn SR, Agnelli G, Goldhaber S, Raskob GE, Comerota AJ (2008) Antithrombotic therapy for venous thromboembolic disease: American College of Chest Physicians Evidence-Based Clinical Practice Guidelines (8th Edition). Chest 133:454S–545S

Kelton JG, Warkentin TE (2008) Heparin-induced thrombocytopenia: a historical perspective. Blood 112:2607–2616

Kinnaird TD, Stabile E, Mintz GS, Lee CW, Canos DA, Gevorkian N, Pinnow EE, Kent KM, Pichard AD, Satler LF, Weissman NJ, Lindsay J, Fuchs S (2003) Incidence, predictors, and prognostic implications of bleeding and blood transfusion following percutaneous coronary interventions. Am J Cardiol 92:930–935

Koch P, Munssinger T, Rupp-John C, Uhl K (2000) Delayed-type hypersensitivity skin reactions caused by subcutaneous unfractionated and low-molecular-weight heparins: tolerance of a new recombinant hirudin. J Am Acad Dermatol 42:612–619

Komericki P, Grims R, Kranke B, Aberer W (2007) Acute generalized exanthematous pustulosis from dalteparin. J Am Acad Dermatol 57:718–721

Krauel K, Furll B, Warkentin TE, Weitschies W, Kohlmann T, Sheppard JI, Greinacher A (2008) Heparin-induced thrombocytopenia–therapeutic concentrations of danaparoid, unlike fondaparinux and direct thrombin inhibitors, inhibit formation of platelet factor 4-heparin complexes. J Thromb Haemost 6:2160–2167

Kukner A, Tore F, Firat T, Terzi EH, Oner H, Balaban YH, Ozogul C (2010) The preventive effect of low molecular weight heparin on CCL-induced necrosis and apoptosis in rat liver. Ann Hepatol 9:445–54

Lassen MR, Borris LC, Jensen HP, Poulsen KA, Ejstrud P, Andersen BS (2000) Dose relation in the prevention of proximal vein thrombosis with a low molecular weight heparin (tinzaparin) in elective hip arthroplasty. Clin Appl Thromb Hemost 6:53–57

Le Templier G, Rodger MA (2008) Heparin-induced osteoporosis and pregnancy. Curr Opin Pulm Med 14:403–407

Lefkou E, Khamashta M, Hampson G, Hunt BJ (2009) Review: low-molecular-weight heparin-induced osteoporosis and osteoporotic fractures: a myth or an existing entity? Lupus 19:3–12

Leizorovicz A, Bara L, Samama MM, Haugh MC (1993) Factor Xa inhibition: correlation between the plasma levels of anti-Xa activity and occurrence of thrombosis and haemorrhage. Haemostasis 23(Suppl 1):89–98

Lim W, Dentali F, Eikelboom JW, Crowther MA (2006) Meta-analysis: low-molecular-weight heparin and bleeding in patients with severe renal insufficiency. Ann Intern Med 144:673–684

Lindhoff-Last E, Piechottka GP, Rabe F, Bauersachs R (2000) Hirudin determination in plasma can be strongly influenced by the prothrombin level. Thromb Res 100:55–60

Lindner T, Cockbain AJ, El Masry MA, Katonis P, Tsiridis E, Schizas C (2008) The effect of anticoagulant pharmacotherapy on fracture healing. Expert Opin Pharmacother 9:1169–1187

Lo GK, Juhl D, Warkentin TE, Sigouin CS, Eichler P, Greinacher A (2006) Evaluation of pretest clinical score (4 T's) for the diagnosis of heparin-induced thrombocytopenia in two clinical settings. J Thromb Haemost 4:759–765

Lobo B, Finch C, Howard A, Minhas S (2008) Fondaparinux for the treatment of patients with acute heparin-induced thrombocytopenia. Thromb Haemost 99:208–214

Lopez LM (2001) Low-molecular-weight heparins are essentially the same for treatment and prevention of venous thromboembolism. Pharmacotherapy 21:56–61, discussion 71S–72S

Lopez S, Torres MJ, Rodriguez-Pena R, Blanca-Lopez N, Fernandez TD, Antunez C, Canto G, de Luque V, Mayorga C (2009) Lymphocyte proliferation response in patients with delayed hypersensitivity reactions to heparins. Br J Dermatol 160:259–265

Louzada ML, Majeed H, Wells PS (2009) Efficacy of low- molecular- weight- heparin versus vitamin K antagonists for long term treatment of cancer-associated venous thromboembolism in adults: a systematic review of randomized controlled trials. Thromb Res 123:837–844

Lubenow N, Kempf R, Eichner A, Eichler P, Carlsson LE, Greinacher A (2002) Heparin-induced thrombocytopenia: temporal pattern of thrombocytopenia in relation to initial use or reexposure to heparin. Chest 122:37–42

Lubenow N, Hinz P, Thomaschewski S, Lietz T, Vogler M, Ladwig A, Junger M, Nauck M, Schellong S, Wander K, Engel G, Ekkernkamp A, Greinacher A (2010) The severity of trauma determines the immune response to PF4/heparin and the frequency of heparin-induced thrombocytopenia. Blood 115:1797–1803

Ludwig RJ, Schindewolf M, Alban S, Kaufmann R, Lindhoff-Last E, Boehncke WH (2005) Molecular weight determines the frequency of delayed type hypersensitivity reactions to heparin and synthetic oligosaccharides. Thromb Haemost 94:1265–1269

Ludwig RJ, Schindewolf M, Utikal J, Lindhoff-Last E, Boehncke WH (2006) Management of cutaneous type IV hypersensitivity reactions induced by heparin. Thromb Haemost 96:611–617

Ludwig RJ, Schindewolf M, Lindhoff-Last E, Boehncke WH (2008) The influence of heparin's molecular weight and the incidence of delayed type hypersensitivity reactions revisited; in response to Grims et al., Br J Dermatol 2007; 157:514–17. Br J Dermatol 158:849–851

Macie C, Forbes L, Foster GA, Douketis JD (2004) Dosing practices and risk factors for bleeding in patients receiving enoxaparin for the treatment of an acute coronary syndrome. Chest 125:1616–1621

Magee KD, Campbell SG, Moher D, Rowe BH (2008) Heparin versus placebo for acute coronary syndromes. Cochrane Database Syst Rev: CD003462

Manoukian SV, Voeltz MD, Eikelboom J (2007) Bleeding complications in acute coronary syndromes and percutaneous coronary intervention: predictors, prognostic significance, and paradigms for reducing risk. Clin Cardiol 30:II24–II34

Martel N, Lee J, Wells PS (2005) Risk for heparin-induced thrombocytopenia with unfractionated and low-molecular-weight heparin thromboprophylaxis: a meta-analysis. Blood 106:2710–2715

Matziolis G, Perka C, Disch A, Zippel H (2003) Effects of fondaparinux compared with dalteparin, enoxaparin and unfractionated heparin on human osteoblasts. Calcif Tissue Int 73:370–379

Melloni C, Alexander KP, Chen AY, Newby LK, Roe MT, Allen LaPointe NM, Pollack CV Jr, Gibler WB, Ohman EM, Peterson ED (2008) Unfractionated heparin dosing and risk of major bleeding in non-ST-segment elevation acute coronary syndromes. Am Heart J 156:209–215

MHRA (2011) Clexane. Summary of product characteristics. http://www.sanofi-aventis.co.uk/products/Clexane_SPC.pdf

Miller PD (2006) Guidelines for the diagnosis of osteoporosis: T-scores vs fractures. Rev Endocr Metab Disord 7:75–89

Mismetti P, Laporte S, Darmon JY, Buchmuller A, Decousus H (2001) Meta-analysis of low molecular weight heparin in the prevention of venous thromboembolism in general surgery. Br J Surg 88:913–930

Monreal M, Lafoz E, Salvador R, Roncales J, Navarro A (1989) Adverse effects of three different forms of heparin therapy: thrombocytopenia, increased transaminases, and hyperkalaemia. Eur J Clin Pharmacol 37:415–418

Monreal M, Lafoz E, Olive A, del Rio L, Vedia C (1994) Comparison of subcutaneous unfractionated heparin with a low molecular weight heparin (Fragmin) in patients with venous thromboembolism and contraindications to coumarin. Thromb Haemost 71:7–11

Muir JM, Andrew M, Hirsh J, Weitz JI, Young E, Deschamps P, Shaughnessy SG (1996) Histomorphometric analysis of the effects of standard heparin on trabecular bone in vivo. Blood 88:1314–1320

Muir JM, Hirsh J, Weitz JI, Andrew M, Young E, Shaughnessy SG (1997) A histomorphometric comparison of the effects of heparin and low-molecular-weight heparin on cancellous bone in rats. Blood 89:3236–3242

Mulloy B (2005) The specificity of interactions between proteins and sulfated polysaccharides. An Acad Bras Cienc 77:651–664

Murphy SA, Gibson CM, Morrow DA, Van de Werf F, Menown IB, Goodman SG, Mahaffey KW, Cohen M, McCabe CH, Antman EM, Braunwald E (2007) Efficacy and safety of the low-molecular weight heparin enoxaparin compared with unfractionated heparin across the acute coronary syndrome spectrum: a meta-analysis. Eur Heart J 28:2077–2086

Nelson-Piercy C (1997) Hazards of heparin: allergy, heparin-induced thrombocytopenia and osteoporosis. Baillieres Clin Obstet Gynaecol 11:489–509

Nieuwenhuis HK, Albada J, Banga JD, Sixma JJ (1991) Identification of risk factors for bleeding during treatment of acute venous thromboembolism with heparin or low molecular weight heparin. Blood 78:2337–2343

Oie CI, Olsen R, Smedsrod B, Hansen JB (2008) Liver sinusoidal endothelial cells are the principal site for elimination of unfractionated heparin from the circulation. Am J Physiol Gastrointest Liver Physiol 294:G520–G528

Palmer AJ, Koppenhagen K, Kirchhof B, Weber U, Bergemann R (1997) Efficacy and safety of low molecular weight heparin, unfractionated heparin and warfarin for thrombo-embolism prophylaxis in orthopaedic surgery: a meta-analysis of randomised clinical trials. Haemostasis 27:75–84

Paus R (1991) Hair growth inhibition by heparin in mice: a model system for studying the modulation of epithelial cell growth by glycosaminoglycans? Br J Dermatol 124:415–422

Pettila V, Kaaja R, Leinonen P, Ekblad U, Kataja M, Ikkala E (1999) Thromboprophylaxis with low molecular weight heparin (dalteparin) in pregnancy. Thromb Res 96:275–282

Pettila V, Leinonen P, Markkola A, Hiilesmaa V, Kaaja R (2002) Postpartum bone mineral density in women treated for thromboprophylaxis with unfractionated heparin or LMW heparin. Thromb Haemost 87:182–186

Pouplard C, Iochmann S, Renard B, Herault O, Colombat P, Amiral J, Gruel Y (2001) Induction of monocyte tissue factor expression by antibodies to heparin-platelet factor 4 complexes developed in heparin-induced thrombocytopenia. Blood 97:3300–3302

Pouplard C, Gueret P, Fouassier M, Ternisien C, Trossaert M, Regina S, Gruel Y (2007) Prospective evaluation of the '4Ts' score and particle gel immunoassay specific to heparin/PF4 for the diagnosis of heparin-induced thrombocytopenia. J Thromb Haemost 5:1373–1379

Pratt DS, Kaplan MM (2000) Evaluation of abnormal liver-enzyme results in asymptomatic patients. N Engl J Med 342:1266–1271

Qian Y, Pan J, Zhou X, Weiser P, Lu H, Zhang L (2010) Molecular mechanism underlines heparin-induced thrombocytopenia and thrombosis. Prog Mol Biol Transl Sci 93:395–421

Rajgopal R, Bear M, Butcher MK, Shaughnessy SG (2008) The effects of heparin and low molecular weight heparins on bone. Thromb Res 122:293–298

Rao SV, Eikelboom J, Steg PG, Lincoff AM, Weintraub WS, Bassand JP, Rao AK, Gibson CM, Petersen JL, Mehran R, Manoukian SV, Charnigo R, Lee KL, Moscucci M, Harrington RA (2009) Standardized reporting of bleeding complications for clinical investigations in acute coronary syndromes: a proposal from the academic bleeding consensus (ABC) multidisciplinary working group. Am Heart J 158(881–886):e1

Rapaport SI, Vermylen J, Hoylaerts M, Saito H, Hirsh J, Bates S, Dahlback B, Poller L (2004) The multiple faces of the partial thromboplastin time APTT. J Thromb Haemost 2:2250–2259

Raskob GE, Hirsh J (2003) Controversies in timing of the first dose of anticoagulant prophylaxis against venous thromboembolism after major orthopedic surgery. Chest 124:379S–385S

Rauova L, Poncz M, McKenzie SE, Reilly MP, Arepally G, Weisel JW, Nagaswami C, Cines DB, Sachais BS (2005) Ultralarge complexes of PF4 and heparin are central to the pathogenesis of heparin-induced thrombocytopenia. Blood 105:131–138

Rauova L, Zhai L, Kowalska MA, Arepally GM, Cines DB, Poncz M (2006) Role of platelet surface PF4 antigenic complexes in heparin-induced thrombocytopenia pathogenesis: diagnostic and therapeutic implications. Blood 107:2346–2353

Riess H, Koppenhagen K, Tolle A, Kemkes-Matthes B, Grave M, Patek F, Drexler M, Siemens HJ, Harenberg J, Weidinger G, Brom J, Haas S (2003) Fixed-dose, body weight-independent subcutaneous low molecular weight heparin Certoparin compared with adjusted-dose intravenous unfractionated heparin in patients with proximal deep venous thrombosis. Thromb Haemost 90:252–259

Riess H, Haas S, Tebbe U, Gerlach HE, Abletshauser C, Sieder C, Rossol S, Pfeiffer B, Schellong SM (2010) A randomized, double-blind study of certoparin vs. unfractionated heparin to prevent venous thromboembolic events in acutely ill, non-surgical patients: CERTIFY Study. J Thromb Haemost 8:1209–1215

Rodger MA, Kahn SR, Cranney A, Hodsman A, Kovacs MJ, Clement AM, Lazo-Langner A, Hague WM (2007) Long-term dalteparin in pregnancy not associated with a decrease in bone mineral density: substudy of a randomized controlled trial. J Thromb Haemost 5:1600–1606

Ruiz-Gimenez N, Suarez C, Gonzalez R, Nieto JA, Todoli JA, Samperiz AL, Monreal M (2008) Predictive variables for major bleeding events in patients presenting with documented acute venous thromboembolism. Findings from the RIETE Registry. Thromb Haemost 100:26–31

Russo MW, Watkins PB (2004) Are patients with elevated liver tests at increased risk of drug-induced liver injury? Gastroenterology 126:1477–1480

Samama MM, Cohen AT, Darmon JY, Desjardins L, Eldor A, Janbon C, Leizorovicz A, Nguyen H, Olsson CG, Turpie AG, Weisslinger N (1999) A comparison of enoxaparin with placebo for the prevention of venous thromboembolism in acutely ill medical patients. Prophylaxis in Medical Patients with Enoxaparin Study Group. N Engl J Med 341:793–800

Sarris E, Tsele E, Bagiatoudi G, Salpigidis K, Stavrianaki D, Kaklamanis L, Siakotos M (2003) Diffuse alopecia in a hemodialysis patient caused by a low-molecular-weight heparin, tinzaparin. Am J Kidney Dis 41:E15

Savi P, Chong BH, Greinacher A, Gruel Y, Kelton JG, Warkentin TE, Eichler P, Meuleman D, Petitou M, Herault JP, Cariou R, Herbert JM (2005) Effect of fondaparinux on platelet activation in the presence of heparin-dependent antibodies: a blinded comparative multicenter study with unfractionated heparin. Blood 105:139–144

Scherer K, Tsakiris DA, Bircher AJ (2008) Hypersensitivity reactions to anticoagulant drugs. Curr Pharm Des 14:2863–2873

Schindewolf M, Ludwig RJ (2010) Need for an increasing awareness for heparin-induced skin lesions. Expert Rev Dermatol 5:1–3

Schindewolf M, Schwaner S, Wolter M, Kroll H, Recke A, Kaufmann R, Boehncke WH, Lindhoff-Last E, Ludwig RJ (2009) Incidence and causes of heparin-induced skin lesions. CMAJ 181:477–481

Schindewolf M, Kroll H, Ackermann H, Garbaraviciene J, Kaufmann R, Boehncke WH, Ludwig RJ, Lindhoff-Last E (2010a) Heparin-induced non-necrotizing skin lesions: rarely associated with heparin-induced thrombocytopenia. J Thromb Haemost 8:1486–1491

Schindewolf M, Scheuermann J, Kroll H, Garbaraviciene J, Hecking C, Marzi I, Wolter M, Kaufmann R, Boehncke WH, Lindhoff-Last E, Ludwig RJ (2010b) Low allergenic potential with fondaparinux: results of a prospective investigation. Mayo Clin Proc 85:913–919

Schulman S, Beyth RJ, Kearon C, Levine MN (2008) Hemorrhagic complications of anticoagulant and thrombolytic treatment: American College of Chest Physicians Evidence-Based Clinical Practice Guidelines (8th Edition). Chest 133:257S–298S

Seitz CS, Brocker EB, Trautmann A (2008) Management of allergy to heparins in postoperative care: subcutaneous allergy and intravenous tolerance. Dermatol Online J 14:4

Senior J (2003) Regulatory perspectives. In: Kaplowitz N, DeLeve LD (eds) Drug-induced liver disease. Marcel Dekker, New York, pp 739–754

Shantsila E, Lip GY, Chong BH (2009) Heparin-induced thrombocytopenia. A contemporary clinical approach to diagnosis and management. Chest 135:1651–1664

Shaughnessy SG, Young E, Deschamps P, Hirsh J (1995) The effects of low molecular weight and standard heparin on calcium loss from fetal rat calvaria. Blood 86:1368–1373

Shaughnessy SG, Hirsh J, Bhandari M, Muir JM, Young E, Weitz JI (1999) A histomorphometric evaluation of heparin-induced bone loss after discontinuation of heparin treatment in rats. Blood 93:1231–1236

Stein PD, Hull RD, Matta F, Yaekoub AY, Liang J (2009) Incidence of thrombocytopenia in hospitalized patients with venous thromboembolism. Am J Med 122:919–930

Suh JS, Aster RH, Visentin GP (1998) Antibodies from patients with heparin-induced thrombocytopenia/thrombosis recognize different epitopes on heparin: platelet factor 4. Blood 91:916–922

Suzuki S, Koide M, Sakamoto S, Yamamoto S, Matsuo M, Fujii E, Matsuo T (1997) Early onset of immunological heparin-induced thrombocytopenia in acute myocardial infarction. Blood Coagul Fibrinolysis 8:13–15

Szolar-Platzer C, Aberer W, Kranke B (2000) Delayed-type skin reaction to the heparin-alternative danaparoid. J Am Acad Dermatol 43:920–922

Taha MO, Goncalves PF, Vidigal RO, Sacchi GP, Pinheiro FG, Brandao LS, Feguri S, Moraes-Filho AM, Ghadie MM, Magalhaes KG, Simoes MJ, Tersariol IL, Monteiro HP, Oliveira-Junior IS, Jurkiewicz A, Caricati-Neto A (2009) Protective effects of heparin on hepatic ischemia and reperfusion lesions in rabbits. Transplant Proc 41:812–815

Tardy-Poncet B, Piot M, Chapelle C, France G, Campos L, Garraud O, Decousus H, Mismetti P, Tardy B (2009) Thrombin generation and heparin-induced thrombocytopenia. J Thromb Haemost 7:1474–1481

Testroote M, Stigter W, de Visser DC, Janzing H (2008) Low molecular weight heparin for prevention of venous thromboembolism in patients with lower-leg immobilization. Cochrane Database Syst Rev: CD006681

Thery C, Simonneau G, Meyer G, Helenon O, Bridey F, Armagnac C, d'Azemar P, Coquart JP (1992) Randomized trial of subcutaneous low-molecular-weight heparin CY 216 (Fraxiparine) compared with intravenous unfractionated heparin in the curative treatment of submassive pulmonary embolism. A dose-ranging study. Circulation 85:1380–1389

TIMI 11A Trial Investigators (1997) Dose-ranging trial of enoxaparin for unstable angina: results of TIMI 11A. The Thrombolysis in Myocardial Infarction. J Am Coll Cardiol 29:1474–1482

Tooher R, Gates S, Dowswell T, Davis LJ (2010) Prophylaxis for venous thromboembolic disease in pregnancy and the early postnatal period. Cochrane Database Syst Rev: CD001689

Trautmann A, Seitz CS (2009) Heparin allergy: delayed-type non-IgE-mediated allergic hypersensitivity to subcutaneous heparin injection. Immunol Allergy Clin North Am 29:469–480

Trautmann A, Seitz CS (2010) The complex clinical picture of side effects to anticoagulation. Med Clin North Am 94:821–834, xii–iii

Trouvin AP, Goeb V (2010) Receptor activator of nuclear factor-kappaB ligand and osteoprotegerin: maintaining the balance to prevent bone loss. Clin Interv Aging 5:345–354

Trujillo-Santos J, Nieto JA, Tiberio G, Piccioli A, Di Micco P, Prandoni P, Monreal M (2008) Predicting recurrences or major bleeding in cancer patients with venous thromboembolism. Findings from the RIETE Registry. Thromb Haemost 100:435–439

Tudhope GR, Cohen H, Meikle RW (1958) Alopecia following treatment with dextran sulphate and other anticoagulant drugs. Br Med J 1:1034–1037

Turcotte JG, Kraft RO, Fry WJ (1965) Heparin reactions in patients with vascular disease. Arch Surg 90:375–379

Turpie AG, Bauer KA, Eriksson BI, Lassen MR (2002) Fondaparinux vs enoxaparin for the prevention of venous thromboembolism in major orthopedic surgery: a meta-analysis of 4 randomized double-blind studies. Arch Intern Med 162:1833–1840

Umlas J, Harken DE (1988) Warfarin-induced alopecia. Cutis 42:63–64

Van de Werf F, Bax J, Betriu A, Blomstrom-Lundqvist C, Crea F, Falk V, Filippatos G, Fox K, Huber K, Kastrati A, Rosengren A, Steg PG, Tubaro M, Verheugt F, Weidinger F, Weis M, Vahanian A, Camm J, De Caterina R, Dean V, Dickstein K, Funck-Brentano C, Hellemans I, Kristensen SD, McGregor K, Sechtem U, Silber S, Tendera M, Widimsky P, Zamorano JL, Aguirre FV, Al-Attar N, Alegria E, Andreotti F, Benzer W, Breithardt O, Danchin N, Di Mario C, Dudek D, Gulba D, Halvorsen S, Kaufmann P, Kornowski R, Lip GY, Rutten F (2008) Management of acute myocardial infarction in patients presenting with persistent ST-segment elevation: the Task Force on the Management of ST-Segment Elevation Acute Myocardial Infarction of the European Society of Cardiology. Eur Heart J 29:2909–2945

van der Wiel HE, Lips P, Huijgens PC, Netelenbos JC (1993) Effects of short-term low-dose heparin administration on biochemical parameters of bone turnover. Bone Miner 22:27–32

van Dongen CJ, MacGillavry MR, Prins MH (2005) Once versus twice daily LMWH for the initial treatment of venous thromboembolism. Cochrane Database Syst Rev: CD003074

Vik A, Brodin E, Sveinbjornsson B, Hansen JB (2007) Heparin induces mobilization of osteoprotegerin into the circulation. Thromb Haemost 98:148–154

Wang YY, Po HL (2006) Enoxaparin-induced alopecia in patients with cerebral venous thrombosis. J Clin Pharm Ther 31:513–517

Warkentin TE (1996) Heparin-induced skin lesions. Br J Haematol 92:494–497

Warkentin TE (2003) Heparin-induced thrombocytopenia: pathogenesis and management. Br J Haematol 121:535–555

Warkentin TE (2006) Think of HIT. Hematology Am Soc Hematol Educ Program: 408–414

Warkentin TE (2007) Clinical picture of heparin-induced thrombocytopenia. In: Warkentin TGA (ed) Heparin-induced thrombocytopenia, 4th edn. Informa Healthcare, New York, pp 21–66

Warkentin TE (2008) Fondaparinux versus direct thrombin inhibitor therapy for the management of heparin-induced thrombocytopenia (HIT) – bridging the River Coumarin. Thromb Haemost 99:2–3

Warkentin TE (2010) Agents for the treatment of heparin-induced thrombocytopenia. Hematol Oncol Clin North Am 24:755–775, ix

Warkentin TE, Greinacher A (2009) Heparin-induced anaphylactic and anaphylactoid reactions: two distinct but overlapping syndromes. Expert Opin Drug Saf 8:129–144

Warkentin TE, Heddle NM (2003) Laboratory diagnosis of immune heparin-induced thrombocytopenia. Curr Hematol Rep 2:148–157

Warkentin TE, Kelton JG (2001) Temporal aspects of heparin-induced thrombocytopenia. N Engl J Med 344:1286–1292

Warkentin TE, Linkins LA (2010) Non-necrotizing heparin-induced skin lesions and the 4T's score. J Thromb Haemost 8:1483–1485

Warkentin TE, Sheppard JI (1999) Generation of platelet-derived microparticles and procoagulant activity by heparin-induced thrombocytopenia IgG/serum and other IgG platelet agonists: a comparison with standard platelet agonists. Platelets 10:319–326

Warkentin TE, Elavathil LJ, Hayward CP, Johnston MA, Russett JI, Kelton JG (1997) The pathogenesis of venous limb gangrene associated with heparin-induced thrombocytopenia. Ann Intern Med 127:804–812

Warkentin TE, Chong BH, Greinacher A (1998) Heparin-induced thrombocytopenia: towards consensus. Thromb Haemost 79:1–7

Warkentin TE, Cook RJ, Marder VJ, Sheppard JA, Moore JC, Eriksson BI, Greinacher A, Kelton JG (2005a) Anti-platelet factor 4/heparin antibodies in orthopedic surgery patients receiving antithrombotic prophylaxis with fondaparinux or enoxaparin. Blood 106:3791–3796

Warkentin TE, Roberts RS, Hirsh J, Kelton JG (2005b) Heparin-induced skin lesions and other unusual sequelae of the heparin-induced thrombocytopenia syndrome: a nested cohort study. Chest 127:1857–1861

Warkentin TE, Sheppard JA, Moore JC, Moore KM, Sigouin CS, Kelton JG (2005c) Laboratory testing for the antibodies that cause heparin-induced thrombocytopenia: how much class do we need? J Lab Clin Med 146:341–346

Warkentin TE, Sheppard JA, Sigouin CS, Kohlmann T, Eichler P, Greinacher A (2006) Gender imbalance and risk factor interactions in heparin-induced thrombocytopenia. Blood 108:2937–2941

Warkentin TE, Greinacher A, Koster A, Lincoff AM (2008a) Treatment and prevention of heparin-induced thrombocytopenia: American College of Chest Physicians Evidence-Based Clinical Practice Guidelines (8th Edition). Chest 133:340S–380S

Warkentin TE, Makris M, Jay RM, Kelton JG (2008b) A spontaneous prothrombotic disorder resembling heparin-induced thrombocytopenia. Am J Med 121:632–636

Warkentin TE, Sheppard JA, Moore JC, Cook RJ, Kelton JG (2009) Studies of the immune response in heparin-induced thrombocytopenia. Blood 113:4963–4969

Warkentin TE, Cook RJ, Marder VJ, Greinacher A (2010) Anti-PF4/heparin antibody formation postorthopedic surgery thromboprophylaxis: the role of non-drug risk factors and evidence for a stoichiometry-based model of immunization. J Thromb Haemost 8:504–512

Zcharia E, Philp D, Edovitsky E, Aingorn H, Metzger S, Kleinman HK, Vlodavsky I, Elkin M (2005) Heparanase regulates murine hair growth. Am J Pathol 166:999–1008

Zikria JC, Ansell J (2009) Oral anticoagulation with factor Xa and thrombin inhibitors: on the threshold of change. Curr Opin Hematol 16:347–356

# Neutralization of Heparin Activity

**Menaka Pai and Mark A. Crowther**

## Contents



M. Pai
Department of Medicine, McMaster University, Hamilton General Hospital, Room 1-270A, 237 Barton Street East, L8L 2X2, Hamilton, ON, Canada
e-mail: mpai@mcmaster.ca

M.A. Crowther (✉)
Department of Pathology and Molecular Medicine, McMaster University, Hamilton, ON, Canada

Hamilton Regional Laboratory Medicine Program, McMaster University, Hamilton, ON, Canada

St. Joseph's Hospital, Room L-208, 50 Charlton Avenue East, Hamilton, ON, Canada L8N 4A6,
e-mail: crowthrm@mcmaster.ca

R. Lever et al. (eds.), *Heparin - A Century of Progress*,
Handbook of Experimental Pharmacology 207,
DOI 10.1007/978-3-642-23056-1_11, 

**Abstract** Heparin is the mainstay in the treatment and prevention of thrombosis in such diverse clinical settings as venous thromboembolism, acute coronary syndrome, cardiopulmonary bypass, and hemodialysis. However, the major complication of heparin – like that of all anticoagulants – is bleeding. Heparin may need to be reversed in the following settings: clinically significant bleeding; prior to an invasive procedure; at the conclusion of a procedure involving extracorporeal circulation (e.g., cardiopulmonary bypass, dialysis). This chapter discusses protamine sulfate, as well as several other agents that are able to neutralize heparin, including their pharmacological properties, indications, dosing, and efficacy.

**Keywords** Heparin • Heparin, Low-Molecular-Weight • Platelet Factor 4 • Protamine Sulfate • Recombinant FVIIa

## 1 Introduction

Heparin is the mainstay in the treatment and prevention of thrombosis in such diverse clinical settings as venous thromboembolism, acute coronary syndrome, cardiopulmonary bypass, and hemodialysis. However, the major complication of heparin – like that of all anticoagulants – is bleeding. Heparin inhibits hemostasis by upregulating the action of antithrombin over 1,000-fold (which in turn inactivates thrombin and factor Xa), impairing platelet function (Fernandez et al. 1986), and increasing capillary permeability (Blajchman et al. 1989). Unfractionated heparin (UFH) is a mixture of sulfated glycosaminoglycans ranging in molecular weight from 3,000 to 30,000 Da, while low-molecular-weight heparin (LMWH) is obtained by depolymerization and/or fractionation of this mixture into a more homogenous subset of lower molecular weight (Hirsh et al. 2008). The anti-Xa effect of LMWH dominates over its antithrombin effect. Heparin is associated with a small, but clinically significant risk of bleeding. The risk of major hemorrhage with UFH or LMWH – bleeding that is intracranial or retroperitoneal, leads directly to death, or results in hospitalization or transfusion – is up to 3% when these agents are used over the short- to medium-term in hospitalized patients (Crowther et al. 2002). This risk depends on the indication for anticoagulation, the use of concomitant medication (including antiplatelet agents and thrombolytics), the heparin dose, and the duration of treatment (Hirsh et al. 2008; Schulman and Bijsterveld 2007; Schulman et al. 2008; Collins et al. 1996; Anonymous 1996, 1997; Klein et al. 1997; Anand et al. 2003; Sun et al. 2008). LMWH is associated with approximately half the risk of bleeding than UFH in the setting of venous thromboembolism, as demonstrated in two meta-analyses and a major pooled analysis of 19 trials (Dolovich et al. 2000; Gould et al. 1999; van Dongen et al. 2004). Conversely, several large trials in patients with acute coronary syndrome suggested that LMWH is associated with a marginally higher bleeding risk (Blazing et al. 2004; Ferguson et al. 2004).

Heparin may need to be reversed in the following settings: clinically significant bleeding; prior to an invasive procedure; at the conclusion of a procedure involving extracorporeal circulation (e.g., cardiopulmonary bypass, dialysis). This chapter discusses several agents that are able to neutralize heparin, including their pharmacological properties, indications, dosing, and efficacy.

## 2 Protamine

### 2.1 *Properties and Pharmacokinetics*

Protamine is a polycationic, highly alkaline protein molecule, predominantly composed of arginine residues. This protein was first described in 1868, when it was found to be present in the spermatozoa of fish (Hirsh et al. 2008). Protamine acts in the nucleus as a DNA stabilizer (similar to a histone) in the haploid phase of spermatogenesis. Today, the major commercial sources of protamine are the sperm ducts of salmon and recombinant technology.

Protamine and UFH interact in a straightforward way, with the large positively charged protamine molecule forming a stable complex with the negatively charged UFH molecule. This ion pair has no anticoagulant effect whatsoever (Hirsh et al. 2008). Protamine itself does have weak anticoagulant activity; however, this effect is clinically insignificant at doses usually administered to patients. After single-dose administration, protamine has a half-life of less than seven minutes (Hirsh et al. 2008). This rapid decline is in contrast to the half-life of UFH, which is 1–2 h, and the half-life of LMWH, which is 4–12 h (Hirsh et al. 2008).

### 2.2 *Efficacy and Indications*

Protamine sulfate is a specific and virtually complete reversal agent for UFH; however, it only partially neutralizes the anticoagulant effect of LMWH (Hirsh et al. 2008; Schulman et al. 2008; van Veen et al. 2005; Patel et al. 2007). This is due to the reduced sulfate charge densities of LMWH, and subsequently decreased protamine binding capacity (Crowther et al. 2002). As the sulfate content of an LMWH goes up, the degree of anti-Xa neutralization by protamine goes up as well (Crowther et al. 2002). Therefore, tinzaparin, which has the highest amount of sulfate per saccharide unit, is more neutralizable than dalteparin, which is in turn more neutralizable than enoxaparin. Protamine sulfate does neutralize the anti-thrombin effect of LMWH (as it does with UFH), so the aPTT and thrombin time may normalize, while the anti-Xa activity remains high (Hirsh et al. 2008).

Although protamine is used for urgent reversal of anticoagulation associated with LMWH, this reversal is not always effective (Schulman and Bijsterveld 2007; Kessler 2004; Holst et al. 1994; Massonnet-Castel et al. 1986; Wiernikowski et al. 2007). In a case series of 15 patients undergoing cardiac surgery, three patients developed abnormal bleeding (Massonnet-Castel et al. 1986). Protamine sulfate, even when given in repeated doses, failed to stop bleeding in two of the patients. There are no well-designed prospective human studies that have demonstrated the clinical efficacy (or inefficacy) of protamine sulfate in controlling bleeding associated with the use of LMWH.

## 2.3 *Dosing*

One milligram of protamine sulfate neutralizes 90 United States Pharmacopeia (USP) units of bovine origin UFH and 115 USP units of porcine origin UFH (Schulman et al. 2008). It is important to take into account the half-life of UFH (~1–2 h) when giving protamine. In the nonextracorporeal circulation setting, protamine sulfate is dosed according to the amount of UFH given, as well as the recency of UFH administration (Table 1) (Hirsh et al. 2008). For example, if a patient has received a bolus of 30,000 U of bovine UFH as a bolus very recently, 300 mg of protamine should be administered. If this dose of UFH was given an hour ago, 150 mg of protamine should be administered, and if it was given 2 h ago, 75 mg of protamine should be administered. If UFH is given as an IV infusion, heparin given in the preceding 2 h should be considered when calculating the dose of protamine sulfate (Hirsh et al. 2008). For example, if a patient were receiving a continuous IV infusion of heparin at 1,000 U/h, the required protamine dose would be about 20 mg. The APTT can be used to assess the effectiveness of protamine sulfate neutralization of the anticoagulant effects of heparin (Hirsh et al. 2008). To avoid hypotensive reactions associated with rapid infusion of large quantities of protamine, doses can be broken into 50 mg aliquots, each administered over 10 min (Schulman and Bijsterveld 2007).

Since protamine has a shorter half-life than UFH, a single-dose regimen may not be sufficient to maintain reversal of heparin anticoagulation (Hirsh et al. 2008; Crowther and Warkentin 2008; Warkentin and Crowther 2002). This phenomenon is called "heparin rebound." Heparin rebound is particularly common in cardiopulmonary bypass, where patients undergo recurrent exposure to small amounts of heparin. In

**Table 1** Protamine sulfate dosing recommendations for unfractionated and low-molecular-weight heparin

| | Dose of protamine sulfate |
|---|---|
| Intravenous bolus of unfractionated heparin | • Administered <1 h ago? 1 mg/100 units of heparin<br>• Administered 1–2 h ago? 1 mg/200 units of heparin<br>• Administered >2 h ago? 1 mg/400 units of heparin |
| Intravenous infusion of unfractionated heparin | • 1 mg/100 units of heparin administered in preceding 2 h |
| Subcutaneous low-molecular-weight heparin | • Administered <8 h ago? 1 mg/100 anti-Xa units of heparin, followed by 0.5 mg/100 anti-Xa units of heparin if bleeding continues<br>• Administered <8 h ago? 1 mg/200 anti-Xa units of heparin, followed by 0.5 mg/200 anti-Xa units of heparin if bleeding continues |

these settings, a maintenance infusion or repeat dosing of protamine may be more effective. The activated clotting time (ACT) should be used in conjunction with the aPTT to determine the degree of residual heparin (Hirsh et al. 2008; Crowther and Warkentin 2008; Warkentin and Crowther 2002). The ACT is less sensitive to heparin than the aPTT, which makes it useful in the setting of extracorporeal bypass, where large amounts of heparin are typically given. The ACT will produce an interpretable result in this setting, whereas the aPTT will likely be prolonged to an uninterpretable degree.

As protamine only partially neutralizes LMWH, and LMWH has a longer half-life than UFH, the suggested dosing regimen is different (Table 1). If LMWH was given within the past 8 h, 1 mg of protamine sulfate per 100 anti-Xa units of LMWH should be administered. (The amount of anti-Xa activity for specific formulations of LMWH should be determined from the manufacturer's product insert.) A second dose of 0.5 mg protamine sulfate per 100 anti-Xa units should be administered if bleeding continues. Smaller doses of protamine sulfate should be given if LMWH administration took place over 8 h ago (Hirsh et al. 2008; Crowther and Warkentin 2008; Warkentin and Crowther 2002).

## 2.4 *Adverse Effects*

Protamine is associated with both minor and severe adverse reactions. Minor adverse reactions, which occur in up to 16% of patients, include pruritis, flushing, urticaria, nausea, leukopenia, and thrombocytopenia (Caplan and Berkman 1976; Mismetti et al. 2005; Nybo and Madsen 2008; Park 2004; Weiler et al. 1990; Weiss et al. 1989; Weiss and Adkinson 1991; Mixon and Dehmer 2004). Severe reactions include hypotension, anaphylactoid reactions, anaphylactic reactions, and catastrophic pulmonary vasoconstriction (Nybo and Madsen 2008). This last reaction causes elevated pulmonary arterial pressures, pulmonary edema, and ultimately, cardiogenic shock. The incidence of severe reactions ranges from 0.2 to 3% (Park 2004; Weiler et al. 1990; Mixon and Dehmer 2004). A prospective study of 243 patients undergoing cardiopulmonary bypass surgery followed by protamine administration showed that 11% had an adverse reaction, while 2% experienced precipitous hypotension immediately after protamine infusion. Neither a positive skin test nor a positive IgE ELISA for antiprotamine antibody was predictive of an adverse reaction (Weiler et al. 1990).

Protamine affects a number of physiological processes, explaining its wide range of adverse effects. On its own, protamine affects platelet aggregation and weakens clot structure (Mochizuki et al. 1998). The heparin–protamine complex activates the classic complement pathway, generating the anaphylatoxins C4a and C5a (Park 2004; Weiler et al. 1990; Mixon and Dehmer 2004). This triggers aggregation and activation of leukocytes and platelets, the release of free oxygen radicals, and the formation of thromboxane A2. The latter may be responsible for pulmonary vasoconstriction (Park 2004; Weiler et al. 1990; Mixon and Dehmer 2004). Protamine also directly activates mast cells, causing histamine release, and stimulates nitric oxide production. These factors may be responsible for its hypotensive effect (Mixon and Dehmer 2004).

Finally, protamine is antigenic, with 15–30% of patients demonstrating IgE and IgG antibodies after administration of this drug (Weiss et al. 1989). Although this does not inevitably lead to an immunological reaction, these patients are at risk of developing anaphylaxis (Weiss et al. 1989). Patients who have previously received protamine sulfate (either for heparin neutralization or in the form of a protamine-containing insulin preparation), have undergone vasectomy, or have an allergy to fish may be at increased risk of having preformed antibodies against protamine sulfate and suffering from allergic reactions (Hirsh et al. 2008; Caplan and Berkman 1976; Stewart et al. 1984). The use of protamine should be undertaken with caution in patients who have had protamine reactions in the past, and they should be pretreated with corticosteroids and antihistamines (Hirsh et al. 2008).

# 3 Platelet Factor 4

## *3.1 Properties*

Platelet factor 4 (PF4) is a small chemokine belonging to the CXC chemokine family. It is synthesized in megakaryocytes, and released from alpha-granules of activated platelets during platelet aggregation (Fig. 1). This 7,800 Da protein has many biological activities, including the inhibition of angiogenesis, chemotactic attraction of neutrophils and monocytes, and maturation effects on megakaryocytes (Mixon and Dehmer 2004; Eisman et al. 1990). PF4 also acts as an endogenous inhibitor of heparin. Unlike protamine, which binds heparin as a simple ion complex, PF4 has several unique features in its primary, secondary, and tertiary structures that are essential for high-affinity heparin binding (Handin and Cohen 1976). Recombinant PF4 has a serum half-life of ~30 min, independent of the dose administered.

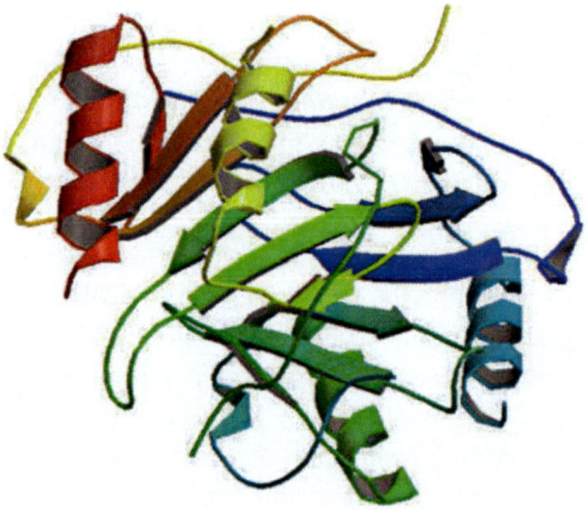

**Fig. 1** Platelet Factor 4. Taken from: http://en.wikipedia.org/wiki/File:PBB_Protein_PF4_image.jpg (Accessed September 1, 2009)

### 3.2 *Efficacy and Indications*

Recombinant PF4 (rPF4) has been tested in a phase one dose-finding trial, as well as compared to protamine in a randomized, blinded phase two trial (Dehmer et al. 1995, 1996). In the latter trial, 81 patients received 5,000 U of UFH in the setting of cardiac catheterization, and were randomized to receive either 1 mg/kg of intravenous recombinant PF4 over 2 min or 50 mg of intravenous protamine over 10 min (Dehmer et al. 1996). Serial measurements of ACT and aPTT (to assess heparin neutralization), hemodynamics, and laboratory tests of liver function were carried out. Both drugs reversed the anticoagulant effect, as measured by the activated coagulation time, and neither were associated with any serious adverse effects. Although the ACT and aPTT were slightly higher in the rPF4 group, there was no difference in clinically significant bleeding.

### 3.3 *Dosing*

rPF4 has been used at a dose of 1 mg/kg, administered intravenously over 2 min, to reverse 5,000 U of UFH in the setting of cardiac bypass. However, this agent is currently not FDA approved for any indication, and its development is no longer being pursued.

### 3.4 *Adverse Effects*

rPF4 has not been associated with any serious adverse effects in human studies, apart from transient chills, headache, back pain, and abdominal pain. All of these resolved without treatment (Mixon and Dehmer 2004). rPF4 is identical to naturally occurring PF4; however, it is not glycosylated (as it is synthesized in *Escherichia coli*). This could make rPF4 more antigenic than the naturally occurring protein, leading to neoantigen formation and (potentially) heparin-induced thrombocytopenia (Mixon and Dehmer 2004).

## 4 Recombinant Human fVIIa

### 4.1 *Properties*

Recombinant human fVIIa (rfVIIa) has been used, off-label, for reversal of heparin's anticoagulant effect, particularly when LMWH has been administered and protamine sulfate neutralization is incomplete and ineffective (Johansson 2008). This drug was initially designed for uncontrollable bleeding in hemophiliacs who had developed inhibitors against replacement coagulation factors. It is currently approved for this indication, as well as factor VII deficiency, in the United States. However, rfVIIa's powerful ability to initiate the coagulation cascade, even in the presence of factor deficiencies and anticoagulants, has made it an attractive option in a number of other settings.

## 4.2 *Efficacy and Indications*

An ex vivo study found that rfVIIa reversed the anticoagulant effects, as measured by thromboelastography, of both heparin and enoxaparin. rfVIIa also significantly decreased the activated partial thromboplastin time in samples containing UFH, though it did not decrease the anti-Xa levels in samples containing enoxaparin (Young et al. 2007). Case reports and case series suggest rfVIIa may be effective in patients with severe bleeding due to LMWH, though further studies are needed to validate its true utility in this setting (Cherfan et al. 2007; Firozvi et al. 2006; Ng et al. 2003).

## 4.3 *Dosing*

Case reports have used a single intravenous bolus of 50–100 U/kg to reverse the anticoagulant effect of heparin (Firozvi et al. 2006).

## 4.4 *Adverse Effects*

rfVIIa is associated with an increased risk of thrombosis, including deep vein thrombosis, pulmonary embolism, and myocardial infarction (Johansson 2008; O'Connell et al. 2006). As these are also the most common indications for heparin, clinicians should be extremely cautious when using rfVIIa for heparin reversal. In individuals with pre-existing thrombosis or thrombotic risk factors, we suggest that rfVIIa be used as an agent of "last resort," and only when heparin-associated bleeding is immediately life threatening. Before using this drug, clinicians should get informed consent from the patient and/or their substitute decision maker, with full disclosure that there is a lack of strong evidence that rfVIIa is safe and effective and that it is known to cause thrombosis.

# 5 Other Agents

Other large basic or polycationic molecules have been hypothesized to neutralize heparin, in a fashion similar to protamine. A 1996 study examined the use of methylene blue and vancomycin as heparin neutralizers; however, both of these drugs were found to be ineffective in reversing heparin in vivo (Kikura et al. 1996). In 1954, hexadimethrine bromide (Polybrene), a quaternary ammonium salt, was introduced into clinical practice (Weiss et al. 1958). During the early years of cardiopulmonary bypass, this agent was used as a highly effective heparin neutralizer. However, hexadimethrine was found to cause life-threatening renal failure at doses greater than 5 mg/kg (Ransdell et al. 1965). Due to its narrow therapeutic window, and the growing use of protamine sulfate, use of hexadimethrine was discontinued in 1962 (Pausescu and Marinescu 1964).

Another drug with promising efficacy, but a poor safety profile, is Heparinase I (Neutralase). This specific heparin-degrading enzyme is synthesized by *Flavobacterium heparinum*. Heparinase I acts by cleaving specific glycosidic linkages within the heparin molecule to create small oligosaccharide fragments, which have no residual activity against thrombin, and only 10–20% residual activity against fXa and none against factor IIa. Heparinase I has a very short half-life of just 5–18 min (Heres et al. 2001). After a 2001 dose-determining trial, heparinase I was compared with protamine in a 2005 randomized, double-blind, non-inferiority trial, in the setting of coronary bypass graft surgery (Heres et al. 2001; Stafford-Smith et al. 2005). About 167 patients received either heparinase I (maximum 35 g/kg) or protamine (maximum 650 mg) for heparin reversal (Stafford-Smith et al. 2005). Both arms displayed comparable efficacy in terms of anticoagulant reversal, as measured by the ACT and clinical assessment. However, the trial was terminated early on advisement of the Data Safety Monitoring Board, as heparinase I was associated with significantly more adverse events, including serious bleeding, transfusions, and increased length of stay.

Extracorporeal heparin removal devices have long been thought to be a safe and useful alternative to protamine, particularly in clinical situations where its use may be contraindicated (e.g., history of anaphylaxis) (Hirsh et al. 2008; Tao et al. 1998). Studies are currently underway to design nanostructured surfaces that bind heparin selectively. These surfaces could be incorporated into a high-efficiency "heparin filter" to decrease heparin concentrations at the conclusion of extracorporeal circulation (Martins et al. 2009; Byun et al. 1995, 1996; Zhang et al. 1998).

More recently, peptides have been developed to bind with high affinity to both UFH and LWMH. Schick et al. (2001) designed a series of peptides with multiple copies, arrange in tandem, of consensus heparin-binding sequences. (The tandem formation was created by including hydrophobic tails or cysteines on these residues to promote multimerization (Schick et al. 2004).) These concatameric peptides were capable of neutralizing both the antithrombin activity of UFH in vitro, and the antifactor Xa activity of UFH and enoxaparin (a LMWH) in vitro and in vivo in rats (Schick et al. 2001). Their neutralizing effect was equivalent to protamine for

UFH and greater than protamine (by 4- to 20-fold) for LMWH. In rats, these peptides were capable of neutralizing 1 U/mL anti-Factor Xa activity rats within 1–2 min, with no effect on blood pressure or heart rate, and very little rebound phenomenon (Schick et al. 2004). The safety and efficacy of concatameric peptides have not yet been confirmed in humans.

## References

Anand SS, Yusuf S, Pogue J, Ginsberg JS, Hirsh J (2003) Relationship of activated partial thromboplastin time to coronary events and bleeding in patients with acute coronary syndromes who receive heparin. Circulation 107(23):2884–2888

Anonymous (1996) Low-molecular-weight heparin during instability in coronary artery disease, Fragmin during Instability in Coronary Artery Disease (FRISC) study group. Lancet 347 (9001):561–568

Anonymous (1997) The International Stroke Trial (IST): a randomised trial of aspirin, subcutaneous heparin, both, or neither among 19435 patients with acute ischaemic stroke. International Stroke Trial Collaborative Group. Lancet 349(9065):1569–1581

Blajchman MA, Young E, Ofosu FA (1989) Effects of unfractionated heparin, dermatan sulfate and low molecular weight heparin on vessel wall permeability in rabbits. Ann N Y Acad Sci 556:245–254

Blazing MA, de Lemos JA, White HD, Fox KA, Verheugt FW, Ardissino D et al (2004) Safety and efficacy of enoxaparin vs unfractionated heparin in patients with non-ST-segment elevation acute coronary syndromes who receive tirofiban and aspirin: a randomized controlled trial. JAMA 292(1):55–64

Byun Y, Yun JH, Han IS, Fu Y, Shanberge JN, Yang VC (1995) A protamine filter for extracorporeal heparin removal. Development, testing, blood compatibility evaluation, and future direction. ASAIO J 41(3):M301–M305

Byun Y, Wang T, Kim JS, Yang VC (1996) Optimization and assessment of the in vivo efficacy of a heparin removing bio-reactor using mathematical modeling. ASAIO J 42(5):M782–M787

Caplan SN, Berkman EM (1976) Letter: protamine sulfate and fish allergy. N Engl J Med 295 (3):172

Cherfan A, Arabi Y, Al Askar A, Al Shimemeri A (2007) Recombinant activated factor VII treatment of retroperitoneal hematoma in a patient with renal failure receiving enoxaparin and clopidogrel. Pharmacotherapy 27(5):755–759

Collins R, MacMahon S, Flather M, Baigent C, Remvig L, Mortensen S et al (1996) Clinical effects of anticoagulant therapy in suspected acute myocardial infarction: systematic overview of randomised trials. BMJ 313(7058):652–659

Crowther MA, Warkentin TE (2008) Bleeding risk and the management of bleeding complications in patients undergoing anticoagulant therapy: focus on new anticoagulant agents. Blood 111 (10):4871–4879

Crowther MA, Berry LR, Monagle PT, Chan AK (2002) Mechanisms responsible for the failure of protamine to inactivate low-molecular-weight heparin. Br J Haematol 116(1):178–186

Dehmer GJ, Fisher M, Tate DA, Teo S, Bonnem EM (1995) Reversal of heparin anticoagulation by recombinant platelet factor 4 in humans. Circulation 91(8):2188–2194

Dehmer GJ, Lange RA, Tate DA, Pirwitz M, Daniel W, Fisher M et al (1996) Randomized trial of recombinant platelet factor 4 versus protamine for the reversal of heparin anticoagulation in humans. Circulation 94(9 Suppl):II347–II352

Dolovich LR, Ginsberg JS, Douketis JD, Holbrook AM, Cheah G (2000) A meta-analysis comparing low-molecular-weight heparins with unfractionated heparin in the treatment of

venous thromboembolism: examining some unanswered questions regarding location of treatment, product type, and dosing frequency. Arch Intern Med 160(2):181–188
Eisman R, Surrey S, Ramachandran B, Schwartz E, Poncz M (1990) Structural and functional comparison of the genes for human platelet factor 4 and PF4alt. Blood 76(2):336–344
Ferguson JJ, Califf RM, Antman EM, Cohen M, Grines CL, Goodman S et al (2004) Enoxaparin vs unfractionated heparin in high-risk patients with non-ST-segment elevation acute coronary syndromes managed with an intended early invasive strategy: primary results of the SYNERGY randomized trial. JAMA 292(1):45–54
Fernandez F, N'guyen P, Van Ryn J, Ofosu FA, Hirsh J, Buchanan MR (1986) Hemorrhagic doses of heparin and other glycosaminoglycans induce a platelet defect. Thromb Res 43(4):491–495
Firozvi K, Deveras RA, Kessler CM (2006) Reversal of low-molecular-weight heparin-induced bleeding in patients with pre-existing hypercoagulable states with human recombinant activated factor VII concentrate. Am J Hematol 81(8):582–589
Gould MK, Dembitzer AD, Doyle RL, Hastie TJ, Garber AM (1999) Low-molecular-weight heparins compared with unfractionated heparin for treatment of acute deep venous thrombosis. A meta-analysis of randomized, controlled trials. Ann Intern Med 130(10):800–809
Handin RI, Cohen HJ (1976) Purification and binding properties of human platelet factor four. J Biol Chem 251(14):4273–4282
Heres EK, Horrow JC, Gravlee GP, Tardiff BE, Luber J Jr, Schneider J et al (2001) A dose-determining trial of heparinase-I (Neutralase) for heparin neutralization in coronary artery surgery. Anesth Analg 93(6):1446–1452
Hirsh J, Bauer KA, Donati MB, Gould M, Samama MM, Weitz JI (2008) Parenteral anticoagulants: American College of Chest Physicians Evidence-Based Clinical Practice Guidelines (8th Edition). Chest 133(6 Suppl):141S–159S
Holst J, Lindblad B, Bergqvist D, Garre K, Nielsen H, Hedner U et al (1994) Protamine neutralization of intravenous and subcutaneous low-molecular-weight heparin (tinzaparin, Logiparin). An experimental investigation in healthy volunteers. Blood Coagul Fibrinolysis 5(5):795–803
Johansson PI (2008) Off-label use of recombinant factor VIIa for treatment of haemorrhage: results from randomized clinical trials. Vox Sang 95(1):1–7
Kessler CM (2004) Current and future challenges of antithrombotic agents and anticoagulants: strategies for reversal of hemorrhagic complications. Semin Hematol 41(1 Suppl 1):44–50
Kikura M, Lee MK, Levy JH (1996) Heparin neutralization with methylene blue, hexadimethrine, or vancomycin after cardiopulmonary bypass. Anesth Analg 83(2):223–227
Klein W, Buchwald A, Hillis SE, Monrad S, Sanz G, Turpie AG et al (1997) Comparison of low-molecular-weight heparin with unfractionated heparin acutely and with placebo for 6 weeks in the management of unstable coronary artery disease. Fragmin in unstable coronary artery disease study (FRIC). Circulation 96(1):61–68
Martins MC, Curtin SA, Freitas SC, Salgueiro P, Ratner BD, Barbosa MA (2009) Molecularly designed surfaces for blood deheparinization using an immobilized heparin-binding peptide. J Biomed Mater Res A 88(1):162–173
Massonnet-Castel S, Pelissier E, Bara L, Terrier E, Abry B, Guibourt P et al (1986) Partial reversal of low molecular weight heparin (PK 10169) anti-Xa activity by protamine sulfate: in vitro and in vivo study during cardiac surgery with extracorporeal circulation. Haemostasis 16 (2):139–146
Mismetti P, Quenet S, Levine M, Merli G, Decousus H, Derobert E et al (2005) Enoxaparin in the treatment of deep vein thrombosis with or without pulmonary embolism: an individual patient data meta-analysis. Chest 128(4):2203–2210
Mixon TA, Dehmer GJ (2004) Recombinant platelet factor 4 for heparin neutralization. Semin Thromb Hemost 30(3):369–377
Mochizuki T, Olson PJ, Szlam F, Ramsay JG, Levy JH (1998) Protamine reversal of heparin affects platelet aggregation and activated clotting time after cardiopulmonary bypass. Anesth Analg 87(4):781–785

Ng HJ, Koh LP, Lee LH (2003) Successful control of postsurgical bleeding by recombinant factor VIIa in a renal failure patient given low molecular weight heparin and aspirin. Ann Hematol 82 (4):257–258

Nybo M, Madsen JS (2008) Serious anaphylactic reactions due to protamine sulfate: a systematic literature review. Basic Clin Pharmacol Toxicol 103(2):192–196

O'Connell KA, Wood JJ, Wise RP, Lozier JN, Braun MM (2006) Thromboembolic adverse events after use of recombinant human coagulation factor VIIa. JAMA 295(3):293–298

Park KW (2004) Protamine and protamine reactions. Int Anesthesiol Clin 42(3):135–145

Patel AA, White CM, Coleman CI (2007) Use of protamine to rapidly reverse anticoagulant effect of unfractionated heparin in patients undergoing percutaneous coronary intervention. Conn Med 71(2):93–95

Pausescu E, Marinescu A (1964) [Neutralization of exogenous heparin by protamine or polybrene in subequivalent quantities after artificial perfusion]. Z Gesamte Inn Med 19(Suppl):43

Ransdell HT Jr, Haller JA Jr, Stowens D Jr, Barton PB Jr (1965) Renal toxicity of polybrene, (hexadimethrine bromide). J Surg Res 5:195–199

Schick BP, Gradowski JF, San Antonio JD, Martinez J (2001) Novel design of peptides to reverse the anticoagulant activities of heparin and other glycosaminoglycans. Thromb Haemost 85(3):482–487

Schick BP, Maslow D, Moshinski A, San Antonio JD (2004) Novel concatameric heparin-binding peptides reverse heparin and low-molecular-weight heparin anticoagulant activities in patient plasma in vitro and in rats in vivo. Blood 103(4):1356–1363

Schulman S, Bijsterveld NR (2007) Anticoagulants and their reversal. Transfus Med Rev 21(1):37–48

Schulman S, Beyth RJ, Kearon C, Levine MN (2008) Hemorrhagic complications of anticoagulant and thrombolytic treatment: American College of Chest Physicians Evidence-Based Clinical Practice Guidelines (8th Edition). Chest 133(6 Suppl):257S–298S

Stafford-Smith M, Lefrak EA, Qazi AG, Welsby IJ, Barber L, Hoeft A et al (2005) Efficacy and safety of heparinase I versus protamine in patients undergoing coronary artery bypass grafting with and without cardiopulmonary bypass. Anesthesiology 103(2):229–240

Stewart WJ, McSweeney SM, Kellett MA, Faxon DP, Ryan TJ (1984) Increased risk of severe protamine reactions in NPH insulin-dependent diabetics undergoing cardiac catheterization. Circulation 70(5):788–792

Sun JC, Whitlock R, Cheng J, Eikelboom JW, Thabane L, Crowther MA et al (2008) The effect of pre-operative aspirin on bleeding, transfusion, myocardial infarction, and mortality in coronary artery bypass surgery: a systematic review of randomized and observational studies. Eur Heart J 29(8):1057–1071

Tao W, Deyo DJ, Brunston RL Jr, Vertrees RA, Zwischenberger JB (1998) Extracorporeal heparin adsorption following cardiopulmonary bypass with a heparin removal device – an alternative to protamine. Crit Care Med 26(6):1096–1102

van Dongen CJ, van den Belt AG, Prins MH, Lensing AW (2004) Fixed dose subcutaneous low molecular weight heparins versus adjusted dose unfractionated heparin for venous thromboembolism. Cochrane Database Syst Rev (4):CD001100

van Veen JJ, Laidlaw S, Maclean RM (2005) Clinical use of protamine for the emergency reversal of anticaogulation with low molecular weight heparin. J Thromb Haemost (Suppl 1). Ref Type: Abstract

Warkentin TE, Crowther MA (2002) Reversing anticoagulants both old and new. Can J Anaesth 49 (6):S11–S25

Weiler JM, Gellhaus MA, Carter JG, Meng RL, Benson PM, Hottel RA et al (1990) A prospective study of the risk of an immediate adverse reaction to protamine sulfate during cardiopulmonary bypass surgery. J Allergy Clin Immunol 85(4):713–719

Weiss ME, Adkinson NF Jr (1991) Allergy to protamine. Clin Rev Allergy 9(3–4):339–355

Weiss WA, Gilman JS, Catenacci AJ, Osterberg AE (1958) Heparin neutralization with polybrene administered intravenously. J Am Med Assoc 166(6):603–607

Weiss ME, Nyhan D, Peng ZK, Horrow JC, Lowenstein E, Hirshman C et al (1989) Association of protamine IgE and IgG antibodies with life-threatening reactions to intravenous protamine. N Engl J Med 320(14):886–892

Wiernikowski JT, Chan A, Lo G (2007) Reversal of anti-thrombin activity using protamine sulfate. Experience in a neonate with a 10-fold overdose of enoxaparin. Thromb Res 120 (2):303–305

Young G, Yonekawa KE, Nakagawa PA, Blain RC, Lovejoy AE, Nugent DJ (2007) Recombinant activated factor VII effectively reverses the anticoagulant effects of heparin, enoxaparin, fondaparinux, argatroban, and bivalirudin ex vivo as measured using thromboelastography. Blood Coagul Fibrinolysis 18(6):547–553

Zhang Y, Singh VK, Yang VC (1998) Poly-L-lysine amplification of protamine immobilization and heparin adsorption. J Biomed Mater Res 42(2):182–187

# Part IV
# Non-anticoagulant Indications for Heparin and Related Compounds

# Non-anticoagulant Effects of Heparin: An Overview

**Rebecca Lever and Clive P. Page**

## Contents



**Abstract** Heparin has long been known to possess biological effects that are unrelated to its anticoagulant activity. In particular, much emphasis has been placed upon heparin, or novel agents based upon the heparin template, as potential anti-inflammatory agents. Moreover, heparin has been reported to possess clinical benefit in humans, including in chronic inflammatory diseases and cancer, that are over and above the expected effects on blood coagulation and which in many cases are entirely separable from this role. This chapter aims to provide an overview of the non-anticoagulant effects that have been ascribed to heparin, from those

R. Lever (✉)
The School of Pharmacy, University of London, 29-39 Brunswick Square, London WC1N 1AX, UK
e-mail: rebecca.lever@pharmacy.ac.uk

R. Lever et al. (eds.), *Heparin - A Century of Progress*,
Handbook of Experimental Pharmacology 207,
DOI 10.1007/978-3-642-23056-1_12, © Springer-Verlag Berlin Heidelberg 2012

involving the binding and inhibition of specific mediators involved in the inflammatory process to effects in whole system models of disease, with reference to the effects of heparin that have been reported to date in human diseases.

**Keywords** Cancer • Inflammation • Inflammatory disease • Non-anticoagulant effects

## 1 Introduction

Since early reports that heparin possesses effects aside of those on coagulation (Jaques 1979), many of which involve modulation of aspects of the inflammatory response (reviewed by Lever and Page 2002), there has been increasing interest in exploiting the non-anticoagulant pharmacology of heparin. Heparin is thought to exert many of its non-anticoagulant actions through binding of proteins such as chemokines and growth factors that are functionally dependent upon heparan sulphate (reviewed in Gallagher 2011, ibid and by Powell et al. 2004; Turnbull et al. 2001). However, the exact structural requirements for the majority of the anti-inflammatory effects of heparin are not fully known; interactions with proteins can vary from strongly sequence specific, such as the binding of antithrombin, to relatively non-specific (Mulloy et al. 1996); hence, the vast array of physiological mediators that is bound by this large and polyanionic molecule (reviewed by Tyrrell et al. 1999). Many well-known heparin-binding proteins are critically involved in the process of inflammation and include cytokines (Gilat et al. 1994), growth factors (Bono et al. 1997), adhesion molecules (Diamond et al. 1995; Koenig et al. 1998; Skinner et al. 1991; Watt et al. 1993), cytotoxic peptides (Fredens et al. 1991) and tissue-degrading enzymes (Redini et al. 1988; Walsh et al. 1991). Heparin, released from mast cells, has been suggested to play a physiological role in regulating the inflammatory response (Page 1991), through binding to and neutralising the activity of such proteins, hence limiting cellular activation and subsequent tissue damage and remodelling. However, whereas the heparin-binding domains of many proteins are either known or can be predicted by modern techniques, the corresponding polysaccharide sequences within the heparin molecule that interact with these sites are yet to be elucidated for the majority of proteins implicated in the inflammatory response. A rationale clearly exists to determine structural requirements within heparin for interactions with specific proteins that, if isolated, may possess useful properties as anti-inflammatory drugs. The use of standard heparin for the treatment of inflammatory diseases is, to an extent, limited by its effects on haemostasis. Therefore, it is important to note that many of the anti-inflammatory actions of heparin are unrelated to anticoagulant activity, as would be expected if these actions are due to interactions with specific mediators implicated in the inflammatory response.

This chapter aims to review the diverse range of effects ascribed to heparin that relate to limitation of the inflammatory response, with discussion of the potential utility of these actions in the treatment of inflammatory diseases.

## 2 Non-anticoagulant Effects of Heparin Relevant to Inflammation

### *2.1 Inflammatory Mediators*

Heparin can inhibit the activation of inflammatory cells (Ahmed et al. 1994; Bazzoni et al. 1992; Brown et al. 2003; Evangelista et al. 1992; Inase et al. 1993; Lever et al. 2007; Piccardoni et al. 1996; Rohrer et al. 1992; Slungaard et al. 1990; Teixeira et al. 1996), an effect that is due in part to the binding and neutralisation of mediators and enzymes released during an inflammatory response (reviewed by Tyrrell et al. 1999) that would otherwise go on to activate these cells. Likewise, certain enzymes and cytotoxic mediators released from these cells, involved in propagation of the inflammatory response and subsequent tissue damage and remodelling, have also been shown to be inhibited by heparin, including elastase (Redini et al. 1988; Walsh et al. 1991), cathepsin G (Redini et al. 1988), eosinophil peroxidase (Pégorier et al. 2006), eosinophil cationic protein (Fredens et al. 1991), major basic protein (Swaminathan et al. 2005), certain cytokines (reviewed by Muramatsu and Muramatsu 2008) and chemokines (reviewed in Shute 2011, ibid and by Handel et al. 2005; Miller and Krangel 1992).

Many growth factors, including basic fibroblast growth factor (Bono et al. 1997) and transforming growth factor-β (McCaffrey et al. 1989; Okona-Mensah et al. 1998), both of which are involved in regulation of smooth muscle proliferation, a factor involved in the tissue remodelling seen in diseases including asthma, atherosclerosis and coronary stenosis, are bound by heparin. A long established property of heparin is that of inhibition of vascular smooth muscle cell proliferation (Clowes and Karnovsky 1977), an effect which is known to be independent of the anticoagulant actions of heparin (Guyton et al. 1980), and which extends to smooth muscle from other sites including the airways (Kanabar et al. 2005; Kilfeather et al. 1995; Okona-Mensah et al. 1998).

Heparin is known to inhibit the degranulation of isolated human mast cells in response to a variety of stimuli, and hence the release of histamine (Dragstedt et al. 1942; Inase et al. 1993). This effect is considered to be due to inhibition of inositol 1,4,5-triphosphate ($IP_3$)-dependent calcium release by heparin. The action of $IP_3$ on the endoplasmic reticulum is potently and competitively blocked by heparin applied to permeabilised cells in vitro (Ghosh et al. 1988), and heparin may be able to prevent the degranulation of other cell types via this mechanism. IgE-mediated degranulation of mast cells in vitro was found to be inhibited by two fractions of heparin, one which lacked anticoagulant activity and was actually the more potent preparation in this respect, evidence that this effect also does not dependent upon the anticoagulant effects of heparin (Ahmed et al. 1997). Cytotoxic effects upon endothelial cells of TNF-α activated eosinophils are markedly inhibited by heparin (Slungaard et al. 1990), as is the homotypic aggregation and chemotaxis of eosinophils in response to complement factor C5a, which is also bound by heparin (Matzner et al. 1984; Teixeira et al. 1996).

Heparin has been shown to bind to the surface of neutrophils (Leculier et al. 1992) and, in these cells, can inhibit their degranulation (Brown et al. 2003; Lever et al. 2007), homotypic aggregation (Bazzoni et al. 1992; Brown et al. 2003; Freischlag et al. 1992; Laghi-Pasini et al. 1984), the production of superoxide anions, the activity of lysosomal enzymes (Bazzoni et al. 1992) and the ability of neutrophils to activate platelets (Bazzoni et al. 1992; Evangelista et al. 1992), again in a manner that is not dependent upon anticoagulant activity. Furthermore, heparin is able to inhibit neutrophil activation in response to thrombin-stimulated platelet products, in addition to inhibiting thrombin-induced platelet aggregation (Piccardoni et al. 1996), and at high concentrations, platelet α-granule secretion is inhibited (Rohrer et al. 1992).

## 2.2 *Cellular Adhesion*

In addition to inhibiting the release of and activity of inflammatory cell-derived mediators, heparin is also effective in limiting the recruitment of these cells to tissues, through modulation of interactions between inflammatory leucocytes and the vascular endothelium. Heparin has been shown to inhibit leucocyte-endothelial adhesion, both in vitro (Bazzoni et al. 1992; Lever et al. 2000; Silvestro et al. 1994; Smailbegovic et al. 2001) and in vivo (Johnson et al. 2004; Lever et al. 2010; Ley et al. 1991; Nelson et al. 1993; Salas et al. 2000; Tangelder and Arfors 1991; Xie et al. 1997), as well as to limit the ultimate accumulation of cells in inflamed tissues, in response to both allergic (Sasaki et al. 1993; Seeds et al. 1995; Seeds and Page 2001; Vancheri et al. 2001) and non-allergic (Johnson et al. 2004; Lever et al. 2010; Nelson et al. 1993; Teixeira and Hellewell 1993; Yanaka et al. 1996) stimuli.

Heparin is known to bind directly to several adhesion molecules expressed during inflammation and the structural requirements for these interactions are becoming increasingly well characterised (e.g. reviewed by Fritzsche et al. 2006). On leucocytes, L-selectin, a molecule involved in early adhesive interactions between inflammatory cells and the vessel wall, is bound by heparin (Koenig et al. 1998) and endothelial heparan sulphate is able to act as an endothelial ligand for this molecule during cell rolling (Giuffrè et al. 1997). The $\beta_2$-integrin adhesion molecule mac-1 (macrophage-1; CD11b/CD18), important for the firm adhesion of leucocytes to endothelium, is also bound by heparin (Diamond et al. 1995; Peter et al. 1999) to an extent that surface immobilised heparin is able to support mac-1-dependent neutrophil adhesion under flow conditions in vitro (Diamond et al. 1995). Therefore, soluble heparin may inhibit mac-1-dependent interactions between leucocytes and the endothelium; the effects of heparin on leucocyte adhesion in vivo having been found to be dependent on such an interaction with mac-1 (Salas et al. 2000). On endothelial cells, heparin binds to P-selectin (Skinner et al. 1991), a selectin adhesion molecule involved in the early sequestration of neutrophils during inflammation. Indeed, the anti-metastatic effects of heparin (vide infra) can be ascribed, at least in part, to inhibition of P- and L-selectin function (Stevenson et al. 2005, 2007). The selectins are a family of glycoprotein adhesion molecules comprising an epidermal growth factor (EGF)-like moiety, repeating sequences mimicking those found on complement binding proteins and an $NH_2$-terminal lectin domain. It is via the lectin domain that these molecules $Ca^{2+}$-dependently bind to carbohydrate structures on the surfaces of interacting cells. Selectins are concerned predominantly with the rolling stages of adhesion, without which firm adhesion and transmigration cannot proceed (Lawrence and Springer 1991). However, despite structural congruencies between the selectins, it is established that heparin is unable to bind to E-selectin (Koenig et al. 1998). This difference is known to rely upon two specific amino-acid residues in the EGF-like domain of the selectins, in that if these residues are altered, E-selectin can be made to bind heparin, and the ability of P-selectin to bind heparin diminished (Revelle et al. 1996). This differential effect may possess physiological significance with respect to the role of endogenous heparin and, possibly, heparan sulphate in the inflammatory process. Indeed, a similar selectivity of binding can be observed amongst key members of the immunoglobulin superfamily adhesion proteins. Heparin has been shown to bind PECAM-1 (platelet endothelial cell adhesion molecule-1; Watt et al. 1993), an IgSF-adhesion molecule thought to be involved in leucocyte transmigration due to its location at intercellular junctions on the endothelium. The homotypic aggregation of PECAM-1-transfected fibroblasts was found to be inhibited by heparin in a manner dependent upon interaction with the second immunoglobulin domain (De Lisser et al. 1993). Similarly, heparin is able to bind directly to neuronal cell adhesion molecule (NCAM), through a heparin-binding region located on the second immunoglobulin domain (Cole et al. 1986), as well as through a further heparin-binding region on the first immunoglobulin domain (Kiselyov et al. 1997), such interactions with heparan sulphate being important for the physiological functioning of this protein in neuronal development (Kallapur and Akeson 1992). However, the IgSF-adhesion molecules intercellular adhesion molecule (ICAM)-1 and ICAM-2, expressed on vascular endothelium and ligands for leucocyte $\beta_2$-integrins, do not appear to be bound by heparin. However, given that heparin can affect the functioning of ICAM-1 indirectly, by binding of mac-1, it is plausible that the cell trafficking associated with physiological immune surveillance, facilitated by, for example, interactions between lymphocyte function-related antigen (LFA-1; CD11a/CD18) and ICAM-1/2 may be spared while those associated with excessive cell recruitment during inflammation may be inhibited. In summary, however, it can be seen that heparin has the potential to interfere with each of the events involved in inflammatory cell recruitment, namely rolling, triggering, adhesion and transmigration.

## 2.3 Heparanase

The ubiquitous distribution of heparan sulphate proteoglycans (HSPG) in mammalian systems provides a clear indication of the physiological importance of these molecules, which possess roles in growth and development, are key structural components of extracellular matrices and are involved in the localisation and bioactivity of a wide array of mediators, including enzymes, growth factors, cytokines and chemokines (reviewed by Powell et al. 2004; Turnbull et al. 2001). The endo-β glucuronidase heparanase (HPSE1) is responsible for the site-selective cleavage of heparan sulphate chains, thus regulating the activity of the wide range of proteins that is functionally dependent upon HSPG. The end of the last century saw the sequencing and cloning of HPSE1 (Fairbanks et al. 1999; Hulett et al. 1999; Kussie et al. 1999; Vlodavsky et al. 1999), the discovery that active HPSE1 exists as a 50 kDa and 8 kDa heterodimer processed from a single, inactive 65 kDa proenzyme (Fairbanks et al. 1999; reviewed by Parish et al. 2001) and the identification of catalytic sites on the enzyme (Hulett et al. 2000),

HPSE1 activity has been demonstrated in immune tissue, including spleen, lymph node, leucocytes and platelets, as well as in endothelial and smooth muscle cells. Moreover, the well-accepted role of HPSE1 in cancer (reviewed by McKenzie 2007) is underscored by the fact that in human tumours, mRNA for HPSE1 is markedly increased with respect to corresponding normal tissues and that HPSE1 activity in tumour cells has been found to correlate positively with metastatic potential (McKenzie et al. 2000).

In the contexts of both inflammatory diseases and cancer, release of HPSE1 by tumour or inflammatory cells facilitates their diapedesis and migration to tissue sites and promotes angiogenesis and tissue remodelling through release or activation of growth factors. The pathophysiological importance of heparanase activity is illustrated by the fact that in tissue sections from inflammatory bowel disease patients, when compared to healthy tissues, areas of extensive GAG disruption are visible on vascular endothelium and basement membrane, which correlate with localised areas of inflammation (Murch et al. 1993) and increased levels of GAG degradation products have been found in the urine of asthmatic patients, which is thought to reflect the breakdown of extracellular matrices as a result of the inflammatory process in the airway (Shute et al. 1997).

Heparin has long been known to be an inhibitor of HPSE1 activity (Bar-Ner et al. 1987), and it is also well established that heparan-degrading enzymes are released by certain leucocytes during the process of diapedesis (Lider et al. 1990; Matzner et al. 1992). Indeed, when heparin is used at low doses in models of inflammatory disease, especially in lymphocyte-driven processes such as allergic encephalomyelitis (Lider et al. 1989; Willenborg and Parish 1988), delayed-type hypersensitivity (DTH) (Sy et al. 1983) and graft versus host reactions (Gorski et al. 1991; Naparstek et al. 1993), leucocyte infiltration into tissues is markedly inhibited. It has further been demonstrated that vascular endothelial cells also secrete heparanase and that exposure of endothelial cells to pro-inflammatory cytokines

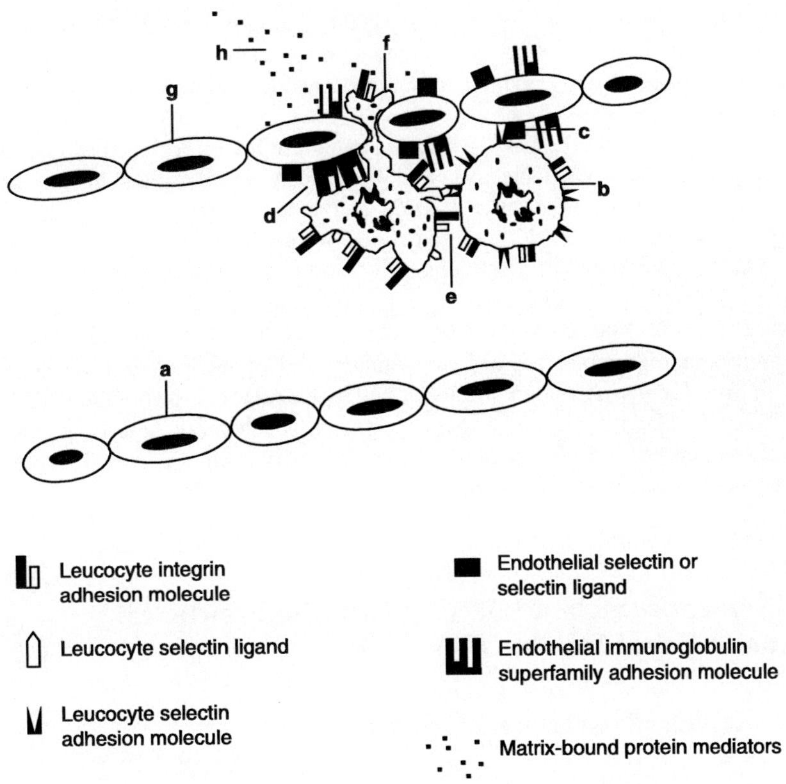


**Fig. 1** Potential sites of action of heparin in the microcirculation. Heparin has the potential to inhibit leucocyte-endothelial adhesion and extravasation through interaction with a variety of molecules with which it is known to bind. Many of the mechanisms involved in leucocyte trafficking are shared by metastatic tumour cells. (**a**) and (**b**) Binding to vascular endothelium and leucocytes, respectively. Heparin may act to augment the general anti-adhesive nature of the endothelium, due to its high net negative charge. (**c**) Inhibition of selectin-mediated adhesive interactions: Heparin binds L-and P-selectins, involved in the early stages of cell adhesion. (**d**) Inhibition of integrin-mediated cell adhesion and extravasation. Heparin binds the leucocyte integrin adhesion molecule mac-1 (CD11b/CD18), involved in firm adhesion of cells to the endothelium. (**e**) Activation of inflammatory cells may be reduced by binding of mediators and adhesion molecules. Platelet–leucocyte interactions may also be inhibited in this manner, as may the actions of cell-derived heparanase on endothelial heparan sulphate. (**f**) Diapedesis across the endothelium may be disrupted through binding of adhesion molecules involved in this process. (**g**) Inhibition of basement membrane and matrix degradation by heparanase and other cell-derived enzymes may limit extravasation of cells. (**h**) Activity of matrix-bound proteins, such as chemokines and growth factors, may be modulated, through inhibition of heparanase digestion of matrix heparan sulphate. This may have consequences both for the movement of extravasated cells to tissue sites and for tissue remodelling

upregulates this secretion (Chen et al. 2004; Edovitsky et al. 2006), further suggesting an important role in inflammation. Moreover, the development of DTH reactions has been found to correlate with endothelial heparanase expression in mice (Edovitsky et al. 2006).

Inhibition of HPSE1 using anti-sense strategies (Edovitsky et al. 2004; Uno et al. 2001), relatively selective inhibitors such as phosphomannopentaose (Basche et al. 2006; Joyce et al. 2005) and indeed heparin (Nakajima et al. 1988; Sciumbata et al. 1996) has been shown to be effective in reducing tumour cell adhesion, migration and subsequent colonisation in tissues and this enzyme is a well-established therapeutic target in the design of new anti-cancer drugs (reviewed by McKenzie 2007). Given the clear similarities that exist between the processes of tumour cell and inflammatory cell diapedesis and the demonstrated role for heparan-degrading enzymes in models of inflammation, HPSE1 inhibition is likely to be an important facet of the overall anti-inflammatory effects of heparin and related molecules and HPSE1 itself a potential target for novel anti-inflammatory drugs (McKenzie 2007).

## 3 Non-anticoagulant Effects of Heparin: Pre-clinical and Clinical Studies

### 3.1 *Acute Inflammatory Reactions*

In animal studies, pre-treatment with heparin has been shown to inhibit eosinophil infiltration into the inflamed lung (Sasaki et al. 1993; Seeds et al. 1993; Seeds et al. 1995) and skin (Teixeira and Hellewell 1993), neutrophil accumulation in the inflamed peritoneal cavity (Nelson et al. 1993; Lever et al. 2010), independently of anticoagulant activity (Seeds and Page 2001; Lever et al. 2010), and to inhibit vascular permeability induced by autacoids (Carr 1979; Jones et al. 2002) and formyl peptide (Jones et al. 2002). Additionally, platelet-activating factor-induced bronchial hyperresponsiveness was inhibited by heparin administration in rabbits (Sasaki et al. 1993) and similar effects have been reported in an allergic sheep model, whereby inhaled heparin was found to inhibit the acute airway response to inhaled allergen (Ahmed et al. 1992), an effect that was shared by very low-molecular-weight and non-anticoagulant heparins (Ahmed et al. 1997), and in guinea-pigs, whereby the protective effect of heparin against airway hyperresponsiveness to methacholine was found to be due to preservation of nitric oxide signalling in the airway (Maarsingh et al. 2004).

Heparin has been found, in a number of models, to protect against ischaemia-reperfusion injury. In a hamster dorsal skin chamber model, leucocyte-endothelial adhesion induced by ischaemia-reperfusion is inhibited by heparin pre-treatment (Becker et al. 1994), as is cardiac muscle damage (Kilgore et al. 1999). Furthermore, administration of heparin subsequent to transient focal cerebral ischaemia in rats was found to reduce the degree of brain injury by inhibiting reperfusion-induced leucocyte accumulation (Yanaka et al. 1996). Interestingly, heparin has recently been suggested as a plausible agent for limitation of the delayed neurological injury that follows subarachnoid haemorrhage (Simard et al. 2010), by virtue of its broad anti-inflammatory effects. Clearly, the potential to promote further haemorrhage is a legitimate concern. Nonetheless, it has been suggested that systemic, sub-anticoagulant doses of heparin may be sufficient to elicit useful effects in this setting (Simard et al. 2010) and, moreover, intra-cisternal administration of heparin has been found to be protective following experimental subarachnoid haemorrhage in rats (Tekkök et al. 1994). However, given that the anti-inflammatory properties of heparin appear largely to be separable from its effects on coagulation, non-anticoagulant heparin species may prove to possess an improved role in this setting.

## 3.2 *Human Inflammatory Diseases*

In addition to having effects in experimental models, heparin has long been considered to be of potential use in human inflammatory disease and was first assessed for this purpose in the 1960s (Dolowitz and Dougherty 1960, 1965), in small, subjectively assessed trials. More recently, in controlled studies, heparin has shown potential in the management of human asthma (Ahmed et al. 1993; Antczak and Kuna 1995; Bowler et al. 1993; Diamant et al. 1996) and chronic obstructive pulmonary disease (Brown et al. 2006), when administered by inhalation, in allergic rhinitis, whereby topical heparin reduces eosinophil recruitment into the nose (Vancheri et al. 2001) and inflammatory bowel disease (Evans et al. 1997; Gaffney et al. 1991, 1995; reviewed by Michell et al. 2001), although meta-analyses of these trials have concluded that there is currently insufficient evidence to support the use of heparin for the treatment of active ulcerative colitis (Chande et al. 2008; Shen et al. 2007).

Importantly, however, in none of these clinical studies was heparin treatment found to elicit significant haemorrhagic side effects, either when administered systemically or locally. Indeed, in a study performed specifically to address the effects of inhaled heparin on coagulation parameters (Bendstrup et al. 1999), it was found that almost 40% of a single inhaled dose of heparin is detectable in the lung 24 h later, with no significant effects on blood coagulation. However, given that the anticoagulant actions of heparin appear not to be necessary for the majority of beneficial effects seen in models of inflammation, it seems likely that novel drugs which retain the anti-inflammatory effects of the parent heparin molecule, without the anticoagulant effects will be useful in the management of inflammatory diseases that have been found to respond positively to the administration of standard heparin; selectively 2,3-O-desulphated heparin, which is currently in clinical trials for human inflammatory diseases, being a good example of such a drug (Fryer et al. 1997).

## 3.3 Cancer

Due to the common use of heparins for the prophylaxis of venous thromboembolism (VTE) in cancer patients, a sizeable body of evidence exists to suggest that heparin confers benefit in the treatment of cancer that is in addition to direct effects on blood coagulation (see Borsig 2010; Engelberg 1999; Hettiarachchi et al. 1999; Mousa 2010; Niers et al. 2007; Smorenburg and Van Noorden 2001; Zacharski and Ornstein 1988; Zacharski et al. 2000). Analysis of trials of heparin treatment in cancer patients indicates an improved rate of survival (Hettiarachchi et al. 1999; Smorenburg et al. 1999) and meta-analyses performed specifically to assess the effects of heparin and LMW heparin treatment on survival in cancer patients have indicated positive effects (Aki et al. 2007; Kuderer et al. 2007).

As discussed previously, the accumulation of metastatic tumour cells in tissue sites, like leucocytes, is dependent upon adhesion to the vascular endothelium and subsequent diapedesis and many similarities exist between the processes utilised by inflammatory cells and tumour cells in this respect (reviewed by Vlodavsky and Friedmann 2001), including a dependency on platelet activation (Borsig et al. 2001; Pitchford et al. 2003). However, the involvement of anticoagulant mechanisms in these effects are less clear than is perhaps the case for many of the anti-inflammatory properties of heparin. Heparin has been demonstrated repeatedly to reduce metastasis of carcinoma cells in animal models (e.g. Alonso et al. 1996; Mousa et al. 2006; Nakajima et al. 1988; Parish et al. 2001; Sciumbata et al. 1996). However, with respect to the contribution of the anticoagulant effects of the drug, it has been suggested that the basis of the anti-metastatic effects of unfractionated heparin lie in inhibition of fibrin deposition around tumour cells, a factor considered to protect the cells from immune attack (Alonso et al. 1996). Nonetheless, many studies have found that fractions of heparin with much reduced anticoagulant activity, or none at all, also inhibit metastasis (e.g. Mousa et al. 2006; Sciumbata et al. 1996). Specific mechanisms thought to be involved in this effect include inhibition of heparanase activity (Engelberg 1999), selectin function (Borsig et al. 2001) and the tissue factor pathway (Amirkhosravi et al. 2007). Tissue factor can promote angiogenesis and metastasis via mechanisms both related and unrelated to plasma coagulation (Mousa et al. 2004). It has been suggested that the effects of heparin and related molecules in models of tumour growth and metastasis rely, at least in part, on the promotion of tissue factor pathway inhibitor (TFPI) release from endothelial cells (Amirkhosravi et al. 2007). Regarding selectin function, clinically relevant levels of LMW heparin, with respect to anticoagulation, have been shown to inhibit experimental metastasis in a manner that correlates with the ability to inhibit P- and L-selectin function; the pentasaccharide fondaparinux, which lacks this ability, was found to be without effect in the same assays, at levels normalised for anticoagulant activity (Stevenson et al. 2005), suggesting that it is not the anticoagulant effects per se of heparin that contribute most significantly to effects on tumour cell metastasis. Moreover, mice deficient in both P- and L-selectin were found to be protected against experimental metastasis and, importantly, in these mice, treatment with heparin conferred no further protection (Stevenson et al. 2007), in contrast to the marked effects seen in wild-type animals in a range of studies.

Protective effects of heparin in cancer models extend beyond the inhibition of metastasis to include those on tumour growth and angiogenesis. Heparin has long been known to be anti-angiogenic and its inhibitory effects on heparanase are again well established to be involved in this effect (reviewed by Vlodavsky et al. 1992). Growth factor-induced endothelial cell proliferation is inhibited by unfractionated and LMW heparins (Takahashi et al. 2005; Khorana et al. 2003; Marchetti et al. 2008). Whilst standard LMW heparins in this respect were found to be more potent than unfractionated heparin (Khorana et al. 2003; Marchetti et al. 2008), ultra-low-molecular-weight species, including the anticoagulant pentasaccharide fondaparinux, were without effect (Marchetti et al. 2008). Moreover, anti-angiogenic and anti-metastatic effects may further be mediated through interference with the chemokine system, which is known to be involved in these phenomena (reviewed by Mehrad et al. 2007).

Therefore, it is likely that heparins inhibit angiogenesis and metastasis via an array of mechanisms, including but by no means limited to heparanase inhibition.

## 3.4 *Wound Healing and Tissue Repair*

Administration of heparin by inhalation has been found to be a viable option for the management of smoke inhalation injury in survivors of fire (reviewed by Cancio 2009; Toon et al. 2010), which reduces the acute lung injury that contributes significantly to morbidity and mortality in these patients, both alone and in combination with N-acetylcysteine (Miller et al. 2009).

Indeed, there are a number of reports of heparin being used, topically or systemically, to treat burns, although there is a lack of effectively controlled studies in this area for clear conclusions to be drawn as to the efficacy of this approach (reviewed by Oremus et al. 2007). However, isolated case reports continue to emerge that suggest heparin is able to promote tissue repair and inhibit inflammation in burns patients (e.g. Ferreira Chacon et al. 2010). Whilst controlled, randomised studies are required to assess the potential utility of heparin in this setting, the demonstrated anti-inflammatory effects in other experimental systems do indicate the potential for a useful effect. Furthermore, in animal models, application of heparin-binding epidermal growth factor-like growth factor (HB-EGF), which is known to be up-regulated both in human burn tissue and during healing of experimental burn tissue (Cribbs et al. 2002), has been shown to promote healing of partial-thickness burn injuries specifically through potentiation of the expression of transforming growth factor-α, another member of the EGF family of growth factors involved in wound repair (Cribbs et al. 1998), and to promote healing of ileal tissue following experimental re-anastomosis surgery (Radulescu et al. 2010). It has recently been reported that tissue localisation of HB-EGF, through binding to HSPG, mediates the function

**Table 1** Conditions (other than thrombosis) in which heparin has been reported to confer benefit

| Condition | Level of evidence |
|---|---|
| Acute respiratory distress syndrome/acute lung injury | Animal models[a,b] |
| | Anecdotal report (human)[c] |
| Allergic encephalomyelitis | Animal models |
| Allergic rhinitis | Controlled trial (human) |
| Arthritis | Anecdotal report (human)[d] |
| Asthma | Animal models |
| | Controlled trials (human) |
| Cancer | Animal models |
| | Some trials (human) |
| | Some meta-analyses |
| Chronic obstructive pulmonary disease | Controlled trial (human) |
| Delayed-type hypersensitivity reactions | Animal models |
| Inflammatory bowel disease | Some controlled trials (human) |
| Interstitial cystitis | Human experimental model of condition[e] |
| | Related molecule (pentosan polysulphate) used clinically[f] |
| Transplant rejection | Animal models |
| Wound healing | Various reports in animals and humans |

[a]Darien et al. (1998),
[b]Li et al. (2009),
[c]Kennedy (1994),
[d]Gaffney and Gaffney (1996),
[e]Lilly and Parsons (1990),
[f]Parsons (1997), for other references, see text

of the growth factor as a juxtacrine inhibitor of cell proliferation and that disruption of this binding allows the released HB-EGF to function as an autocrine mitogen (Prince et al. 2010). Therefore, it is possible that soluble heparin at the site of injury could act competitively to release this growth factor from HS binding sites and promote its participation in tissue repair. Moreover, topically applied heparin has been found to promote effective tissue repair in rabbit trachea, in a model of tissue healing following airway surgery (Sen et al. 2009), further suggesting that the immunomodulatory effects of heparin may be useful in the specific situation of tissue repair following localised injury.

## 4 Endogenous Heparin

The physiological role of mast cell-derived heparin is incompletely understood. Mast cells are classically associated with allergic and inflammatory responses and, whilst heparin is known to be important in the storage of histamine and certain granule proteins within the mast cell (Forsberg et al. 1999; Humphries et al. 1999), it would appear incongruous that a potent anticoagulant should be biosynthesised solely for this purpose. However, the localisation of mast cells close to vessels of the microcirculation and, indeed, their more recent description in pathological tissue sites such as atheromas and tumours, suggests that endogenous heparin may play an important role in regulation of physiological and pathophysiological responses. Despite the fact that malignancy tends in general to present as a prothrombotic state, it has been observed in malignant melanoma that blood specifically within tumours fails to clot. It was shown in both murine and human melanoma that mast cells infiltrate these tumours in large numbers. Moreover, inhibition of endogenous heparin activity in mouse models, with protamine, heparinase or genetic N-deacetylase/N-sulphotransferase (NDST)-2 deficiency, was found both to restore thrombosis within tumours and lead to increased tumour size, suggesting that mast-cell-derived heparin plays an inhibitory role in melanoma progression (Samoszuk et al. 2003). The same investigators found mast cell-derived heparin to inhibit the growth of a human breast cancer cell line when in co-culture with fibroblasts, but not in their absence, the potential significance of which is illustrated by their histological observations of degranulating mast cells in fibrous tissues of a range of human tumour types (Samoszuk et al. 2005).

Heparin released from degranulating rat mast cells was found to inhibit the proliferation of aortic smooth muscle cells more potently than commercial heparins of lower molecular weights than the mast cell species isolated; in particular, high-molecular-weight proteoglycans released from the mast cell granules were found to confer the majority of the anti-proliferative effect, although GAGs subsequently purified from these proteoglycans were also found to possess inhibitory activity (Wang and Kovanen 1999). The results of this study and others (reviewed by Kovanen 2009) suggest that mast cells per se, and mast cell-derived heparin in particular, may possess a protective or regulatory role in the process of atherogenesis. Indeed, endogenous heparin deficiency has been suggested as a potential contributing factor in atherosclerosis, through loss of heparin-derived anti-inflammatory and anticoagulant mechanisms relevant to this process (reviewed by Engelberg 2001). Furthermore, a small study compared serum levels of immunoglobulin E (IgE), endogenous heparin-like material and thrombin–antithrombin (TAT) complexes between healthy controls and patients with a recent history of myocardial infarction. Levels of heparin-like material were significantly lower in the patient cohort and, in both groups, levels correlated positively with IgE concentrations, suggesting mast cells to be the source of the heparin. Moreover, TAT complex levels were negatively correlated with heparin concentrations (Wladyslaw 2002). In a separate study, reduced activity of heparin-like material was demonstrated in the blood of patients with peripheral arterial occlusive disease, all of whom showed an increased coagulation tendency in comparison with healthy controls, in a manner that correlated with disease severity (Shankar et al. 2008).

With respect to allergic inflammation, there have been mixed suggestions as to the role of mast cell-derived heparin. On the one hand, given the known anti-inflammatory effects of heparin, it has been proposed that heparin released from degranulating mast cells during the allergic asthmatic response functions to limit the extent of subsequent inflammation in response to the initial stimulus, and that inhibition of mast cell degranulation by agents such as $\beta_2$-adrenoceptor agonists (Green et al. 1993) may thus impair this protective function (Page 1991).

By contrast, others have suggested that heparin from mast cells contributes to allergic inflammation: Heparin proteoglycans, isolated from a murine mastocytoma line (Brunnee et al. 1997) and from human lung (Noga et al. 1999), were found to initiate the contact system via factor XII activation, and it was suggested that in allergic reactions, therefore, mast-cell-derived heparin can act as a suitable surface to promote this cascade and contribute to the generation of kinins to participate in the allergic response. However, a recent small study compared plasma levels of endogenous heparin-like material between a group of controlled asthmatic patients and healthy controls and found these levels to be reduced in the patient group, and the authors conclude that the lack of heparin-associated anti-inflammatory activity in these individuals may be a significant factor in the disease process (Davids et al. 2010).

## 5 Implications and Conclusions

It is clear from the large number of studies that have been published over recent decades, from basic science investigations to clinical observations in human disease states, that heparin and related molecules possess considerable promise for the treatment of a range of conditions that are not specifically associated with disorders of blood coagulation. However, the potential utility of heparin as it stands, for many of these indications, may be limited by two key factors: lack of selectivity of action, including the need to dissociate specific non-anticoagulant effects from anticoagulant activity, and inconvenience of the parenteral route of administration.

With respect to the first point, it seems likely that novel heparin-based modalities will emerge that isolate specific activities of the parent molecule. Aside from the archetypal example of the antithrombin-binding pentasaccharide, it is established that selectivity of protein binding exists within GAG chains (reviewed by Taylor and Gallo 2006) and structural requirements within heparin for the binding of certain growth factors, enzymes and adhesion molecules have been identified (reviewed by Casu et al. 2008; Karnovsky et al. 1989). Moreover, the knowledge that minimum GAG chain lengths exist for the effective binding of specific proteins, including thrombin and platelet factor 4 (PF4), has already been exploited in the development and use of anticoagulant heparins with more predictable effects (reviewed by Gray et al. 2008; Petitou et al. 2009). Indeed, we have found that very low-molecular-weight heparin preparations, of defined and homogenous chain length, have differential and size-dependent effects on neutrophil degranulation (Lever et al. 2007), and a simple approach such as molecular weight limitation could alone go some way towards removing activities that are unwanted for a specific indication. From a physiological perspective, the fact that NDST-2-deficient mice, which are unable to synthesise heparin and thus produce serglycin proteoglycans decorated with chondroitin only, possess defined and limited abnormalities in the granular storage function of mast cells (Forsberg et al. 1999; Humphries et al. 1999), whereas mice that lack this proteoglycan completely display a severe loss of mast cell function (Abrink et al. 2004), suggests a fundamental specificity of function within the GAG element of the GAG-protein interaction, as opposed to a simple requirement for negative charge. Apropos of the physiological role(s) of heparin specifically, the fact that commercially available heparin has been standardised based upon its anticoagulant activity, and given the heterogeneity of naturally occurring heparin chains, it is not necessarily the case that alternative and unrelated activities will be present in a manner that correlates with the former. Furthermore, it is possible that certain activities may be lost or selected out in the processing of raw heparin for clinical use as an anticoagulant. Full characterisation of the structural features within heparin that account for specific non-anticoagulant effects could, in the future, allow standardisation of native heparin for these activities.

Regarding the route of administration of heparin, there has long been interest in developing heparin-based anticoagulants that do not require parenteral administration and with respect to, for example, the management of chronic inflammatory diseases, the need for convenient and palatable methods of drug delivery is arguably an even greater issue. In some circumstances, such as inflammatory diseases of the lung, local administration of heparin by inhalation is an option but where systemic effects are required, an efficient and predictable drug absorption profile becomes necessary. Absorption of unmodified, unfractionated and LMW heparins has been reported following oral administration in rats (Hiebert 2002; Hiebert et al. 2008; Pinel et al. 2004), and in rats and humans when administered with the absorption-enhancing delivery agent sodium N-[8(2-hydroxybenzoyl)amino]caprylate (SNAC) (Baughman et al. 1998; Berkowitz et al. 2003; Gonze et al. 2000; Pineo et al. 2001). Similarly, augmentation of heparin absorption via the pulmonary (Bai and Ahsan 2009, 2010; Bai et al. 2010; Qi et al. 2004; Rawat et al. 2008; Yang et al. 2004c) and nasal (Mustafa et al. 2004; Yang et al. 2004a, b, 2006) routes has been described, when the drug is co-administered with delivery systems including polyethyleneimines, cyclodextrins, alkylmaltosides, alkanoylsucroses, poly-L-arginine and within PEGylated nanocarriers. Moreover, heparin was administered successfully in validation studies of needle-free injection devices (Baer et al. 1996; Hollingsworth et al. 2000; Wagner et al. 2004), designed to reduce the pain and inconvenience associated with the regular administration of substances such as insulin, presenting a possible alternative to conventional subcutaneous injection of heparins. In all of these studies, measurement of coagulation parameters was used to assess the efficacy of heparin delivery. However, the fact that a robust and well-characterised effect of heparin can be measured, following the administration of standard heparins by non-standard routes, is promising with respect to the potential delivery of heparin species with alternative pharmacological profiles.

## References

Abrink M, Grujic M, Pejler G (2004) Serglycin is essential for maturation of mast cell secretory granule. J Biol Chem 279:40897–40905

Ahmed T, Abraham WM, D'Brot J (1992) Effects of inhaled heparin on immunologic and nonimmunologic bronchoconstrictor responses in sheep. Am Rev Respir Dis 145:566–570

Ahmed T, Garrigo J, Danta I (1993) Preventing bronchoconstriction in exercise-induced asthma with inhaled heparin. N Engl J Med 329:90–95

Ahmed T, Syriste T, Mendelssohn R, Sorace D, Mansour E, Lansing M, Abraham WM, Robinson MJ (1994) Heparin prevents antigen-induced airway hyperresponsiveness: interference with IP3-mediated mast cell degranulation? J Appl Physiol 76:893–901

Ahmed T, Campo C, Abraham MK, Molinari IF, Abraham WM, Ashkin D, Syriste T, Andersson LO, Svahn CM (1997) Inhibition of antigen-induced acute bronchoconstriction, airway hyperresponsiveness, and mast cell degranulation by a nonanticoagulant heparin – comparison with a low molecular weight heparin. Am J Respir Crit Care Med 155:1848–1855

Aki EA, van Doormaal FF, Barba M, Kamath G, Kim SY, Kuipers S, Middeldorp S, Yosuico V, Dickinson HO, Schünemann HJ (2007) Parenteral anticoagulation for prolonging survival in patients with cancer who have no other indication for anticoagulation. Cochrane Database Syst Rev 3:CD006652

Alonso DF, Bertolesi GE, Farias EF, Eijan AM, Joffe EBD, De Cidre LL (1996) Antimetastatic effects associated with anticoagulant properties of heparin and chemically modified heparin species in a mouse mammary tumor model. Oncol Rep 3:219–222

Amirkhosravi A, Meyer T, Amaya M, Davila M, Mousa SA, Robson T, Francis JL (2007) The role of tissue factor pathway inhibitor in tumor growth and metastasis. Semin Thromb Hemost 33:643–652

Antczak M, Kuna P (1995) Heparin inhibits allergen induced airway response in asthmatics. Results of a double blind placebo-controlled, crossover study. J Allergy Clin Immunol 95:386 (Abstract)

Baer CL, Bennett WM, Folwick DA, Erickson RS (1996) Effectiveness of a jet injection system in administering morphine and heparin to healthy adults. Am J Crit Care 5:42–48

Bai S, Ahsan F (2009) Synthesis and evaluation of pegylated dendrimeric nanocarrier for pulmonary delivery of low molecular weight heparin. Pharm Res 26:539–548

Bai S, Ahsan F (2010) Inhalable liposomes of low molecular weight heparin for the treatment of venous thromboembolism. J Pharm Sci 99:4554–4564

Bai S, Gupta V, Ahsan F (2010) Inhalable lactose-based dry powder formulations of low molecular weight heparin. J Aerosol Med Pulm Drug Deliv 23:97–104

Bar-Ner M, Eldor A, Wasserman L, Matzner Y, Cohen IR, Fuks Z, Vlodavsky I (1987) Inhibition of heparanase-mediated degradation of extracellular matrix heparan sulfate by non-anticoagulant heparin species. Blood 70:551–557

Basche M, Gustafson DL, Holden SN, O'Bryant CL, Gore L, Witta S, Schultz MK, Morrow M, Levin A, Creese BR, Kangas M, Roberts K, Nguyen T, Davis K, Addison RS, Moore JC, Eckhardt SG (2006) A phase I biological and pharmacologic study of the heparanase inhibitor PI-88 in patients with advanced solid tumors. Clin Cancer Res 12:5471–5480

Baughman RA, Kapoor SC, Agarwal RK, Kisicki J, Catella-Lawson F, FitzGerald GA (1998) Oral delivery of anticoagulant doses of heparin. A randomized, double-blind, controlled study in humans. Circulation 98:1610–1615

Bazzoni G, Nuñez AB, Mascellani G, Bianchini P, Dejana E, Del Maschio A (1992) Effect of heparin, dermatan sulfate, and related oligo-derivatives on human polymorphonuclear leukocyte functions. J Lab Clin Med 121:268–275

Becker M, Menger MD, Lehr HA (1994) Heparin-released superoxide dismutase inhibits postischemic leukocyte adhesion to venular endothelium. Am J Physiol 267:H925–930

Bendstrup KE, Chambers CB, Jensen JI, Newhouse MT (1999) Lung deposition and clearance of Inhaled $^{99m}$Tc-heparin in healthy volunteers. Am J Respir Crit Care Med 160:1653–1658

Berkowitz SD, Marder VJ, Kosutic G, Baughman RA (2003) Oral heparin administration with a novel drug delivery agent (SNAC) in healthy volunteers and patients undergoing elective total hip arthroplasty. J Thromb Haemost 1:1914–1919

Bono F, Rigon P, Lamarche I, Savi P, Salel V, Herbert J-M (1997) Heparin inhibits the binding of basic fibroblast growth factor to cultured human aortic smooth-muscle cells. Biochem J 326:661–668

Borsig L (2010) Antimetastatic activities of heparins and modified heparins: experimental evidence. Thromb Res 125:S66–S71
Borsig L, Wong R, Feramisco J, Nadeau DR, Varki NM, Varki A (2001) Heparin and cancer revisited: mechanistic connections involving platelets, P-selectin, carcinoma mucins, and tumor metastasis. Proc Natl Acad Sci USA 98:3352–3357
Bowler SD, Smith SM, Laverombe PS (1993) Heparin inhibits the immediate response to antigen in the skin and lungs of allergic subjects. Am Rev Respir Dis 147:160–163
Brown RA, Lever R, Jones NA, Page CP (2003) Effects of heparin and related molecules upon neutrophil aggregation and elastase release in vitro. Br J Pharmacol 139:845–853
Brown RA, Allegra L, Matera MG, Page CP, Cazzola M (2006) Additional clinical benefit of enoxaparin in COPD patients receiving salmeterol and fluticasone propionate in combination. Pulm Pharmacol Ther 19:419–424
Brunnee T, Reddigan SR, Shibayama Y, Kaplan AP, Silverberg M (1997) Mast cell derived heparin activates the contact system: a link to kinin generation in allergic reactions. Clin Exp Allergy 27:653–663
Cancio LC (2009) Airway management and smoke inhalation injury in the burn patient. Clin Plast Surg 36:555–567
Carr J (1979) The anti-inflammatory action of heparin: Heparin as an antagonist to histamine, bradykinin and prostaglandin $E_1$. Thromb Res 16:507–516
Casu B, Vlodavsky I, Sanderson RD (2008) Non-anticoagulant heparins and inhibition of cancer. Pathophysiol Haemost Thromb 36:195–203
Chande N, MacDonald JW, Macdonald JK (2008) Unfractionated or low-molecular weight heparin for induction of remission in ulcerative colitis. Cochrane Database Syst Rev 2:CD006774
Chen G, Wang D, Vikramadithyan R, Yagyu H, Saxena U, Pillarisetti S, Goldberg IJ (2004) Inflammatory cytokines and fatty acids regulate endothelial cell heparanase expression. Biochemistry 43:4971–4977
Clowes AW, Karnovsky MJ (1977) Suppression by heparin of smooth muscle cell proliferation in injured arteries. Nature 265:625–626
Cole GJ, Loewy A, Glaser L (1986) Neuronal cell–cell adhesion depends on interactions of N-CAM with heparin-like molecules. Nature 320:445–447
Cribbs RK, Luquette MH, Besner GE (1998) Acceleration of partial-thickness burn wound healing with topical application of heparin-binding EGF-like growth factor (HB-EGF). J Burn Care Rehabil 19:95–101
Cribbs RK, Harding PA, Luquette MH, Besner GE (2002) Endogenous production of heparin-binding EGF-like growth factor during murine partial-thickness burn wound healing. J Burn Care Rehabil 23:116–125
Darien BJ, Fareed J, Centgraf KS, Hart AP, MacWilliams PS, Clayton MK, Wolf H, Kruse-Elliott KT (1998) Low molecular weight heparin prevents the pulmonary hemodynamic and pathomorphologic effects of endotoxin in a porcine acute lung injury model. Shock 9:274–281
Davids H, Ahmed A, Oberholster A, van der Westhuizen C, Mer M, Havlik I (2010) Endogenous heparin levels in the controlled asthmatic patient. S Afr Med J 100:307–308
De Lisser HM, Horng CY, Newman PJ, Muller WA, Buck CA, Albelda SM (1993) Platelet/endothelial cell adhesion molecule-1 (CD31)-mediated cellular aggregation involves cell surface glycosaminoglycans. J Biol Chem 268:16037–16046
Diamant Z, Timmers MC, Van Der Veen H, Page CP, Van Der Meer FJM, Sterk PJ (1996) Effect of inhaled heparin on allergen-induced early and late asthmatic responses in patients with atopic asthma. Am J Respir Crit Care Med 153:1790–1795
Diamond MS, Alon R, Parkos CA, Quinn MT, Springer TA (1995) Heparin is an adhesive ligand for the leukocyte integrin Mac-1 (CD11b/CD18). J Cell Biol 130:1473–1482
Dolowitz DA, Dougherty TF (1960) The use of heparin as an anti-inflammatory agent. Laryngoscope 70:873–874

Dolowitz DA, Dougherty TF (1965) The use of heparin in the control of allergies. Ann Allergy 23:309–313

Dragstedt CA, Wells JA, Rocha E, Silva M (1942) Inhibitory effect of heparin upon histamine release by trypsin, antigen, and protease. Proc Soc Exp Biol Med 51:191–192

Edovitsky E, Elkin M, Zcharia E, Peretz T, Vlodavsky I (2004) Heparanase gene silencing, tumor invasiveness, angiogenesis, and metastasis. J Natl Cancer Inst 96:1194–1195

Edovitsky E, Lerner I, Zcharia E, Peretz T, Vlodavsky I, Elkin M (2006) Role of endothelial heparanase in delayed-type hypersensitivity. Blood 107:3609–3616

Engelberg H (1999) Actions of heparin that may affect the malignant process. Cancer 85:257–272

Engelberg H (2001) Endogenous heparin activity deficiency: the 'missing link' in atherogenesis? Atherosclerosis 159:253–260

Evangelista V, Piccardoni P, Maugeri N, De Gaetano G, Cerletti C (1992) Inhibition by heparin of platelet activation induced by neutrophil-derived cathepsin G. Eur J Pharmacol 216: 401–405

Evans RC, Wong VS, Morris AI, Rhodes JM (1997) Treatment of corticosteroid-resistant ulcerative colitis with heparin – a report of 16 cases. Aliment Pharmacol Ther 11:1037–1040

Fairbanks MB, Mildner AM, Leone JW, Cavey GS, Mathews WR, Drong RF, Slightom JL, Bienkowski MJ, Smith CW, Bannow CA, Heinrikson RL (1999) Processing of the human heparanase precursor and evidence that the active enzyme is a heterodimer. J Biol Chem 274: 29587–29590

Ferreira Chacon JM, Mello de Andrea LM, Blanes L, Ferreira LM (2010) Effects of topical application of 10,000 IU heparin on patients with perineal dermatitits and second-degree burns treated in a public pediatric hospital. J Tissue Viability 19:150–158

Forsberg E, Pejler G, Ringvall M, Lunderius C, Tomasini-Johansson B, Kusche-Gullberg M, Eriksson I, Ledin J, Hellman L, Kjellén L (1999) Abnormal mast cells in mice deficient in a heparin-synthesizing enzyme. Nature 400:773–776

Fredens K, Dahl R, Venge P (1991) In vitro studies of the interaction between heparin and eosinophil cationic protein. Allergy 46:27–29

Freischlag JA, Colburn MD, Quinones-Baldrich WJ, Moore WS (1992) Heparin, urokinase, and ancrod alter neutrophil function. J Vasc Surg 16:565–572

Fritzsche J, Alban S, Ludwig RJ, Rubant S, Boehncke W-H, Schumacher G, Bendas G (2006) The influence of various structural parameters of semisynthetic sulfated polysaccharides on the P-selectin inhibitory capacity. Biochem Pharmacol 72:474–485

Fryer A, Huang YC, Rao G, Jacoby D, Mancilla E, Whorton R, Piantadosi CA, Kennedy T, Hoidal J (1997) Selective O-desulfation produces nonanticoagulant heparin that retains pharmacological activity in the lung. J Pharmacol Exp Ther 282:208–219

Gaffney A, Gaffney P (1996) Rheumatoid arthritis and heparin. Br J Rheumatol 35:808

Gaffney PR, O'Leary J, Doyle CT, Gaffney A, Hogan J, Smew F, Annis P (1991) Response to heparin in patients with ulcerative colitis. Lancet 337:238–239

Gaffney PR, Doyle CT, Gaffney A, Hogan J, Hayes DP, Annis P (1995) Paradoxical response to heparin in 10 patients with ulcerative colitis. Am J Gastroenterol 90:220–223

Gallagher JT (2011) Heparan sulphate: a heparin in miniature. In: Lever R, Mulloy B, Page CP (eds) Heparin – a century of progress. Springer, Heidelberg

Ghosh TK, Eis PS, Mullaney JM, Ebert CL, Gill DL (1988) Competitive, reversible and potent antagonism of inositol 1,4,5-triphosphate-activated calcium release by heparin. J Biol Chem 263:11075–11079

Gilat D, Hershkoviz R, Mekori YA, Vlodavsky I, Lider O (1994) Regulation of adhesion of CD4+ T lymphocytes to intact or heparinase-treated subendothelial extracellular matrix by diffusible or anchored RANTES and MIP-1α. J Immunol 153:4899–4906

Giuffrè L, Cordey A-S, Monai N, Tardy Y, Schapira M, Spertini O (1997) Monocyte adhesion to activated aortic endothelium: role of L-selectin and heparan sulfate proteoglycans. J Cell Biol 136:945–956

Gonze MD, Salartash K, Sternbergh WC, Baughman RA, Leone-Bay A, Money SR (2000) Orally administered unfractionated heparin with carrier agent is therapeutic for deep venous thrombosis. Circulation 101:2658–2661

Gorski A, Lao M, Gradowska L, Nowaczyk M, Wasik M, Lagodzinski Z (1991) New strategies of heparin treatment used to prolong allograft survival. Transplant Proc 23:2251–2252

Gray E, Mulloy B, Barrowcliffe TW (2008) Heparin and low-molecular-weight heparin. Thromb Haemost 99:807–818

Green WF, Konnaris K, Woolcock AJ (1993) Effect of salbutamol, fenoterol, and sodium cromoglycate on the release of heparin from sensitized human lung fragments challenged with *Dermatophagoides pteronyssinus* allergen. Am J Respir Cell Mol Biol 8:518–521

Guyton JR, Rosenberg RD, Clowes AW, Karnovsky (1980) Inhibition of rat arterial smooth muscle cell proliferation by heparin: in vivo studies with anticoagulant and nonanticoagulant heparin. Circ Res 46:625–634

Handel TM, Johnson Z, Crown SE, Lau EK, Proudfoot AE (2005) Regulation of protein function by glycosaminoglycans – as exemplified by chemokines. Annu Rev Biochem 74:385–410

Hettiarachchi RJ, Smorenburg SM, Ginsberg J, Levine M, Prins MH, Büller HR (1999) Do heparins do more than just treat thrombosis? The influence of heparins on cancer spread. Thromb Haemost 82:947–952

Hiebert LM (2002) Oral heparins. Clin Lab 48:111–116

Hiebert LM, Ping T, Wice SM (2008) Enhanced antithrombotic effects of unfractionated heparin in rats after repeated oral doses and its relationship to endothelial heparin concentration. Br J Pharmacol 153:1177–1184

Hollingsworth SJ, Hoque K, Linnard D, Corry DG, Barker SG (2000) Delivery of low molecular weight heparin for prophylaxis against deep vein thrombosis using a novel, needle-less injection device (J-Tip). Ann R Coll Surg Engl 82:428–431

Hulett MD, Freeman C, Hamdorf BJ, Baker RT, Harris MJ, Parish CR (1999) Cloning of mammalian heparanase: an essential enzyme involved in tumor invasion and metastasis. Nat Med 5:803–809

Hulett MD, Hornby JR, Ohms SJ, Zuegg J, Freeman C, Gready JE, Parish CR (2000) Identification of active-site residues of the pro-metastatic endoglycosidase heparanase. Biochemistry 39:15659–15667

Humphries DE, Wong GW, Friend DS, Gurish MF, Qiu WT, Huang C, Sharpe AH, Stevens RL (1999) Heparin is essential for the storage of specific granule proteases in mast cells. Nature 400:769–772

Inase N, Schreck RE, Lazarus SC (1993) Heparin inhibits histamine release from canine mast cells. Am J Physiol 264:L387–L390

Jaques LB (1979) Heparins – anionic polyelectrolyte drugs. Pharmacol Rev 31:99–167

Johnson Z, Kosco-Vilbois MH, Herren S, Cirillo R, Muzio V, Zaratin P, Carbonatto M, Mack M, Smailbegovic A, Rose M, Lever R, Page C, Wells TN, Proudfoot AE (2004) Interference with heparin binding and oligomerization creates a novel anti-inflammatory strategy targeting the chemokine system. J Immunol 173:5776–5785

Jones H, Paul W, Page CP (2002) The effects of heparin and related molecules on vascular permeability and neutrophil accumulation in rabbit skin. Br J Pharmacol 135:469–479

Joyce JA, Freeman C, Meyer-Morse N, Parish CR, Hanahan D (2005) A functional heparan sulfate mimetic implicates both heparanase and heparan sulfate in tumor angiogenesis and invasion in a mouse model of multistage cancer. Oncogene 24:4037–4051

Kallapur SG, Akeson RA (1992) The neural cell adhesion molecule (NCAM) heparin binding domain binds to cell surface heparan sulfate proteoglycans. J Neurosci Res 33:538–548

Kanabar V, Hirst SJ, O'Connor BJ, Page CP (2005) Some structural determinants of the antiproliferative effect of heparin-like molecules on human airway smooth muscle. Br J Pharmacol 146:370–377

Karnovsky MJ, Wright TC Jr, Castellot JJ Jr, Choay J, Lormeau JC, Petitou M (1989) Heparin, heparan sulfate, smooth muscle cells, and atherosclerosis. Ann NY Acad Sci 556:268–281

Kennedy TP (1994) Use of heparin to inhibit interleukin-8. International patent application, WO94/18989

Khorana AA, Sahni A, Altland OD, Francis CW (2003) Heparin inhibition of endothelial cell proliferation and organization is dependent on molecular weight. Arterioscler Thromb Vasc Biol 23:2110–2115

Kilfeather SA, Tagoe S, Perez AC, Okona-Mensah K, Matin R, Page CP (1995) Inhibition of serum-induced proliferation of bovine tracheal smooth muscle cells in culture by heparin and related glycosaminoglycans. Br J Pharmacol 114:1442–1446

Kilgore KS, Tanhehco EJ, Naylor KB, Lucchesi BR (1999) Ex vivo reversal of heparin-mediated cardioprotection by heparinase after ischemia and reperfusion. J Pharmacol Exp Therapeut 290:1041–1047

Kiselyov VV, Berezin V, Maar TE, Soroka V, Edvardsen K, Schousboe A, Bock E (1997) The first immunoglobulin-like neural cell adhesion molecule (NCAM) domain is involved in double-reciprocal interaction with the second immunoglobulin-like NCAM domain and in heparin binding. J Biol Chem 272:10125–10134

Koenig A, Norgard-Sumnicht K, Linhardt R, Varki A (1998) Differential interactions of heparin and heparan sulfate glycosaminoglycans with the selectins. J Clin Invest 101:877–889

Kovanen PT (2009) Mast cells in atherogenesis: actions and reactions. Curr Atheroscler Rep 11:214–219

Kuderer NM, Khorana AA, Lyman GH, Francis CW (2007) A meta-analysis and systematic review of the efficacy and safety of anticoagulants as cancer treatment: impact on survival and bleeding complications. Cancer 110:1149–1161

Kussie PH, Hulmes JD, Ludwig DL, Patel S, Navarro EC, Seddon AP, Giorgio NA, Bohlen P (1999) Cloning and functional expression of a human heparanase gene. Biochem Biophys Res Commun 261:183–187

Laghi-Pasini F, Pasqui AL, Ceccatelli L, Capecchi PL, Orrico A, Di Perri T (1984) Heparin inhibition of polymorphonuclear leukocyte activation in vitro. A possible pharmacological approach to granulocyte-mediated vascular damage. Thromb Res 35:527–537

Lawrence MB, Springer TA (1991) Leukocytes roll on a selectin at physiologic flow rates: distinction from and prerequisite for adhesion through integrins. Cell 65:859–873

Leculier C, Benzerara O, Couprie N, Francina A, Lasne Y, Archimbaud E, Fiere D (1992) Specific binding between human neutrophils and heparin. Br J Haematol 81:81–85

Lever R, Page CP (2002) Novel drug development opportunities for heparin. Nat Rev Drug Discov 1:140–148

Lever R, Hoult JRS, Page CP (2000) The effects of heparin and related molecules upon the adhesion of human polymorphonuclear leucocytes to vascular endothelium in vitro. Br J Pharmacol 129:533–540

Lever R, Lo WT, Faraidoun M, Amin V, Brown RA, Gallagher J, Page CP (2007) Size-fractionated heparins have differential effects on human neutrophil function in vitro. Br J Pharmacol 151:837–843

Lever R, Smailbegovic A, Page CP (2010) Locally available heparin modulates inflammatory cell recruitment in a manner independent of anticoagulant activity. Eur J Pharmacol 630:137–144

Ley K, Cerrito M, Arfors K-E (1991) Sulfated polysaccharides inhibit leukocyte rolling in rabbit mesentery venules. Am J Physiol 260:H1667–H1673

Li L-F, Huang C-C, Lin H-C, Tsai Y-H, Quinn DA, Liao S-K (2009) Unfractionated heparin and enoxaparin reduce high-stretch ventilation augmented lung injury: a prospective, controlled animal experiment. Crit Care 13:R108

Lider O, Baharav E, Mekori YA, Miller T, Naparstek Y, Vlodavsky I, Cohen IR (1989) Suppression of experimental autoimmune diseases and prolongation of allograft survival by treatment of animals with low doses of heparin. J Clin Invest 83:752–756

Lider O, Mekori YA, Miller T, Bar-Tana R, Vlodavsky I, Baharav E, Cohen IR, Naparstek Y (1990) Inhibition of T lymphocyte heparanase by heparin prevents T cell migration and T cell-mediated immunity. Eur J Immunol 20:493–499

Lilly JD, Parsons CL (1990) Bladder surface glycosaminoglycans is a human epithelial permeability barrier. Surg Gynaecol Obstet 171:493–496

Maarsingh H, de Boer J, Kauffman HF, Zaagsma J, Meurs H (2004) Heparin normalizes allergen-induced nitric oxide deficiency and airway hyperresponsiveness. Br J Pharmacol 142:1293–1299

Marchetti M, Vignoli A, Russo L, Balducci D, Pagnoncelli M, Barbui T, Falanga A (2008) Endothelial capillary tube formation and cell proliferation induced by tumor cells are affected by low molecular weight heparins and unfractionated heparin. Thromb Res 121:637–645

Matzner Y, Marx G, Drexler R, Eldor A (1984) The inhibitory effect of heparin and related glycosaminoglycans on neutrophil chemotaxis. Thromb Haemost 52:134–137

Matzner Y, Vlodavsky I, Bar-Ner M, Ishai-Michaeli R, Tauber AI (1992) Subcellular localization of heparanase in human neutrophils. J Leukoc Biol 51:519–524

McCaffrey TA, Falcone DJ, Brayton CF, Agarwal LA, Welt FG, Weksler BB (1989) Transforming growth factor-β activity is potentiated by heparin via dislocation of the transforming growth factor-β/$\alpha_2$-macroglobulin inactive complex. J Cell Biol 109:441–448

McKenzie EA (2007) Heparanase: a target for drug discovery in cancer and inflammation. Br J Pharmacol 151:1–14

McKenzie E, Tyson K, Stamps A, Smith P, Turner P, Barry R, Hircock M, Patel S, Barry E, Stubberfield C, Terrett J, Page M (2000) Cloning and expression profiling of hpa2, a novel mammalian heparanase family member. Biochem Biophys Res Commun 276:1170–1177

Mehrad B, Keane MP, Strieter RM (2007) Chemokines as mediators of angiogenesis. Thromb Haemost 97:755–762

Michell NP, Lalor P, Langman MJ (2001) Heparin therapy for ulcerative colitis? Effects and mechanisms. Eur J Gastroenterol Hepatol 13:449–456

Miller MD, Krangel MS (1992) Biology and biochemistry of the chemokines: a family of chemotactic and inflammatory cytokines. Crit Rev Immunol 12:17–46

Miller AC, Rivero A, Ziad S, Smith DJ, Elamin EM (2009) Influence of nebulized unfractionated heparin and N-acetylcysteine in acute lung injury after smoke inhalation injury. J Burn Care Res 30:249–256

Mousa SA (2010) Heparin and low-molecular weight heparins in thrombosis and beyond. Methods Mol Biol 663:109–132

Mousa SA, Fareed J, Iqbal O, Kaiser B (2004) Tissue factor pathway inhibitor in thrombosis and beyond. Methods Mol Med 93:133–155

Mousa SA, Linhardt R, Francis JL, Amirkkhosravi A (2006) Anti-metastatic effect of a non-anticoagulant low-molecular-weight heparin versus the standard low-molecular weight heparin, enoxaparin. Thromb Haemost 96:816–821

Mulloy B, Crane DT, Drake AF, Davies DB (1996) The interaction between heparin and polylysine: a circular dichroism and molecular modelling study. Braz J Med Biol Res 29:721–729

Muramatsu T, Muramatsu H (2008) Glycosaminoglycan-binding cytokines as tumor markers. Proteomics 8:3350–3359

Murch SH, MacDonald TT, Walker-Smith JA, Levin M, Lionetti P, Klein NJ (1993) Disruption of sulphated glycosaminoglycans in intestinal inflammation. Lancet 341:711–714

Mustafa F, Yang T, Khan MA, Ahsan F (2004) Chain length-dependent effects of alkylmaltosides on nasal absorption of enoxaparin. J Pharm Sci 93:675–683

Nakajima M, Irimura T, Nicolson GL (1988) Heparanases and tumor metastasis. J Cell Biochem 36:157–167

Naparstek E, Slavin S, Weiss L, Sidi H, Ohana M, Reich S, Vlodavsky I, Cohen IR, Naparstek Y (1993) Low-dose heparin inhibits acute graft versus host disease in mice. Bone Marrow Transplant 12:185–189

Nelson RM, Cecconi O, Roberts WG, Aruffo A, Linhardt RJ, Bevilacqua MP (1993) Heparin oligosaccharides bind L- and P-selectin and inhibit acute inflammation. Blood 82:3253–3258

Niers TM, Klerk CP, DiNisio M, Van Noorden CJ, Büller HR, Reitsma PH, Richel DJ (2007) Mechanisms of heparin induced anti-cancer activity in experimental cancer models. Crit Rev Oncol Hematol 33:358–368

Noga O, Brunnee T, Schaper C, Kunkel G (1999) Heparin, derived from the mast cells of human lungs is responsible for the generation of kinins in allergic reactions due to the activation of the contact system. Int Arch Allergy Immunol 120:310–316

Okona-Mensah KB, Shittu E, Page C, Costello J, Kilfeather SA (1998) Inhibition of serum and transforming growth factor beta (TGF-β1)-induced DNA synthesis in confluent airway smooth muscle by heparin. Br J Pharmacol 125:599–606

Oremus M, Hanson M, Whitlock R, Young E, Archer C, Dal Cin A, Gupta A, Raina P (2007) A systematic review of heparin to treat burn injury. J Burn Care Res 28:794–804

Page CP (1991) One explanation of the asthma paradox: inhibition of natural anti-inflammatory mechanism by beta 2-agonists. Lancet 337:717–720

Parish CR, Freeman C, Hulett MD (2001) Heparanase: a key enzyme involved in cell invasion. Biochim Biophys Acta 1471:M99–M108

Parsons CL (1997) Epithelial coating techniques in the treatment of interstitial cystitis. Urology 49:100–104

Pégorier S, Wagner LA, Gleich GJ, Pretolani M (2006) Eosinophil-derived cationic proteins activate the synthesis of remodeling factors by airway epithelial cells. J Immunol 177:4861–4869

Peter K, Schwarz M, Conradt C, Nordt T, Moser M, Kübler W, Bode C (1999) Heparin inhibits ligand binding to the leukocyte integrin Mac-1 (CD11b/CD18). Circulation 100:1533–1539

Petitou M, Nancy-Portebois V, Dubreucq G, Motte V, Meuleman D, de Kort M, van Boeckel CA, Vogel GM, Wisse JA (2009) From heparin to EP217609: the long way to a new pentasaccharide-based neutralisable anticoagulant with an unprecedented pharmacological profile. Thromb Haemost 102:804–810

Piccardoni P, Evangelista V, Piccoli A, De Gaetano G, Walz A, Cerletti C (1996) Thrombin-activated platelets release two NAP-2 variants that stimulate polymorphonuclear leukocytes. Thromb Haemost 76:780–785

Pinel C, Wice SM, Hiebert LM (2004) Orally administered heparins prevent arterial thrombosis in a rat model. Thromb Haemost 91:919–926

Pineo GF, Hull RD, Marder VJ (2001) Orally active heparin and low-molecular-weight heparin. Curr Opin Pulm Med 7:344–348

Pitchford SC, Yano H, Lever R, Riffo-Vasquez Y, Ciferri S, Rose MJ, Giannini S, Momi S, Spina D, O'connor B, Gresele P, Page CP (2003) Platelets are essential for leukocyte recruitment in allergic inflammation. J Allergy Clin Immunol 112:109–118

Powell AK, Yates EA, Fernig DG, Turnbull JE (2004) Interactions of heparin/heparan sulfate with proteins: appraisal of structural factors and experimental approaches. Glycobiology 14:17R–30R

Prince RN, Schreiter ER, Zou P, Wiley HS, Ting AY, Lee RT, Lauffenburger DA (2010) The heparin-binding domain of HB-EGF mediates localization to sites of cell–cell contact and prevents HB-EGF proteolytic release. J Cell Sci 123:2308–2318

Qi Y, Zhao G, Liu D, Shriver Z, Sundaram M, Sengupta S, Venkataraman G, Langer R, Sasisekharan R (2004) Delivery of therapeutic levels of heparin and low-molecular-weight heparin through a pulmonary route. Proc Natl Acad Sci USA 101:9867–9872

Radulescu A, Zhang HY, Chen CL, Chen Y, Zhou Y, Yu X, Otabor I, Olson JK, Besner GE (2010) Heparin-binding egf-like growth factor promotes intestinal anastomotic healing. J Surg Res doi:10.1016/j.jss.2010.06.036

Rawat A, Yang T, Hussain A, Ahsan F (2008) Complexation of a poly-L-arginine with low molecular weight heparin enhances pulmonary absorption of the drug. Pharm Res 25:936–948

Redini F, Tixier JM, Petitou M, Choay J, Robert L, Hornebeck S (1988) Inhibition of leucocyte elastase by heparin and its derivatives. Biochem J 252:515–519

Revelle BM, Scott D, Beck PJ (1996) Single amino acid residues in the E- and P-selectin epidermal growth factor domains can determine carbohydrate binding specificity. J Biol Chem 271:16160–16170

Rohrer MJ, Kestin AS, Ellis PA, Barnard MR, Rodino L, Breckwoldt WL, Li J-M, Michelson AD (1992) High-dose heparin supresses platelet α-granule secretion. J Vasc Surg 15:1000–1009
Salas A, Sans M, Soriano A, Reverter JC, Anderson DC, Piqué JM, Panés J (2000) Heparin attenuates TNF-alpha induced inflammatory response through a CD11b dependent mechanism. Gut 47:88–96
Samoszuk M, Corwin M, Yu H, Wang J, Nalcioglu O, Su MY (2003) Inhibition of thrombosis in melanoma allografts in mice by endogenous mast cell heparin. Thromb Haemost 90:351–360
Samoszuk M, Kanakubo E, Chan JK (2005) Degranulating mast cells in fibrotic regions of human tumors and evidence that mast cell heparin interferes with the growth of tumor cells through a mechanism involving fibroblasts. BMC Cancer 5:121
Sasaki M, Herd CM, Page CP (1993) Effect of heparin and a low-molecular weight heparinoid on PAF-induced airway responses in neonatally immunized rabbits. Br J Pharmacol 110:107–112
Sciumbata T, Caretto P, Pirovano P, Pozzi P, Cremonesi P, Galimberti G, Leoni F, Marcucci F (1996) Treatment with modified heparins inhibits experimental metastasis formation and leads, in some animals, to long-term survival. Invasion Metastasis 16:132–143
Seeds EAM, Page CP (2001) Heparin inhibits allergen-induced eosinophil infiltration into guinea-pig lung via a mechanism unrelated to its anticoagulant activity. Pulm Pharmacol Ther 14:111–119
Seeds EAM, Hanss J, Page CP (1993) The effect of heparin and related proteoglycans on allergen and PAF-induced eosinophil infiltration. J Lipid Mediat 7:269–278
Seeds EAM, Horne AP, Tyrrell DJ, Page CP (1995) The effect of inhaled heparin and related glycosaminoglycans on allergen-induced eosinophil infiltration in guinea-pigs. Pulm Pharmacol 8:97–105
Sen S, Meteoglu I, Ogurlu M, Sen S, Derinceoz OO, Barutca S (2009) Topical heparin: a promising agent for the prevention of tracheal stenosis in airway suregery. J Surg Res 157:e23–e29
Shankar VK, Handa A, Hands L (2008) Endogenous heparin activity is decreased in peripheral arterial occlusive disease. J Vasc Surg 47:1033–1038
Shen J, Ran ZH, Tong JL, Xiao SD (2007) Meta-analysis: the utility and safety of heparin in the treatment of active ulcerative colitis. Aliment Pharmacol Ther 26:653–663
Shute J (2011) Glycosaminoglycan and chemokine/growth factor interactions. In: Lever R, Mulloy B, Page CP (eds) Heparin – a century of progress. Springer, Heidelberg
Shute JK, Parmar J, Holgate ST, Howart PH (1997) Urinary glycosaminoglycan levels are increased in acute severe asthma – a role for eosinophil-derived gelatinase B? Int Arch Allergy Immunol 113:366–367
Silvestro L, Viano I, Macario M, Colangelo D, Montrucchio G, Panico S, Fantozzi R (1994) Effects of heparin and its desulfated derivatives on leukocyte-endothelial adhesion. Semin Thromb Hemost 20:254–258
Simard JM, Schreibman D, Aldrich EF, Stallmeyer B, Le B, James RF, Beaty N (2010) Unfractionated heparin: multitargeted therapy for delayed neurological deficits induced by subarachnoid hemorrhage. Neurocrit Care 13(3):439–449
Skinner MP, Lucas CM, Burns GF, Chesterman CN, Berndt MC (1991) GMP-140 binding to neutrophils is inhibited by sulfated glycans. J Biol Chem 266:5371–5374
Slungaard A, Vercellotti GM, Walker G, Nelson RD, Jacob HS (1990) Tumor necrosis factor-α/cachectin stimulates eosinophil oxidant production and toxicity towards human endothelium. J Exp Med 171:2025–2041
Smailbegovic A, Lever R, Page CP (2001) The effects of heparin on the adhesion of human peripheral blood mononuclear cells to human stimulated umbilical vein endothelial cells. Br J Pharmacol 134:827–836
Smorenburg SM, Van Noorden CJ (2001) The complex effects of heparins on cancer progression and metastasis in experimental studies. Pharmacol Rev 53:93–105
Smorenburg SM, Hettiarachchi RJ, Vink R, Büller HR (1999) The effects of unfractionated heparin on survival in patients with malignancy – a systematic review. Thromb Haemost 82:1600–1604

Stevenson JL, Choi SH, Varki A (2005) Differential metastasis inhibition by clinically relevant levels of heparins – correlation with selectin inhibition, not antithrombotic activity. Clin Cancer Res 11:7003–7011

Stevenson JL, Varki A, Borsig L (2007) Heparin attenuates metastasis mainly due to inhibition of P- and L-selectin, but non-anticoagulant heparins can have additional effects. Thromb Res 120:S107–S111

Swaminathan GJ, Myszka DG, Katsamba PS, Ohnuki LE, Gleich GJ, Acharya KR (2005) Eosinophil-granule major basic protein, a C-type lectin, binds heparin. Biochemistry 44:14152–14158

Sy MS, Schneeberger E, McCluskey R, Greene MI, Rosenberg RD, Benacerraf B (1983) Inhibition of delayed-type hypersensitivity by heparin depleted of anticoagulant activity. Cell Immunol 82:23–32

Takahashi H, Ebihara S, Okazaki T, Asada M, Sasaki H, Yamaya M (2005) A comparison of the effects of unfractionated heparin, dalteparin and danaparoid on vascular endothelial growth factor-induced tumour angiogenesis and heparanase activity. Br J Pharmacol 146:333–343

Tangelder GJ, Arfors K-E (1991) Inhibition of leukocyte rolling in venules by protamine and sulfated polysaccharides. Blood 7:1565–1571

Taylor KR, Gallo RL (2006) Glycosaminoglycans and their proteoglycans@host-associated molecular patterns for initiation and modulation of inflammation. FASEB J 20:9–22

Teixeira MM, Hellewell PG (1993) Suppression by intradermal administration of heparin of eosinophil accumulation but not oedema formation in inflammatory reactions in guinea-pig skin. Br J Pharmacol 110:1496–1500

Teixeira MM, Rossi AG, Hellewell PG (1996) Adhesion mechanisms involved in C5a-induced eosinophil homotypic aggregation. J Leukoc Biol 59:389–396

Tekkök IH, Tekkök S, Ozcan OE, Erbengi T, Erbengi A (1994) Preventive effect of intracisternal heparin for proliferative angiopathy after experimental subarachnoid haemorrhage in rats. Acta Neurochir (Wien) 127:112–117

Toon MH, Maybauer MO, Greenwood JE, Maybauer DM, Fraser JF (2010) Management of acute smoke inhalation injury. Crit Care Resusc 12:53–61

Turnbull J, Powell A, Guimond S (2001) Heparan sulfate: decoding a dynamic multifunctional cell regulator. Trends Cell Biol 11:75–82

Tyrrell DJ, Horne AP, Holme KR, Preuss JM, Page CP (1999) Heparin in inflammation: potential therapeutic applications beyond anticoagulation. Adv Pharmacol 46:151–208

Uno F, Fujiwara T, Takata Y, Ohtani S, Katsuda K, Takaoka M, Ohkawa T, Naomoto Y, Nakajima M, Tanaka N (2001) Antisense-mediated suppression of human heparanase gene expression inhibits pleural dissemination of human cancer cells. Cancer Res 61:7855–7860

Vancheri C, Mastruzzo C, Armato F, Tomaselli V, Magrì S, Pistorio MP, LaMicela M, D'amico L, Crimi N (2001) Intranasal heparin reduces eosinophil recruitment after nasal allergen challenge in patients with allergic rhinitis. J Allergy Clin Immunol 108:703–708

Vlodavsky I, Friedmann Y (2001) Molecular properties and involvement of heparanase in cancer metastasis and angiogenesis. J Clin Invest 108:341–347

Vlodavsky I, Eldor A, Haimovitz-Friedman A, Matzner Y, Ishai-Michaeli R, Lider O, Naparstek Y, Cohen IR, Fuks Z (1992) Expression of heparinase by platelets and circulating cells of the immune system: possible involvement in diapedesis and extravasation. Invasion Metastasis 12:112–127

Vlodavsky I, Friedmann Y, Elkin M, Aingorn H, Atzmon R, Ishai-Michaeli R, Bitan M, Pappo O, Peretz T, Michal I, Spector L, Pecker I (1999) Mammalian heparanase: gene cloning, expression and function in tumor progression and metastasis. Nat Med 5:793–802

Wagner S, Dues G, Sawitzky D, Frey P, Christ B (2004) Assessment of the biological performance of the needle-free injector INJEX using the isolated porcine forelimb. Br J Dermatol 150:455–461

Walsh RL, Dillon TJ, Scicchitano R, McLennan G (1991) Heparin and heparan sulphate are inhibitors of human leucocyte elastase. Clin Sci 81:341–346

Wang Y, Kovanen PT (1999) Heparin proteoglycans released from rat serosal mast cells inhibit proliferation of rat aortic smooth muscle cells in culture. Circ Res 84:74–83
Watt SM, Williamson J, Genevier H, Fawcett J, Simmons DL, Hatzfield A, Nesbitt SA, Coombe DR (1993) The heparin binding PECAM-1 adhesion molecule is expressed by CD34+ hematopoietic precursor cells with early myeloid and B-lymphoid cell phenotypes. Blood 82:2649–2663
Willenborg DO, Parish CR (1988) Inhibition of allergic encephalomyelitis in rats by treatment with sulphated polysaccharides. J Immunol 140:3401–3405
Wladyslaw S (2002) Endogenous heparin – a protective marker in patients with myocardial infarction. Coron Artery Dis 13:423–426
Xie X, Thorlacius H, Raud J, Hedqvist P, Lindbom L (1997) Inhibitory effect of locally administered heparin on leukocyte rolling and chemoattractant-induced firm adhesion in rat mesenteric venules in vivo. Br J Pharmacol 122:906–910
Yanaka K, Nose T, Hindman BJ (1996) Heparin ameliorates brain injury by inhibiting leukocyte accumulation. Stroke 27:2146–2147
Yang T, Hussain A, Paulson J, Abbruscato TJ, Ahsan F (2004a) Cyclodextrins in nasal delivery of low-molecular-weight heparins: in vivo and in vitro studies. Pharm Res 21:1127–1136
Yang T, Mustafa F, Ahsan F (2004b) Alkanoylsucroses in nasal delivery of low molecular weight heparins: in-vivo absorption and reversibility studies in rats. J Pharm Pharmacol 56:53–60
Yang T, Mustafa F, Bai S, Ahsan F (2004c) Pulmonary delivery of low molecular weight heparins. Pharm Res 21:2009–2016
Yang T, Hussain A, Bai S, Khalil IA, Harashima H, Ahsan F (2006) Positively charged polyethylenimines enhance nasal absorption of the negatively charged drug, low molecular weight heparin. J Control Release 115:289–297
Zacharski LR, Ornstein DL (1988) Heparin and cancer. Thromb Haemost 80:10–23
Zacharski LR, Ornstein DL, Mamourian AC (2000) Low-molecular-weight heparin and cancer. Semin Thromb Hemost 26:69–77

# Glycosaminoglycan and Chemokine/Growth Factor Interactions

**Janis Shute**

## Contents



**Abstract** Heparin and glycosaminoglycans (GAGs) related structurally to heparin, notably heparan sulphate, bind to most, if not all, chemokines and many growth factors. The chemokine and growth factor interactions with GAGs localise the peptide mediators to specific sites in tissues and influence their stability and function. This chapter discusses the nature of these interactions and the effect on the function of a number of chemokines (PF-4, interleukin-8, RANTES and SDF-1) and growth factors (FGF, HGF, VEGF) in normal physiology and the disease setting. Novel therapeutic interventions that target chemokine and growth factor interactions with GAGs are also discussed.

**Keywords** Chemokine • Glycosaminoglycan • Growth factor • Therapeutic intervention

J. Shute (✉)
Institute of Biomedical and Biomolecular Sciences, University of Portsmouth, Portsmouth PO1 2UP, UK
e-mail: jan.shute@port.ac.uk

R. Lever et al. (eds.), *Heparin - A Century of Progress*,
Handbook of Experimental Pharmacology 207,
DOI 10.1007/978-3-642-23056-1_13, 

# 1 General Introduction

Glycosaminoglycans (GAGs) are a family of linear polysaccharides, which interact with a wide range of proteins, including chemokines and growth factors, to regulate both physiological and pathological processes (Handel et al. 2005). The GAG polysaccharide chains are made up of sulphated and non-sulphated derivatives of repeating disaccharide units comprising a uronic acid moiety (D-glucuronic acid or L-iduronic acid) and an amino sugar (glucosamine or galactosamine). The family consists of heparin (H), heparan sulphate (HS), hyaluronic acid (HA), chondroitin sulphate (CS), dermatan sulphate (DS) and keratan sulphate (KS), in which the disaccharide atypically comprises galactose linked with a galactosamine residue. Following their synthesis, the GAG chains are modified by N- and O-sulphotransferase activity and epimerisation of the glucuronic acid residue into iduronic acid (Capila and Linhardt 2002; Gandhi and Mancera 2008), giving rise to highly variable structures. Hyaluronic acid is the only member that is not sulphated and not synthesised on or found covalently linked with the core protein of a proteoglycan.

Best characterised are the binding interactions of chemokines and growth factors with heparin and heparan sulphate, although roles for chondroitin/dermatan sulphate and hyaluronic acid in the regulation of chemokine and growth factor activity have also been identified. The principal disaccharide units found in heparan sulphate and heparin and their ordered assembly in HS chains are illustrated in Fig. 1.

The major disaccharide units of HS and heparin are N-acetylated and N-sulphated derivatives of glucosamine linked with glucuronic acid and iduronic acid, respectively. The difference is quantitative. Heparin consists mainly of tri-sulphated disaccharides with ~90% iduronic acid as the uronate component; however, variations in the composition and sulphation pattern lead to microheterogeneity in heparin structure. At physiological pH, all carboxylic acid and sulphate groups are deprotonated, and GAGs therefore have high negative charge density, with heparin having the highest negative charge of any molecule in biology. HS is structurally related but is much less substituted with sulphate groups than heparin, and carries less charge. However, in contrast to the linear, uniform, high charge density of heparin, HS has a unique multidomain structure, which brings sulphated sugars into complex clusters [the composite sulphated regions, Fig. 1 (Gallagher 2006)]. Most protein binding is to the S-domains but the T-domains, which contribute 20–25% of the polymer chain, also contribute to binding specificity. The differential sulphation pattern in the S and T domains, and the conformational flexibility of the iduronate residue, are proposed to contribute to both the structural and functional diversity of the HS chains (Capila and Linhardt 2002; Gandhi and Mancera 2008). A polymer of just six disaccharide units has the potential to display more than 12 billion different sequences, greater diversity than seen amongst hexapeptides or a 6-mer of DNA (Handel et al. 2005), indicating the vast potential of specific GAG sequences to interact with specific ligands.

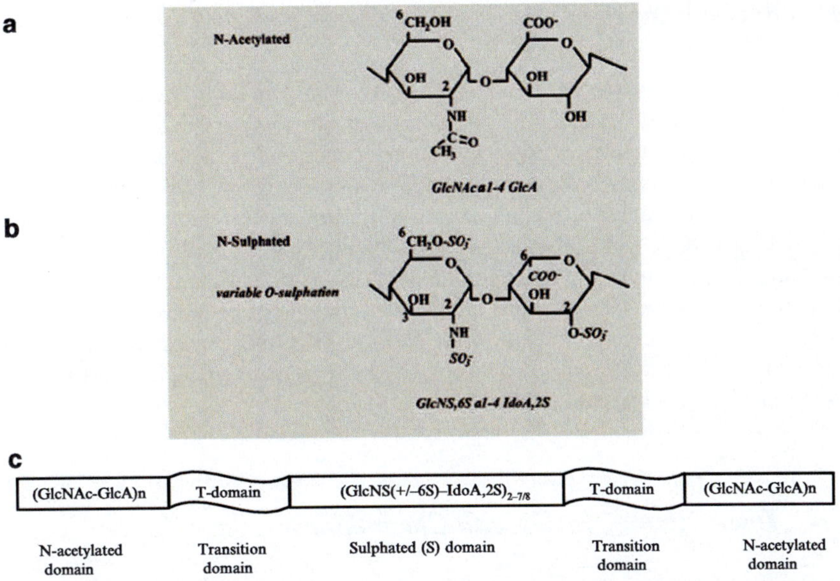


**Fig. 1** The major disaccharide repeating units in (**a**, **b**) heparan sulphate and (**b**) heparin and (**c**) their organisation into composite sulphated regions (CSRs) in heparan sulphate. The major disaccharide units of HS are N-acetylated (**a**) and N-sulphated (**b**) derivatives of glucosamine linked with glucuronic acid and iduronic acid, respectively, which may be further O-sulphated. Heparin consists mainly of tri-sulphated disaccharides (**b**) with iduronic acid-2-O-sulphate as the major uronate component. (**c**) The CSRs of HS have average length of 14–16 disaccharides and comprise the heavily modified S-domains, the areas of highest sulphate density, plus the flexible NAc/NS transition (T) domains (GlcNAc(+/−6S)-GlcA-GlcNS(+/−6S)-IdoA) and are separated by the largely unmodified N-acetylated domains. Heparin is analogous to the S-domains and lacks the domain structure of HS [reproduced with permission from Gallagher J.T., 2006, Biochem Soc Trans, 34:438–441 (http://www.biochemsoctrans.org)]

GAGs, with the exception of HA, are synthesised on a core protein, and found in storage granules, on cell surfaces and in the extracellular matrix as proteoglycans. Heparin is unique in that it is released from mast cells following protease-mediated cleavage of the serglycin core protein as a single long polysaccharide chain of approximately 100 kDa attached to a small peptide. The polysaccharide chain is subsequently cleaved by an endoglucuronidase to GAG chains of average molecular weight 15 kDa (range 5–40 kDa). HS chains are generally longer with average molecular weight 30 kDa (range 5–50 kDa) and remain attached to the core proteins on which they are synthesised. Heparan sulphate containing proteoglycans (HSPGs; agrin, collagen XVIII, perlecan) are major components of the extracellular matrix, with essential roles in the regulation of a wide range of physiological processes (Bishop et al. 2007). HSPGs are also found widely distributed at cell surfaces as two major families, the transmembrane syndecans and the GPI-linked glypicans. HSPGs interact with a wide range of proteins, predominantly, but not exclusively, through the HS chains, and most ECM molecules contain binding sites for HS (Handel et al. 2005; Bishop et al. 2007).

## 2 Glycosaminoglycan Interactions with Proteins

In view of the negative charge of the GAGs, it was initially thought that non-specific electrostatic interactions were responsible for the binding of proteins to the GAG side chains of proteoglycans in tissues and on cell surfaces. Subsequently, investigations have demonstrated a high degree of specificity in these interactions, and that GAG-binding to consensus sequences in proteins is a function of GAG structure, composition and chain length as well as charge (Capila and Linhardt 2002; Gandhi and Mancera 2008). Clusters of basic amino acid residues [arginine (R), lysine (K), histidine (H)], either contiguous or brought together not only by protein folding, interact with specific GAG sequences predominantly through ionic interaction, but also through hydrogen-bonding (Capila and Linhardt 2002; Gandhi and Mancera 2008).

### *2.1 Interaction of GAGs with Chemokines*

The chemokines are a family of structurally and functionally related chemoattractant cytokines with roles in the selective recruitment and activation of cells during inflammation, angiogenesis and cancer metastasis (Zlotnik et al. 2006). There are at least 46 chemokines with molecular mass of 8–14 kDa that signal their effects through 18 members of the G-protein-coupled seven-helix transmembrane receptors (GPCRs). The chemokines typically contain four cysteine residues in positions conserved through evolution and are classified into CC, CXC, C or CX3C, depending on the position of the first two cysteines near the amino terminus. Residues of the relatively disordered N-terminal domain are involved in receptor ligation, which is characterised by cross-reactivity as most chemokines can bind to several receptors and most receptors can be activated by several chemokines (Zlotnik et al. 2006). Despite low sequence homologies, chemokines have similar tertiary structures and most are highly basic proteins. In addition, most, if not all, chemokines demonstrate binding with heparin/heparan sulphate, and this includes MIP-1$\alpha$ and MIP-1$\beta$, which have acidic pI, indicating the specificity of these interactions (Handel et al. 2005).

The following are examples of ways in which GAG-binding plays an important role in the biological function of chemokines.

#### 2.1.1 Platelet Factor-4

Platelet factor-4 (PF-4) is released rapidly by activated platelets from pre-formed $\alpha$-granule stores and was the first of the chemokines to be isolated and sequenced. The heparin-binding and neutralising (pro-coagulant) activity of PF-4 (CXCL4) was suspected as long ago as 1948, and this property was used in the affinity purification and sequencing of PF-4 in 1977 (Brandt et al. 2000), more than ten years before the description of IL-8 and the recognition of the existence of a family of structurally and functionally related peptides, including PF-4 (Yoshimura et al. 1987). PF-4 is a member of a subset of CXC chemokines that are identified by the lack of an ELR (Glu-Leu-Arg) tripeptide motif at the N-terminal adjacent to the CXC grouping (Brandt et al. 2000). The presence of the ELR grouping is essential for the binding of $ELR^+$ chemokines to high-affinity CXCR1 and CXCR2 receptors. Thus, $ELR^-$ chemokines are unable to interact with these chemokine receptors, but interact with the CXCR3 chemokine receptor, or the CXCR3B spliced variant on microvascular endothelial cells in the case of PF-4 (Gleissner et al. 2008). In addition, PF-4 is proposed to bind to HSPG on the luminal surface of vascular endothelial cells (Stringer and Gallagher 1997) and CSPG on the surface of platelets and neutrophils (Brandt et al. 2000; Gleissner et al. 2008; Xiao et al. 2008).

Under physiological conditions of ionic strength and pH, PF-4 exists in active form mainly as a tetramer and binds in this form with high affinity in a 1:1 ratio with heparin and HS (Stringer and Gallagher 1997). A HS oligosaccharide of high molecular mass, equivalent to 21 disaccharides, was proposed to wrap around the PF-4 tetramer, binding and neutralising a ring of positive charges. The structural motif was identified as two S-domains (see above) enriched in N-sulphated disaccharides and 2-O-sulphate residues, separated by an N-acetylated region, in which the 2-O-sulphate but not the N-sulphated residues were important for binding interactions with basic lysine residues. The release of high concentrations of PF-4 from activated platelets and binding to HSPG at sites of vascular injury is proposed to neutralise the anticoagulant activity of endothelial HSPG. In addition, PF-4 has been shown to bind preferentially to the endothelium at sites of active angiogenesis, where it limits endothelial cell proliferation and angiogenesis via inhibition of FGF-2 and VEGF activity. The mechanisms involved include the displacement of growth factors from HSPG co-receptors on the endothelium (see below) and binding of PF-4 to growth factors in heterodimeric complexes, limiting their binding to specific receptors and subsequent neovascularisation and tumour growth (Brandt et al. 2000; Belpeiro et al. 2000).

Treatment of patients with heparin for anticoagulation results in PF-4–heparin complexes and an increase in PF-4 concentration in the blood, as PF-4 is released from low affinity endothelial HSPG binding sites. In 1–5% of patients, PF-4–heparin complexes induce an immune response and the generation of anti-PF-4 antibodies, leading to a dangerous fall in platelet count and, paradoxically, an increased risk of venous thrombosis, a condition known as heparin-induced thrombocytopenia (HIT) (Poncz 2005). It is suggested that at a certain ratio of PF-4:heparin, the heparin chain induces a conformational change in the PF-4 tetramer to a less folded state eliciting pathogenic IgG HIT antibodies that recognise the neoepitopes exposed as a result of heparin binding (Mikhailov et al. 1999). Alternatively, it is suggested that the antigen is generated by two PF-4 tetramers brought in close proximity by heparin binding (Greinacher 2009). Macromolecular antibody-heparin-PF-4 HIT immune complexes cross-link platelet FcγIIa (CD32a) receptors, triggering platelet activation and aggregation, thrombocytopenia and a pro-coagulant response (Xiao et al. 2008; Greinacher 2009).

PF-4 is also reported to bind to a high-molecular-weight CSPG with an unidentified core protein on neutrophils (Brandt et al. 2000; Gleissner et al. 2008). A role for leukocyte surface CSPG in neutrophil activation during the development of HIT (Xiao et al. 2008) has been indicated. In this model, CSPG enhances the binding of PF-4 and PF-4 immune complexes to the neutrophil surface, and activation of CD32a on the same or adjacent cells. Thus, while PF-4 per se is not a neutrophil chemoattractant, neutrophil activation via this mechanism may contribute to thrombosis and inflammation in patients mounting an immune response to heparin-PF-4.

### 2.1.2 Interleukin-8

Interleukin-8 (IL-8, or CXCL8) is an $ELR^+$ CXC chemokine that interacts with CXCR1 and CXCR2 receptors on neutrophils to activate diverse signalling pathways (Waugh and Wilson 2008) to induce neutrophil survival and recruitment during inflammatory responses. It is also an angiogenic cytokine (Belpeiro et al. 2000), mediating its effect in tumour microvessels either directly through increased expression of endothelial CXCR1 and CXCR2, or indirectly through effects on neutrophil recruitment and activation (De Larco et al. 2004), to induce endothelial cell proliferation, survival and migration and significantly influencing the tumour microenvironment (Waugh and Wilson 2008). The secretion of IL-8 from cancer cells and the expression of CXCR1 and CXCR2 by cancer cells indicate that IL-8 signalling is also likely to induce the transcription of multiple genes involved in regulation of the cell cycle, migration, invasion and survival, each response independently associated with cancer progression (Waugh and Wilson 2008).

The recruitment of leukocytes from the vasculature into tissues at sites of infection and inflammation occurs in a multi-step process that involves the capture and selectin-mediated rolling of cells on the endothelial surface, the chemokine-induced activation of integrin-mediated adhesion and leukocyte arrest, followed by transendothelial migration and migration through the basement membrane (Ley et al. 2007). The importance of HSPG in these processes is indicated in Fig. 2 (Parish 2006). Rolling is initiated by endothelial P-selectin interacting with its ligand (PSGL1) on leukocytes, and supported by leukocyte L-selectin binding to HS side chains of endothelial syndecan proteoglycans. Chemokines, including IL-8, are synthesised by endothelial cells in response to pro-inflammatory cytokines such as IL-1 and TNF-α, or by sub-endothelial cells such as the tissue macrophages and, following HSPG-mediated transcytosis, are presented bound to HSPG on the luminal endothelial cell surface to GPCRs on leukocytes (Parish 2006; Colditz et al. 2007). Leukocytes are guided by a concentration gradient of HSPG-bound chemokines across the endothelial cell layer, through the basement membrane and into the tissue. During this process, leukocyte-derived enzymes, heparanase,

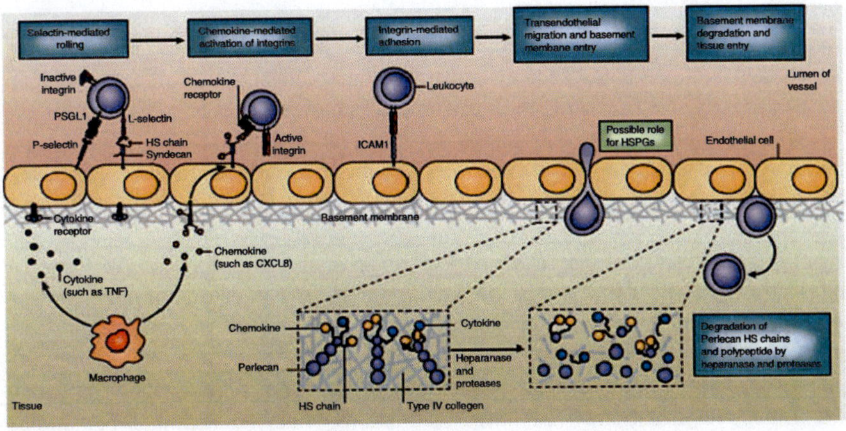


**Fig. 2** The importance of HS at different stages of leukocyte entry into sites of inflammation [Reprinted by permission from Macmillan Publishers Ltd: Nat Rev Immunol, Parish CR, 6: 633–643, 2006 (http://www.nature.com)]

sulphatase, matrix metalloproteases and reactive oxygen species solubilise basement membrane HSPG and depolymerise HS, releasing cytokines and growth factors involved in tissue remodelling and repair, so that there is normally no collateral tissue damage during an inflammatory response (Parish 2006).

While chemokines, including IL-8 (Rajarathnam et al. 1994), bind to specific receptors and are active as monomers in vitro, all CXC chemokines form dimers (e.g. IL-8, SDF-1) or tetramers (e.g. PF-4, NAP-2) and oligomerisation maybe enhanced in the presence of HS, although some chemokines form dimers and tetramers in the absence of HS (Handel et al. 2005; Lortat-Jacob et al. 2002). Molecular modelling studies predict four different modes of binding of heparin to the chemokines, depending on the way they fold and oligomerise (Lortat-Jacob et al. 2002). Specifically, dimeric IL-8 binding to HS was found to require a minimum of 18 monosaccharide units consisting of two N-sulphated regions of six monosaccharides linking two IL-8 monomers through a flexible N-acetylated spacer (Spillman et al. 1998). The spatial separation of the GAG-binding domain from the receptor binding domain suggests that IL-8 bound to GAGs in tissues is able to activate neutrophil IL-8 receptors (Kuschert et al. 1998).

The binding of IL-8 to endothelial cell surfaces was first proposed by Rot (1992) as a mechanism to promote neutrophil adhesion and emigration from the blood, preventing the dissipation of chemoattractant gradients by dilution of IL-8 in the circulation, while soluble IL-8 was proposed to inhibit adhesion and migration through receptor desensitisation. IL-8 binding sites on endothelial cells in tissues (Rot et al. 1996) were subsequently identified as the GAGs HS and CS in lung tissue (Frevert et al. 2003). In free solution, chemokines are monomers at physiological concentrations. However, binding to GAGS mediates the oligomerisation of chemokines, including IL-8, in tissues (Frevert et al. 2003) and on endothelial cell surfaces (Hoogewerf et al. 1997), and was proposed to immobilise the chemokine gradient for directed leukocyte migration. In addition, IL-8 binding to leukocyte cell-surface GAGs was proposed to increase the local concentration of the chemokine for presentation to specific high-affinity receptors on the target cell. The binding of IL-8 to endothelial cell surfaces can be competed with soluble GAGs, which inhibit the binding of IL-8 in the rank order heparin > heparan sulphate > dermatan sulphate > chondroitin sulphate (Kuschert et al. 1999). Thus, soluble GAGs bind IL-8, but inhibit the biological activity since the soluble IL-8/GAG complex is unable to bind the receptor (Kuschert et al. 1999; Ramdin et al. 1997). This is believed to be an electrostatic effect in view of the acidic nature of both the N-terminus of the receptor and the GAG (Kuschert et al. 1999).

IL-8 binds to the HS chains of syndecan-1 (Marshall et al. 2003) and to both the HS and core protein of syndecan-2 (Halden et al. 2004) on endothelial cell surfaces. The functional importance of these interactions was demonstrated by studies showing inhibition of transendothelial neutrophil migration in vitro following the plasmin-mediated shedding of endothelial syndecan-1/IL-8 complexes (Marshall et al. 2003) and impaired L-selectin and chemokine-mediated neutrophil trafficking during inflammatory responses in vivo as a result of endothelial cell heparan sulphate deficiency induced by inhibition of the gene coding for expression of N-acetyl glucosamine N-deacetylase-N-sulfotransferase-1, the enzyme responsible for adding sulphate groups to heparan sulphate chains, in mice (Wang et al. 2005). These effects are cell-type specific and a deficiency of HS on neutrophils did not affect inflammatory responses in mice (Wang et al. 2005), indicating that endothelial HS dominates in this system. However, deficiencies of syndecan-1 on neutrophils *increased* integrin-dependent neutrophil adhesion to the endothelium (Masouleh et al. 2009), indicating a role for the syndecan core protein and/or chondroitin sulphate side chains in limiting neutrophil adhesion.

With respect to the regulation of chemokine function by HS, both pro-inflammatory and anti-inflammatory roles have also been demonstrated for *soluble* syndecan-1 in the airway (Li et al. 2002; Xu et al. 2005). In a mouse model of bleomycin-induced acute lung injury, matrilysin-shed epithelial syndecan-1 supported CXC chemokine-mediated neutrophil trafficking into the alveolar space (Li et al. 2002), while soluble syndecan-1 binding to CC chemokines attenuated allergic lung inflammation (Xu et al. 2005).

In summary, IL-8 binding to HSPG on vascular endothelial cell surfaces is proposed to support neutrophil recruitment by stabilising the chemoattractant gradient, and protect the chemokine from proteolysis. Changes in the expression of HSPG under inflammatory conditions are likely to influence the binding and activity of IL-8 and other chemokines. For example, increased expression of HS in lung tissue of patients with cystic fibrosis (Solic et al. 2005) and the induction of an IL-8 binding site on endothelial syndecan-3 in rheumatoid synovium (Patterson et al. 2005) are likely to contribute to the sustained IL-8 mediated inflammatory response in these diseases.

### 2.1.3 Stromal Cell-Derived Factor-1

Stromal cell-derived factor-1 (SDF-1, or CXCL12) is an ELR$^-$ CXC chemokine and constitutively expressed agonist of the CXCR4 receptor, which has multiple roles in embryonic development. Post-natally, SDF-1 is a lymphocyte chemoattractant and plays prominent roles in inflammation, angiogenesis, wound healing and the metastatic potential of a number of tumours (Lortat-Jacob 2009; Rueda et al. 2008). Alternative splicing from the *Cxcl12* gene gives rise to six different isoforms (α, β, γ, δ, ε and φ) of the chemokine. The major α isoform comprises the core protein of 68 amino acids, found in all isoforms, containing the CXCR4 binding site and canonical HS-binding region, responsible for heparin/HS-binding and chemokine dimerisation, and is a target for inhibition by heparin (Murphy et al. 2007). The other isoforms differ in their C-terminal domain, which fine tunes GAG binding. Of particular interest is the extended 30 amino acid C-terminal domain of CXCL12γ. This intrinsically disordered domain is highly enriched in basic amino acids and has four overlapping BBXB (where B is a basic amino acid and X is any other) HS-binding domains that increase the capacity of CXCL12γ to interact with HS. This structural modification is responsible for high affinity and stable binding to extracellular HS structures and the increased functional capacity of CXCL12γ, compared to CXCL12α, to sustain lymphocyte recruitment in vivo (Rueda et al. 2008).

Other functions of the interaction of this chemokine with HS have been identified. Endothelial bound CXCL12, but not the soluble chemokine, induces a conformational change in lymphocyte function-associated antigen 1 (LFA-1) from a bent (inactive) to an extended (active) form that primes LFA-1 for binding to its ligand, intercellular adhesion molecule 1 (ICAM1), on endothelial cells thereby promoting stable cell adhesion (Shamri et al. 2005). Binding to HS also protects CXCL12α from proteolytic inactivation by CD26/DPP IV, which mediates the selective removal of the N-terminal dipeptide (Lortat-Jacob 2009).

In addition to the binding of CXCL12 to endothelial HSPG, HS-dependent binding of CXCL12 to syndecan-4 on HeLa cells, and on T-cells and macrophages, has been demonstrated (Charnaux et al. 2005). The binding of CXCL12 to syndecan-4 on macrophages facilitates chemokine binding to the CXCR4 receptor. However, on HeLa cells, HS-dependent binding of CXCL12 directly to syndecan-4 activates signal transduction pathways, with the proteoglycan behaving as a CXCL12 receptor (Charnaux et al. 2005) and indicating a role beyond chemokine presentation.

### 2.1.4 RANTES

RANTES (released on activation, normal T-cell expressed and secreted) is a member of the CC chemokine family (CCL5) and binds three GPCRs, CCR1, CCR3 and CCR5, inducing the recruitment of T-lymphocytes, monocytes and other leukocytes into tissues. The binding of RANTES, and other CC chemokines such as MCP-1 and MIP-1α, to endothelial cell surfaces is mediated by GAGs, which induce oligomerisation of the chemokines, increasing their concentration and enhancing their interaction with high-affinity GPCR chemokine receptors (Hoogewerf et al. 1997). RANTES formed oligomers of 60 kDa average molecular weight in the presence of heparin (Hoogewerf et al. 1997). Extensive oligomerisation of RANTES (>MIP-1α > MCP-1 > IL-8) in the presence of soluble heparin may reflect its high positive (+5) charge density at neutral pH (Martin et al. 2001). The affinity of chemokine binding to GAGs on HUVECs (RANTES > MCP-1 > IL-8 > MIP-1α) reflected the rank order in the affinity of the chemokines for heparin, and all could be competed with soluble GAGs. RANTES binding to HUVECs was competed by soluble GAGs in the rank order heparin > dermatan sulphate > chondroitin sulphate > heparan sulphate > chondroitin sulphate and, unusually, dermatan sulphate interacted with RANTES more strongly than heparan sulphate (Kuschert et al. 1999), which might reflect a non-specific charge interaction (Yamaguchi et al. 2006). The specific interaction of RANTES with heparin is mediated through a BBXB cluster of basic residues, (44) RKNR(47), common in heparin-binding proteins (Proudfoot et al. 2001). Mutations in this sequence to form the (44)AANA(47) variant do not affect chemotactic activity in vitro (Proudfoot et al. 2003). However, these GAG-binding mutants are unable to form higher order oligomers, and since in vivo activity requires a minimum tetrameric structure, were inactive in vivo. These studies demonstrated that both GAG-binding and oligomerisation are essential for the in vivo activity of RANTES (Proudfoot et al. 2003). Binding of RANTES to heparan sulphate induces a significant conformational change, which is believed to be a pre-requisite for oligomarisation, and receptor activation in vivo (Rek et al. 2009a). More recently, a second cluster of basic amino acid residues, (55)KKWVR(59), was found to be essential for in vivo activity but not GAG-binding (Segerer et al. 2009) and may be involved in binding other sulphated oligosaccharides, including those exposed on the cell surface by sulphatides, a class of glycosphingolipids.

The functional significance of RANTES–proteoglycan interactions are further indicated by the release of RANTES, as well as MIP-1α and MIP-1β, in high molecular weight complexes from T-lymphocytes (Wagner et al. 1998). In addition, RANTES has been shown to bind in a GAG-dependent manner to syndecan-1 and syndecan-4 on primary normal macrophages and human hepatoma cell lines (Charni et al. 2009). This interaction is involved in chemokine-induced tumour cell migration in vitro and is inhibited by heparin, indicating a new therapeutic approach for hepatocellular carcinoma (Charni et al. 2009).

## 2.2 *Targeting GAG–Chemokine Interactions as a Novel Anti-inflammatory Strategy*

Heparin has been demonstrated in vitro and in vivo, in animal models and clinical studies, to have an array of anti-inflammatory and anti-cancer effects (discussed elsewhere in this volume). In part, these are likely to be mediated by interference of heparin with GAG/chemokine interactions (see above) since soluble heparin competes with endothelial GAGs for chemokine binding, while the soluble chemokine/heparin complex is unable to bind and activate its specific receptor. In view of the potential side effects of heparin, other approaches are also of interest. Recent studies have shown that a non-anticoagulant fully sulphated synthetic hexasaccharide found in heparin inhibits CCL21 (but not CXCL12 or CCL19)-induced lymphocyte chemotaxis in vitro (de Paz et al. 2007), but only when presented in multivalent form attached to polyamidoamine (PAMAM) dendrimers. It was predicted from these results that the functionalised dendrimers would inhibit the formation of chemokine gradients and cell recruitment in vivo.

GAG mimetics synthesised by grafting carboxylate and sulphate groups onto a dextran backbone in extents similar to those found in heparin have been shown to inhibit RANTES-induced migration and invasion of human hepatatoma cells (Sutton et al. 2007). GAG mimetics directly bind to RANTES, which could result in inhibition of chemokine binding to GAGs carried on hepatoma cell-surface proteoglycans. GAG mimetics based on synthetic sulphated linked cyclitol structures have also been developed that have been used to probe the heparin/HS-binding specificity of a number of HS-binding proteins and may have therapeutic value by specifically blocking the activity of certain HS-binding proteins (Freeman et al. 2005). These HS mimetics comprise sulphated cyclitol-based pseudo-disaccharides linked by a flexible spacer of variable chain length. The length of the spacer, >7 carbon atoms, critically determined the ability of the mimetic to inhibit IL-8 binding to heparin/HS, presumably to allow each of the linked disaccharides to interact with the HS-binding site of the IL-8 dimer.

Other approaches in development involve mutational modifications in chemokine structures with a view to interfering with the interaction of native chemokines with endothelial GAGs. In the first example, the (44)AANA(47) RANTES variant is not only unable to bind to endothelial GAGs and recruit cells in vivo, but also has unexpected anti-inflammatory properties and was able to significantly inhibit inflammatory cell recruitment into the peritoneum, airway and CNS in mouse models of inflammation (Johnson et al. 2004). Intraperitoneal wild-type RANTES in mice induced recruitment of a heterogeneous population of leukocytes, consisting of macrophages, neutrophils, lymphocytes, eosinophils and mast cells. However, inflammatory cell recruitment was nearly completely abrogated by prior administration of the (44)AANA(47) RANTES variant. The effect was not due to receptor desensitisation, but to the ability of the variant to sequester endogenous RANTES, forming a non-functional heterodimer that is unable to form the higher order oligomers that are necessary for the biological activity of RANTES in vivo. Altering the GAG-binding site therefore rendered the variant RANTES a dominant negative inhibitor of endogenous RANTES activity that was as effective as heparin. In addition, small molecules that bind to the chemokine and interfere with chemokine binding to GAGs are currently being sought (Proudfoot et al. 2008).

In an experimental autoimmune encephalomyelitis (EAE) animal model of multiple sclerosis the P8A-MCP-1 obligate monomer of MCP-1 (CCL2), which is unable to recruit cells in vivo (Proudfoot et al. 2003) also had unexpected anti-inflammatory properties (Proudfoot et al. 2008). In this case, the variant retains GAG-binding properties and is believed to displace endogenous JE [the mouse analogue of MCP-1 (Zlotnik et al. 2006)] from vascular endothelial cells, but is then unable to oligomerise and mediate inflammatory cell recruitment.

Further protein-engineering approaches have led to the development of chemokine-based GAG antagonists (Rek et al. 2009b). These molecules are engineered to have *enhanced* GAG-binding affinity combined with Met-derived N-terminal amino acids to create a GAG antagonist that is able to displace wild-type RANTES from its endothelial HSPG co-receptor, but is unable to induce monocyte recruitment (Brandner et al. 2009). Similarly, an IL-8-based therapeutic has been engineered with these combined properties, although in this case the receptor activating properties were deleted with six N-terminal amino acids (Bedke et al. 2010). This molecule binds endothelial HS with high affinity, competing off the wild-type IL-8, but is unable to induce leukocyte recruitment. Thus, by disabling both chemokine/GAG and chemokine/receptor interactions, this molecule has been shown to inhibit neutrophil recruitment in vitro and in vivo in an animal model of acute experimental renal allograft damage (Bedke et al. 2010).

Finally, a small molecule antagonist of heparan sulphate, surfen (MW 372.4), has been described that interacts electrostatically with heparin > dermatan sulphate > heparan sulphate > chondroitin sulphate, blocks FGF binding (see below) and may be useful therapeutically in treating inflammatory diseases, tumour growth and angiogenesis (Schuksz et al. 2008).

## 2.3 *Interaction of GAGs with Growth Factors*

HSPGs bind a number of growth factors in the extracellular matrix and at cell surfaces (Taipale and Keski-Oja 1997). The HS-dependent binding and regulation of growth factor activity is a function of syndecans at cell surfaces (Alexopoulou et al. 2007) and of perlecan in the basement membrane (Taipale and Keski-Oja 1997). While the selective HS-binding of growth factors may involve preferred saccharide sequences and sulphation patterns (Rek et al. 2009a, b), it appears that degree of sulphation and relatively non-specific charge interactions, in addition to domain organisation (Fig. 1), may also have been involved (Kreuger et al. 2006). The following are examples of the interaction of growth factors with HS, although FGF, HGF and VEGF also bind to CS and DS side chains of proteoglycans (Malavaki et al. 2008).

### 2.3.1 Fibroblast Growth Factors

Fibroblast growth factors (FGFs) play a role in tissue patterning and organogenesis in embryogenesis, and induce proliferation, differentiation, motility, adhesion, survival and apoptosis at both embryonic and adult stages. There are 18 mammalian

FGFs (FGF1-FGF10 and FGF16-FGF23) that bind and activate a family of four FGFR tyrosine kinase receptors in a HS-dependent manner (Beenken and Mohammadi 2009). (FGF11-FGF14 are now known as FGF homologous factors (FHFs) that do not activate FGFRs and FGF15 is the mouse orthologue of human FGF19.) Alternative splicing of FGFR1-3 yields FGFR1-3b isoforms on epithelia and FGFR1-3c isoforms on mesenchymal cells. In normal physiology, FGFs synthesised by the epithelium will activate mesenchymal receptors, and vice versa (Beenken and Mohammadi 2009). FGF1 (acidic) and FGF2 (basic) growth factors lack signal peptides and are secreted in an unconventional manner that appears to be dependent on extracellular HSPG to trap secreted FGF and drive membrane translocation of the protein (Nickel 2007). FGF-FGFR binding specificity, or pairing, is determined by primary sequence variation of the N-and C-terminal tails of FGFs and FGFRs, and by the spatial and temporal expression patterns of FGFs, FGFRs and HSPG.

Both FGF and FGFR bind to HS (and heparin) and HS increases the affinity of FGF for FGFR and stabilises the FGF–FGFR complex (Beenken and Mohammadi 2009; Harmer 2006). HS and heparin induce dimerisation of FGF and, following the binding of FGF to the extracellular domains of FGFR, FGFRs dimerise and the tyrosine kinase domains of the two receptors phosphorylate each other, and induce cell signalling (Harmer 2006). The stoichiometry of the FGF–FGFR–HS ternary complex (2:2:1 or 2:2:2) is a matter of current debate, and may depend on GAG chain length, octasaccharide and larger heparin fragments supporting the dimerisation of FGF-2 and the 2:2:1 configuration (Goodger et al. 2008). It is suggested that the binding of FGF to HS and the formation of ternary complexes with FGF receptors is determined by the length and overall charge density of sulphated domains, rather than specific sulphation patterns (Kreuger et al. 2006). Further, it has been suggested that multiple FGF complexes form on a single HS chain initiating aggregation of multiple FGFRs to generate a signalling plaque (Harmer 2006).

Cell-surface syndecans regulate HS-dependent FGF activity and syndecan-4, through its cytoplasmic domain, is also involved in endothelial cell signalling (Alexopoulou et al. 2007). Proteolytic shedding of syndecan-1 ectodomains during wound repair generates an inhibitor of FGF-2. However, subsequent processing of the HS chains with a heparanase that cleaved less sulphated domains, converted the shed ectodomain from an inhibitor to a potent FGF-2 activator by releasing heparin-like S-domains, which increase FGF-2 bioavailability (Kato et al. 1998).

Each tissue or cell-type produces specific repertoires of HS structures, contributing to the specificity of the cellular response, and expression of these structures is altered in disease settings, such as inflammation and cancer (Bishop et al. 2007; Kreuger et al. 2006), altering the context and activity of the growth factor. High-affinity binding to HS generally renders the FGFs paracrine ligands, acting locally to the source of their expression, with the exception of FGF19, FGF21, and FGF23, which have reduced HS-binding affinity and act as endocrine ligands. Thus, in the extracellular matrix, HS acts as a storage reservoir for FGFs, determining the radius of diffusion, and also protects the FGFs from proteolysis.

### 2.3.2 Hepatocyte Growth Factor/Scatter Factor

Hepatocyte growth factor (HGF) is a mesenchymal factor that stimulates growth, motility and morphogenesis to neural, epithelial and endothelial cells through activation of its specific tyrosine kinase receptor MET (Birchmeier et al. 2003). HGF binds with high affinity to MET, and also to heparin, HS and DS. HGF binds to MET in the absence of HS, with the GAGs acting as co-factors in the activation of MET. Unlike the situation for FGF-2, there is little evidence of a requirement for GAG interaction with MET. The binding of HGF to MET has low GAG sequence specificity, but the affinity of binding increases with increasing sulphate density (Catlow et al. 2008). As for FGF, it appears that length and overall charge density of sulphated domains, rather than specific dispositions and linear sequences of sulphate groups, may dictate the affinity of HGF binding to HS.

### 2.3.3 Vascular Endothelial Growth Factor

Vascular endothelial growth factor (VEGF) is an endothelial cell mitogen and is the major factor involved in physiological and pathological angiogenesis (Ruhrberg 2003). Alternative mRNA splicing gives rise to a number of VEGF isoforms, differing in the presence or absence of short C-terminal heparin-binding domains. These domains mediate VEGF interaction with HSPG and neuropilin co-receptors. $VEGF_{165}$ is the most abundantly expressed variant that interacts with HSPGs in an active form. The longer $VEGF_{189}$ variant is sequestered in an inactive form in the extracellular matrix, while the short $VEGF_{110}$ is active, but lacks heparin-binding domains.

$VEGF_{165}$ is secreted as a disulphide-linked homodimer with two identical HS-binding domains. Heparin and HS bind both $VEGF_{165}$ and VEGFR, and interactions of $VEGF_{165}$ with HS regulate the diffusion, half-life and affinity of the growth factor for its specific receptors. Cell-surface HSPGs enhance the affinity of dimeric $VEGF_{165}$ binding to its receptors. The interaction is mediated by two S-domains in a single HS chain, in which the presence of a 6-O sulphate group is a critical factor in determining HS affinity for VEGF (Robinson et al. 2006).

## 2.4 *Targeting Growth Factor/HS Interactions in Angiogenesis*

VEGF, HGF and FGF2 all mediate angiogenesis and their heparin-binding properties have been applied to novel strategies for the controlled delivery of these growth factors for tissue regeneration purposes (Joung et al. 2008). Conversely, heparin mimetics such as the sulphated linked cyclitols (Freeman et al. 2005), suramin (Beenken and Mohammadi 2009) and PI-88 (phosphomannopentaose sulphate)

(Khachigian and Parish 2004) and its analogues (Ferro et al. 2007) that bind growth factors and interfere with HS and receptor binding have anti-angiogenic effects and may be useful therapeutics in a range of cancers.

## References

Alexopoulou AN, Multhaupt HAB, Couchman JR (2007) Syndecans in wound healing, inflammation and vascular biology. Int J Biochem Cell Biol 39:505–528

Bedke J, Nelson PJ, Kiss E et al (2010) A novel CXCL8 protein-based antagonist in acute experimental renal allograft damage. Mol Immunol 47(5):1047–1057

Beenken A, Mohammadi M (2009) The FGF family: biology, pathophysiology and therapy. Nat Rev Drug Discov 8:235–253

Belpeiro JA, Keane MP, Arenberg DA et al (2000) CXC chemokines in angiogenesis. J Leukoc Biol 68:1–8

Birchmeier C, Birchmeier W, Gheradi E, van de Woude GF (2003) MET, metastasis, motility and more. Nat Rev Mol Cell Biol 4:915–925

Bishop JR, Schuksz M, Esko JD (2007) Heparan sulphate proteoglycans fine-tune mammalian physiology. Nature 446:1030–1037

Brandner B, Rek A, Diedrichs-Mohring M et al (2009) Engineering the glycosaminoglycan-binding affinity, kinetics and oligomerization behaviour of RANTES: a tool for generating chemokine-based glycosaminoglycan antagonists. Prot Eng Des Select 22:367–373

Brandt E, Ludwig A, Petersen F, Flad H-D (2000) Platelet-derived CXC chemokines: old players in new games. Immunol Rev 177:204–216

Capila I, Linhardt RJ (2002) Heparin-protein interaction. Angew Chem Int Ed Engl 41:390–412

Catlow KR, Deakin JA, Wei Z et al (2008) Interactions of hepatocyte growth factor/scatter factor with various glycosaminoglycans reveal an important interplay between the presence of iduronate and sulphate density. J Biol Chem 283:5235–5248

Charnaux N, Brule S, Hamon M et al (2005) Syndecan-4 is a signalling molecule for stromal cell-derived factor-1 (SDF-1)/CXCL12. FEBS J 272:1937–1951

Charni F, Friand V, Haddad O et al (2009) Syndecan-1 and syndecan-4 are involved in RANTES/CCL5-induced migration and invasion of human hepatoma cells. Biochim Biophys Acta 1790:1314–1326

Colditz IG, Schneider MA, Pruenster M, Rot A (2007) Chemokines at large: in-vivo mechanisms of their transport, presentation and clearance. Thromb Haemost 97:688–693

De Larco JE, Wuertz BRK, Furcht LT (2004) The potential role of neutrophils in promoting the metastatic phenotype of tumours releasing interleukin-8. Clin Cancer Res 10:4895–4900

de Paz JL, Moseman EA, Noti C et al (2007) Profiling heparan-chemokine interactions using synthetic tools. ACS Chem Biol 2:735–744

Ferro V, Dredge K, Liu L et al (2007) PI-88 and novel heparan sulphate mimetics inhibit angiogenesis. Semin Thromb Hemost 33:557–568

Freeman C, Liu L, Banwell MG et al (2005) Use of sulphated linked cyclitols as heparan sulphate mimetics to probe the heparin/heparan sulphate binding specificity of proteins. J Biol Chem 280:8842–8849

Frevert CW, Kinsella MG, Vathanaprida C et al (2003) Binding of interleukin-8 to heparan sulfate and chondroitin sulfate in lung tissue. Am J Respir Cell Mol Biol 28:464–472

Gallagher JT (2006) Multiprotein signalling complexes: regional assembly on heparan sulphate. Biochem Soc Trans 34:438–441

Gandhi NS, Mancera RL (2008) The structure of glycosaminoglycans and their interactions with proteins. Chem Biol Drug Des 72:455–482

Gleissner CA, von Hundelhausen P, Ley K (2008) Platelet chemokines in vascular disease. Arterioscler Thromb Vasc Biol 28:1920–1927
Goodger SJ, Robinson CJ, Murphy KJ et al (2008) Evidence that heparin saccharides promote FGF2 mitogenesis through two distinct mechanisms. J Biol Chem 283:13001–13008
Greinacher A (2009) Heparin-induced thrombocytopenia. J Thromb Haemost 7:9–12
Halden Y, Rek A, Atzenhofer W et al (2004) Interleukin-8 binds to syndecan-2 on human endothelial cells. Biochem J 377:533–538
Handel TM, Johnson Z, Crown SE et al (2005) Regulation of protein function by glycosaminoglycans – as exmplified by chemokines. Annu Rev Biochem 74:385–410
Harmer NJ (2006) Insights into the role of heparan sulphate in fibroblast growth factor signalling. Biochem Soc Trans 34:442–445
Hoogewerf AJ, Kushert GSV, Proudfoot AEI et al (1997) Glycosaminoglycans mediate cell surface oligomerisation of chemokines. Biochemistry 36:13570–13578
Johnson Z, Kosco-Vilbois MH, Herren S et al (2004) Interference with heparin binding and oligomerisation creates a novel anti-inflammatory strategy targeting the chemokine system. J Immunol 173:5776–5785
Joung YK, Bae JW, Park KD (2008) Controlled release of heparin-binding growth factors using heparin-containing particulate systems for tissue regeneration. Expert Opin Drug Deliv 5: 1173–1184
Kato M, Wang H, Kainulainen V et al (1998) Physiological degradation converts the soluble syndecan-1 ectodomain from an inhibitor to a potent activator of FGF-2. Nat Med 4:691–697
Khachigian LM, Parish CR (2004) Phosphomannopentaose sulphate (PI-88): heparan sulphate mimetic with clinical potential in multiple vascular pathologies. Cardiovasc Drug Rev 22:1–6
Kreuger J, Spillman D, Li J-P, Lindahl U (2006) Interactions between heparan sulfate and proteins: the concept of specificity. J Cell Biol 174:323–327
Kuschert GS, Hoogewerf AJ, Proudfoot AE et al (1998) Identification of a glycosaminoglycan binding surface on human interleukin-8. Biochemistry 37:11193–11201
Kuschert GSV, Coulin F, Power CA et al (1999) Glycosaminoglycans interact selectively with chemokines and modulate receptor binding and cellular responses. Biochemistry 38: 12959–12968
Ley K, Laudanna C, Cybulsky MI, Nourshargh S (2007) Getting to the site of inflammation: the leukocyte adhesion cascade updated. Nat Rev Immunol 7:678–689
Li Q, Park PW, Wilson CL, Parks WC (2002) Matrilysin shedding of syndecan-1 regulates chemokine mobilization and transepithelial efflux of neutrophils in acute lung injury. Cell 111:635–646
Lortat-Jacob H (2009) The molecular basis and functional implications of chemokine interactions with heparan sulphate. Curr Opin Struct Biol 19:543–548
Lortat-Jacob H, Grosdidier A, Imberty A (2002) Structural diversity of heparan sulphate binding domains in chemokines. Proc Natl Acad Sci USA 99:1229–1234
Malavaki C, Mizumoto N, Karamanos N, Sugahara K (2008) Recent advances in the structural study of functional chondroitin and dermatan sulphate in health and disease. Conn Tiss Res 49:133–139
Marshall LJ, Ramdin LS, Brooks T et al (2003) Plasminogen activator inhibitor-1 supports IL-8-mediated neutrophil transendothelial migration by inhibition of the constitutive shedding of endothelial IL-8/heparan sulfate/syndecan-1 complexes. J Immunol 171:2057–65
Martin L, Blanpain C, Garnier P et al (2001) Structural and functional analysis of the RANTES-glycosaminoglycans interactions. Biochemistry 40:6303–6318
Masouleh BK, Ten Dam GB, Wild MK et al (2009) Role of heparan sulphate proteoglycan syndecan-1 (CD138) in delayed type hypersensitivity. J Immunol 182:4985–4993
Mikhailov D, Young HC, Linhardt RJ, Mayo KH (1999) Heparin dodecasaccaride binding to platelet factor-4 and growth-related protein-alpha. Induction of a partially folded state and implications for heparin-induced thrombocytopenia. J Biol Chem 274:25317–25329

Murphy JW, Cho Y, Sachpatzidis A et al (2007) Structural and functional basis of CXCL12 (stromal cell-derived factor-1 alpha) binding to heparin. J Biol Chem 282:10018–10027
Nickel W (2007) Unconventional secretion: an extracellular trap for export of fibroblast growth factor 2. J Cell Sci 120:2295–2299
Parish CR (2006) The role of heparan sulphate in inflammation. Nat Rev Immunol 6:633–643
Patterson AM, Gardner L, Shaw J et al (2005) Induction of a CXCL8 binding site on endothelial syndecan-3 in rheumatoid synovium. Arthritis Rheum 52:2331–2342
Poncz M (2005) Mechanistic basis of heparin-induced thrombocytopenia. Semin Thorac Cardiovasc Surg 17:73–79
Proudfoot AE, Fritchley S, Borlat F et al (2001) The BBXB motif of RANTES is the principal site for heparin binding and controls receptor selectivity. J Biol Chem 276:10620–10626
Proudfoot AE, Handel TM, Johnson Z et al (2003) Glycosaminoglycan binding and oligomerization are essential for the in vivo activity of certain chemokines. Proc Natl Acad Sci USA 100:1885–1890
Proudfoot AEI, de Souza ALS, Muzio V (2008) The use of chemokine antagonists in EAE models. J Neuroimmunol 198:27–30
Rajarathnam K, Sykes BD, Kay CM et al (1994) Neutrophil activation by monomeric interleukin-8. Science 264:90–92
Ramdin L, Perks B, Sheron N, Shute JK (1997) Regulation of interleukin-8 binding and function by heparin and alpha2-macroglobulin. Clin Exp Allergy 28:616–624
Rek A, Brandner B, Geretti E, Kungl AJ (2009a) A biophysical insight into the RANTES-glycoaminoglycan interaction. Biochim Biophys Acta 1794:577–582
Rek A, Krenn E, Kungl AJ (2009b) Therapeutically targeting protein-glycan interactions. Br J Pharmacol 157:686–694
Robinson CJ, Mulloy B, Gallagher JT, Stringer SE (2006) $VEGF_{165}$-binding sites within heparan sulphate encompass two highly sulphated domains and can be liberated by K5 lyase. J Biol Chem 281:1731–1740
Rot A (1992) Endothelial cell binding of NAP-1/IL-8: role in neutrophil emigration. Immunol Today 13:291–294
Rot A, Hub E, Middleton J et al (1996) Some aspects of IL-8 pathophysiology III: chemokine interaction with endothelial cells. J Lukoc Biol 59:39–44
Rueda P, Balabanian K, Lagane B et al (2008) The CXCL12γ chemokine displays unprecedented structural ad functional properties that make it a paradigm of chemoattractant proteins. PLoS One 3:e2543
Ruhrberg C (2003) Growing and shaping the vascular tree: multiple roles for VEGF. Bioessays 25:1052–1060
Schuksz M, Fuster MM, Brown JR et al (2008) Surfen, a small molecule antagonist of heparan sulphate. Proc Natl Acad Sci USA 105:13075–13080
Segerer S, Johnson Z, Rek A et al (2009) The basic residue cluster $^{55}$KKWVR$^{59}$ in CCL5 is required for in vivo biologic function. Mol Immunol 46:2533–2538
Shamri R, Grabovsky V, Gauguet JM et al (2005) Lymphocyte arrest requires instantaneous induction of an extended LFA-1 conformation mediated by endothelium-bound chemokines. Nat Immunol 6:497–506
Solic N, Wilson J, Wilson SJ, Shute JK (2005) Endothelial activation and increased heparan sulfate expression in cystic fibrosis. Am J Respir Crit Care Med 72:892–898
Spillman D, Witt D, Lindahl U (1998) Defining the interleukin-8 binding domain of heparan sulphate. J Biol Chem 273:15487–15493
Stringer SE, Gallagher JT (1997) Specific binding of the chemokine platelet factor 4 to heparan sulphate. J Biol Chem 272:20508–20514
Sutton A, Friand V, Papy-Garcia D et al (2007) Glycosaminoglycans and their synthetic mimetics inhibit RANTES-induced migration and invasion of human hepatoma cells. Mol Cancer Ther 6:2948–2958
Taipale J, Keski-Oja J (1997) Growth factors in the extracellular matrix. FASEB J 11:51–59

Wagner L, Yang OO, Garcia-Zepeda EA et al (1998) Beta-chemokines are released from HIV-1-specific cytolytic T-cell granules complexed to proteoglycans. Nature 391(6670):908–911
Wang L, Fuster M, Sriramarao P, Esko JD (2005) Endothelial heparan sulphate deficiency impairs L-selectin- and chemokine-mediated neutrophil trafficking during inflammatory responses. Nat Immunol 6:902–910
Waugh DJJ, Wilson C (2008) The interleukin-8 pathway in cancer. Clin Cancer Res 14:6735–6741
Xiao Z, Visentin GP, Dayananda K, Neelamegham S (2008) Immune complexes formed following the binding of anti-platelet factor 4 (CXCL4) antibodies to CXCL4 stimulate human neutrophil activation and cell adhesion. Blood 112:1091–1100
Xu J, Park PW, Kheradmand F, Corry DB (2005) Endogenous attenuation of allergic lung inflammation by syndecan-1. J Immunol 174:5758–5765
Yamaguchi K, Tamaki H, Fukui S (2006) Detection of oligosaccharide ligands for hepatocyte growth factor/scatter factor (HGF/SF), keratinocyte growth factor (KGF/FGF-7), RANTES and heparin cofactor II by neoglycolipid microarrays of glycosaminoglycan-derived oligosaccharide fragments. Glycoconj J 23:513–523
Yoshimura T, Matsushima K, Tanaka S et al (1987) Purification of a human monocyte-derived neutrophil chemotactic factor that has peptide sequence similarity to other host-defense cytokines. Proc Natl Acad Sci USA 84:9233–9237
Zlotnik A, Yoshie O, Nomiyama H (2006) The chemokine and chemokine receptor superfamilies and their molecular evolution. Genome Biol 7:243

# Glycosaminoglycans and Neuroprotection

**B. Dudas and K. Semeniken**

## Contents



**Abstract** Glycosaminoglycans (GAGs) are basic building blocks of the ground substance of the extracellular matrix and present at the cellular level as an important component of the glycocalyx covering the cell membrane. In addition to the general role of GAGs in maintaining the integrity of the cell and extracellular matrix by retaining water, certain GAGs exhibit anticoagulant and neuroprotective properties and serve as cell-surface receptors for various molecules. Although heparin, a highly sulfated GAG, has been used as a drug for more than 70 years due to its anticoagulant attributes, the neuroprotective properties of GAGs came into focus only in recent years. The discovery of some of the roles GAGs play in the pathomechanism of numerous neurodegenerative disorders as well as shedding light on the neuroprotective properties of these compounds in animal studies raised the possibility that GAGs may provide an entirely new avenue in the treatment of neurodegenerative diseases. Indeed, some GAGs were successfully used to improve the cognitive function of patients with various neurodegenerative

B. Dudas (✉)
Neuroendocrine Organization Laboratory, Lake Erie College of Osteopathic Medicine, 1858 West Grandview Blvd, Erie, PA 1509, USA
e-mail: bdudas@lecom.edu

R. Lever et al. (eds.), *Heparin - A Century of Progress*,
Handbook of Experimental Pharmacology 207,
DOI 10.1007/978-3-642-23056-1_14, © Springer-Verlag Berlin Heidelberg 2012

conditions (Ban et al. (1991, 1992); Conti et al. (1989a, b); Passeri and Cucinotta, (1989); Santini (1989). Although the mechanism by which the GAGs exhibit neuroprotective properties is not entirely clear, there is a general consensus that the major factors of the neuroprotective attributes of GAGs include the impact of GAGs on amyloidogenesis and the regulatory action of GAGs in the apoptotic pathway.

**Keywords** Alzheimer's disease · Apoptosis · Glycosaminoglycans · Neuroprotection · Oxidative stress · Parkinson's disease · Prion · Proteoglycans

## 1 Structure of the Glycosaminoglycans

Glycosaminoglycans (GAGs) are long unbranched polysaccharide chains composed of disaccharide subunits formed by hexose (D-galactose), hexosamine (D-glucosamine or D-galactosamine) and hexuronic acid (D-glucuronic acid or L-iduronic acid) components. According to the disaccharide subunits, GAGs can be classified as chondroitin sulfate (CS), dermatan sulfate (DS), keratan sulfate (KS), heparan sulfate (HS), heparin, and hyaluronic acid (HA) (Fig. 1). The disaccharide components of GAGs are sulfated to various degrees with the exception of HA that is the only nonsulfated GAG. The sulfate and carboxylic groups are responsible for the negative charge of the GAGs. Indeed, heparin has one of the highest negative charge density of any known biological molecule.

GAG chains, with the exception of HA, may be covalently linked with core proteins forming proteoglycans (PGs, Fig. 2). The GAG-protein linkage usually

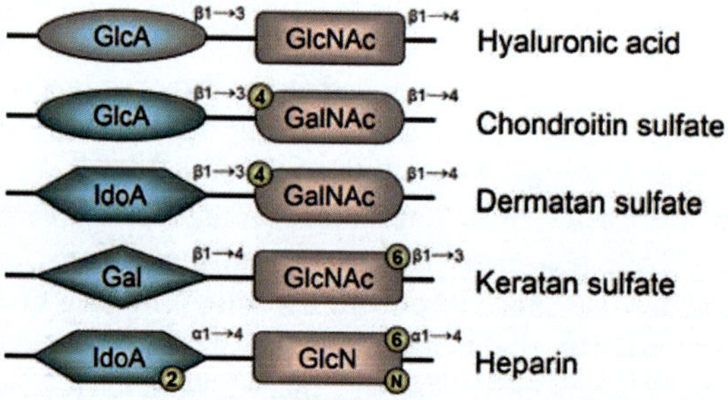


**Fig. 1** Structure of GAGs. GAGs are long unbranched polysaccharide chains composed of repeating disaccharide subunits formed by D-glucuronic acid (GlcA), L-iduronic acid (IdoA), D-galactose (Gal), N-acetyl-D-glucosamine (GlcNAc) or N-acetyl-D-galactosamine (GalNAc) components. Sulfate groups are denoted by spheres on the saccharide backbone. The numbers on the spheres mark the position of the sulfate groups in the sugar ring. The *N* indicates sulfation on the nitrogen at the number 2 position. The acetal linkages are indicated above the bonds. Heparin is a fractionated form of heparan sulfate

**Fig. 2** General structure of most of the PGs. GAG side chains are covalently attached to a centrally located core protein. The terminal part of the core protein is often anchored to HA chains

involves the hydroxyl groups of serine residues, but threonine or asparagine residues may also be involved. The ability of GAGs to interact with various proteins is crucial for the neuroprotective and anticoagulant properties of these molecules.

## 2 GAGs and Amyloidogenesis

There is a growing body of evidence that GAGs and PGs play a significant role in the pathomechanism of neurodegenerative diseases. One of the major histological hallmarks of neurodegenerative disorders is undoubtedly the deposition of abnormal proteins usually in the form of insoluble fibrils with a secondary structure of β-pleated sheets. In turn, these aggregated proteins are believed to be the primary trigger of neuronal loss. Although there are structural similarities between the peptides forming the protein deposits in the central nervous system, the primary structures of these molecules are rather heterogeneous. The term "amyloidosis," literally meaning deposition of starch-like substance, was first used by Virchow in the nineteenth century in order to describe these pathological conditions.

Since most of these accumulated proteins have affinity to, and often colocalize with, PGs and GAGs (Guiroy et al. 1991; Van et al. 1993), it is conceivable that these molecules play an active role in the process of protein aggregation. Indeed, GAGs and PGs appear to facilitate the polymerization and accumulation of abnormal proteins with cytotoxic nature. In contrast, certain low-molecular-weight (LMW) GAGs appear to competitively inhibit the detrimental effect of the endogenous GAGs and PGs and thus may be considered as useful adjuncts in the treatment of neurodegenerative disorders (Caughey and Raymond 1993; Leveugle et al. 1994). Among the numerous neurodegenerative diseases, the role of GAGs was intensively studied in the pathogenesis of Alzheimer's disease (AD), Parkinson's disease (PD), and transmissible spongiform encephalopathies (TSEs), also known as prion diseases.

## 2.1 GAGs and AD

The major histological hallmark of AD is the extra- and intracellular deposition of abnormal proteins. A 40–42 amino acid protein, amyloid β (Aβ), is deposited extracellularly in the wall of the cerebral vasculature and senile plaques in the brain of AD patients. In addition to the plaques, the hyperphosphorylated form of the normally present tau protein is deposited intracellularly as paired helical filaments (PHFs) that are responsible for the formation of flame-like neurofibrillary tangles (NFTs) in the cytoplasm of the affected neurons. There is a general consensus that the action of these deposited proteins may be a major factor in triggering the cholinergic cell loss that is responsible for the majority of the clinical symptoms of AD including cognitive deficits and behavioral changes.

PGs and GAGs appear to play a role in the protein deposition characteristic of AD. The levels of heparan sulfate proteoglycans (HSPGs) are significantly higher in the hippocampus and the superior frontal gyrus of patients suffering from senile dementia of Alzheimer's type compared to healthy individuals at the same age (Shimizu et al. 1997). GAGs can be found in senile plaques and intracellular tangles of AD patients (Diaz-Nido et al. 2002; Guiroy et al. 1991). Chondroitin sulfate proteoglycans (CSPGs) and HSPGs have high affinity for the Aβ, accelerate fibril formation, and maintain fibril stability by protecting the fibrils against proteolysis in vitro (Castillo et al. 1997; Gupta-Bansal et al. 1995) and in tissue cultures (Shaffer et al. 1995). LMW GAGs appear to counteract the amyloidogenic nature of the PGs, and thus, exhibit neuroprotective attributes, possibly by competing for the PG-binding sites. Shorter GAG chains inhibit the Aβ-PG-binding (Leveugle et al. 1994; Snow et al. 1995). LMW anionic sulfonate or sulfate compounds significantly reduced splenic amyloid progression in mice and interfered with HS-stimulated Aβ fibril aggregation in vitro (Kisilevsky et al. 1995). Novel GAG precursors appear to act as anti-amyloidogenic agents (Kisilevsky et al. 2003, 2004, 2007; Kisilevsky and Szarek 2002). Moreover, GAGs and other sulfated compounds were reported to inhibit the aggregation and toxicity of Aβ itself (Pollack et al. 1995). Since the sulfated dye Congo red has the same effect (Pollack et al. 1995; Sadler et al. 1995), it is conceivable that sulfate groups play a crucial role in the amyloid-PG and amyloid-GAG binding.

In addition to influencing the formation of the amyloid deposits, GAGs appear to play a significant role in the formation of NFT as well. The microtubule-associated tau protein is the major constituent of NFT, and it is able to polymerize in vitro forming fibrillar deposits similar to PHF (Goedert et al. 1996). GAGs, including HS, colocalize with tau protein in AD and other neurodegenerative diseases (Diaz-Nido et al. 2002; DeWitt et al. 1993; Spillantini et al. 1999). In vitro, GAGs induce the assembly of tau protein into filaments that resemble PHF (Arrasate et al. 1997; Goedert et al. 1996; Hasegawa et al. 1997; Perez et al. 1996). Since HA, a nonsulfated GAG, does not affect the aggregation of tau, sulfate groups appear to be essential for the development of PHF (Arrasate et al. 1997).

The fragments of Aβ and tau protein, which are capable to form polymerized deposits in the presence of GAGs, include sequences with similar cationic motifs containing histidine and lysine. Since the substitution of these amino acids in these sequences of tau protein appears to prevent GAG binding, these cationic motifs were suggested to have a crucial role in the tau protein aggregation triggered by GAGs (Diaz-Nido et al. 2002). Consequently, certain GAGs may inhibit the deposition of tau by binding to these sequences without triggering the aggregation of the protein into fibrils (Diaz-Nido et al. 2002). Indeed, administration of C3 (Neuroparin), a low-molecular-weight GAG manufactured from heparin by γ-irradiation, significantly reduces Aβ(25–35)-induced tau-2 immunoreactivity (Dudas et al. 2002).

In addition to the effect on tau protein, C3 exhibits neuroprotective properties in animal models that simulate cholinergic deficits characteristic of Alzheimer's disease. In rat, C3 administration protects against cholinergic lesion induced by cholinotoxin AF64A (Rose et al. 2003) in a dose-dependent and time-dependent manner (Dudas et al. 2003; Rose et al. 2004).

## 2.2 *GAGs and PD*

PD is the one of the most common neurodegenerative diseases. The underlying cause of the symptoms of PD is the loss of the dopaminergic neurons in the substantia nigra. Histologically, one of the major hallmarks of PD is the appearance of spherical protein deposits called Lewy bodies in the cytoplasm of the affected dopaminergic neurons. Lewy bodies have been suggested to be a key factor responsible for the dopaminergic neuronal loss. Lewy bodies, however, are not restricted to the brain tissue of PD patients; they have been observed in numerous neurodegenerative disorders including AD, dementia with Lewy bodies and neurodegeneration with brain iron accumulation 1 (NBIA1, Hallervorden-Spatz syndrome).

The major constituent of Lewy bodies is an abnormally folded form of the soluble, unfolded protein alpha-synuclein that is normally present in the presynaptic terminals. Under pathological conditions, alpha-synuclein may undergo conformational change and form β-pleated sheets that eventually result in fibrillary deposition constituting the Lewy bodies.

Whether GAGs may be involved in the formation of Lewy bodies is still unclear. GAGs have been shown to be associated with amyloid deposits in most neurodegenerative diseases with amyloidosis, and GAGs appear to play a pivotal role in the aggregation, deposition, and neurotoxicity of amyloid itself. Previous studies revealed that heparin, HS, and other highly sulfated polymers (dextran sulfate) stimulate the formation of alpha-synuclein fibrils (Cohlberg et al. 2002). Since KS has a negligible effect on alpha-synuclein fibrillization, this effect exerted by certain GAGs appears to be quite specific (Cohlberg et al. 2002). There may be other proteins involved in the GAG-modulated alpha-synuclein deposition.

Agrin, a synaptogenic protein that colocalizes with alpha-synuclein in Lewy bodies of the substantia nigra of PD brain, also binds to alpha-synuclein in a HS-dependent manner, induces conformational changes of alpha-synuclein resulting in beta-pleated sheet structure, and enhances its insolubility (Liu et al. 2005).

Although HSPGs are the major components of senile plaques, the colocalization of GAGs with Lewy bodies remains contradictory. Incorporation of fluorescein-labeled heparin into the fibrils demonstrated that the heparin is integrated into the fibrils (Cohlberg et al. 2002). Heparinase sensitive basic fibroblast growth factor (bFGF) binding sites, which are used to locate HSPGs, were observed in the intraneuronal inclusions of PD patients (Perry et al. 1992). In contrast to these findings, previous studies were unable to identify HSPGs or GAGs in Lewy bodies indicating that alpha-synuclein fibrillization and stabilization may also occur independently of the presence of HSPGs or GAGs (van et al. 2004).

GAGs are suggested to play a role in alpha-synuclein deposition via affecting the aggregation of beta-synuclein. Beta-synuclein is a homolog of alpha-synuclein and they colocalize in presynaptic terminals of neurons in numerous brain regions, including the dopaminergic neurons of the substantia nigra. Although beta-synuclein does not form fibrils and appears to inhibit alpha-synuclein aggregation, it can trigger rapid fibrillation of alpha-synuclein in the presence of GAGs (Yamin et al. 2005).

## 2.3 *GAGs and TSEs*

TSEs, also known as prion diseases, are rare, fatal neurodegenerative diseases affecting both animals and humans. In human, TSEs include Creutzfeldt–Jakob disease (CJD), Gerstmann–Straussler–Scheinker syndrome (GSS), fatal insomnia, and kuru. Due to the transmissible nature of TSEs and the consequent public health concerns, these diseases have been intensively studied in the past 15 years.

Prions are infectious agents that are made of the misfolded form of a protein that is normally present in the body. The normally present cellular form, the $PrP^C$, is a 209 amino acid protein located on the surface of the cell. The function of $PrP^C$ has never been fully revealed, although it is known to have high affinity to copper (II) ions. The misfolded form of the protein, $PrP^{Sc}$, which is identical to the prion itself, appears to be responsible for the pathological lesions characteristic for TSEs. Unlike $PrP^C$ that is mostly composed of protein chains with $\alpha$-helical structure, the $PrP^{Sc}$ has $\beta$-pleated sheet secondary structure, and it is highly resistant to proteases. $PrP^{Sc}$ accumulates in the brain during the disease forming amyloid plaques that disrupt normal tissue structure. Moreover, $PrP^{Sc}$ is able to convert normal $PrP^C$ to abnormal $PrP^{Sc}$, and thus it is responsible for the infectious characteristic of the prion as well as the rapid progression of the disease.

GAGs appear to play a pivotal role in the pathogenesis of TSEs. GAGs are secreted in the urine of animals and humans infected with prion, and in the urine of mice ablated for the PrP gene, suggesting that both the presence of $PrP^{Sc}$ or the absence of $PrP^{C}$ may alter the metabolism of GAGs (Mayer-Sonnenfeld et al. 2005). Indeed, sulfated GAGs have high affinity to PrP. Cultured cells express saturable and specific surface binding sites for PrP, many of which are GAGs (Shyng et al. 1995). HS serves as a cell-surface receptor for prions (Horonchik et al. 2005; Vana et al. 2007) and HSPGs have been shown to be associated with amyloid deposits in the mouse model of scrapie (McBride et al. 1998) and in numerous TSEs (Guiroy et al. 1991). $PrP^{Sc}$ binding to cultured cell lines could be inhibited by heparin (Hijazi et al. 2005) and binding of $PrP^{Sc}$ to cells missing GAGs on the cell surface was significantly reduced (Hijazi et al. 2005). Preincubation of scrapie brain homogenate with heparin prior to intraperitoneal inoculation into normal hamsters results in a significant delay in manifestation of the prion disease (Hijazi et al. 2005). The affinity of heparin and HS to $PrP^{C}$ has been confirmed in vitro (Warner et al. 2002). Sulfated dyes, such as Congo red, have a similar affinity to PrP (Caughey et al. 1994).

The sequence of PrP that is responsible for the GAG binding has not been precisely located yet. Free Congo red blocks heparin binding to PrP and vice versa suggesting that these molecules compete for the same binding site (Caughey et al. 1994). Similar to the Aβ and tau protein, PrP contains cationic sequences containing histidine and lysine. These sequences, similar to the ones of the tau protein, are believed to be the primary GAG binding sites of PrP (Diaz-Nido et al. 2002). Recombinant human PrP binds GAGs including CS, HA, and heparin via the N-terminus (Pan et al. 2002). The role of the N-terminus in GAG binding is supported by the finding that recombinant mutant human prion protein rPrP(8OR) that binds more monoclonal antibodies that are specific for the N-terminus of $rPrP^{C}$ than wild-type recombinant normal human $rPrP^{C}$ itself, indicating that the N-terminus of rPrP(8OR) is more exposed, also binds more GAGs than $rPrP^{C}$ (Yin et al. 2007; Yin et al. 2006). To complicate the picture even more, several regions of the prion protein have been identified as potential binding sites for GAGs (Warner et al. 2002). Interestingly, PrP-GAG binding appears to be modulated by the presence of $Cu^{2+}$ and $Zn^{2+}$ ions (Pan et al. 2002; Warner et al. 2002).

By binding to the PrP, GAGs appear to modulate the polymerization of PrP into protease resistant fibrils that ultimately get deposited in the brain tissue (Caughey 1994). Pentosan polysulfate (Caughey and Raymond 1993; Priola et al. 1994), and Congo red (Caughey and Race 1994; Priola and Caughey 1994; Priola et al. 1994) inhibits the accumulation of $PrP^{Sc}$ in cell culture. Moreover, GAGs appear to affect $PrP^{C}$ expression itself. Pentosan sulfate and related compounds rapidly and dramatically reduces the amount of $PrP^{C}$ on the cell surface by stimulating endocytosis of $PrP^{C}$, thus causing a redistribution of the protein from the plasma membrane to the interior of the cell (Shyng et al. 1995). On the other hand, interaction of PrP with endogenous sulfated GAGs or PGs appears to be crucial for $PrP^{Sc}$ accumulation (Caughey 1994). Certain GAGs and $Cu^{2+}$ have been shown to promote the aggregation of recombinant human PrP (Yu et al. 2008a) and HS and pentosan polysulfate stimulated $PrP^{Sc}$ formation (Wong et al. 2001). Since hyaluronic acid, which is a nonsulfated GAG, does not affect PrP polymerization, sulfate groups may be essential for this effect (Perez et al. 1998).

The ambiguous nature of GAGs on the PrP polymerization and deposition, coupled with the fact that $PrP^{Sc}$ deposits are known to contain sulfated GAGs, can be resolved by suggesting that LMW GAGs may competitively block an interaction between PrP and endogenous GAGs that is essential for the polymerization and accumulation of $PrP^{Sc}$ (Caughey et al. 1994; Caughey and Raymond 1993; Yin et al. 2007). Indeed, administration of HS can increase the concentration of PrP in normal neuroblastoma cells, whereas LMW heparin does not (Gabizon et al. 1993). Unlike HS, LMW heparin can inhibit the synthesis of $PrP^{Sc}$ in scrapie-infected cells and reverse their phenotype back to normal (Gabizon et al. 1993). Heparan mimetic biopolymers that are synthetic dextran derivatives with different degrees of sulfation, eliminate $PrP^{Sc}$ from prion-infected cells much more effectively than pentosan sulfate (Schonberger et al. 2003) by attenuating the conversion of $PrP^{C}$ into $PrP^{Sc}$. These molecules also prevent the uptake of prion rods by cultured cells, possibly by blocking the interaction of $PrP^{Sc}$ with a putative GAG cellular receptor, probably HS. The anti-$PrP^{Sc}$ effect of heparan mimetics correlates with the degree of sulfation (Schonberger et al. 2003). In contrast, agents inducing lysosomal accumulation of endogenous GAGs, such as tilorone, reduce the clearance of the $PrP^{Sc}$ from the infected cells. Since prolonged administration of tilorone to mice prior to the prion infection results in significant delay in disease onset, it is conceivable that GAGs may form complexes with $PrP^{Sc}$ and thus impair the clearance of $PrP^{Sc}$ from the cells by further stabilizing its conformation. In turn, overstabilized $PrP^{Sc}$ molecules have been demonstrated to exhibit reduced converting activity that may be responsible for the delayed onset of the disease (Mayer-Sonnenfeld et al. 2008).

It is a general consensus that the neuroprotective role of certain GAGs in prion diseases is based on the competition of these molecules with the endogenous GAGs for the PrP-binding sites (Caughey and Race 1994). Certain GAGs and Congo red inhibit the cytotoxic $PrP^{Sc}$ accumulation by interfering with the interaction of endogenous GAGs or PGs with PrP (Caughey et al. 1994). GAGs including heparin, KS, and CS inhibit the neurotoxicity of amyloid fibrils possibly via inhibition of the polymerization of the PrP peptide (Perez et al. 1998). Mice infected with bovine spongiform encephalopathy treated with pentosan polysulfate live significantly longer than controls (Larramendy-Gozalo et al. 2007). Moreover, HS mimetics attenuate prion propagation in scrapie-infected cells, impede $PrP^{Sc}$ accumulation in scrapie- and BSE-infected mice, and significantly prolong the survival time of scrapie-infected hamsters (Adjou et al. 2003).

## 3 GAGs and Apoptosis

In addition to the beneficial and detrimental effects of GAGs on the deposition of abnormal proteins in neurodegenerative diseases, GAGs may exert neuroprotective properties via different mechanisms. Since apoptotic processes often result in neuronal loss characteristic for neurodegenerative disorders, it is possible that the neuroprotective effects of GAGs are exerted via the modulation of apoptotic processes. Indeed, previous studies reported that LMW heparin derivative C3 suppressed AF64A-induced activity of caspase-3 (Dudas 2009).

### 3.1 *The Process of Apoptosis*

Apoptosis is one of the most complex and sophisticated signaling pathways of the cells (Fig. 3). Apoptosis, unlike necrosis, is a controlled cell death, a natural process

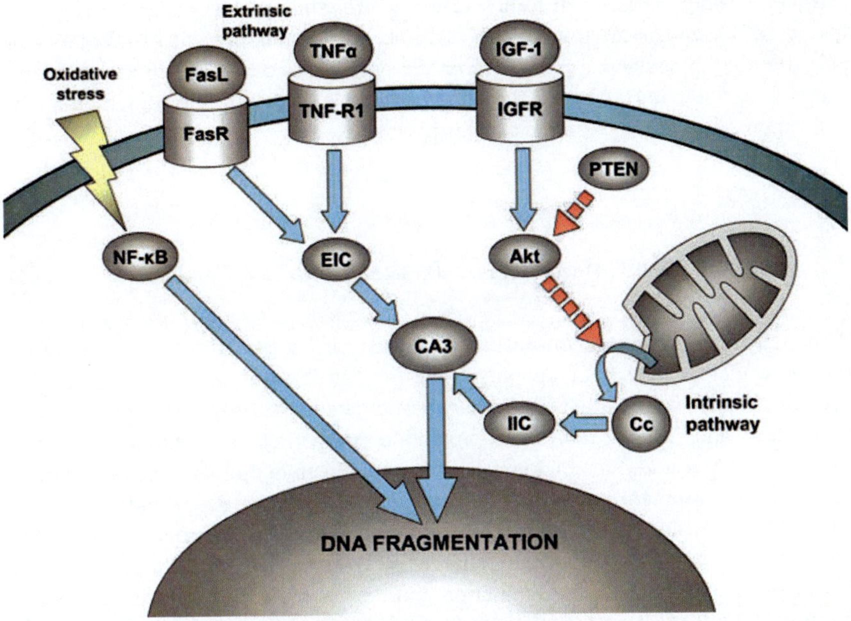


**Fig. 3** Some of the apoptotic signaling pathways that may play a role in the neuroprotection exerted by GAGs and PGs. Apoptosis is a natural process that ensures that damaged cells are packaged and removed by the surrounding cells. This mechanism prevents inflammation that is characteristic for necrotic processes. Apoptosis is triggered by external stimuli (extrinsic pathway) and internal events (intrinsic pathway). The extrinsic pathway is activated by binding "death ligands" (FasL, TNFα) to specific "death receptors" (FasR, TNF-R1). The ligand-receptor binding leads to activation of initiator caspases of the external pathway (EIC). The intrinsic pathway of apoptosis is triggered by intracellular lesion, typically a mitochondrial damage. This results in release of cytochrome-c (Cc) from damaged mitochondria that in turn activates initiator caspases of the internal pathway (IIC). The extrinsic and intrinsic pathways merge at caspase-3 (CA3) that is activated by initiator caspases (EIC and IIC). Numerous additional factors participate in the caspase activation and the subsequent DNA fragmentation. A trophic factor, IGF-1, binds to its receptor (IGFR) and activates Akt, a protein kinase, that can prevent cytochrome-c release by maintaining the integrity of the mitochondrial membrane. Akt is supressed by PTEN, a phosphatase. Oxidative stress modulates the activation of NF-κB that in turn can modulate DNA fragmentation in the nucleus. Subsequent processes result in multiple events, including packaging the cell into small compact units that are removed by the neighboring cells

that involves the packaging and removal of damaged cells by surrounding tissue, thus preventing inflammation generally caused by the necrotic processes. Apoptosis may be triggered by external stimuli (extrinsic pathway) and internal events (intrinsic pathway). The extrinsic pathway is initiated outside of the cell, usually when the conditions of the extracellular environment determine that the cell must die. The specific trigger for the extrinsic pathway is the binding of ligands to specific "death receptors" found on the surface of cells, which leads to activation of cysteinyl proteases called initiator caspases. In contrast, the intrinsic pathway of apoptosis is initiated within the cell. The extrinsic and intrinsic pathways merge at caspase-3, which is activated by initiator caspases of the external and internal pathways. The processes that follow result in numerous events, including degradation of DNA and packaging of the cell into small units are easily taken up by neighboring phagocytic cells. Apoptotic processes can be modulated via suppressing/inducing various factors that participate in the initiation, augmentation, and suppression of the extrinsic and intrinsic pathways.

## *3.2 Modulatory Role of GAGs in the Apoptotic Processes*

Although it is a common consensus that GAGs play a crucial role in modulating the molecular processes of apoptosis, the published data are rather ambiguous. Previous studies suggest that GAGs may participate in the suppression of apoptotic processes, and thus exert neuroprotective properties in neurodegenerative disorders. Heparin and its analogs attenuate apoptosis in epithelial cells (Belmiro et al. 2009), glomerular cells (Ishikawa and Kitamura 1999), and in human trophoblasts (Hills et al. 2006). LMW heparin derivative, certoparin limits the apoptotic process in cardiac and renal tissues (Deepa and Varalakshmi 2006). Heparin suppresses DNA fragmentation and the consequent apoptosis in primary culture of adult rat hepatocytes (Maeda et al. 1993). Ultra low-molecular-weight heparin also exerts a protective effect on glutamate-induced apoptosis in cortical cells (Yu et al. 2008b).

Heparin and its derivatives are not the only GAGs exhibiting suppressive properties toward apoptotic processes. High-molecular-weight (HMW) HA decreases apoptosis in human epithelial cells (Haider et al. 2003; Pauloin et al. 2008, 2009), in granulosa cells (Kaneko et al. 2000; Tunjung et al. 2009), and chondrocytes (Campo et al. 2009; Echigo et al. 2006; Grishko et al. 2009; Lisignoli et al. 2001; Maneiro et al. 2004; Takahashi et al. 2000; Zhou et al. 2008). Transforming growth factor-β (TGF-β)-stimulated HA production reduces apoptosis in human fibroblasts exposed to oxidative stress (Campo et al. 2008c). Moreover, HA has an important effect on tumor cell survival by suppressing the apoptotic processes of the cells. HA induces chemoresistance in nonsmall-cell lung cancer (Ohashi et al. 2007) and antagonizes dexamethasone-induced apoptosis of malignant multiplex myeloma (MM) cells

(Vincent et al. 2003). Since HA is the major component of the bone marrow, this latter finding could account for the accumulation of MM cells in the bone marrow of patients with MM and why these cells escape conventional chemotherapy (Vincent et al. 2003). In addition to the anti-apoptotic effects of HA, HS, and CS reduces apoptosis in skin fibroblasts submitted to oxidative stress (Yue et al. 2009). CS attenuates apoptosis in the animal models of acute hepatitis (Campo et al. 2008e), pancreatitis (Campo et al. 2008d), and arthritis (Campo et al. 2008a) in mice. The anti-apoptotic properties of GAGs are diverse; for example, HA, HS, and CS but not KS or DS exerted anti-apoptotic effects on lipopolysaccharide-treated chondrocytes (Campo et al. 2009).

In contrast to these findings, several studies reported augmentation of the apoptotic processes by GAGs, indicating that GAGs may be valuable therapeutic agents in tumor cell elimination. Heparin induces apoptosis in oral squamous carcinoma cells (Ueda et al. 2009), melanoma cells (Bae et al. 2009; Berry et al. 2004), lymphoblasts obtained from acute lymphoid leukemia (ALL) patients (Erduran et al. 1999, 2004, 2007), human peripheral blood neutrophils (Manaster et al. 1996) nasopharyngeal carcinoma cells (Li et al. 2001, 2002), and human hepatoma cells (Karti et al. 2003). Dalteparin, an LMW heparin, increases apoptosis in lung adenocarcinoma cell line in vitro (Chen et al. 2008). On the other hand, heparin has no significant anti-proliferative and apoptotic effects on colon cancer cells (Uzun et al. 2009).

Similar to heparin and its derivatives, HA can also induce apoptosis. LMW HA has been shown to suppress survival and proliferation of colorectal carcinoma cells by inducing apoptosis (Alaniz et al. 2009). HMW HA induces apoptosis in activated T cells (Ruffell and Johnson 2008) and in macrophage cells, particularly at high concentrations (Sheehan et al. 2003, 2004). Small chains of HA (6–18 sugar units), but not large polymers, attenuate many types of cancer cells by triggering apoptosis while leaving normal cells unaffected (Toole et al. 2008). HA oligosaccharides have potent antitumor effects on osteosarcoma cell lines (Hosono et al. 2007). Glioma-associated HA increases apoptosis in dendritic cells (Yang et al. 2002). HA, especially when fragmented, augments apoptosis of the synovial cells from rheumatoid arthritis patients (Fujii et al. 2001).

Similar to GAGs, certain PGs can also modulate apoptotic processes. Versican, a large CSPG, protects cells from oxidative stress-induced apoptosis (Wu et al. 2005). Decorin, a PG with CS and DS side chains, contributes to prevention of apoptosis in endothelial cells (Schonherr et al. 1999). In contrast, lumican, a keratan sulfate proteoglycan (KSPG), inhibits melanoma growth and progression by inducing apoptosis (Brezillon et al. 2007; Vuillermoz et al. 2004) and stromal apoptosis is downregulated in the lumican-null mouse (Vij et al. 2005). Neoglycans, carbodiimide-modified GAGs, inhibit cancer cell proliferation by inducing apoptosis (Pumphrey et al. 2002). Moreover, articular chondrocytes from animals with a DS storage disease undergo a high rate of apoptosis and release nitric oxide and inflammatory cytokines (Simonaro et al. 2001).

### 3.3 *Molecular Mechanisms of GAG- and PG-Modulated Apoptosis*

The mechanism by which GAGs influence apoptosis is not entirely clear and involves numerous components of the apoptotic pathway (Fig. 3). However, previous studies indicate that the major processes that may trigger apoptosis include oxidative stress and apoptotic receptor–ligand binding. Thus, revealing the role of GAGs in these processes is crucial for understanding the mechanisms of the modulatory function of GAGs on apoptosis.

Oxidative stress is one of the major factors triggering apoptosis (Ganguly et al. 2002). Oxidative stress can be defined as an imbalance between reactive oxygen species (ROS), reactive nitrogen species (RNS), and antioxidant defence mechanisms. ROS/RNS result in cell death via lipid peroxidation (LPO) and oxidation of nucleic acids and proteins. Mild oxidative stress appears to cause apoptosis by activating nuclear transcription factors such as NF-κB, and consequently, factors including antioxidants that inhibit NF-κB also suppress subsequent apoptosis. Indeed, HA is reported to block LPO and suppress the activity of NF-κB and caspases, and thus attenuate apoptosis (Campo et al. 2008c). Heparin analogs have also been reported to suppress NF-κB (Belmiro et al. 2009). GAGs including HS and CS reduce LPO and intracellular ROS levels, inhibit cytochrome-c release and activation of caspases-9 and -3, and block NF-κB activation (Campo et al. 2008b, d, e; Yue et al. 2009). HMW HA decreases benzalkonium chloride-induced oxidative stress and the consequent apoptosis in human epithelial cell lines (Pauloin et al. 2008). On the other hand, sustained oxidative stress attenuates TNF-induced NF-κB activation (Wu et al. 2009), emphasizing the complex role of NF-κB in the oxidative stress-induced apoptosis.

Since HMW HA decreases UVB-induced apoptosis in human epithelial corneal cells without significant effect on ROS and 8-hydroxy-2′-deoxyguanosine (8-oxo-dG) release, other mechanisms may play a role in the suppression of apoptosis in addition to the LPO-induced pathway (Pauloin et al. 2009). Indeed, GAGs appear to exert attenuation of apoptotic processes by inhibiting ligand–receptor interactions that normally trigger apoptosis. Heparin analogs decrease the production of tumor necrosis factor-α (TNFα) and TGF-β and thus attenuate epithelial cell apoptosis (Belmiro et al. 2009). Certoparin suppresses TNFα in the cardiac and renal tissues (Deepa & Varalakshmi 2006). Lumican decreases cell proliferation and aids Fas receptor (FasR)–Fas ligand (FasL) mediated apoptosis (Vij et al. 2004). Apoptosis of stromal cells is downregulated in the lumican-null mouse possibly via disruption of FasR–FasL signaling (Vij et al. 2005). Heparin-like agents stimulate FasR-agonistic antibody-induced apoptosis (Manero et al. 2004). Moreover, interaction of CD44 with HA fragments enhances expression of FasR and the consequent FasR-mediated apoptosis of synovial cells (Fujii et al. 2001).

CD44 is a principal cell-surface receptor for HA. HMW HA induces apoptosis in activated T cells via CD44, independently of FasR and caspase activation. Cells expressing a CD44 mutant unable to bind HA were resistant to apoptosis (Ruffell and Johnson 2008). The interaction of CD44 with fragmented HA on rheumatoid synovial cells induces expression of FasR on the cells, which leads to FasR-mediated apoptosis of synovial cells by the interaction of FasL on the surface of T cells. On the other hand, CD44-ligand binding on the surface of lung cancer cells reduces FasR expression and FasR-mediated apoptosis, resulting in increased cell survival (Yasuda et al. 2001, 2002). Anti-FasR-induced apoptosis in chondrocytes is reduced by HA binding to CD44 (Lisignoli et al. 2001).

Other cellular factors may also play roles in the modulatory function of GAGs in apoptosis. The protein kinase Akt signaling pathway is one of the most potent inhibitors of cell death, and consequently, Akt is elevated in large number of cancers. Akt is activated by various trophic factors including insulin-like growth factor 1 (IGF-1). Although the mechanisms by which Akt attenuates apoptosis have not been completely elucidated yet, Akt appears to prevent cytochrome-c release from the mitochondria by maintaining the integrity of the mitochondrial membrane following an apoptotic insult, thus inhibiting the intrinsic pathway of apoptosis. Indeed, heparin induces apoptosis in oral squamous carcinoma cells via suppression of Akt (Ueda et al. 2009). LMW but not HMW HA has been shown to suppress survival and proliferation of colorectal carcinoma cells via Akt signaling mechanism (Alaniz et al. 2009). HA oligomers stimulate expression of PTEN, a phosphatase that suppresses Akt function (Ghatak et al. 2002) and consequently may participate in the suppression of tumor cell survival. These data indicate that Akt signaling pathway may play a pivotal role in the GAG-modulated apoptotic processes. The ambiguous effects of GAGs on the Akt signaling and other apoptotic pathways indicate that the structural properties of GAGs, including molecular weight, may determine whether these molecules exert neuroprotective or apoptotic effects.

## 4 Conclusion

The neuroprotective role of GAGs has been extensively studied in the past ten years. Among the major factors leading to neurodegeneration/neuronal loss, the deposition of cytotoxic proteins and apoptosis received considerable attention due to their role in neurodegenerative disorders. Indeed, a massive amount of data indicates that GAGs and related PGs play a crucial role in these processes. The neuroprotection of GAGs exerted in the animal models of neuronal lesions as well as their pivotal role in the pathogenesis of neurodegenerative disorders raised the possibility that GAGs may be valuable therapeutic adjuncts in the treatment of various neurodegenerative disorders. Future studies should aim to (1) develop standardized compounds that can be used in the therapy of AD, PD, prion diseases and other neurodegenerative conditions, as well as (2) elucidate the mechanism(s) by which GAGs exert their neuroprotective attributes.

## References


Adjou KT, Simoneau S, Sales N et al (2003) A novel generation of heparan sulfate mimetics for the treatment of prion diseases. J Gen Virol 84:2595–2603

Alaniz L, Rizzo M, Malvicini M et al (2009) Low molecular weight hyaluronan inhibits colorectal carcinoma growth by decreasing tumor cell proliferation and stimulating immune response. Cancer Lett 278:9–16

Arrasate M, Perez M, Valpuesta JM et al (1997) Role of glycosaminoglycans in determining the helicity of paired helical filaments. Am J Pathol 151:1115–1122

Bae KH, Moon CW, Lee Y et al (2009) Intracellular delivery of heparin complexed with chitosan-g-poly(ethylene glycol) for inducing apoptosis. Pharm Res 26:93–100

Ban TA, Morey LC, Santini V (1991) Clinical investigations with ateroid in old-age dementias. Semin Thromb Hemost 17(Suppl 2):161–163

Ban TA, Morey LC, Fjetland OK et al (1992) Early manifestations of dementing illness: treatment with glycosaminoglycan polysulfate. Prog Neuropsychopharmacol Biol Psychiatry 16:661–676

Belmiro CL, Castelo-Branco MT, Melim LM et al (2009) Unfractionated heparin and new heparin analogues from ascidians (chordate-tunicate) ameliorate colitis in rats. J Biol Chem 284:11267–11278

Berry D, Lynn DM, Sasisekharan R et al (2004) Poly(beta-amino ester)s promote cellular uptake of heparin and cancer cell death. Chem Biol 11:487–498

Brezillon S, Venteo L, Ramont L et al (2007) Expression of lumican, a small leucine-rich proteoglycan with antitumour activity, in human malignant melanoma. Clin Exp Dermatol 32:405–416

Campo GM, Avenoso A, Campo S et al (2008a) Chondroitin-4-sulphate inhibits NF-kB translocation and caspase activation in collagen-induced arthritis in mice. Osteoarthritis Cartilage 16:1474–1483

Campo GM, Avenoso A, Campo S et al (2008b) NF-kB and caspases are involved in the hyaluronan and chondroitin-4-sulphate-exerted antioxidant effect in fibroblast cultures exposed to oxidative stress. J Appl Toxicol 28:509–517

Campo GM, Avenoso A, Campo S et al (2008c) The antioxidant effect exerted by TGF-1beta-stimulated hyaluronan production reduced NF-kB activation and apoptosis in human fibroblasts exposed to FeSo4 plus ascorbate. Mol Cell Biochem 311:167–177

Campo GM, Avenoso A, Campo S et al (2008d) Chondroitin-4-sulphate reduced oxidative injury in caerulein-induced pancreatitis in mice: the involvement of NF-kappaB translocation and apoptosis activation. Exp Biol Med (Maywood) 233:741–752

Campo GM, Avenoso A, Campo S et al (2008e) The antioxidant activity of chondroitin-4-sulphate, in carbon tetrachloride-induced acute hepatitis in mice, involves NF-kappaB and caspase activation. Br J Pharmacol 155:945–956

Campo GM, Avenoso A, Campo S et al (2009) Glycosaminoglycans modulate inflammation and apoptosis in LPS-treated chondrocytes. J Cell Biochem 106:83–92

Castillo GM, Ngo C, Cummings J et al (1997) Perlecan binds to the beta-amyloid proteins (A beta) of Alzheimer's disease, accelerates A beta fibril formation, and maintains A beta fibril stability. J Neurochem 69:2452–2465

Caughey B (1994) Scrapie-associated PrP accumulation and agent replication: effects of sulphated glycosaminoglycan analogues. Philos Trans R Soc Lond B Biol Sci 343:399–404

Caughey B, Race RE (1994) Scrapie-associated PrP accumulation and its inhibition: revisiting the amyloid-glycosaminoglycan connection. Ann N Y Acad Sci 724:290–295

Caughey B, Raymond GJ (1993) Sulfated polyanion inhibition of scrapie-associated PrP accumulation in cultured cells. J Virol 67:643–650

Caughey B, Brown K, Raymond GJ et al (1994) Binding of the protease-sensitive form of PrP (prion protein) to sulfated glycosaminoglycan and congo red [corrected]. J Virol 68:2135–2141

Chen X, Xiao W, Qu X et al (2008) The effect of dalteparin, a kind of low molecular weight heparin, on lung adenocarcinoma A549 cell line in vitro. Cancer Invest 26:718–724

Cohlberg JA, Li J, Uversky VN et al (2002) Heparin and other glycosaminoglycans stimulate the formation of amyloid fibrils from alpha-synuclein in vitro. Biochemistry 41:1502–1511

Conti L, Placidi GF, Cassano GB (1989a) Ateroid in the treatment of dementia: results of a clinical trial. Mod Probl Pharmacopsychiatry 23:76–84

Conti L, Re F, Lazzerini F et al (1989b) Glycosaminoglycan polysulfate (Ateroid) in old-age dementias: effects upon depressive symptomatology in geriatric patients. Prog Neuropsychopharmacol Biol Psychiatry 13:977–981

Deepa PR, Varalakshmi P (2006) Influence of a low-molecular-weight heparin derivative on the nitric oxide levels and apoptotic DNA damage in adriamycin-induced cardiac and renal toxicity. Toxicology 217:176–183

DeWitt DA, Silver J, Canning DR et al (1993) Chondroitin sulfate proteoglycans are associated with the lesions of Alzheimer's disease. Exp Neurol 121:149–152

Diaz-Nido J, Wandosell F, Avila J (2002) Glycosaminoglycans and beta-amyloid, prion and tau peptides in neurodegenerative diseases. Peptides 23:1323–1332

Dudas B (2009) Low molecular weight glycosaminoglycans and apoptosis: potential treatment of neurodegenerative disorders including Alzheimer's Disease. Int J Med Biol Front 12:337–348

Dudas B, Cornelli U, Lee JM et al (2002) Oral and subcutaneous administration of the glycosaminoglycan C3 attenuates Abeta(25–35)-induced abnormal tau protein immunoreactivity in rat brain. Neurobiol Aging 23:97–104

Dudas B, Rose M, Hanin I (2003) Dose-dependent effect of cholinotoxin AF64A on the cholinergic elements of the cingulum bundle in rat. Brain Res 965:295–298

Echigo R, Mochizuki M, Nishimura R et al (2006) Suppressive effect of hyaluronan on chondrocyte apoptosis in experimentally induced acute osteoarthritis in dogs. J Vet Med Sci 68:899–902

Erduran E, Tekelioglu Y, Gedik Y et al (1999) Apoptotic effects of heparin on lymphoblasts, neutrophils, and mononuclear cells: results of a preliminary in vitro study. Am J Hematol 61:90–93

Erduran E, Tekelioglu Y, Gedik Y et al (2004) In vitro determination of the apoptotic effect of heparin on lymphoblasts using DNA analysis and measurements of Fas and Bcl-2 proteins by flow cytometry. Pediatr Hematol Oncol 21:383–391

Erduran E, Deger O, Albayrak D et al (2007) In vitro investigation of the apoptotic effect of heparin on lymphoblasts by using flow cytometric DNA analysis and fluorometric caspase-3 and -8 activities. DNA Cell Biol 26:803–808

Fujii K, Fujii Y, Hubscher S et al (2001) CD44 is the physiological trigger of Fas up-regulation on rheumatoid synovial cells. J Immunol 167:1198–1203

Gabizon R, Meiner Z, Halimi M et al (1993) Heparin-like molecules bind differentially to prion-proteins and change their intracellular metabolic fate. J Cell Physiol 157:319–325

Ganguly BJ, Tonomura N, Benson RM et al (2002) Hyperbaric oxygen enhances apoptosis in hematopoietic cells. Apoptosis 7:499–510

Ghatak S, Misra S, Toole BP (2002) Hyaluronan oligosaccharides inhibit anchorage-independent growth of tumor cells by suppressing the phosphoinositide 3-kinase/Akt cell survival pathway. J Biol Chem 277:38013–38020

Goedert M, Jakes R, Spillantini MG et al (1996) Assembly of microtubule-associated protein tau into Alzheimer-like filaments induced by sulphated glycosaminoglycans. Nature 383:550–553

Grishko V, Xu M, Ho R et al (2009) Effects of hyaluronic acid on mitochondrial function and mitochondria-driven apoptosis following oxidative stress in human chondrocytes. J Biol Chem 284:9132–9139

Guiroy DC, Yanagihara R, Gajdusek DC (1991) Localization of amyloidogenic proteins and sulfated glycosaminoglycans in nontransmissible and transmissible cerebral amyloidoses. Acta Neuropathol 82:87–92

Gupta-Bansal R, Frederickson RC, Brunden KR (1995) Proteoglycan-mediated inhibition of A beta proteolysis. A potential cause of senile plaque accumulation. J Biol Chem 270:18666–18671

Haider AS, Grabarek J, Eng B et al (2003) In vitro model of "wound healing" analyzed by laser scanning cytometry: accelerated healing of epithelial cell monolayers in the presence of hyaluronate. Cytometry A 53:1–8

Hasegawa M, Crowther RA, Jakes R et al (1997) Alzheimer-like changes in microtubule-associated protein Tau induced by sulfated glycosaminoglycans. Inhibition of microtubule binding, stimulation of phosphorylation, and filament assembly depend on the degree of sulfation. J Biol Chem 272:33118–33124

Hijazi N, Kariv-Inbal Z, Gasset M et al (2005) PrPSc incorporation to cells requires endogenous glycosaminoglycan expression. J Biol Chem 280:17057–17061

Hills FA, Abrahams VM, Gonzalez-Timon B et al (2006) Heparin prevents programmed cell death in human trophoblast. Mol Hum Reprod 12:237–243

Horonchik L, Tzaban S, Ben-Zaken O et al (2005) Heparan sulfate is a cellular receptor for purified infectious prions. J Biol Chem 280:17062–17067

Hosono K, Nishida Y, Knudson W et al (2007) Hyaluronan oligosaccharides inhibit tumorigenicity of osteosarcoma cell lines MG-63 and LM-8 in vitro and in vivo via perturbation of hyaluronan-rich pericellular matrix of the cells. Am J Pathol 171:274–286

Ishikawa Y, Kitamura M (1999) Inhibition of glomerular cell apoptosis by heparin. Kidney Int 56:954–963

Kaneko T, Saito H, Toya M et al (2000) Hyaluronic acid inhibits apoptosis in granulosa cells via CD44. J Assist Reprod Genet 17:162–167

Karti SS, Ovali E, Ozgur O et al (2003) Induction of apoptosis and inhibition of growth of human hepatoma HepG2 cells by heparin. Hepatogastroenterology 50:1864–1866

Kisilevsky R, Szarek WA (2002) Novel glycosaminoglycan precursors as anti-amyloid agents part II. J Mol Neurosci 19:45–50

Kisilevsky R, Lemieux LJ, Fraser PE et al (1995) Arresting amyloidosis in vivo using small-molecule anionic sulphonates or sulphates: implications for Alzheimer's disease. Nat Med 1:143–148

Kisilevsky R, Szarek WA, Ancsin J et al (2003) Novel glycosaminoglycan precursors as anti-amyloid agents, part III. J Mol Neurosci 20:291–297

Kisilevsky R, Szarek WA, Ancsin J et al (2004) Novel glycosaminoglycan precursors as antiamyloid agents: part IV. J Mol Neurosci 24:167–172

Kisilevsky R, Ancsin JB, Szarek WA et al (2007) Heparan sulfate as a therapeutic target in amyloidogenesis: prospects and possible complications. Amyloid 14:21–32

Larramendy-Gozalo C, Barret A, Daudigeos E et al (2007) Comparison of CR36, a new heparan mimetic, and pentosan polysulfate in the treatment of prion diseases. J Gen Virol 88:1062–1067

Leveugle B, Scanameo A, Ding W et al (1994) Binding of heparan sulfate glycosaminoglycan to beta-amyloid peptide: inhibition by potentially therapeutic polysulfated compounds. Neuroreport 5:1389–1392

Li HL, Ye KH, Zhang HW et al (2001) Effect of heparin on apoptosis in human nasopharyngeal carcinoma CNE2 cells. Cell Res 11:311–315

Li HL, Ren XD, Zhang HW et al (2002) Synergism between heparin and adriamycin on cell proliferation and apoptosis in human nasopharyngeal carcinoma CNE2 cells. Acta Pharmacol Sin 23:167–172

Lisignoli G, Grassi F, Zini N et al (2001) Anti-Fas-induced apoptosis in chondrocytes reduced by hyaluronan: evidence for CD44 and CD54 (intercellular adhesion molecule 1) invovement. Arthritis Rheum 44:1800–1807

Liu IH, Uversky VN, Munishkina LA et al (2005) Agrin binds alpha-synuclein and modulates alpha-synuclein fibrillation. Glycobiology 15:1320–1331

Maeda S, Kimura H, Koga N et al (1993) Cell density-dependent DNA fragmentation and its suppression by heparin in primary culture of adult rat hepatocytes. Biochem Biophys Res Commun 195:270–275

Manaster J, Chezar J, Shurtz-Swirski R et al (1996) Heparin induces apoptosis in human peripheral blood neutrophils. Br J Haematol 94:48–52

Maneiro E, de Andres MC, Fernandez-Sueiro JL et al (2004) The biological action of hyaluronan on human osteoartritic articular chondrocytes: the importance of molecular weight. Clin Exp Rheumatol 22:307–312

Manero F, Ljubic-Thibal V, Moulin M et al (2004) Stimulation of Fas agonistic antibody-mediated apoptosis by heparin-like agents suppresses Hsp27 but not Bcl-2 protective activity. Cell Stress Chaperones 9:150–166

Mayer-Sonnenfeld T, Zeigler M, Halimi M et al (2005) The metabolism of glycosaminoglycans is impaired in prion diseases. Neurobiol Dis 20:738–743

Mayer-Sonnenfeld T, Avrahami D, Friedman-Levi Y et al (2008) Chemically induced accumulation of GAGs delays PrP(Sc) clearance but prolongs prion disease incubation time. Cell Mol Neurobiol 28:1005–1015

McBride PA, Wilson MI, Eikelenboom P et al (1998) Heparan sulfate proteoglycan is associated with amyloid plaques and neuroanatomically targeted PrP pathology throughout the incubation period of scrapie-infected mice. Exp Neurol 149:447–454

Ohashi R, Takahashi F, Cui R et al (2007) Interaction between CD44 and hyaluronate induces chemoresistance in non-small cell lung cancer cell. Cancer Lett 252:225–234

Pan T, Wong BS, Liu T et al (2002) Cell-surface prion protein interacts with glycosaminoglycans. Biochem J 368:81–90

Passeri M, Cucinotta D (1989) Ateroid in the clinical treatment of multi-infarct dementia. Mod Probl Pharmacopsychiatry 23:85–94

Pauloin T, Dutot M, Warnet JM et al (2008) In vitro modulation of preservative toxicity: high molecular weight hyaluronan decreases apoptosis and oxidative stress induced by benzalkonium chloride. Eur J Pharm Sci 34:263–273

Pauloin T, Dutot M, Joly F et al (2009) High molecular weight hyaluronan decreases UVB-induced apoptosis and inflammation in human epithelial corneal cells. Mol Vis 15:577–583

Perez M, Valpuesta JM, Medina M et al (1996) Polymerization of tau into filaments in the presence of heparin:the minimal sequence required for tau–tau interaction. J Neurochem 67:1183–1190

Perez M, Wandosell F, Colaco C et al (1998) Sulphated glycosaminoglycans prevent the neurotoxicity of a human prion protein fragment. Biochem J 335(Pt 2):369–374

Perry G, Richey P, Siedlak SL et al (1992) Basic fibroblast growth factor binds to filamentous inclusions of neurodegenerative diseases. Brain Res 579:350–352

Pollack SJ, Sadler II, Hawtin SR et al (1995) Sulfated glycosaminoglycans and dyes attenuate the neurotoxic effects of beta-amyloid in rat PC12 cells. Neurosci Lett 184:113–116

Priola SA, Caughey B (1994) Inhibition of scrapie-associated PrP accumulation. Probing the role of glycosaminoglycans in amyloidogenesis. Mol Neurobiol 8:113–120

Priola SA, Caughey B, Raymond GJ et al (1994) Prion protein and the scrapie agent: in vitro studies in infected neuroblastoma cells. Infect Agents Dis 3:54–58

Pumphrey CY, Theus AM, Li S et al (2002) Neoglycans, carbodiimide-modified glycosaminoglycans: a new class of anticancer agents that inhibit cancer cell proliferation and induce apoptosis. Cancer Res 62:3722–3728

Rose M, Dudas B, Cornelli U et al (2003) Protective effect of the heparin-derived oligosaccharide C3, on AF64A-induced cholinergic lesion in rats. Neurobiol Aging 24:481–490

Rose M, Dudas B, Cornelli U et al (2004) Glycosaminoglycan C3 protects against AF64A-induced cholinotoxicity in a dose-dependent and time-dependent manner. Brain Res 1015:96–102

Ruffell B, Johnson P (2008) Hyaluronan induces cell death in activated T cells through CD44. J Immunol 181:7044–7054

Sadler II, Hawtin SR, Tailor V et al (1995) Glycosaminoglycans and sulphated polyanions attenuate the neurotoxic effects of beta-amyloid. Biochem Soc Trans 23:106S

Santini V (1989) A general practice trial of Ateroid 200 in 8,776 patients with chronic senile cerebral insufficiency. Mod Probl Pharmacopsychiatry 23:95–100
Schonberger O, Horonchik L, Gabizon R et al (2003) Novel heparan mimetics potently inhibit the scrapie prion protein and its endocytosis. Biochem Biophys Res Commun 312:473–479
Schonherr E, O'Connell BC, Schittny J et al (1999) Paracrine or virus-mediated induction of decorin expression by endothelial cells contributes to tube formation and prevention of apoptosis in collagen lattices. Eur J Cell Biol 78:44–55
Shaffer LM, Dority MD, Gupta-Bansal R et al (1995) Amyloid beta protein (A beta) removal by neuroglial cells in culture. Neurobiol Aging 16:737–745
Sheehan KM, DeLott LB, Day SM et al (2003) Hyalgan has a dose-dependent differential effect on macrophage proliferation and cell death. J Orthop Res 21:744–751
Sheehan KM, DeLott LB, West RA et al (2004) Hyaluronic acid of high molecular weight inhibits proliferation and induces cell death in U937 macrophage cells. Life Sci 75:3087–3102
Shimizu H, Sato S, Ohishi H et al (1997) Proteoglycans in senile dementia of Alzheimer's type. Nippon Ronen Igakkai Zasshi 34:461–467
Shyng SL, Lehmann S, Moulder KL et al (1995) Sulfated glycans stimulate endocytosis of the cellular isoform of the prion protein, PrPC, in cultured cells. J Biol Chem 270:30221–30229
Simonaro CM, Haskins ME, Schuchman EH (2001) Articular chondrocytes from animals with a dermatan sulfate storage disease undergo a high rate of apoptosis and release nitric oxide and inflammatory cytokines: a possible mechanism underlying degenerative joint disease in the mucopolysaccharidoses. Lab Invest 81:1319–1328
Snow AD, Kinsella MG, Parks E et al (1995) Differential binding of vascular cell-derived proteoglycans (perlecan, biglycan, decorin, and versican) to the beta-amyloid protein of Alzheimer's disease. Arch Biochem Biophys 320:84–95
Spillantini MG, Tolnay M, Love S et al (1999) Microtubule-associated protein tau, heparan sulphate and alpha-synuclein in several neurodegenerative diseases with dementia. Acta Neuropathol 97:585–594
Takahashi K, Hashimoto S, Kubo T et al (2000) Effect of hyaluronan on chondrocyte apoptosis and nitric oxide production in experimentally induced osteoarthritis. J Rheumatol 27:1713–1720
Toole BP, Ghatak S, Misra S (2008) Hyaluronan oligosaccharides as a potential anticancer therapeutic. Curr Pharm Biotechnol 9:249–252
Tunjung WA, Yokoo M, Hoshino Y et al (2009) Effect of hyaluronan to inhibit caspase activation in porcine granulosa cells. Biochem Biophys Res Commun 382:160–164
Ueda K, Inoue S, Zhang Y et al (2009) Heparin induces apoptosis through suppression of AKt in oral squamous cell carcinoma cells. Anticancer Res 29:1079–1088
Uzun Y, Akdogan E, Ozdemir F et al (2009) The effects of heparin on DLD-1 colon cancer cell line. Bratisl Lek Listy 110:3–6
Van GD, David G, Lammens M et al (1993) Heparan sulfate expression patterns in the amyloid deposits of patients with Alzheimer's and Lewy body type dementia. Dementia 4:308–314
van HJ, de Vos RA, Steur EN et al (2004) Absence of heparan sulfate proteoglycans in Lewy bodies and Lewy neurites in Parkinson's disease brains. J Alzheimers Dis 6:469–474
Vana K, Zuber C, Nikles D et al (2007) Novel aspects of prions, their receptor molecules, and innovative approaches for TSE therapy. Cell Mol Neurobiol 27:107–128
Vij N, Roberts L, Joyce S et al (2004) Lumican suppresses cell proliferation and aids Fas-Fas ligand mediated apoptosis: implications in the cornea. Exp Eye Res 78:957–971
Vij N, Roberts L, Joyce S et al (2005) Lumican regulates corneal inflammatory responses by modulating Fas-Fas ligand signaling. Invest Ophthalmol Vis Sci 46:88–95
Vincent T, Molina L, Espert L et al (2003) Hyaluronan, a major non-protein glycosaminoglycan component of the extracellular matrix in human bone marrow, mediates dexamethasone resistance in multiple myeloma. Br J Haematol 121:259–269
Vuillermoz B, Khoruzhenko A, D'Onofrio MF et al (2004) The small leucine-rich proteoglycan lumican inhibits melanoma progression. Exp Cell Res 296:294–306

Warner RG, Hundt C, Weiss S et al (2002) Identification of the heparan sulfate binding sites in the cellular prion protein. J Biol Chem 277:18421–18430
Wong C, Xiong LW, Horiuchi M et al (2001) Sulfated glycans and elevated temperature stimulate PrP(Sc)-dependent cell-free formation of protease-resistant prion protein. EMBO J 20:377–386
Wu Y, Wu J, Lee DY et al (2005) Versican protects cells from oxidative stress-induced apoptosis. Matrix Biol 24:3–13
Wu M, Bian Q, Liu Y et al (2009) Sustained oxidative stress inhibits NF-kappaB activation partially via inactivating the proteasome. Free Radic Biol Med 46:62–69
Yamin G, Munishkina LA, Karymov MA et al (2005) Forcing nonamyloidogenic beta-synuclein to fibrillate. Biochemistry 44:9096–9107
Yang T, Witham TF, Villa L et al (2002) Glioma-associated hyaluronan induces apoptosis in dendritic cells via inducible nitric oxide synthase: implications for the use of dendritic cells for therapy of gliomas. Cancer Res 62:2583–2591
Yasuda M, Tanaka Y, Fujii K et al (2001) CD44 stimulation down-regulates Fas expression and Fas-mediated apoptosis of lung cancer cells. Int Immunol 13:1309–1319
Yasuda M, Nakano K, Yasumoto K et al (2002) CD44: functional relevance to inflammation and malignancy. Histol Histopathol 17:945–950
Yin S, Yu S, Li C et al (2006) Prion proteins with insertion mutations have altered N-terminal conformation and increased ligand binding activity and are more susceptible to oxidative attack. J Biol Chem 281:10698–10705
Yin S, Pham N, Yu S et al (2007) Human prion proteins with pathogenic mutations share common conformational changes resulting in enhanced binding to glycosaminoglycans. Proc Natl Acad Sci USA 104:7546–7551
Yu S, Yin S, Pham N et al (2008a) Ligand binding promotes prion protein aggregation – role of the octapeptide repeats. FEBS J 275:5564–5575
Yu TG, Zhang QZ, Zhang ZG et al (2008b) Protective effect of ultra low molecular weight heparin on glutamate-induced apoptosis in cortical cells. Yonsei Med J 49:486–495
Yue XL, Lehri S, Li P et al (2009) Insights on a new path of pre-mitochondrial apoptosis regulation by a glycosaminoglycan mimetic. Cell Death Differ 16:770–781
Zhou PH, Liu SQ, Peng H (2008) The effect of hyaluronic acid on IL-1beta-induced chondrocyte apoptosis in a rat model of osteoarthritis. J Orthop Res 26:1643–1648

# Part V
# Heparin-Like Entities

# Heparan Sulphate: A Heparin in Miniature

**J.T. Gallagher**

## Contents



**Abstract** Heparan sulphate (HS), discovered in 1948 in heparin by-products, only emerged slowly from the shadow of heparin. Its inauspicious beginning was followed by the gradual realisation that HS was a separate entity with distinctive features. Both HS and heparin follow a common biosynthetic route but while heparin reaches full maturity, HS holds on to some of its youthful traits. The novel design and complex patterning of sulphation in HS enable it fulfil key roles in many, diverse biological processes.

**Keywords** Heparan sulphate • S-domain • NA-domain • Interferon • Fibroblast growth factor

J.T. Gallagher (✉)
Paterson Institute for Cancer Research, Iduron Ltd., University of Manchester, Manchester M20 4BX, UK
e-mail: jgallagher@picr.man.ac.uk

R. Lever et al. (eds.), *Heparin - A Century of Progress*,
Handbook of Experimental Pharmacology 207,
DOI 10.1007/978-3-642-23056-1_15, 

# 1 Heparan Sulphate: The Beginning

It is very apt that a chapter on heparan sulphate (HS) should appear in a book dedicated to heparin. Historically, the early discovery of HS in 1948 was largely due to the biomedical interest in the structure of heparin, which was already in extensive use as an anticoagulant. The discarded by-products of heparin preparations were known to contain heparin-like polymers with low sulphation and low-anticoagulant activity that could be separated from heparin on the basis of their high solubility in barium salts. The question posed by Jorpes and Gardell in their 1948 paper in the Journal of Biological Chemistry was whether the soluble fractions "are distinct chemical entities per se or only mixtures of one or several different polysaccharides and the trisulphuric acid ester." Working with extracts from ox liver and lung they discovered that the soluble material had a similar backbone structure to heparin but contained just one sulphate group per disaccharide in contrast to the trisulphated disaccharides in heparin. They named the polysaccharide *heparin monosulfuric acid*. Jorpes and Gardell discussed the possibility that this was a degradation product of heparin brought about by the action of sulphatases and suggested that one way to resolve this issue and to address the question of its unique chemical identity would be to try and isolate heparin monosulphuric acid in tissues with a low mast cell content and therefore likely to be deficient in heparin.

# 2 Early Developments in the Analysis of Heparan Sulphate

The Jorpes/Gardell paper attracted considerable interest amongst carbohydrate biochemists. Their data suggested the existence of a separate but related molecular entity from heparin but the possibility that the new low-sulphated heparin was either a breakdown product or an intermediate in heparin biosynthesis could not be ruled out. The potential physiological significance of the Jorpes/Gardell heparin-like substance (variously renamed as heparin monosulphate, heparitin sulphate and now heparan sulphate) emerged when Dorfman and colleagues (Knecht et al. 1967) published data on polysaccharides extracted from human aorta and from pathological material accumulated in Hurler tissues. These sulphated polymers had characteristics that more closely resembled heparan sulphate than heparin. Subsequently, the pioneering work of Cifonelli using nitrous acid scission (Cifonelli 1968) and the isolation and characterisation of heparin and HS-degrading bacterial lyases (Hovingh and Linker 1970; Silva et al. 1976) revealed many of the distinguishing structural features of HS including the important finding that sulphated disaccharides tended to occur in clusters separated by N-acetyl-rich regions that were largely devoid of sulphate groups. The picture that emerged was of a polysaccharide chain composed of the same basic repeating unit as heparin (i.e. α/β 1,4 glucosamine – uronic acid repeats) but with a novel molecular design in which the sulphated regions, though similar to heparin, lacked the density and uniformity of sulphation that characterised the heparin polymer.

## 3 Biological Dimensions Revealed

Throughout the early years of study, it was widely believed that in common with most other glycosaminoglycans (GAGs) HS was principally an extracellular component with important roles in maintaining the structural integrity, hydration and elasticity of connective tissues. Indeed, these are vital functions of GAGs but the potential for HS to directly influence cell physiology emerged in the early seventies from the papers of Kramer who showed for the first time that HS was detected in trypsin extracts of cultured cells, strongly suggesting an association with the plasma membrane by an attached protein component (Kraemer 1971a, b). In agreement with earlier indications from Linker's work, Kramer also showed that the degree of sulphation of HS varied from one cell type to another. The strategic location of HS at the interface of the cell and its microenvironment was highly suggestive of important roles in cell:cell and cell:ECM interactions. Following up on Kramer's work the isolation of hydrophobic HS-proteoglycans from rat liver membranes (Oldberg et al. 1979) was a significant advance that ultimately lead to the discovery of the two major families of plasma membrane HS-proteoglycans (HSPGs), the syndecans and the glypicans by the research groups of Bernfield (Saunders et al. 1989) and David (David et al. 1990), respectively. Despite these important advances, the functions of HSPGs remained elusive.

The work of Couchman and Woods on cell adhesion gave the first clear indication that membrane HSPGs were involved in conveying information across the cell membrane (for review, see Couchman et al. 2001). They demonstrated that the formation of focal adhesions on fibronectin required the co-ordinated binding of both cell surface integrins and HSPGs to distinct sites on fibronectin substrates, the speed of the response being indicative of a process elicited by trans-membrane signalling. Then in 1991 two papers, one by Yayon and colleagues and the other by Rapraeger and co-workers, revealed a critical role for HS in FGF-signalling (Rapraeger et al. 1991; Yayon et al. 1991). The term co-receptor was coined to describe the proposed role of cell surface HS in binding FGF, and transferring it to its signal transducing receptor. Drawing on the model of the allosteric activation of anti-thrombin by heparin, it was suggested that HS induced a conformational change in FGF2 that was essential for receptor recognition These landmark publications enabled HS to finally "find its voice" in the wider community of cell biologists. The significance of HS was further emphasised when genetic manipulation of model organisms, initially in Drosophila, but later in mice, Xenopus and *C. elegans*, revealed essential roles for HS and its associated core proteins in various critical embryonic signalling pathways mediated by often structurally unrelated growth factors and morphogenic proteins (Bellache et al. 1998; Bornemann et al. 2004; Bullock et al. 1998; Esko and Selleck 2002; Han et al. 2004; Kramer and Yost 2002; Lin and Perrimon 1999; Nakato et al. 2002). We are now beginning to understand the molecular basis of some of the key HS–protein interactions in development but the diversity of these interactions in relation to both the proteins involved and the pathways regulated seems to defy a unifying theory of the mode of action of HS in controlling biological systems.

# 4 Heparan Sulphate: An Ordered Polymeric Structure

Over the past 20 years or so biochemical analysis of the structure of HS has provided compelling evidence for an ordered polymeric structure with a unique design that sets HS apart from heparin and from other sulphated GAGs (Fig. 1). The defining feature of HS first recognised in the work of Linker and Cifonelli and later refined by high-resolution separation methods, sequencing strategies (Merry et al. 1999; Turnbull et al. 1999; Venkataraman et al. 1999) and a clearer understanding of the specificities of bacterial heparinases, (Desai et al. 1993), is its domain structure in which sulphated regions called S-domains or NS-domains are disposed at fairly regular intervals along the GAG chain separated by N-acetylated regions (NA domains) of very low sulphation (Gallagher and Lyon 2000; Turnbull and Gallagher 1990, 1991a). The S-domains are remote from HS-core proteins with a long stretch of about nine or ten N-acetylated disaccharides positioned between the most proximal N-sulphate group and the GAG–protein linkage region (Turnbull and Gallagher 1991b). This proximal N-sulphate does not correspond to the first downstream S-domain, which is about 15 disaccharides from the core protein. In addition to the S-domains, HS contains a second type of sulphated motif composed of alternating N-acetylated and N-sulphated disaccharide units. These mixed sequences, named NA/NS domains, reside at the interface of S-domains and NA-regions; they ensure a smooth transition from non-sulphated to highly sulphated sections of the chain (Murphy et al. 2004). Although rare in heparin, NA/NS domains comprise about 25% of a typical HS chain. Various lines of evidence

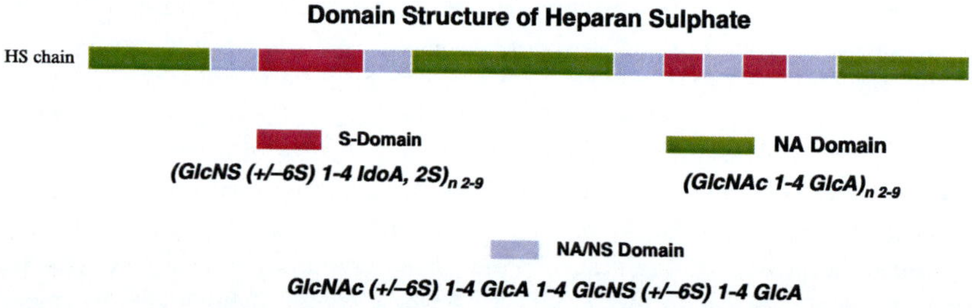


**Fig. 1** The domain structure of heparan sulphate. Heparan sulphate is composed of hypervariable sulphated domains (S-domains) separated by long N-acetylated regions (NA domains) largely devoid of sulphate groups. S-domains and NA domains are adjoined by transition zone sequences composed of alternating N-acetylated and N-sulphated disaccharides (NA/NS domains). S-domains range in length from 2 to 9 disaccharide units. The predominant sequence in S-domains is GlcNS – IdoA,2S modified to varying degrees by 6-O-sulphation of the GlcNS residues. The NA/NS domains are modified by 6-O-sulphate groups principally on GlcNAc but also to a lesser degree on GlcNS. In the S- and NA/NS domains, the densities and patterns of 6-sulphation vary between HS extracted from different cells and tissues. Sequence variability in HS is enhanced in certain polymer species by 3-O-sulphation of GlcNS, 2-O-sulphation of glucuronate residues and by the presence of N-unsubstituted glucosamine residues

now indicate that the structural distinctions between HS species from different cells and tissues take the form of variations in pattern and density of sulphation superimposed on a common design principle (Fig. 1).

## 5 Biosynthesis of HS

How is the domain structure of HS formed during biosynthesis and what factors regulate polymer sulphation? These are very important and challenging questions that are addressed in more detail in Carlsson and Kjellén (2011). Earlier work on this topic has been reviewed in detail by Lindahl et al. (1989). A few points will be discussed here.

HS is initially formed as a non-sulphated precursor named heparan or N-acetylheparosan on core proteins primed by the common GAG-linkage sequence of Xyl-Gal-Gal GlcA (Zhang et al., 1995). The linkage sequence is attached via the reducing-end xylose to specific serine residues in peptide motifs that strongly favour polymerisation of heparan rather than chondroitin-type GAG chains. A series of sequential and stepwise modifications then convert heparan to HS. In these respects, the biosynthesis of HS and heparin is very similar. What appears to set the heparan precursor along the pathway to HS rather than heparin is the first modification step, the targeted conversion of ~40–50% of GlcNAc residues to GlcNS by the N-deacetylase/N-sulphotransferase enzymes (NDSTs). NDST1 appears to be mainly responsible for HS synthesis, whereas NDST2 plays the major role in the synthesis of heparin. Mechanistically, it is unclear how the action of the NDST1 is confined to particular regions of the heparan chain, but it is assumed that the domain structure is established at this primary conversion step. The substrate specificities of all the polymer modifying enzymes that act after NDST1 restrict their activities to the regions of N-sulphation. However, there are interesting differences in modification patterns between the NA/NS and nascent S-domains. In S-domains, the presence of N-sulphated disaccharide sequences strongly favours the epimeriszation of GlcA to IdoA, and this step is tightly coupled to C2 sulphation of the newly formed IdoA residue. In contrast, iduronate occurs with very low frequency in the NA/NS regions and it is not sulphated.

Following epimerisation and C2 sulphation, the final steps in HS biosynthesis are carried out by two multigene families, the 6-OSTs and the 3-OSTs, that transfer sulphate groups to C6 and C3 of GlcNS and GlcNAc. These late-stage modifications considerably amplify the informational content of HS. C6 sulphation is particularly crucial for "imprinting" the molecular features that distinguish different species of HS. Unlike the tight association of 2-O-sulphate groups with S-domains, C6 sulphation is common in both NA/NS and S-domains; it is a key modification step on which many of the biological properties of HS depend (see below). Despite its low abundance, C3 sulphation is very important functionally. In addition to its precise location in the AT-binding sequence, a C3 sulphate group, in association with a rare N-unsubstituted glucosamine residue, is present in the novel HS-binding site for the gD surface protein of HSV (Shukla et al. 1999). Moreover, on the basis of specific recognition by the HS4C3 antibody, a C3-sulphate is present in a transient, developmentally regulated HS motif that marks a population of mesodermal cells with high haemangioblast potential (Baldwin et al. 2008).

## 6 Editing the Sulphation Patterns of HS: The Role of Endosulphatases (Sulfs)

Any description of HS-sulphation patterns would be incomplete without reference to the relatively recent, striking discovery of the HS-endosulphatase (Sulfs). Sulfs (initially named Qsulf) were first identified in quail embryos as genes essential for the hedgehog and wingless signalling pathways (Dhoot et al. 2001). Despite some sequence similarities to lysosomal 6-O-sulphatases, the Sulfs are cell surface rather than lysosonal enzymes acting efficiently at physiological pH (Frese et al. 2009). Sulfs remove 6-sulphates specifically from N-sulphated glucosamine residues targeting the S-domains rather than the NA/NS regions and effecting a partial rather than complete depletion 6-sulphate groups (Ai et al. 2003, 2006; Lamanna et al. 2006; Viviano et al. 2004). Thus, they can be considered as antagonists of the 6-OSTs; no other sulphate groups in HS are subject to selective removal on the cell surface (Fig. 2). Sulfs rapidly and dramatically modulate the recognition properties of HS. At present, it is not known how the synthesis, intracellular transport and catalytic activities of the Sulfs are regulated but genetic modulation of their expression levels in cultured cells provides clear evidence that they play an important role in finely tuning 6-sulphation patterns on the cell surface (Ai et al. 2003; Lamanna et al. 2006). Sulfs clearly effect cell responses to growth factors and morphogens. They dampen mitogenic and motility properties of FGFs and VEGF that are potentiated by 6-sulphates (Uchimura et al. 2006) but, in a notable twist to the activity repertoire of HS, Sulf-mediated depletion of 6-O-sulphates enhances the activity of Wnt proteins apparently by diminishing their affinity for cell surface

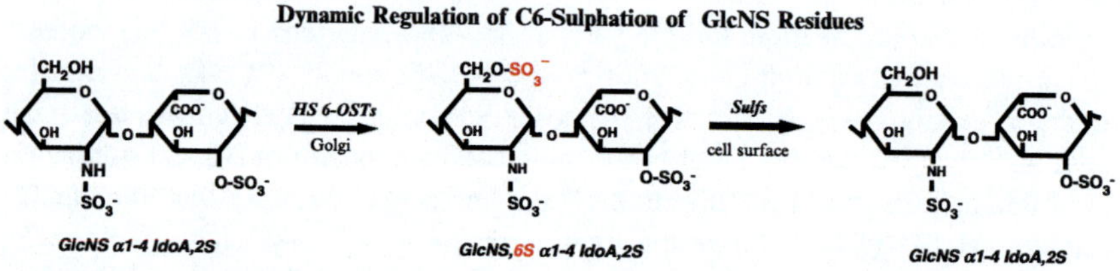


**Fig. 2** Dynamic regulation of HS-sulphation at C6. Sulphation at C6 of GlcNS residues is regulated by the balance of activities of the HS 6-O-sulphatransferases (6-OSTs) and the Sulfs, endosulphatases that edit the 6-sulphation patterning of HS in the S-domains. The dynamic regulation of 6-sulphation reflects the importance of this functional group in dictating many of the binding properties of HS

HS and facilitating ligand delivery to the Wnt signalling receptor frizzled (Ai et al. 2003). Sulfs also indirectly enhance the morphogenic effects of BMP4 by displacing the BMP4 inhibitor Noggin from its binding site on HS (Viviano et al. 2004).

# 7 Diverse Activities of HS

A vast array of effector proteins and peptides bind to HS and make up the so-called HS "interactome" (Ori et al., 2011). Invariably, when a protein binds to HS, its activity is modified in some way but as already discussed generally-applicable mechanistic explanation for the mode of action of HS is not immediately apparent. Some HS-protein interactions are quite specific, others less so (for reviews, see Bulow and Hobert 2006; Casu and Lindahl 2001; Gallagher 2001; Lindahl and Li 2009; Lyon and Gallagher 1998). For many proteins (e.g. FGF1 and FGF2), an HS-activator is essential for signalling through high-affinity transmembrane receptor (Fig. 3). HS directs the assembly of FGF–FGF receptor signalling complexes, although the architecture and stoichiometry of the complexes is still a matter of debate (see below). The growth and motility factor HGF/SF that binds HS with low specificity despite a high dependency on GAGs for activity may require HS to stabilise an active configuration of its disulphide-bonded subunits (Deakin et al 2009; Lyon and Gallagher 1998).

A proximity-type mechanism perhaps best describes the way HS regulates Interferon gamma (IFNγ), a dimeric cytokine essential for innate and acquired immunity. In the active structure of IFNγ, two identical monomers are combined in a compact antiparallel orientation with unstructured C-termini extending on opposite sides of the protein. The HS interaction sites are in these unstructured regions (Lortat-Jacob and Grimaud 1991); in consequence, IFNγ binds over an extensive segment of HS, about 20 disaccharides in length, with each C-terminus

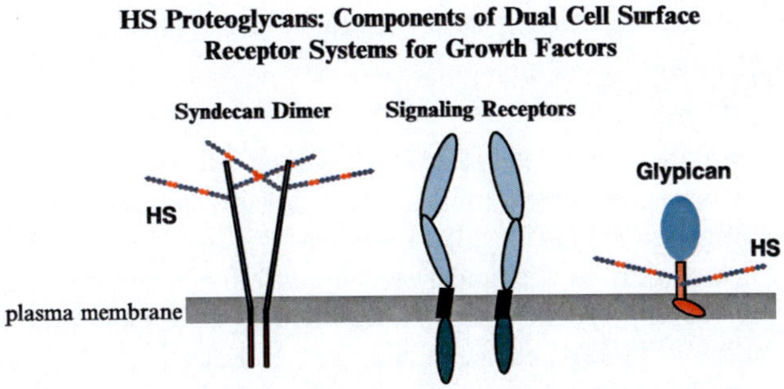


**Fig. 3** HS-proteoglycans: components of the dual receptor system for growth factors. The two major cell surface HS-proteoglycans are the dimeric, trans-membrane syndecans and the GPI-anchored glypicans. In the syndecans, the HS chains are positoned towards the N-terminus, whereas in the glypicans they are located in the stem region close to the cell surface. Both proteoglycans can function as co-receptors for growth factors, cytokines, etc., which bind to HS chains and are then transferred to signalling receptors. In some cases such as the FGFs, HS forms part of the signalling complex (Fig. 4), whereas in others such as the Wnt proteins, the protein may dissociate from HS in close proximity to its receptor (see text for details)

bound to individual S-domains bridged by a flexible N-acetylated "linker" sequence (Lortat-Jacob et al. 1995). The binding sites for HS and the IFNγ-receptor partially overlap and the protein–GAG complex on the cell surface is inactive. However, the cytokine is in close proximity to its receptor. The spontaneous release of IFNγ, which is likely to be influenced by changes in the microenvironment, will determine the level of activation of the IFNγ receptor. In an analogous manner to IFNγ, the large family of multimeric chemokines exploit the high pericellular concentrations of HS as a means of attachment and accumulation on cell surfaces of vascular tissues at sites of infection and inflammation. Chemokines are chemo-attractants that stimulate the migration and extravasation of circulating neutrophils and lymphocytes. However, unlike IFNγ the chemokine–GAG complex is active and high local concentrations of protein enhance receptor recognition and signalling (Lortat-Jacob 2009; Proudfoot 2006). Recombinant chemokines defective in receptor binding but with increased GAG-affinity are being developed as antagonists of endogenous counterparts for treatment of inflammation and autoimmune diseases (Potzinger et al. 2006).

The influence of HS on ligand concentration and diffusion are of critical importance in embryogenesis. HS plays a critical role in setting-up diffusion gradients of morphogenic proteins such as Wnt/Wg, Hedgehog and BMP across fields of differentiating cells responsive to different concentrations of inducer (Han et al. 2004; Gallagher and Lyon 2000; Lindahl and Li 2009). Template and catalytic functions of HS are evident in the regulation of enzymes involved in blood coagulation and lipid metabolism (Bishop et al. 2007; Spillmann et al. 2006). HS elicits the fusogenic activity of the gD protein of HSV essential for viral entry into the cell (Shukla et al. 1999). It is not known whether this novel mechanism is exploiting a normal physiological process mediated by HS.

## 8 Some Issues of Binding Specificity: The Problem of the FGFs

Following the first reports in 1991 that FGF2 was an HS-dependent growth factor, several groups began to search for its active site in the GAG chain – despite considerable progress the issue is still unresolved. The first FGF2 binding sequence to be published was a low abundance, high-affinity S-domain isolated from skin fibroblast HS and named Oligo-H that bound to immobilised FGF2 with comparable affinity to the parent molecule (Turnbull et al. 1992). Oligo-H was a 14mer sequence, the longest S-domain in fibroblast HS that consisted of an internal repeat motif of N- and 2-O-sulphated disaccharides in the sequence:

$$\text{GlcA} - \text{GlcNS} - [\text{IdoA,2S} - \text{GlcNS}]_5 - \text{IdoA} - \text{GlcNAc}$$

Soon after the publication of the Oligo–H sequence, a second paper described a minimal N-sulphated pentasaccharide for FGF2 binding that contained just one 2-O-sulphate group (Maccarana et al. 1993).

HexA – GlcNS – HexA – GlcNS – IdoA,2S (where HexA is GlcA or IdoA).

The pentasaccharide was compatible with crystal structures of FGF in complex with heparin that revealed a GAG-binding site that accommodated five sugar residues (Faham et al. 1996). Both sequences were devoid of 6-O-sulphates and at the time it seemed reasonable to assume that the long, high-affinity Oligo-H was the active site that in principle could bind two FGFs, whereas short minimal binding sequences that are common in HS would assist in capture and transfer of FGF2 to the high-affinity Oligo-H site. This simple notion was soon dispelled when assays of growth factor activity revealed that activation of FGFs signalling by structures of the Oligo-H type required additional modification by at least one, possibly two (the question is still unresolved) 6-O-sulphate groups (Guimond et al. 1993; Pye et al. 1998). One proposal for the requirement for 6-O-sulphates was that HS active sites for FGF2 contain two distinct sites, one for the ligand and the other for the FGF-receptor with 6-sulphates needed for receptor binding (Guimond et al. 1993). In this operational mode, HS would be acting as a template for docking ligand and receptor in an analogous manner to the way in which heparin aligns anti-thrombin and thrombin A, second possibility was that 6-O-sulphates influenced the local geometry S-domains. Although not formally proven, it is probable that S-domains adopt a helical structure similar to heparin (Mulloy and Forster 2000). FGF2 (and also FGF1) bind in a co-operative manner to active saccharide sequences, the binding of the first FGF creating a higher affinity site for interaction with a second FGF (Goodger et al. 2008; Robinson et al., 2005). Crystal structures have shown that two FGFs are able to bind in close proximity on opposite faces of the saccharide helix to form a trans-dimer that binds two FGF-receptors in a putative signalling complex (Pellegrini et al. 2000). An alternative model of the FGF-signalling mechanism is more compatible with a directional template; this model, also derived from high-resolution crystallography, depicts two half complexes of 1:1:1 FGF/FGF-receptor/heparin that assemble in a symmetrical manner with each heparin saccharide interacting with an FGF monomer and a receptor (Schlessinger et al. 2000) (Fig. 4). Recent biochemical data and cell activation studies provide support for both models (Goodger et al. 2008). It seems that cells may have evolved more than one solution to the problem of FGF-signalling (Jastrebova et al. 2006).

Although the FGF–HS interaction has dominated the sequence-activity debate, there have been many quite detailed studies on other HS-binding growth factors and ECM proteins, and these have been thoroughly reviewed in several recent publications (Kreuger et al. 2006; Lamanna et al. 2007; Lindahl and Li 2009). Although in general there tends to be a good correlation between level of sulphation and binding affinities, the strict cell/organ specific regulation of HS structure suggest that a high degree of selectivity may prevail at cell surfaces. It seems inconceivable (Ledin et al., 2004) that the tight regulation of HS synthesis has

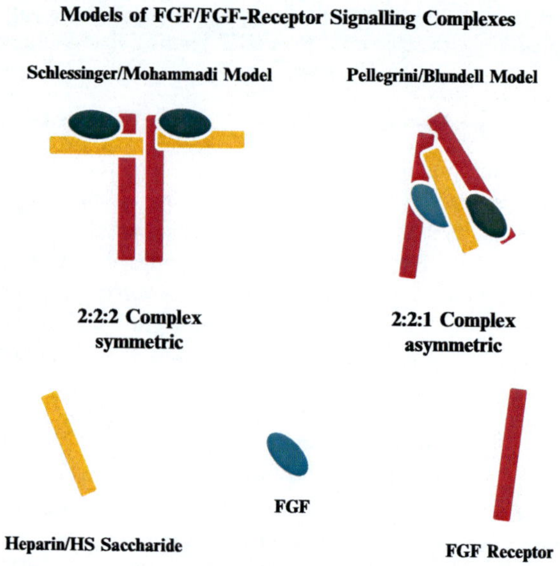


**Fig. 4** Models of the crystal structures of FGF-signalling complexes. The diagrams are based on two crystal structures of putative FGF-signalling complexes. The symmetrical model (a) of Schlessinger and Mohammadi was formed between FGF2, FGFR1 and a heparin decasaccharide (dp10); the 2:2:2 stiochiometry of the complex was stabilised by each saccharide binding to one of the ligand-receptor pairs. In contrast, the asymmetrical Pellegrini/Blundell model formed from FGF1, FGFR2 and a dp10 heparin saccharide has the heparin at the core of the complex with two FGFs binding in close proximity in a trans-dimeric orientation in a favourable disposition for docking of two receptors. Biochemical data suggesting co-operative binding of FGFs to a single heparin dp10 supports the asymmetrical model, but it is likely that both types of architecture will form on the cell surface

evolved without any clear purpose. Control of sulphation patterns and densities will enable cells to be selective in their interactions with protein effectors, binding and activating only those that are essential for specific cellular functions.

## 9 Concluding Remarks

HS has come a long way since its discovery over 50 years ago in what was essentially discarded by-products of methods for the production of heparin. The two polymers have maintained a close kinship with similar methods of analysis being used to study them and with the more abundant heparin being an invaluable source of oligosaccharides for revealing some of the potential binding and activating properties of HS. Heparin in various molecular forms, including synthetic compounds (Petitou and Van Boeckel 2004), continues to dominate in the sphere of anticoagulant drugs but HS has emerged as the main GAG involved in cell regulation. Its vast repertoire of client proteins has been discovered only over the past 20 years or so. It seems likely that in time structural motifs in HS will have important biomedical applications especially if sequence specific protein targets can be identified. Refinements in HS analysis and creative ideas for assembling saccharide arrays should bring this goal closer to realisation (Shi and Zaia, 2009., Yang et al., 2011).

## References


Ai X, Do AT, Lozynska O, Kusche-Gullberg M, Lindahl U, Emerson CP Jr (2003) Qsulf1 remodels the 6-O sulfation states of cell surface heparan sulfate proteoglycans to promote Wnt signaling. J Cell Biol 162:341–351

Ai X, Do AT, Kusche-Gullberg M, Lindahl U, Lu K, Emerson CP Jr (2006) Substrate specificity and domain functions of extracellular heparan sulfate 6-O-endosulfatases, Qsulf1 and Qsulf2. J Biol Chem 281:4969–4976

Baldwin RJ, ten Dam GB, van Kuppevelt TH, Lacaud G, Gallagher JT, Kouskoff V et al (2008) A developmentally regulated heparan sulfate epitope defines a subpopulation with increased blood potential during mesodermal differentiation. Stem Cells 26:3108–3118

Bellache Y, The I, Perrimon N (1998) Tout-velu is a Drosophila homologue of the putative tumour suppressor Ext-1 and is neede for Hh diffusion. Nature 394:85–88

Bishop JR, Schuksz M, Esko JD (2007) Heparan sulphate proteoglycans fine-tune mammalian physiology. Nature 446:1030–1037

Bornemann DJ, Duncan JE, Staatz W, Selleck S, Warrior R (2004) Abrogation of heparan sulfate synthesis in Drosophila disrupts the Wingless, Hedgehog and Decapentaplegic signaling pathways. Development 131:1927–1938

Bullock SL, Fletcher JM, Beddington RS, Wilson VA (1998) Renal agenesis in mice homozygous for a gene trap mutation in the gene encoding heparan sulfate 2-sulfotransferase. Genes Dev 12:1894–1906

Bulow HE, Hobert O (2006) The molecular diversity of glycosaminoglycans shapes animal development. Annu Rev Cell Dev Biol 22:375–407

Carlsson P, Kjellén L (2011) Heparin biosynthesis. In: Lever R, Mulloy B, Page CP (eds) Heparin – a century of progress. Springer, Heidelberg

Casu B, Lindahl U (2001) Structure and biological interactions of heparin and heparan sulfate. Adv Carbohydr Chem Biochem 57:159–206

Cifonelli JA (1968) Reaction of heparitin sulphate with nitrous acid. Carbohydr Res 8:233–242

Couchman JR, Chen L, Woods A (2001) Syndecans and cell adhesion. Int Rev Cytol 207:113–150

David G, Lories V, Decock B, Marynen P, Cassiman JJ, Van den Berghe H (1990) Molecular cloning of a phosphatidylinositol-anchored membrane heparan sulfate proteoglycan from human lung fibroblasts. J Cell Biol 111:3165–3176

Deakin JA, Blaum BS, Gallagher JT, Uhrin D, Lyon M (2009) The binding properties of minimal oligosaccharides reveal a common heparan sulphate/dermatan sulphate binding site in hepatocyte growth factor/scatter factor that can accommodate a wide variety of sulphation patterns. J Biol Chem 284:6311–6322

Desai UR, Wang HM, Linhardt RJ (1993) Substrate specificity of the heparin lyases from *Flavobacterium heparinum*. Arch Biochem Biophys 306:461–468

Dhoot GK, Gstafsson MK, Ai X, Sun W, Standford DM, Emerson CP (2001) Regulation of Wnt signalling and embryo patterning by an extracellular sulfatase. Science 293:1663–1666

Esko JD, Selleck SB (2002) Order out of chaos: assembly of ligand binding sites in heparan sulfate. Annu Rev Biochem 71:435–471

Faham S, Hileman RE, Fromm JR, Linhardt RJ, Ress DC (1996) Heparin structure and interactions with basic fibroblast growth factor. Science 23:1116–1120

Frese MA, Milz F, Dick M, Lamanna WC, Dierks T (2009) Characterisation of the human sulfatase sulf1 and its high affinity heparin/heparan sulphate interaction domain. J Biol Chem 284:28033–28044

Gallagher JT (2001) Heparan sulfate: growth control with a restricted sequence menu. J Clin Invest 108:357–361

Gallagher JT, Lyon M (2000) Heparan sulphate; molecular structure and interactions with growth factors and morphogens. In: Iozzo RV (ed) Proteoglycans, structure, biology and molecular interactions. Marcel Dekker, New York

Goodger SJ, Robinson CJ, Murphy KJ, Gasiunas N, Harmer NJ, Blundell TL, Gallagher JT (2008) Evidence that heparin saccharides promote FGF2 mitogenesis through two distinct mechanisms. J Biol Chem 283:13001–13008

Guimond S, Maccarana M, Olwin BB, Lindahl U, Rapraeger AC (1993) Activating and inhibitory heparin sequences for FGF-2 (basic FGF). Distinct requirements for FGF-1, FGF-2, and FGF-4. J Biol Chem 268:23906–23914

Han C, Belenkaya TY, Khodoun M, Tauchi M, Lin X (2004) Distinct and collaborative roles of Drosophila EXT family proteins in morphogen signalling and gradient formation. Development 131:1563–1575

Hovingh P, Linker A (1970) The enzymatic degradation of heparin and heparitin sulphate. J Biol Chem 245:6170–6175

Jastrebova N, Vanwildemeersch M, Rapraeger AC, Gimenez-Gallego G, Lindahl U, Spillmann D (2006) Heparan sulfate-related oligosaccharides in ternary complex formation with fibroblast growth factors 1 and 2 and their receptors. J Biol Chem 281:26884–26892

Jorpes JE, Gardell S (1948) On heparin monosulfuric acid. J Biol Chem 176:267–275

Knecht J, Cifonelli JA, Dorfman A (1967) Structural studies on heparitin sulphate of normal and Hurler tissues. J Biol Chem 242:4652–4661

Kraemer PM (1971a) Heparan sulfates of cultured cells. 1. Membrane-associated and cell sap species in Chinese hamster cells. Biochemistry 10:1437–1445

Kraemer PM (1971b) Heparan sulfates of cultured cells. 11 Acid soluble and precipitable species of different cell lines. Biochemistry 10:1445–1451

Kramer KL, Yost HJ (2002) Ectodermal syndecan-2 mediates left-right axis formation in migrating mesoderm as a cell-nonautonomous Vg1 cofactor. Dev Cell 2:115–124

Kreuger J, Spillmann D, Li JP, Lindahl U (2006) Interactions between heparan sulfate and proteins: the concept of specificity. J Cell Biol 174:323–327

Lamanna WC, Baldwin RJ, Pavda M, Kalus I, ten Dam G, van Kuppvelt TH, Gallagher JT, Von Figura K, Dierks T, Merry CL (2006) Heparan sulphate 6-O-endosulphatases; discrete in vivo activities and functional co-operativity. Biochem J 400:63–73

Lamanna WC, Kalus I, Padva M, Baldwin RJ, Merry CL, Dierks T (2007) The heparanome – the enigma of encoding and decoding heparan sulfate sulfation. J Biotechnol 129:290–307

Ledin J, Staatz W, Li JP, Gotte M, Selleck S, Kjellen L, Spillmann D (2004) Heparan sulphate structure in mice with genetically-modified heparan sulphate production. J. Biol. Chem. 279: 42732–42741

Lin X, Perrimon N (1999) *Dally* cooperates with Drosophila Frizzled 2 to transducer wingless signalling. Nature 400:281–284

Lindahl U, Li JP (2009) Interactions between heparan sulphate and proteins-design and functional implications. Int Rev Cell Mol Biol 276:105–159

Lindahl U, Kusche M, Lidholt K, Oscarsson LG (1989) Biosynthesis of heparin and heparan sulfate. Ann N Y Acad Sci 556:36–50

Lortat-Jacob H (2009) The molecular basis and functional implications of chemokine interactions with heparan sulphate. Curr Opin Struct Biol 19:543–548

Lortat-Jacob H, Grimaud JA (1991) Interferon γ binds to heparan sulphate by a cluster of amino acids located in the C-terminus part of the molecule. FEBS Lett 280:152–154

Lortat-Jacob H, Turnbull JE, Grimaud JA (1995) Molecular organization of the interferon γ-binding domain in heparan sulphate. Biochem J 310:497–505

Lyon M, Gallagher JT (1998) Biospecific sequences and domains in heparan sulphate and the regulation of cell growth and adhesion. Matrix Biol 17:485–493
Maccarana M, Casu B, Lindahl U (1993) Minimal sequence in heparin/heparan sulfate required for binding of basic fibroblast growth factor. J Biol Chem 268:23898–23905
Merry CL, Lyon M, Deakin JA, Hopwood JJ, Gallagher JT (1999) Highly sensitive sequencing of the sulphated domains of heparan sulphate. J Biol Chem 274:18455–18462
Mulloy B, Forster MJ (2000) Conformation and dynamics of heparin and heparan sulfate. Glycobiology 10:1147–1156
Murphy K, Merry CL, Lyon M, Thompson J, Roberts I, Gallagher JT (2004) A new model for the domain structure of heparan sulphate based on the novel specificity of K5 lyase. J Biol Chem 279:27239–27245
Nakato H, Fox B, Selleck SB (2002) *Dally*, a Drosophila member of the glypican family of integral membrane proteoglycans, affects cell cycle progression and morphogenesis by a cyclin A mediated process. J Cell Sci 115:123–130
Oldberg A, Kjellen L, Hook M (1979) Cell surface heparan sulphate. Isolation and characterisation form rat liver membranes. J Biol Chem 254:8505–8510
Ori A, Wilkinson MC, Fernig DG (2011) A systems biology approach for investigation of the heparin/heparan sulphate interactome. J. Biol. Chem. 286;19892–19904
Pellegrini L, Burke DF, von Delft F, Mulloy B, Blundell TL (2000) Crystal structure of fibroblast growth factor receptor ectodomain bound to ligand and heparin. Nature 407:1029–1034
Petitou M, van Boeckel CA (2004) A synthetic antithrombin III binding pentasaccharide is now a drug! What comes next? Angew Chem Int Ed Engl 43:3118–3133
Potzinger H, Geretti E, Brander B, Wabitsch V, Piccinini AM, Rek A, Kungl A (2006) Developing chemokine mutants with improved proteoglycan affinity and knocked-out GPCR as anti-inflammatory recombinant drugs. Biochem Soc Trans 34:435–437
Proudfoot AE (2006) The biological relevance of chemokine-proteoglycan interactions. Biochem Soc Trans 34:422–426
Pye D, Vives R, Turnbull JE, Hyde P, Gallagher JT (1998) Heparan sulphate oligosaccharides require 6-O sulphation for promotion of basic fibroblast factor mitogeneic activity. J Biol Chem 273:1183–1192
Rapraeger A, Kufka A, Olwin BB (1991) Requirement for heparan sulphate for bFGF-mediated growth and myoblast differentiation. Science 253:1705–1708
Robinson CJ, Harmer NJ, Godger SJ, Blundell TL, Gallagher JT, (2005) Cooperative dimerisation of fibroblast growth factor 1 upon a single heparin saccharide may drive the formation of a 2:2:1 FGF1:FGFR2c:heparin ternary complex. J.Biol.Chem. 280:42274–42282
Saunders S, Jalkanen M, O'farrell S, Bernfield M (1989) Molecular cloning of syndecan, an integral membrane proteoglycan. J Cell Biol 108:1547–1556
Schlessinger J, Plotnikov AN, Ibrahimi OA, Eliseenkova AV, Yeh BK, Yayon A, Linhardt RJ, Mohammadi M (2000) Crystal structure of a ternary FGF–FGFR–heparin complex reveals a dual role for heparin in FGFR binding and dimerization. Mol Cell 6:743–747
Shi X, Zaia J (2009) Organ specific heparan sulphate structural phenotypes. J. Biol. Chem. 284;11806–11814
Shukla D, Liu J, Blaiklock P, Shworak NW, Bai X, Esko JD, Cohen GH, Eisenberg RJ, Rosenberg RD, Spear RJ (1999) A novel role for 3-O-sulfated heparan sulfate in herpes simplex virus 1 entry. Cell 99:13–22
Silva ME, Dietrich CP, Nader HB (1976) On the structure of heparitin sulfates. Analysis of products formed from heparitin sulfates and heparin by two heparitinases and heparinise from Flavobacterium heparinum. Biochem Biophys Acta 437:129–141
Spillmann D, Lookene A, Olivecrona G (2006) Isolation and characterization of low sulfated heparan sulfate sequences with affinity for lipoprotein lipase. J Biol Chem 281:23405–23413
Turnbull JE, Gallagher JT (1990) Molecular organisation of heparan sulphate from human skin fibroblasts. Biochem J 265:715–724

Turnbull JE, Gallagher JT (1991a) Distribution of iduronate-2-sulphate residues in heparin sulphate; evidence for an ordered polymeric structure. Biochem J 273:553–559
Turnbull JE, Gallagher JT (1991b) Sequence analysis of heparan sulphate indicates defined locations of N-sulphated glucosamine and iduronate-2-sulphate residues proximal to the protein-linkage region. Biochem J 277:297–303
Turnbull JE, Fernig DG, Ye Y, Wilkinson MC, Gallagher JT (1992) Identification of the basic fibroblast factor binding sequence in heparan sulfate. J Biol Chem 267:10337–10341
Turnbull JE, Hopwood JJ, Gallagher JT (1999) A strategy for rapid sequencing of heparan sulphate and heparin saccharides. Proc Natl Acad Sci USA 96:2698–2703
Uchimura K, Morimoto-Tomita M, Bistrup A, Li J, Lyon M, Gallagher J, Werb Z, Rosen S (2006) Hsulf-2, an extracellular endoglucosamine-6-sulfatase, selectively mobilises heparin- bound growth factors and chemokines: effects on VEGF, FGF-1, and SDF-1. BMC Biochem 7:2
Venkataraman G, Shriver Z, Raman R, Sasisekharan R (1999) Sequencing complex polysaccharides. Science 286:537–542
Viviano BL, Paine-Saunders S, Gasiunas N, Gallagher J, Saunders S (2004) Domain-specific modification of heparan sulfate by Qsulf1 modulates the binding of the bone morphogenetic protein antagonist Noggin. J Biol Chem 279:5604–5611
Yang B, Solakyildirim K, Chang Y, Linhardt RJ (2011) Hyphenated techniques for the analysis of heparin and heparan sulphate. Chemistry and Materials Science 341:541–557
Yayon A, Klagsbrun M, Esko JD, Leder P, Ornitz DM (1991) Cell surface, heparin-like molecules are required for binding of basic fibroblast growth factor to its high affinity receptor. Cell 64:841–848
Zhang L, David G, Esko JD (1995) Repetitive ser-gly sequences enhance heparan sulphate assembly in proteoglycans. J.Biol. Chem. 270; 27127–27135

# Heparin Mimetics

**Deirdre R. Coombe and Warren C. Kett**

## Contents



**Abstract** Explorations of the therapeutic potential of heparin mimetics, anionic compounds that are analogues of glycosaminoglycans (GAGs), have gone hand-in-hand with the emergence of understanding as to the role of GAGs in many essential biological processes. A myriad of structurally different heparin mimetics have been prepared and examined in many diverse applications. They range in complexity from heterogeneous polysaccharides that have been chemically sulphated to well-defined compounds, designed in part to mimic the natural ligand, but with binding specificity and potency increased by conjugation to non-carbohydrate pharmacophores.

D.R. Coombe (✉)
Molecular Immunology, School of Biomedical Sciences, Curtin Health Innovation Research Institute, Curtin University of Technology, Level 3 MRF Building, Rear 50 Murray Street, Perth, WA 6000, Australia
e-mail: d.coombe@curtin.edu.au

R. Lever et al. (eds.), *Heparin - A Century of Progress*,
Handbook of Experimental Pharmacology 207,
DOI 10.1007/978-3-642-23056-1_16, 

The maturity of the field is illustrated by the seven heparin mimetics that have achieved marketing approval and there are several more in late-stage clinical development. An overview of the structural determinants of heparin mimetics is presented together with an indication of their activities. The challenges in developing heparin mimetics as drugs, specificity and potential toxicity issues, are highlighted. Finally, the development path of three structurally very different mimetics, PI-88®, GMI-1070 and RGTAs, each of which is in clinical trials, is described.

**Keywords** Heparin • Glycosaminoglycan • Heparin-mimetics • PI-88 • RGTAs • GMI-1070

## Abbreviations

| | |
|---|---|
| AT | Antithrombin |
| CMBDS | Carboxymethyl benzylamide sulfonate dextrans |
| COPD | Chronic obstructive pulmonary disease |
| d.p. | Degree of polymerization |
| FGF | Fibroblast growth factor |
| GAG | Glycosaminoglycans |
| HSV | Herpes simplex virus |
| MMP | Matrixmetallo proteinase |
| PECAM-1 | Platelet endothelial cell adhesion molecule |
| RANTES/CCL5 | Regulated on Activation Normal T Cell Expressed and Secreted/ (C-C motif) ligand 5 |
| RGTA | Regenerating agents |
| SDF | Stromal-derived factor |
| Sialyl $Le^a$ | Sialyl Lewis A |
| Sialyl $Le^x$ | Sialyl Lewis X |
| TFPI | Tissue factor pathway inhibitor |
| TGF-β | Transforming growth factor-β |
| VEGF | Vascular endothelial growth factor |

## 1 Introduction

Earlier chapters have described the involvement of heparin, heparan sulphates and other GAGs in regulating biological processes. Although studies on the biological function of these molecules initially focused on their anticoagulation activity, in more recent years a greater understanding of the extent of their involvement in a range of fundamental biological processes essential for normal mammalian development and physiology has emerged. Taking place simultaneously with the structure–function studies of natural GAGs have been similar studies exploring highly sulphated compounds that are analogues of GAGs. Collectively such compounds are referred to as heparin mimetics. They are synthetic, or at least semi-synthetic (i.e. chemically transformed natural products) compounds that are structurally distinct from naturally occurring GAGs. A heparin mimetic may also be characterized in terms of whether it performs similar functions as heparin (or another GAG), e.g. it binds to the heparin-binding site upon a protein. Chapters dedicated to several types of heparin mimetics appear in this volume; hence, the reader will be referred to appropriate chapters in preference to discussing these mimetics at length.

Investigations with heparin mimetics for therapeutic use have predominantly been directed towards medical needs outside of the antithrombotic and anticoagulant fields. The development of heparin mimetics closely followed the development of heparin itself, with the preparation and investigation of sulphate esters of malto-oligosaccharides, amylose, cellulose and other polysaccharides appearing in the late 1940s. Since that time, 7 heparin mimetics have been approved for marketing by the various regulatory authorities throughout the world. These include sucrose octasulphate (Sucralfate®), pentosan polysulphate (SP54®, Elmiron®), dextran sulphate, Hirudoid® (an oversulphated heparin), Macugen®, Suramin® and Cacipliq20®. Although this suggests that the development of heparin mimetics has surpassed that of heparin and the various low-molecular-weight heparins (LMWHs), it is noteworthy that these drugs do not serve high value markets and hence total sales per year of all the mimetics amount to a fraction of the billions earned by heparin, LMWH and Arixtra® (a totally synthetic version of the antithrombin (AT)-binding pentasaccharide) in the blockbuster anticoagulant market. However, this situation may begin to change with several of the mimetics described below.

Reasons for developing new heparin mimetics are multiple and varied. The core of almost every drug development program is the search for compounds that display a higher relative potency and selectivity (specificity) of action and this is also true for heparin mimetics. Heparin binds many proteins in addition to AT and has a diversity of biological activities, which give rise to heparin's well-recognized polypharmacy. Thus, a prime consideration for mimetic design is the need to curtail heparin's polypharmacy and in particular its anticoagulant activity. Clearly, in the context of treating many diseases patient anticoagulation is undesirable. However, occasionally anticoagulation may contribute to heparin's efficacy in treating conditions, where coagulation is not the underlying cause. For example, anticoagulants, including heparin, have been investigated as a treatment for acute respiratory distress syndrome (MacLaren and Stringer 2007). Heparin's anti-inflammatory activity and its anticoagulation activity are both likely to contribute to any therapeutic effect for this complex disease. Thus, the goal for heparin mimetic design is the removal of unwanted activities and the maximization of activities with therapeutic benefit for the disease being targeted.

In the main, heparin mimetics have lower anticoagulant activity than heparin, and this potentially gives access to a therapeutic dosage window. The polypharmacy of heparin is due in part to its enormous heterogeneity and the diversity of structures within it. Thus, heparin mimetics with reduced heterogeneity, ranging from a mixture of more narrowly defined structures to the extreme of a single chemical entity, may be expected to display a greater specificity of action. Reduced heterogeneity may also give rise to a greater potency, particularly if the mimetic has been designed with structure–activity relationships in mind.

Few organizations can dedicate the considerable resources required to synthesize GAG fragments for exploratory research, thus using heparin mimetics that are readily synthesized is a more accessible entry point for many groups. In addition, given the ultimate goal of developing and marketing a drug, the protracted multi-step chemical synthesis of heparin fragments is technically challenging and very costly, particularly if manufacturing at industrial scales. Depending on dose, treatment regimen and number of patients, the scale of synthesis required may exceed 1,000 kg/year and this is especially challenging for heparin hexasaccharides, octasaccharides or larger. Heparin fragments of this size are frequently required before nM binding affinities can be measured. Marketing a fully synthetic heparin fragment as a drug may not be commercially viable, because the cost of production and development may demand a price that the market cannot support. Thus, heparin mimetics, which are more readily synthesized, may offer a clearer route to negotiating these technical hurdles and achieving a price competitive drug in the market.

## *1.1 Structural Diversity of Heparin Mimetics*

Given the inspiration for heparin mimetics, it is no surprise that most of the mimetics created to date are carbohydrate based, be they polysaccharides or oligosaccharides. Generally, anionic charge is introduced by sulphation, although phosphorylation and carboxylation have also been used. Non-carbohydrate-based heparin mimetics have also been explored and there is a growing subset of mimetics that are a combination of carbohydrates and aglycones. Many of the heparin mimetics that have been synthesized are presented in Table 1, where they are classified according to the starting material and the type of modifications introduced. Before describing these classes, we would like to highlight particular structural features encompassed by the diversity of mimetic structures and a rationalization as to why they have been synthesized and their activity investigated.

1. Size/molecular weight – Variations in the size (e.g. degree of polymerization) of the mimetic is a mainstay of this field of investigation. Historically, observing an effect with a polysaccharide mimetic of a certain size has inspired investigations of smaller analogues in order to increase the selectivity of action and possibly also the potency. However, reducing the size of the starting material for a heparin mimetic can have pronounced effects on a number of other parameters (e.g. sulphation reactions, shape and heterogeneity). Thus, additional factors should be considered in structure–function analyses, as an alteration in the size of the starting material is likely to have complex and possibly unpredicted

**Table 1** Classes of heparin mimetics

| Mimetic class | Examples of class | References |
|---|---|---|
| I Polysaccharides | | |
| | Dextran | Christensen et al. (2001) |
| | Pentosan | Parsons (1997) |
| | Chitin | Kumar et al. (2004) and Jayakumar et al. (2007) |
| | Cellulose | Usher et al. (1995) and Baumann et al. (2000) |
| | Inulin (and other polysaccharides) | Pahi et al. (2005) |
| | Other glucans (Pullalan, phycarin, curdlan) | Fritzsche et al. (2006) and Smelcerovic et al. (2008) |
| | Alginate | Freeman et al. (2008) |
| | Galactans | Burgermeister et al. (2002) |
| | Colominic acid | Yang et al. (1996) |
| | Pectin | Pahi et al. (2005) and Cipriani et al. (2009) |
| | Polymerized sugar monomers | Hou et al. (2004) and Chaikof et al. (2007) |
| | RGTA | Rouet et al. (2006); Barritault and Caruelle (2009) and Barritault and Caruelle (2004) |
| Ia GAGs | | |
| | Oversulphated Heparins (e.g. Hirudoid) | Livaoglu et al. (2009) and Casu et al. (2001) |
| | Desulphated GAGS | Kennedy (2009), Kennedy (1999), Hara et al. (2007) and Casu et al. (2005) |
| | Oxidized GAGs | Casu et al. (2004, 2008) |
| | K5 polysaccharide (heparosan) | Chapter 18 in this volume and Oreste and Zoppetti (2005) |
| | Carboxyl modified heparins | Ponedel'kina et al. (2006) |
| | N-acyl heparins | Fernandez et al. (2006) |
| | Lipophilic modified heparins | Kim et al. (2007) and Hoarau et al. (2007) |
| II Oligosaccharides | | |
| | Sucrose octasulphate | |
| | Malto- | Nair et al. (1977) and Parish and Cowden (2000) |
| | Linear homo-oligosaccharides (e.g. PI-88) | Parish and Cowden (2000), Mousa et al. 2006 and Fugedi et al. (1998) |
| | Chito- | Wall et al. (2001) |
| | Laminarin oligos | Katsuraya et al. (1999) and Hoffman et al. (1995) |
| | Galacto- | Kasbauer et al. (2001) |
| | Stachyose series | Nair et al. (1978) |
| | Cyclodextrins | Crandall et al. (2007) |
| | Hyaluronic oligos | Suzuki et al. (2001) |
| | Alginate oligos | Zhao et al. (2006) |

(continued)

**Table 1** (continued)

| Mimetic class | Examples of class | References |
|---|---|---|
| | Lentinan oligos | Wang and Zhang (2009) and Kong et al. (2008) |
| | Trestatin and other trehaloses | Wessel (1997) |
| | Oversulphated heparin disaccs (Ivax) | Duong et al. (2008) |
| | K5 oligosaccharides | Presta et al. (2005) |
| III Oligosaccharide-aglycone conjugates | | |
| | Aprosulate | Papoulias et al. (1993) and Raake et al. (1994) |
| | Maltodapoh | Martin et al. (1999) |
| | GMI-1070 and family | Magnani (2009) |
| | Linked glycosyl-1-amines | Fugedi and Peto (1998) |
| | Napthol maltose dimers | Wessel et al. (2005) and Chucholowski et al. (1996) |
| | Cyclitols | Freeman et al. (2005) |
| | Heparin oligosaccharide dimers | Rele et al. (2004) |
| | Galacto dendrimers | Kensinger et al. (2004) |
| | Sulphated oligosaccharide conjugates | Shoji et al. (1994) and Vogel (2006) |
| | PG-545 and family (oligosaccharide lipophilic conjugates) | Dredge et al. (2009) |
| IV Non-carbohydrate-based, sulfated mimetics | | |
| | Suramin | De Clercq (2009a) |
| | Aurine tricarboxylate | Myskiw et al. (2007) and De Clercq (2009b) |
| | Polyvinyl sulphate, poly(anetholesulfonic acid), poly(2-acrylamido-2-methyl-propanesulfonic acid) | Bugatti et al. (2007) |
| | Aptamers e.g. Macugen | Lee et al. (2005) and Nimjee et al. (2009) |
| | Napthalene disulfonate dendrimer (VivaGel SPL7013) | O'Loughlin et al. (2009) |
| | Napthalene sulphonate polymer (PRO2000) | Nunn et al. (2009) |
| | Peptidespe | Maynard and Hubbell (2005) and Ueki et al. (2001) |
| | Lignins | Raghuraman et al. (2007) and Henry et al. (2009) |
| V Non-sulfate anionic groups | | |
| | Various modified chitins | Jayakumar et al. (2008) and Kumar et al. (2004) |
| | Bisphosphonates | Renner et al. (2006) |

ramifications on the three-dimensional configuration of the mimetic and hence on its biological activity. To simply assume, the overall structure is maintained but it is shorter, or longer, may not always be valid.

2. Degree or density of sulphation – Whilst disaccharides such as sucrose octasulphate carry 4 anionic groups per monosaccharide, typically larger oligosaccharides and polysaccharides approach a theoretical maximum of 2–3

anionic groups per monosaccharide. A practical limitation with large oligosaccharides and polysaccharides is that sulphation reactions do not proceed to completion and a heterogeneous product results. For example, for the pseudo-nonasaccharide (containing nine carbon rings), trestatin A, the average degree of sulphation achieved was only 21–22 per molecule out of a possible 27 hydroxyl groups (Wessel 1997). Techniques to control the degree of sulphation include the choice of starting material (e.g. xylans and chitin have only 2 sulphatable groups per monosaccharide), selective sulphation (e.g. primary hydroxyls and amines can be selectively sulphated) or the use of limiting reaction conditions (e.g. time, heat or amount of sulphating reagent). The latter technique increases the structural heterogeneity (number of isomers).

3. Orientation of sulphates – Stereochemistry defines a particular carbohydrate and nature supplies a wide diversity of carbohydrates. It is this diversity which affords a route to exploring the influence of sulphate orientation on biological activity, merely by careful selection of the starting material to be sulphated. A major factor determining the three-dimensional structures assumed by polysaccharides is the rotations that are permissible about the bonds that form the glycosidic linkages. In addition, the well-recognized conformational flexibility of iduronate residues within heparin might prompt the exploration of other residues that also display flexibility in their ring conformations under certain circumstances.
4. Pattern of sulphation – With group selective sulphation procedures and/or judicious choice of starting material, it is possible to examine the influence of sulphation pattern upon activity. For example, 3,6 and 2,3 and 2,6 disulphated chitin/chitosan derivatives have been prepared. As a primary aim of heparin mimetics has been to maintain simplicity of synthesis, such studies are generally limited to group modifications and not selective sulphation of a particular site in one residue but not in others. The latter is still predominantly the domain of complicated protection/de-protection synthetic schemes. Whilst simplicity of synthesis has been a motivating factor historically, the future of heparin mimetics will likely encompass more selective sulphation schemes.
5. Linkage patterns – Exploration of the influence of anomers (e.g. cello- vs. malto-oligosaccharides) or of linkage site (e.g. 1,4- vs. 1,6- linked sugars) is readily accessible by the choice of starting material. Perhaps, the most thorough investigation of the influence of this parameter is evident in the discovery of PI-88 (Parish and Cowden 2000), in which many series of homo-oligosaccharides varying in monosaccharide and linkage pattern were prepared and examined.
6. Flexibility and "stiffness" of the backbone – Carbohydrates possess a relatively ordered structure limited by the conformation of the rings and the angle of the glycosidic bonds. The allowed rotations about a glycosidic linkage are such that only limited bending in a polysaccharide chain is possible. In essence, polysaccharide chains are relatively "stiff". More flexibility can be introduced by chemically modifying the polysaccharide, for example by opening some of the monosaccharide rings (e.g. glycol splitting of heparins). This approach appears

particularly suitable when different domains of the mimetic straddle binding patches on the protein. Moreover, by using synthetic, non-carbohydrate polymers of varying degrees of flexibility to link short carbohydrate chains, it is possible to produce mimetics where flexibility is controlled. Conjugates such as aprosulate and its analogues possess acyclic, ring-opened monosaccharide derivatives. Frequently, enthalpy of binding arguments is used when considering whether greater flexibility or rigidity is preferable. However, this should be settled experimentally. Interestingly, the conformational flexibility of certain monosaccharide residues can allow the protein binding site to force a residue to adopt a conformation that is not the most energetically favoured when the mimetic is free in solution. There is evidence that this occurs when heparin binds fibroblast growth factor (FGF)-1 and FGF-2 (Raman et al. 2003) and when heparin octasaccharides bind AT (Guerrini et al. 2006).

7. Heterogeneity – Heterogeneous mixtures are more likely to display multiple functions than a structurally similar homogeneous product. However, a more practical concern is the significant technical challenges associated with the reproducible synthesis of a heterogenous mixture and its characterization so that reproducibility can be confirmed. This is a real issue when the mimetic enters clinical development and a regulatory authority requires proof of reproducible synthesis. To put the problem of heterogeneity into perspective, consider the example of trestatin A. The distribution of degree of sulphation can be approximated with a Poisson distribution (assuming that sulphation is a stochastic event), by rounding the average degree of sulphation to 22. Thus, as seen in Fig. 1a, the degree of sulphation for extends from 13 to 27. Moreover, the number of isomers for each degree of sulphation is a simple combinatorial function (Fig. 1b) indicating that there are 81,000 theoretical isomers with 22 sulphates.
8. Conjugates with aglycones – Only a few of these that have entered human trials, but the precedent has been set by the following: heparin conjugated to aglycone thrombin inhibitors under development by Endotis, a selectin inhibitor in Phase III clinical trials under the sponsorship of Glycomimetics and the RGTA (dextran) derivatives. Non-ionic interactions contribute significantly to some heparin–protein interactions (Fernandez et al. 2006). For example, we have seen a role for hydrogen bonding in the interactions of platelet endothelial cell adhesion molecule (PECAM-1) with heparin fragments (Gandhi et al. 2008). An emerging strategy of incorporating other pharmacophores (e.g. investigational or existing drugs) into the mimetic is illustrated by the development of the dual antithrombin and thrombin inhibitor EP217609 (Petitou et al. 2009) and GMI-1070 (see below). Thus, this strategy is likely to be a fertile area for future investigation.

The number and diversity of heparin mimetics prepared and tested in biological systems is such that only a selection of the mimetics presented in each class in Table 1 will be described.

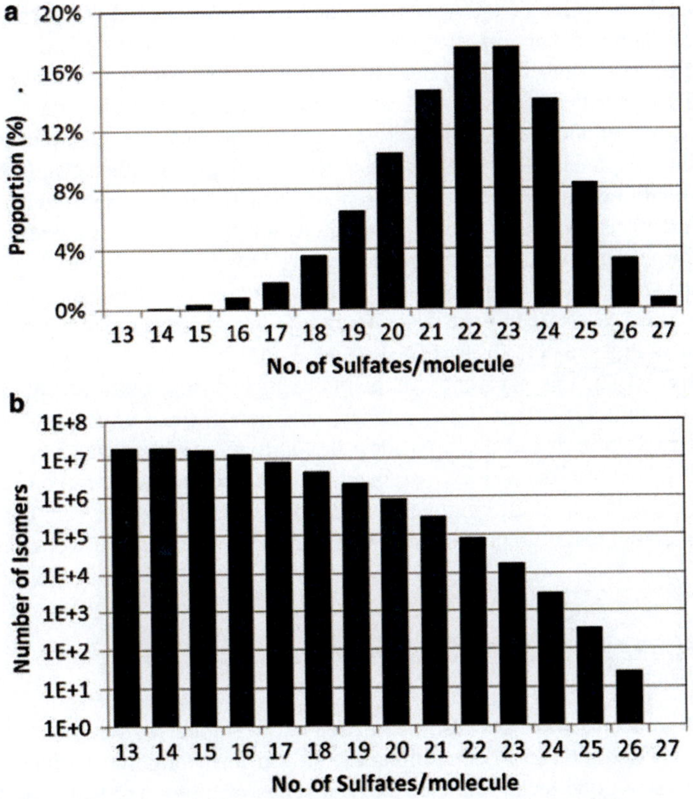


**Fig. 1** (**a**) The calculated Poisson distribution for sub-species of sulphated Trestatin A. An average number of 22 sulfates was used for the calculations (Wessel et al. 2005). (**b**) The number of theoretical isomers for a Trestatin A molecule of each degree of sulphation (Note the logarithmic Y-scale)

# 2 Classes of Heparin Mimetics

## *2.1 Modified Polysaccharides*

From surveying the literature, it appears that the most common polysaccharides have been chemically sulphated. A selection of these is presented in Table 1. Typically, the sulphation of large oligosaccharides and polysaccharides does not proceed to completion. Other technical challenges are also encountered during the sulphation process due to solubility problems of the polysaccharide in the sulphating mixture and the potential degradation of the polysaccharide chain. As a result, like heparin, the polysaccharide derivatives are polydisperse and heterogenous.

*GAGs*: A sub-class of this group are chemically modified GAGs. Of note is a 2,3-de-O-sulphated heparin that has entered clinical trials for treating chronic obstructive pulmonary disease under the sponsorship of ParanGenix (clinical trial ID NCT00457951). The same derivative has been described for treating asthma (Kennedy 1999). Other GAG derivatives such as chemically oversulphated chondroitin and sulphated K-5 polysaccharide are the subject of other chapters in this book. An interesting perspective is the replacement of N-sulfo groups in heparin by N-acyl groups drawn from aliphatic, aromatic and heterocyclic moieties with the goal of producing mimetics that bind a selective group of heparin-binding proteins (Fernandez et al. 2006). The incorporation of aglycones (e.g. lipophilic groups) into heparin is also motivated by the desire to change the pharmacokinetic properties of heparins to improve oral absorption (Kim et al. 2007).

Finally, within this class are compounds which do not quite fit either of the above two sub-classes, such as synthetic polymers formed from sulphated and polymerizable mono- and disaccharides. In addition, over the last 15 years or so, a family of dextran derivatives has been prepared containing a mixture of both anionic and aglycone units conjugated to the polysaccharide. These are the RGTA family of compounds and one of these, Cacipliq20® is marketed for wound healing.

## 2.2 *Synthetically Sulphated Oligosaccharides*

Many oligosaccharides have also been chemically sulphated. Without exception, if the polysaccharide has been sulphated, so too have oligosaccharides derived from the same polysaccharide. Generally, a series of oligosaccharides of discrete size are examined, thereby reducing heterogeneity. Although a notable exception is the phosphosulfomannoligosaccharide, PI-88. The starting material for PI-88 is isolated after hydrolysis of yeast mannan and contains a mix of di- to hexasaccharides. The heterogeneity is exacerbated by the subsequent chemical sulphation. PI-88 is heterogeneous both in the size of its core saccharides and in its sulphation patterns (Cochran et al. 2003). It is currently in phase III clinical trials, sponsored by Progen Industries Ltd. Presumably, the ease of obtaining quantities of starting material suitable for large-scale production at a commercially viable price contributed to the selection of PI-88 for clinical development. Currently, very few other oligosaccharides offer this feature. Notably, several families of oligosaccharides were examined prior to selecting PI-88 (Parish and Cowden 2000).

## 2.3 *Oligosaccharide-Aglycone Conjugates*

Frequently biological activity is proportional to oligosaccharide size, but the lack of ready access to larger oligosaccharides as starting materials may have inspired several groups to form larger *pseudo*-oligosaccharides by conjugating together two or more smaller oligosaccharides. The conjugation is achieved by non-glycosidic linkages as this greatly simplifies the chemistry. An early success was Aprosulate, a conjugate of lactobionic acid, which was submitted for Phase II clinical trials as an anticoagulant. Unfortunately, Aprosulate was associated with liver toxicity and development was discontinued. As yet, no other conjugate has been developed to an equivalent stage. Variations on this theme include the extension to higher order dendrimers to take advantage of avidity effects. The selectin inhibitor, GMI-1070, is an example of a heterobifunctional conjugate of two separate pharmacophores, an oligosaccharide and a sulfonated aglycone.

### 2.4 Non-carbohydrate-Based Sulphated Mimetics

The quintessential example is suramin, a hexasulphated naphthalene derivative. Many other sulphated and sulphonated (here, we adopt the convention whereby a sulphonate is characterized by a carbon-sulfur bond as opposed to sulphates, which have a carbon-oxygen-sulphur bond system) polymers have also been examined as heparin mimetics. Aptamers (RNA or DNA) have been designed to bind to the heparin-binding sites of proteins and one, Macugen, has gained marketing approval. Two napthalene sulfonate compounds show promise in clinical trials. The first, VivaGel (SPL 7013), a vaginal microbicide created by StarPharma is a dendrimer with 32 pendant naphthalene disulfonate groups (i.e. 64 sulfonates). The second (PRO2000) is also a vaginal microbicide and was examined in a recently completed Phase III study of HIV infection.

### 2.5 Anionic Groups Other than Sulphates

The observation that a phosphorylated analogue of the AT activating pentasaccharide (Boeckel and Petitou 1993) had diminished activity raised the possibility that not all anionic groups are created equal. Even more intriguing was the discovery that the reverse situation sometimes occurred, for example synthetic heparin-like antithrombotics with perphosphorylated thrombin-binding domains had greater antithrombin activity compared to structurally related conjugates with persulphated thrombin-binding domains (Basten et al. 1998; Buijsman et al. 1999). Thus, it seems that phosphate substitution for sulphate groups (and vice versa) is another tool medicinal chemists can add to their armory; however, the merit of the substitution needs to be evaluated on a case-by-case basis. Polycarboxylates can also exhibit similar activities to the stronger acid sulphates and indeed polyacrylic acid can pontentiate AT activity (Monien and Desai 2005). In the RGTA family, the observed selectivity of biological activities is implicitly related to the mix of carboxylate and sulphate groups.

The type of anionic groups may also alter the in vivo metabolism of the mimetic. For example, not only are C-sulfonates inherently more chemically stable than their O-sulphate ester counterparts, but they also appear to be much more resistant to degradation in vivo. The toxicity of poly(vinyl sulfate) and poly(anethole sulfate) is ascribed to a lack of metabolism or clearance by filtration (Islam and Linhardt 2003).

## 3 Applications of Mimetics

The intended targets of heparin mimetics reflect the proteins that have also been investigated for heparin binding; they include the suite of mammalian heparin-binding proteins, viral proteins and infective bacteria. Many of these applications with heparin and GAGs are discussed in other chapters in this volume. In addition, comprehensive reviews of the activities of particular GAGs and heparin mimetics also exist (Hassan 2007; Urbinati et al. 2008; Gunay and Linhardt 1999).

Here, heparin mimetics will be examined in the context of the aims of drug development. Any drug in development is ultimately judged by regulatory authorities according to safety and efficacy. Since safety places an upper limit on dosing, perhaps this requirement can be restated, as "is the drug efficacious at a safe dosing level"? This brings us back to why heparin mimetics have been created, that is, to curtail the promiscuity and polypharmacy of heparin. For a particular mimetic, this might be achieved by increasing potency so that dosage can be decreased to the point that deleterious side effects are insignificant. Alternatively, undesired interactions can be eliminated if binding selectivity is increased, thereby decreasing the number of proteins with which the mimetic interacts. Most desirable is a combination of both approaches.

When assessing the activities of any drug lead, it is advisable to determine the potential interactions that might be problematic for a certain disease application, and to investigate these early in development. This is the essence of the "fail early" paradigm adopted by the pharmaceutical industry. The long history of clinical use of heparin and some heparin mimetics provides vital information as to where toxicity effects might be encountered (Alban 2011). Thus, if mimetics are being developed for a disease indication that does not involve the coagulation cascade, it is usual to screen for anticoagulant activity at an early stage. However, any detrimental activities should be put into the context of an effective therapeutic dose. For example, to synthesize a mimetic with 10% of the anticoagulant activity of heparin is encouraging, but if in clinical studies it is found intravenous dosages of 300 mg are required for efficacy (i.e. approximately ten times an anticoagulant dose of heparin), then there is potential for bleeding complications. Other toxicity issues such as heparin/mimetic-induced thrombocytopenia (occurs when the interaction of heparin/mimetic with platelet factor 4 exposes novel antigenic sites), urticaria, liver toxicity (which seemingly led to the abandonment of aprosulate) and activation of the contact system (which occurred as a result of oversulphated chondroitin sulphate contamination of heparin batches) would now be considered amongst the prime toxicity issues to be addressed early in the development of a mimetic.

A delivery route that largely confines the drug to the target organ is a way of decreasing possible systemic toxicity. Large doses of the orally delivered mimetics sucrose octasulphate and pentosan are tolerated, with the recommended daily dose of pentosan for treatment of interstitial cystitis being 300 mg. Only a small proportion of patients suffer rectal bleeding at this dose (Parsons 1997). The aptamer Macugen®, designed to slow vision loss in patients with wet, age-related, macular degeneration, is given as an ophthalmic injection and so is contained. The promising treatments of the naphthalene sulphonate vaginal microbicides VivaGel®, PRO2000® and even UsherCell® (cellulose sulphate, now abandoned for preventing HIV infection) show a good safety profile with little systemic absorption. There is also a long history of topical heparinoids and Hemoclar® (pentosan) (Vecchio and Frisinghelli 2008). Numerous clinical trials involving inhaled heparins and heparin derivatives for treating respiratory disorders such as asthma and chronic obstructive pulmonary disease (COPD) point to the fact that this is a safe route of administration for preparations (e.g. heparin) that have well known side effects when similar doses are administered parenterally (Rose and Page 2004). There are several companies investigating heparin mimetics for these disease indications (ParanGenix, Ivax and Glycan Biosciences), which is a striking concentration in one area given the number of companies developing heparin mimetics overall. Finally, heparin mimetics, such as pentosan and novel mimetics (Larramendy-Gozalo et al. 2007) are also being investigated as treatment of prion diseases, with intraventricular delivery to bypass the blood–brain barrier.

## 4 The Road to Clinical Development: Three Case Studies

### *4.1 PI-88*

PI-88, now renamed Muparfostat, arose out of a discovery program searching for a heparanase inhibitor performed as a collaboration between researchers at Australian National University and Progen Industries Ltd. (now Progen Pharmaceuticals Ltd.). PI-88 is in phase II and III clinical trials for liver cancer and metastatic melanoma, and other indications are being explored. It is a mixture of predominantly pentamanno-(60%) and tetramanno-oligosaccharides (30%) (Fig. 2). Despite this small size, it displays a polypharmacy evident by high-affinity interactions with several angiogenesis-stimulating growth factors (Cochran et al. 2003), the Slit-2 protein that regulates axonal regeneration following injury (Lau and Margolis 2009) as well as heparanase. The affinity of PI-88 interactions with certain growth factors exceeds the affinity of those growth factors with heparin or heparan sulphate, for example its affinity for vascular endothelial growth factor (VEGF) is sub-nanomolar. In addition, PI-88 targets the mucin-like regions of glycoprotein C and

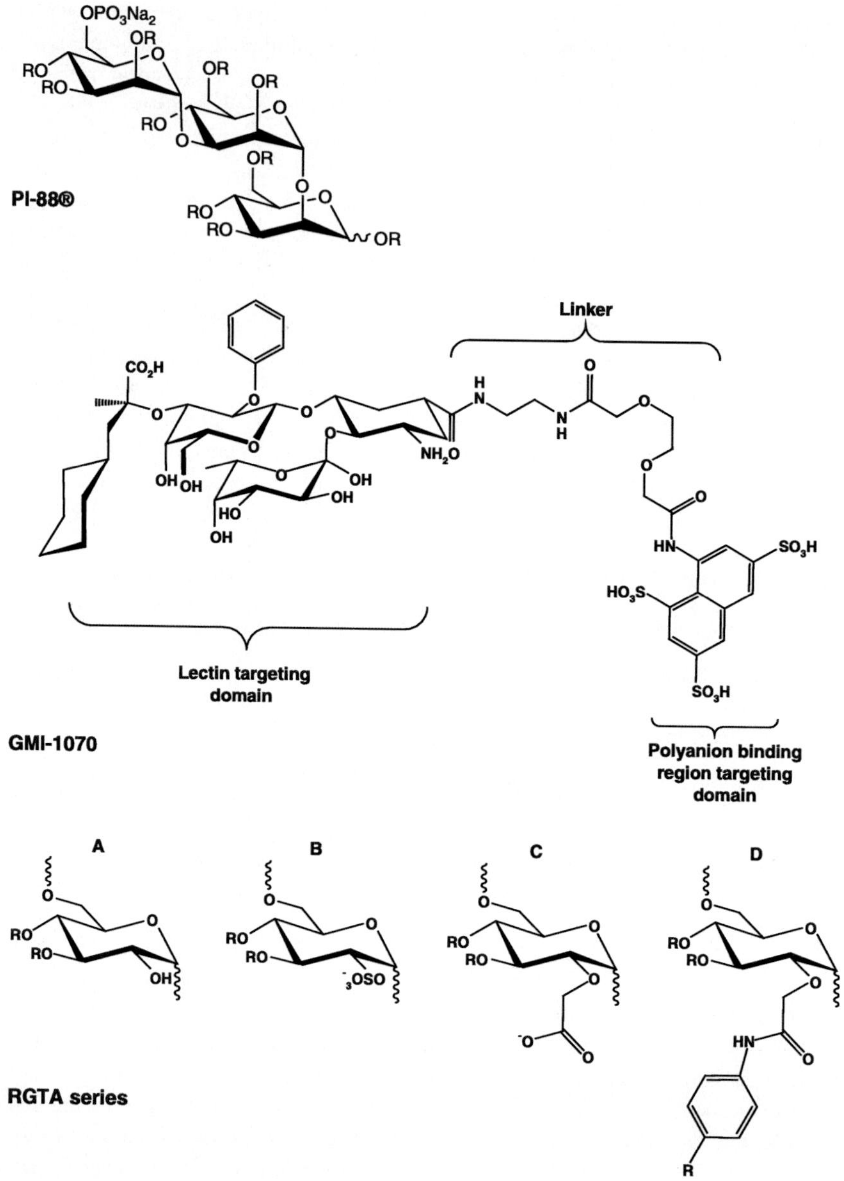


**Fig. 2** Schematic representations of PI-88 (*upper*), GMI-1070 (*middle*) and RGTA. For PI-88, R represents either H or a sulphate group. RGTA is a random distribution of the four monomeric units shown, which are derivatized dextran monomer units and R represents either H, a sulphate group, a carboxymethylgroup or a carboxylmethylamide group. For OTR4120, the proportions of A, B, C and D are A < 1%, B = 67%, C = 32% and D = 0% (i.e. it is a mixture of carboxymethyl and sulphate substitutions only)

glycoprotein G of herpes simplex viruses (HSV) type 1 and type 2, respectively to inhibit HSV infection of cells and cell-to-cell spread (Adamiak et al. 2007), it also inhibits cell infection by flaviviruses, most likely by binding to flavivirus protein E on the virus surface (Lee et al. 2006) and has anti-malarial activity, inhibiting the invasion of *Plasmodium falciparum* merozoites into erythrocytes, a critical part of the parasites life-cycle (Adams et al. 2006). Heparin and other sulphated polysaccharides have similarly been shown experimentally to inhibit cell infection by these viruses and the malaria parasite. PI-88 does have anticoagulant properties (Khachigian and Parish 2004) arising from heparin cofactor II activation and the stimulation of tissue factor pathway inhibitor (TFPI) release from endothelial cell surfaces (Demir et al. 2001). PI-88 is generally well tolerated but the most common toxicity issues are thrombocytopenia and thrombosis, injection site haemorrhage and other bleeding events. Nevertheless, in separate clinical trials of hepatocellular carcinoma and prostate cancer doses of 160 mg/day (Liu et al. 2009; Khasraw et al. 2009) and up to 250 mg/day of PI-88 were tolerated (Chow et al. 2008). These dosage levels are 16-fold greater than recommended therapeutic doses of Arixtra® and at least fivefold higher than recommended doses for heparin (Hirsh and Raschke 2004). Progen Pharmaceuticals Ltd. have formulated second-generation versions of PI-88 with improved selectivity (markedly less in vitro anticoagulant activity than PI-88) and activity. These new mimetics are based on anomerically pure, fully sulphated oligosaccharide glycosides modified by a lipophilic moiety at the reducing terminus (Dredge et al. 2009).

## 4.2 RGTAs

Precursors of RGTAs were first synthesized in the 1980s and were called carboxymethyl benzylamide sulfonate dextrans (CMDBS). The native dextran polymer was modified by first carboxymethlyation of hydroxyl groups on D-glycosyl units, benzylamidation of the carboxylic groups, then sulphonation of phenyl rings, with sulphates also appearing on free hydroxyl groups during this step (Fig. 2). Many derivatives were synthesized, purified and characterized for reproducibility. They were found to have a range of biological activities, many mimicking those of heparin and the levels of the various activities varied with the composition of the polymer. Their ability to stimulate wound healing in various in vivo models led to the suggestion that these compounds could offer a new approach to tissue repair (Logeart-Avramoglou and Jozefonvicz 1999). Numerous publications describing the activities of the different functionalized dextrans, now renamed RGTAs (for ReGeneraTing Agents) followed. RGTA11, which acts synergistically with FGF-2 in various wound healing and muscle regeneration models, was studied extensively. Interactions with different growth factors could be tailored according to the RGTA structure, for example RG-1192, which contains benzylamide groups binds FGF-2 with nM affinity equating that of heparin, but its activity on Type III collagen biosynthesis in the presence of FGF-2 did not mirror that of heparin. In contrast,

RG-1503, which has no benzylamide groups interacts with TGF-β and not with FGF-2 (Alexakis et al. 2004). Second-generation RGTAs were made by a synthesis procedure using 2-methyl-2-butene as a neutral acid scavenger to produce a non-destructive sulphation protocol. OTR4120, the active ingredient in Cacipliq20®, a marketed topical product for wound healing, arose from this procedure. OTR4120 mimics heparin in its global content of carboxylate and sulphate groups and the seven mainly represented residues are believed to be arranged in a 15 sugar unit sequence statistically repeated along the molecule (Papy-Garcia et al. 2005). This compound has multiple activities; it binds VEGF and modulates angiogenesis, it binds the chemokines stromal-derived factor-1 (SDF-1) and RANTES/CCL5, it is an anticoagulant and it assists healing of chronic wounds, burned skin and full-thickness excisional wounds through mechanisms that are not well defined but seem to involve regulation of collagen synthesis, activation of matrix metalloproteinases (MMP)-2 and -9 and VEGF activities (Liu et al. 2009). The anticoagulant activity of OTR4120 is 1/10 that of heparin and is mediated through direct thrombin inhibition, which it binds with nM affinity (Charef et al. 2007).

## 4.3 An Inhibitor of Selectins: GMI-1070

The selectins (E-, P- and L-selectin) play a key role in the early stages of an inflammatory response. Their function of initiating the tethering and rolling of leucocytes on endothelial cells lining blood vessels is a prerequisite for leucocyte adhesion and passaging across the vessel wall into adjacent inflamed tissue. Antagonists of selectin activities could be valuable therapeutics for diseases where extravasation of leucocytes is a key part of the pathology; examples include the cell infiltration associated with reperfusion injuries, various inflammatory diseases (e.g. rheumatoid arthritis, inflammatory bowel disease, asthma) and certain cancers. The discovery of the selectin inhibitor GMI-1070 illustrates how processes of rational drug design led to the building of a heterobifunctional conjugate. A range of GAGs and heparin mimetics are known to inhibit the selectins (Magnani 2004). However, Magnani and coworkers believed these structures lacked the necessary specificity to be effective selectin antagonists in vivo, and they focused upon the structure of the carbohydrate recognition domain of the selectins to engineer the desired specificity. Sequential refinement of structure–function studies guided by extensive characterization and modelling of the natural ligands sialyl Lewis −A and −X (sialyl Le$^a$ and sialyl Le$^x$, respectively) led to the glycan being substituted with 4 critical aglycones. A further key element is the orientation of the carboxylate group. This compound possessed three orders of magnitude higher activity than sialyl Le$^x$. Magnani and coworkers then sought to exploit the polyanion binding domain (Fig. 2), to further improve selectivity and potency. Coupling of a naphthalene trisulphonic acid through a small flexible linker led to GMI-1070 (Fig. 2) an inhibitor of all three selectins, which has now entered phase I clinical trials (Ernst and Magnani 2009). GMI-1070 is presently being developed as a treatment for multiple myeloma and vaso-occlusive crisis associated with sickle cell disease. This represents the culmination of 15 years of development from when the first patent was filed (Magnani et al. 2000), and several more years are required before marketing approval could be expected.

## 5 Concluding Remarks

Heparin mimetics occupy an under exploited space in drug discovery, they are intermediate in size between small molecules and biological drugs like antibodies. They have the advantage of being very stable and are inherently water soluble. They are ideal for blocking the binding of growth factors to their receptors because they usually occupy extended binding sites on their target proteins. It is now being recognized that absolute specificity to one protein as occurs with monoclonal antibody drugs is not always the most effective way to treat complex diseases. One part of the disease pathway may be blocked but often the molecular redundancy of the pathology is such that symptoms are not alleviated. Of the three heparin mimetic drug examples we presented in detail two structurally heterogeneous mimetics displayed multiple activities, whilst the third arose from sequential structure–function studies, is homogeneous and probably has very restricted activities. We believe that the design of future heparin mimetics will favour structural homogeneity over the heterogeneous products of the past. Homogeneous heparin mimetics designed so that desirable activities are maximized and delivered in a way that largely restricts the drug to the site of disease may be more effective than monoclonal antibody drugs in treating complex diseases and at a more favourable cost. It will be fascinating to watch the developments in this field over the next few years as new synthetic approaches (e.g. conjugates with aglycones) produce heparin mimetics that match the criteria required for a drug.

## References

Adamiak B, Ekblad M, Bergström T, Ferro V, Trybala E (2007) Herpes simplex virus type 2 glycoprotein G is targeted by the sulfated oligo- and polysaccharide inhibitors of virus attachment to cells. J Virol 81:13424–13434

Adams Y, Freeman C, Schwartz-Albiez R, Ferro V, Parish CR, Andrews KT (2006) Inhibition of *Plasmodium falciparum* growth in vitro and adhesion to chondroitin-4-sulfate by the heparan sulfate mimetic PI-88 and other sulfated oligosaccharides . Antimicrob Agents Chemother 50:2850–2852

Alban S (2011) Adverse effects of heparin. In: Lever R, Mulloy B, Page CP (eds) Heparin – a century of progress. Springer, Heidelberg

Alexakis C, Mestries P, Garcia S, Petit E, Barbier V, Papy-Garcia D, Sagot M-A, Barritault D, Caruelle JP, Kern P (2004) Structurally different RGTAs modulate collagen-type expression by cultured aortic smooth muscle cells via different pathways involving fibroblast growth factor-2 or transforming growth factor-β1 . FASEB J 18:1147–1159

Barritault D, Caruelle J-P (2004) Biocompatible polymers, process for their preparation and compositions containing them, Patent No. US 6689741

Barritault DP, Caruelle J-P (2009) Biocompatible polymers, process for their preparation and compositions containing them, Patent No. US 2009/0012040 A1

Basten JE, Dreef-Tromp CM, de Wijs B, van Boeckel CA (1998) In vitro evaluation of synthetic heparin-like conjugates comprising different thrombin binding domains. Bioorg Med Chem Lett 8:1201–1206

Baumann H, Richter A, Klemm D, Faust V (2000) Concepts for preparation of novel regioselective modified cellulose derivatives sulfated, aminated, carboxylated and acetylated for hemocompatible ultrathin coatings on biomaterials. Macromol Chem Phys 201:1950–1962

van Boeckel CAA, Petitou M (1993) The unique antithrombin III binding domain of heparin: a lead to new synthetic antithrombotics. Angew Chem Int Ed Engl 32:1671–1690

Bugatti A, Urbinati C, Ravelli C, De Clercq E, Liekens S, Rusnati M (2007) Heparin-mimicking sulfonic acid polymers as multitarget inhibitors of human immunodeficiency virus type 1 Tat and gp120 proteins. Antimicrob Agents Chemother 51:2337–2345

Buijsman RC, Basten JE, Dreef-Tromp CM, van der Marel GA, van Boeckel CA, van Boom JH (1999) Synthesis of heparin-like antithrombotics having perphosphorylated thrombin binding domains. Bioorg Med Chem 7:1881–1890

Burgermeister J, Paper DH, Vogl H, Linhardt RJ, Franz G (2002) LaPSvS1, a (1– > 3)-beta-galactan sulfate and its effect on angiogenesis in vivo and in vitro. Carbohydr Res 337: 1459–1466

Casu B, Guerrini M, Guglieri S, Naggi A, Perez M, Torri G, Cassinelli G, Ribatti D, Carminati P, Giannini G, Penco S, Pisano C, Belleri M, Rusnati M, Presta M (2004) Undersulfated and glycol-split heparins endowed with antiangiogenic activity. J Med Chem 47:838–848

Casu B, Naggi A, Torri G (2001) Glycosaminoglycans having high antithrombotic activity, Patent No. US 6197943

Casu B, Vlodavsky I, Sanderson RD (2008) Non-anticoagulant heparins and inhibition of cancer . Pathophysiol Haemost Thromb 36:195–203

Casu BM, Torri GM, Naggi AMM, Giannini GP, Pisano CP, Penco SM (2005) Derivatives of partially desulphated glycosaminoglycans endowed with antiangiogenic activity and devoid of anticoagulating effect, Patent No. US 2005/0222084 A1

Chaikof EL, Grande D, Baskaran S (2007) Glycopolymers and free radical polymerization methods, Patent No. US 7244830

Charef S, Petit E, Barritault D, Courty J, Caruelle JP (2007) Effects on coagulation of a synthetic heparin mimetic given intraperitoneally or orally. J Biomed Mater Res 83A:1024–1031

Chow LQM, Gustafson DL, O'Bryant CL, Gore L, Basche M, Holden SN, Morrow MC, Grolnic S, Creese BR, Roberts KL, Davis K, Addison R, Eckhardt SG (2008) A phase I pharmacological and biological study of PI-88 and docetaxel in patients with advanced malignancies . Cancer Chemother Pharmacol 63:65–74

Christensen ND, Reed CA, Culp TD, Hermonat PL, Howett MK, Anderson RA, Zaneveld LJ (2001) Papillomavirus microbicidal activities of high-molecular-weight cellulose sulfate, dextran sulfate, and polystyrene sulfonate. Antimicrob Agents Chemother 45:3427–3432

Chucholowski A, Fingerle J, Iberg N, Marki HP, Muller R, Pech M, Rouge M, Schmid G, Tschopp T, Wessel HP (1996) Sulfuric acid esters of sugar alcohols, Patent No. US5521160

Cipriani TR, Gracher AH, de Souza LM, Fonseca RJ, Belmiro CL, Gorin PA, Sassaki GL, Iacomini M (2009) Influence of molecular weight of chemically sulfated citrus pectin fractions on their antithrombotic and bleeding effects. Thromb Haemost 101:860–866

Cochran S, Li C, Fairweather JK, Kett WC, Coombe DR, Ferro V (2003) Probing the interactions of phosphosulfomannans with angiogenic growth factors by surface plasmon resonance. J Med Chem 46:4601–4608

Crandall IE, Szarek WA, Vlahakis JZ, Xu Y, Vohra R, Sui J, Kisilevsky R (2007) Sulfated cyclodextrins inhibit the entry of Plasmodium into red blood cells. Implications for malarial therapy. Biochem Pharmacol 73:632–642

De Clercq E (2009a) Antiviral drug discovery: ten more compounds, and ten more stories (part B). Med Res Rev 29:571–610

De Clercq E (2009b) The next ten stories on antiviral drug discovery (part E): advents, advances, and adventures. Med Res Rev. doi:10.1002/med.20179

Demir M, Iqbal O, Hoppensteadt DA, Piccolo P, Ahmad S, Schultz CL, Linhardt RJ, Fareed J (2001) Anticoagulant and antiprotease profiles of a novel natural heparinomimetic mannopentaose phosphate sulfate (PI-88). Clin Appl Thromb Hemost 7:131–140

Dredge K, Hammond E, Davis K, Li C, Liu L, Johnstone K, Handley P, Wimmer N, Gonda T, Gautam A, Ferro V, Bytheway I (2009) The PG500 series: novel heparan sulfate mimetics as potent angiogenesis and heparanase inhibitors for cancer therapy. Invest New Drugs. doi:10.1007/s10637-009-9245-5

Duong M, Cockcroft D, Boulet LP, Ahmed T, Iverson H, Atkinson DC, Stahl EG, Watson R, Davis B, Milot J, Gauvreau GM, O'Byrne PM (2008) The effect of IVX-0142, a heparin-derived hypersulfated disaccharide, on the allergic airway responses in asthma. Allergy 63: 1195–1201

Ernst B, Magnani JL (2009) From carbohydrate leads to glycomimetic drugs. Nat Rev Drug Discov 8:661–677

Fernandez C, Hattan CM, Kerns RJ (2006) Semi-synthetic heparin derivatives: chemical modifications of heparin beyond chain length, sulfate substitution pattern and N-sulfo/N-acetyl groups. Carbohydr Res 341:1253–1265

Freeman C, Liu L, Banwell MG, Brown KJ, Bezos A, Ferro V, Parish CR (2005) Use of sulfated linked cyclitols as heparan sulfate mimetics to probe the heparin/heparan sulfate binding specificity of proteins. J Biol Chem 280:8842–8849

Freeman I, Kedem A, Cohen S (2008) The effect of sulfation of alginate hydrogels on the specific binding and controlled release of heparin-binding proteins. Biomaterials 29:3260–3268

Fritzsche J, Alban S, Ludwig RJ, Rubant S, Boehncke WH, Schumacher G, Bendas G (2006) The influence of various structural parameters of semisynthetic sulfated polysaccharides on the P-selectin inhibitory capacity. Biochem Pharmacol 72:474–485

Fugedi P, Peto C (1998) Saccharopeptides and derivatives thereof, Patent No. US 5849709

Fugedi P, Tyrrell DJ, Tressler RJ, Stack RJ, Ishihara M (1998) Sulfated maltooligosaccharides with heparin-like properties, Patent No. US 5739115

Gandhi NS, Coombe DR, Mancera RL (2008) Platelet endothelial cell adhesion molecule-1 (PECAM-1) and its interactions with glycosaminoglycans I. Molecular modelling studies. Biochemistry 47:4851–4862

Guerrini M, Guglieri S, Beccati D, Torri G, Viskov C, Mourier P (2006) Conformational transitions induced in heparin octasaccharides by binding with antithrombin III. Biochem J 399:191–198

Gunay NS, Linhardt RJ (1999) Heparinoids: structure, biological activities and therapeutic applications. Planta Med 65:301–306

Hara ST, Yoshida KO, Ishihara MI (2007) Use of desulfated heparin, Patent No. EP 1371666/B1

Hassan HH (2007) Chemistry and biology of heparin mimetics that bind to fibroblast growth factors. Mini Rev Med Chem 7:1206–1235

Henry BL, Connell J, Liang A, Krishnasamy C, Desai UR (2009) Interaction of antithrombin with sulfated, low molecular weight lignins: opportunities for potent, selective modulation of antithrombin function. J Biol Chem 284:20897–20908

Hirsh J, Raschke R (2004) Heparin and low-molecular-weight heparin: the Seventh ACCP Conference on Antithrombotic and Thrombolytic Therapy. Chest 126:188S–203S

Hoarau DQ, Boustta MP, Braud CCLL, Vert MCLL (2007) Amphiphilic heparin derivative formed by coupling a heparin with a bile acid, Patent No. US 2007/0141158 A1

Hoffman R, Paper DH, Donaldson J, Alban S, Franz G (1995) Characterisation of a laminarin sulphate which inhibits basic fibroblast growth factor binding and endothelial cell proliferation. J Cell Sci 108(Pt 11):3591–3598

Hou S, Sun XL, Dong CM, Chaikof EL (2004) Facile synthesis of chain-end functionalized glycopolymers for site-specific bioconjugation. Bioconjug Chem 15:954–959
Islam T, Linhardt R (2003) Chemistry, biochemistry, and pharmaceutical potentials of glycosaminoglycans and related saccharides. In: Wong C (ed) Carbohydrate-based drug discovery. Wiley, Weinheim, pp 407–427
Jayakumar R, Nwe N, Tokura S, Tamura H (2007) Sulfated chitin and chitosan as novel biomaterials. Int J Biol Macromol 40:175–181
Jayakumar R, Selvamurugan N, Nair SV, Tokura S, Tamura H (2008) Preparative methods of phosphorylated chitin and chitosan – an overview. Int J Biol Macromol 43:221–225
Kasbauer CW, Paper DH, Franz G (2001) Sulfated beta-(1– > 4)-galacto-oligosaccharides and their effect on angiogenesis. Carbohydr Res 330:427–430
Katsuraya K, Nakashima H, Yamamoto N, Uryu T (1999) Synthesis of sulfated oligosaccharide glycosides having high anti-HIV activity and the relationship between activity and chemical structure. Carbohydr Res 315:234–242
Kennedy T (1999) Methods of treating asthma with o-desulfated heparin, Patent No. US 5990097
Kennedy T (2009) Methods of treating acute exacerbations of chronic obstructive pulmonary disease, Patent No. US 2009/0054374 A1
Kensinger RD, Yowler BC, Benesi AJ, Schengrund CL (2004) Synthesis of novel, multivalent glycodendrimers as ligands for HIV-1 gp120. Bioconjug Chem 15:349–358
Khachigian LM, Parish CR (2004) Phosphomannopentaose sulfate (PI-88): heparan sulfate mimetic with clinical potential in multiple vascular pathologies. Cardiovasc Drug Rev 22:1–6
Khasraw M, Pavlakis N, McCowatt S, Underhill C, Begbie S, de Souza P, Boyce A, Parnis F, Lim V, Harvie R, Marx G (2009) Multicentre phase I/II study of PI-88, a heparanase inhibitor in combination with docetaxel in patients with metastatic castrate-resistant prostate cancer. Ann Oncol. doi:10.1093/annonc/mdp524
Kim SK, Lee DY, Kim CY, Nam JH, Moon HT, Byun Y (2007) A newly developed oral heparin derivative for deep vein thrombosis: non-human primate study. J Control Release 123:155–163
Kong FHD, Ning JHD, Gu JS (2008) Type of oligosaccharides and their sulfate derivatives, Patent No. US 7365191
Kumar MN, Muzzarelli RA, Muzzarelli C, Sashiwa H, Domb AJ (2004) Chitosan chemistry and pharmaceutical perspectives. Chem Rev 104:6017–6084
Larramendy-Gozalo C, Barret A, Daudigeos E, Mathieu E, Antonangeli L, Riffet C, Petit E, Papy-Garcia D, Barritault D, Brown P, Deslys JP (2007) Comparison of CR36, a new heparan mimetic, and pentosan polysulfate in the treatment of prion diseases. J Gen Virol 88: 1062–1067
Lau E, Margolis RU (2009) Inhibitors of slit protein interactions with the heparan sulfate proteoglycan glypican-1: potential agents for the treatment of spinal cord injury. Clin Exp Pharmacol Physiol. doi:10.1111/j.1440-1681.2009.05318.x
Lee E, Pavy M, Young N, Freeman C, Lobigs M (2006) Antiviral effect of the heparan sulfate mimetic, PI-88, against dengue and encephalitic flaviviruses. Antiviral Res 69:31–38
Lee JH, Canny MD, De Erkenez A, Krilleke D, Ng YS, Shima DT, Pardi A, Jucker F (2005) A therapeutic aptamer inhibits angiogenesis by specifically targeting thc heparin binding domain of VEGF165. Proc Natl Acad Sci USA 102:18902–18907
Liu C-J, Lee P-H, Lin D-Y, Wu C-C, Jeng L-B, Lin P-W, Mok K-T, Lee W-C, Yeh H-Z, Ho M-C, Yang S-S, Lee C-C, Yu M-C, Hu R-H, Peng C-Y, Lai K-L, Chang SS-C, Chen P-J (2009) Heparanase inhibitor PI-88 as adjuvant therapy for hepatocellular carcinoma after curative resection: a randomized phase II trial for safety and optimal dosage. J Hepatol 50:958–968
Livaoglu M, Kerimoglu S, Sonmez B, Livaoglu A, Karacal N (2009) The effect of Hirudoid on random skin-flap survival in rats. J Plast Reconstr Aesthet Surg. doi:10.1016/j.bjps.2009.03.018
Logeart-Avramoglou D, Jozefonvicz J (1999) Carboxymethyl benzylamide sulfonate dextrans (CMDBS), a family of biospecific polymers endowed with numerous biological properties: a review. J Biomed Mater Res 48:578–590

MacLaren R, Stringer KA (2007) Emerging role of anticoagulants and fibrinolytics in the treatment of acute respiratory syndrome. Pharmacotherapy 27:860–873
Magnani JL (2004) The discovery, biology, and drug development of sialyl Lea and sialyl Lex. Arch Biochem Biophys 426:122–131
Magnani JL, Woodlark R, Butcher EC, Berg EL (2000) Methods for the inhibition of cancer metastasis mediated by endothelial adhesion molecules, Patent No. US 6121233
Magnani JLTCRGMD (2009) Pan-selectin inhibitor with enhanced pharmacokinetic activity, Patent No. WO 2009/126556 A1
Martin DJ, Toce JA, Anevski PJ, Tollefsen DM, Abendschein DR (1999) Anticoagulant and antithrombotic activity of maltodapoh, a novel sulfated tetrasaccharide. J Pharmacol Exp Ther 288:516–521
Maynard HD, Hubbell JA (2005) Discovery of a sulfated tetrapeptide that binds to vascular endothelial growth factor . Acta Biomater 1:451–459
Monien BH, Desai UR (2005) Antithrombin activation by nonsulfated, non-polysaccharide organic polymer. J Med Chem 48:1269–1273
Mousa SA, Feng X, Xie J, Du Y, Hua Y, He H, O'Connor L, Linhardt RJ (2006) Synthetic oligosaccharide stimulates and stabilizes angiogenesis: structure-function relationships and potential mechanisms. J Cardiovasc Pharmacol 48:6–13
Myskiw C, Deschambault Y, Jefferies K, He R, Cao J (2007) Aurintricarboxylic acid inhibits the early stage of vaccinia virus replication by targeting both cellular and viral factors. J Virol 81:3027–3032
Nair VG, Joseph JP, Bernstein S (1977) Complement inhibitors, Patent No. US 4021544
Nair VG, Joseph JP, Bernstein S (1978) Polygalactosido-sucrose Poly(H-)sulfate salts, Patent No. US 4098995
Nimjee SM, Oney S, Volovyk Z, Bompiani KM, Long SB, Hoffman M, Sullenger BA (2009) Synergistic effect of aptamers that inhibit exosites 1 and 2 on thrombin. RNA 15:2105–2111
Nunn A, McCormack S, Crook AM, Pool R, Rutterford C, Hayes R (2009) Microbicides Development Programme: design of a phase III trial to measure the efficacy of the vaginal microbicide PRO 2000/5 for HIV prevention. Trials 10:99
O'Loughlin J, Millwood IY, McDonald HM, Price CF, Kaldor JM, Paull JR (2009) Safety, tolerability, and pharmacokinetics of SPL7013 Gel (VivaGel(R)): a dose ranging, phase I study. Sex Transm Dis. doi:10.1097/OLQ.0b013e3181bc0aac
Oreste PM, Zoppetti GM (2005) Anticoagulant and antithrombotic LMW-glycosaminoglycans derived from K5 polysaccharide and process for their preparation, Patent No. US 2005/0027117 A1
Pahi V, Ferrer J, Verbruggen A, Milano V, Roura R, Verson A (2005) New therapeutic use for a group of sulphated polysaccharides, Patent No. WO 2005/084610 A3R4
Papoulias UE, Wyld PJ, Haas S, Stemberger A, Jeske W, Hoppensteadt D, Kammereit A (1993) Phase I – study with aprosulate, a new synthetic anticoagulant . Thromb Res 72:99–108
Papy-Garcia D, Barbier-Chassefière V, Rouet V, Kerros M-E, Klochendler C, Tournaire M-C, Barritault D, Caruelle JP, Petit E (2005) Nondegradative sulfation of polysaccharides. Synthesis and structure characterization of biologically active heparan sulfate mimetics. Macromolecules 38:4647–4654
Parish CR, Cowden WB (2000) Preparation and use of sulfated oligosaccharides, Patent No. US 6143730
Parsons CL (1997) Epithelial coating techniques in the treatment of interstitial cystitis. Urology 49:100–104
Petitou M, Nancy-Portebois V, Dubreucq G, Motte V, Meuleman D, de Kort M, van Boeckel CA, Vogel GM, Wisse JA (2009) From heparin to EP217609: the long way to a new pentasaccharide-based neutralisable anticoagulant with an unprecedented pharmacological profile. Thromb Haemost 102:804–810
Ponedel'kina I, Odinokov V, Lukina E, Tyumkina T, Khalilov L, Dzhemilev U (2006) Chemical modification of heparin. Russ J Bioorg Chem 32:472–477

Presta M, Oreste P, Zoppetti G, Belleri M, Tanghetti E, Leali D, Urbinati C, Bugatti A, Ronca R, Nicoli S, Moroni E, Stabile H, Camozzi M, Hernandez GA, Mitola S, Dell'Era P, Rusnati M, Ribatti D (2005) Antiangiogenic activity of semisynthetic biotechnological heparins: low-molecular-weight-sulfated Escherichia coli K5 polysaccharide derivatives as fibroblast growth factor antagonists. Arterioscler Thromb Vasc Biol 25:71–76

Raake W, Klauser RJ, Elling H, Zeiller P, Meinetsberger E (1994) Derivatives of bis-aldonic acid amides and their anticoagulant and antithrombotic properties. Semin Thromb Hemost 20: 176–181

Raghuraman A, Riaz M, Hindle M, Desai UR (2007) Rapid and efficient microwave-assisted synthesis of highly sulfated organic scaffolds. Tetrahedron Lett 48:6754–6758

Raman R, Venkataraman G, Ernst S, Sasisekharan V, Sasisekharan R (2003) Structural specificity of heparin binding in the fibroblast growth factor family of proteins. Proc Natl Acad Sci USA 100:2357–2362

Rele SM, Iyer SS, Baskaran S, Chaikof EL (2004) Design and synthesis of dimeric heparinoid mimetics. J Org Chem 69:9159–9170

Renner C, Piehler J, Schrader T (2006) Arginine- and lysine-specific polymers for protein recognition and immobilization. J Am Chem Soc 128:620–628

Rose MJ, Page C (2004) Glycosaminoglycans and the regulation of allergic inflammation. Curr Drug Targets Inflamm Allergy 3:221–225

Rouet V, Meddahi-Pelle A, Miao HQ, Vlodavsky I, Caruelle JP, Barritault D (2006) Heparin-like synthetic polymers, named RGTAs, mimic biological effects of heparin in vitro. J Biomed Mater Res A 78:792–797

Shoji T, Ikushima N, Kaname Katsuraya S, Nahoko Takahashi C, Fusayo Kobayashi S, Toshiyuki Uryu T, Takashi Yoshida T, Naoki Yamamoto T, Hideki Nakashima T (1994) Sulfated tocopheryl oligosaccharides and antiviral agents including the same as active ingredients, Patent No. US 5280111

Smelcerovic A, Knezevic-Jugovic Z, Petronijevic Z (2008) Microbial polysaccharides and their derivatives as current and prospective pharmaceuticals. Curr Pharm Des 14:3168–3195

Suzuki A, Toyoda H, Toida T, Imanari T (2001) Preparation and inhibitory activity on hyaluronidase of fully O-sulfated hyaluro-oligosaccharides. Glycobiology 11:57–64

Ueki M, Watanabe S, Ishii Y, Okunaka O, Uchino K, Saitoh T, Higashi K, Nakashima H, Yamamoto N, Ogawara H (2001) Synthesis and anti-HIV activity of nonatyrosine N- and O1-9-decasulfate. Bioorg Med Chem 9:477–486

Urbinati C, Chiodelli P, Rusnati M (2008) Polyanionic drugs and viral oncogenesis: a novel approach to control infection, tumor-associated inflammation and angiogenesis. Molecules 13:2758–2785

Usher TC, Patel N, Tele CG (1995) Process for preparing soluble alkali metal salts of cellulose sulfate using chlorosulfuric acid and pyridine, Patent No. US 5378828

Vecchio C, Frisinghelli A (2008) Topically applied heparins for the treatment of vascular disorders: a comprehensive review. Clin Drug Investig 28:603–614

Vogel G (2006) Pulmonary administration of an antithrombotic compound, Patent No. WO 2006/ 037771 A1

Wall D, Douglas S, Ferro V, Cowden W, Parish C (2001) Characterisation of the anticoagulant properties of a range of structurally diverse sulfated oligosaccharides. Thromb Res 103: 325–335

Wang X, Zhang L (2009) Physicochemical properties and antitumor activities for sulfated derivatives of lentinan. Carbohydr Res 344:2209–2216

Wessel H (1997) Heparinoid mimetics, vol 187, Topics in current chemistry. Springer, Heidelberg, pp 215–239

Wessel HP, Chucholowski A, Fingerle J, Iberg N, Märki HP, Müller R, Pech M, Pfister-Downar M, Rouge M, Schmid G, Tschopp T (2005) From glycosaminoglycans to heparinoid mimetics with antiproliferative activity. In: Yves C (ed) Carbohydrate mimics. Wiley, Weinheim, pp 417–431

Yang DW, Ohta Y, Yamaguchi S, Tsukada Y, Haraguchi Y, Hoshino H, Amagai H, Kobayashi I (1996) Sulfated colominic acid: an antiviral agent that inhibits the human immunodeficiency virus type 1 in vitro. Antiviral Res 31:95–104

Zhao H, Liu H, Chen Y, Xin X, Li J, Hou Y, Zhang Z, Zhang X, Xie C, Geng M, Ding J (2006) Oligomannurarate sulfate, a novel heparanase inhibitor simultaneously targeting basic fibroblast growth factor, combats tumor angiogenesis and metastasis. Cancer Res 66:8779–8787

# Hyaluronic Acid

## Perspectives in lung diseases

**Luigi Allegra, Sabrina Della Patrona, and Giuseppe Petrigni**

**Contents**



**Abstract** Hyaluronic acid (HA) is a non-sulphated glycosaminoglycan. It is a natural polymer characterised by a coiled linear chain in particularly well-hydrated configuration composed of repeating disaccaride units. In mammals, its molecular

L. Allegra (✉)
Università degli Studi, IRCCS Fondazione Ca' Granda, Ospedale Policlinico, Via Francesco Sforza 35, 20122 Milano, Italy
e-mail: luigi.allegra@unimi.it

R. Lever et al. (eds.), *Heparin - A Century of Progress*,
Handbook of Experimental Pharmacology 207,
DOI 10.1007/978-3-642-23056-1_17, 

weight can be extremely wide, ranging from 20 to 4,000 kDa. High molecular mass forms are provided with anti-inflammatory properties. A unique characteristic of HA is hydration (up to 6,000 molecules water/molecule of HA) with a major role in the regulation of fluid balance in the interstitium, a fundamental activity on the amorphous colloidal matrix gluing connective cell and fibers, and many other biological functions including lubrication, solute transport and microcirculatory exchange. HA has been widely used in the treatment of eye, ear, joint and skin disorders; in the last 15 years HA has been also proposed successfully in the treatment of a number of lung diseases in vitro, experimental animals and humans. In particular, inhaled HA at relatively high molecular weight has been proven to prevent bronchoconstiction induced in asthmatics by direct and indirect challenges such as inhalation of methacholine, inhalation of ultrasonically nebulised distilled water, muscular exercise. More recently, in patrients affected by chronic obstructive pulmonary diseases, we have demonstrated that repeated administrations of inhaled HA (daily, for 8 weaks) induce significant increase in bronchial patency as well as progressive lung deflation with decrease of residual volume. In conclusion there are elements that can let us state that is perhaps time to change the focus to connective tissue and extracellular matrix substances such as HA, in order to prevent and treat chronic lung diseases.

**Keywords** Hyalurnic acid • High molecular weight • Lung diseases

## 1 Background

### *1.1 Discovery and Nomenclature*

*Hyaluronic acid* (*HA*) was first described in eye vitreous humour (Meyer and Palmer 1934). In the last 40 years, HA has been widely used in the treatment of eye, ear, joint and skin disorders, leading to >6,000 papers. For medical use, HA has been obtained by extraction from bovine joints, cockscomb and human hair, although more recently (Goh 1998) HA has been produced by *Staphylococcus equi* spp. *Zooepidermicus* (biological synthesis).

*Hyaluronan*, strictly speaking, is used as a generic name for any salt of hyaluronic acid when the specific salt is not indicated (Balasz et al. 1986).

### *1.2 Structure*

HA is a natural polysaccharide polymer characterised by a linear chain, in a flexible coiled and particularly well-hydrated configuration (Cleland and Wang 1970), composed of repeating disaccharide units, each unit formed by glucuronic acid (GlcA) and N-acetyl-glucosamine (GlcNAc), linked in the repeating disaccharide [→4)-β-D-GlcA-(1→3)-β-D-GlcNAc-(1].

Its molecular weight (mw) can range from 20 to 4,000 kDa. HA exists in both a high molecular mass form (1–6 × $10^6$ Da) and a polydisperse lower molecular mass form (0.1–0.5 × $10^6$ Da), the latter predominating under inflammatory conditions (Poole and Dieppe 1994) and the former being associated with anti-inflammatory properties (Dahl 2008).

Polymerisation of HA is regulated by three *HA synthases*, $HAS_1$, $HAS_2$ and $HAS_3$ (Itano et al. 1999), through the joining of the glycosidic residues to the reducing chain extremity. HA is metabolised by *hyaluronidases* (HYAL), mainly by $HYAL_1$ and $HYAL_2$, present in most tissues, including lung (Dentener et al. 2005).

## 1.3 Content and Concentration

In mammals, the maximum concentration of the ubiquitous HA molecules (in particular, as intercellular space filler of the intercellular matrix) is found in eye vitreous humour, synovial liquid and loose connective tissue (Fraser et al. 1997), with a maximum content in skin, intestine and lung (Hallgren et al. 1985; Monzon et al. 2006). Quantitatively, skin, intestine and lung (together with a relatively high content in pleura and bronchial secretions) contain more than 50% of the total HA in the body. HA is drained by the lymphatic vessels and catabolised in the local lymph nodes (Fraser et al. 1988) and liver (Smedrson et al. 1990).

## 1.4 Glycosaminoglycans

HA molecules belong to the family of glycosaminoglycans, which also includes heparin, heparan, chondroitin sulphate A and C, dermatan sulphate and keratan sulphate.

HA is the exception in not being sulphated. All of these GAGs have been extensively investigated for possible therapeutic indications in several diseases.

## 1.5 Proteoglycans

HA (as well as other glycosaminoglycans) can interact in different tissues with several different HA-binding proteins, which have been called *hyaladerins* or *hyaldherins* (Toole 1990). These complex macromolecules vary in the different tissues (aggrecan, versican, hyaluronectin, neurocan, decorin, lumican, laminin, fibronectin, fibrinogen). One of the properties of HA is the influence on the extracellular matrix through the organisation of proteoglycans within such structures, to provide load-bearing functions.

Some of these hyaladherins are provided with the function of HA receptors and are located on the surface of different cells in different tissues. Among these surfaces receptors are the well-characterised family (with several isoforms) of *CD44 receptors* (Haynes et al 1991; Underhill 1992).

The effects of HA are mainly exerted through interactions with CD44 receptors, which is the main receptor type mediating HA signalling. In addition to CD44, a non-CD44 surface receptor called *RHAMM* (Receptor for the HA-Mediated Motility) (Yang et al. 1993) has also been described, which is of particular interest in the respiratory system due to its activity on ciliary beating in bronchi (Scuri et al. 2007).

Reports on a functional role of HA in chronic inflammatory diseases are sometimes contradictory, but several authors are in favour of the hypothesis that possible differences depend on low-molecular-weight fragments of HA being correlated with inflammatory events and high-molecular-weight HA molecules having anti-inflammatory properties (Poole and Dieppe 1994; Scuri and Abraham 2003; Vitanzo and Sennett 2006; Allegra et al. 2008; Dahl 2008).

# 2 Properties

## 2.1 *Hydration: A Major Property*

A *unique characteristic* of HA, which relates to its variable functions, is hydration (up to ~6,000 molecules $H_2O$/molecule of HA) (Bhattacharya et al. 1989; Turino 2003). *Verbatim* From Turino (2003): "*HA's major role is most likely the regulation of the fluid balance in the interstitium...In several structures the concentration of HA determines water content*".

Mature tissues and organs have less water concentration than in foetal or early life. Synthesis of HA is an early protective response to injury and connective cell activation. Tissue repair requires an environment with water and fibre scaffolds to rebuild structures and elimination of low-molecular-weight HA fragments from tissue compartment is enhanced by an increased interstitial water fluid (Bhattacharya et al. 1989; Turino 2003).

A remarkable characteristic of HA is its rapid and marked increase in content and concentration in response to different kinds of physical injuries to tissue (King et al. 1991; Goldberg et al. 1993) or cytokine attack due to endotoxin activity (Blackwood et al. 1983), thus stimulating and facilitating defensive or reparatory cell movements within the matrix.

### 2.2 *Biological Functions*

Apart from the unique capacity to link and retain a relevant number of water molecules in the interfibrillar spaces, contributing a fundamental part of the amorphous colloidal matrix gluing connective cells and fibres, HA has other diverse biological functions including migration and proliferation (Papakonstantinou et al. 1998), tissue morphogenesis (Laurent and Reed 1991), embryonic development, cell growth, differentiation and ovulation (Toole 1991), disease progression (Toole 2004), lubrication, solute transport and microcirculatory exchanges (due to its influence on interstitial volume, hydraulic conduction and macromolecular diffusion) (Reed and Laurent 1992).

## 3 Hyaluronic Acid in Lung Diseases: Novel Perspectives

### 3.1 *Glycosaminoglycans in Respiratory Diseases: Background*

In the last two decades, *heparins and heparan* have also been investigated for their interesting benefits in COPD (Brown et al. 2006) and asthma, with studies performed in vitro (Ahmed et al. 1991; Ahmed et al. 1994), experimental animals (Ahmed et al. 1992) and in humans (Ahmed et al. 1993; Garrigo et al. 1996; Tranfa et al. 2001). As the clinical anticoagulant properties of GAGs are a major limitation of more widespread use, this has led to an attempt to identify GAG-like structures not exhibiting anticoagulant activity (Fryer et al. 1997; Ahmed et al. 1999).

*HA* has also been demonstrated to have beneficial properties in obstructive respiratory diseases in experimental animals (Cantor et al. 1995; Scuri et al. 2001; Scuri and Abraham 2003; Turino 2003) and in humans, both ex vivo (Klagas et al. 2009) and in vivo, following subcutaneous (Venge et al. 1996) or aerosol (Allegra et al. 2006, 2008; Petrigni and Allegra 2006) administration.

The remainder of this review is focused on the most recent promising observations regarding the possible clinical use of HA in treating respiratory disturbances such as those induced by asthma and chronic obstructive pulmonary disease (COPD).

### 3.2 *Controversies*

As in diseases of other organs (Vitanzo and Sennett 2006), reports on the functional role of HA in chronic inflammatory lung diseases are sometimes controversial. This may be attributed to the fact that most studies on GAG expression in chronic inflammatory lung diseases are hindered by the lack of healthy lung tissue being used as the basic control condition (Klagas 2009). Some controversy surrounds the pro- and anti-inflammatory activity of HA, which can be easily explained when we consider the positive experimental (Scuri and Abraham 2003) and clinical (Venge et al 1996; Petrigni and Allegra 2006; Allegra et al. 2008) effects of higher molecular weight HA (anti-inflammatory) versus the neutral or negative effects of HA preparations characterised by lower molecular weight HA or fragments (Kunz et al. 2006; Allegra et al. 2008), which are devoid of anti-inflammatory properties or may even have pro-inflammatory effects.

## 3.3 HA in the "Dynamic" Matrix of Pulmonary Parenchyma

"*In the lung HA works as a major matrix substance in which fibers and fibrous constituents of the matrix (such as elastin and collagen) are embedded*": a statement by Turino pronounced in his "classic" J Burns Amberson lecture (Turino 1986).

Specifically, HA is structurally an integral part of the microfibrils of collagen (Laurent 1970; Keity et al. 1992) and, more importantly, of elastin fibres (Baccarani-Contri et al. 1990), which has been demonstrated immunohistochemically (Turino 2003). This fact (with others) supports the hypothesis that HA protects lung tissue matrix against elastic fibre degradation and works as barrier function against the destructive activities of elastases.

## 3.4 Content in the Lung

In mammalian lung, HA content is 15–150 μg/g dry weight (with some variation amongst species. In the normal human adult lung, the total HA content is ~160 mg, declining with age (Schmid et al. 1982) and (rapidly) in certain diseases (Hallgren 1985) including asthma and COPD (Klagas 2009).

## 3.5 HA: The "Exclusion" Principle

The physico-chemical properties of HA derive from its molecular structure (in solution: casual winding of a linear chain) and its hydration. HA occupies a volume up to 1,000-fold greater than that of other organic materials (Granger 1981), such that contiguous molecules are trapped by HA >1 mg/kg (Bothner and Wyk 1987) with consequent influence on biomechanical forces and water balance.

Colloidal HA, on account of its exceptionally high hydrating properties, occupies a considerable volume in tissues. For the principle of "volume exclusion", the simultaneous occupation of the same space by two different bodies is impossible. As a consequence, once HA is driven into bronchi and lung and/or reaches interstitial space and connective tissue, the possibility for the same tissue volumes to be invaded by pathogenic agents, inflammatory cells and macromolecules is limited since the available spaces are consistently reduced by the more "physiological" presence of hydrated HA. Such mechanical properties help to explain the adjuvant effect of aerosolised HA in bronchopulmonary diseases, where it reduces severity. It has been calculated, for example, that HA solutions of 5–15 mg/ml "exclude", respectively, 25–75% albumin in the bronchopulmonary tissue fluid volume (Bert and Pierce 1984). Due to such mechanical "exclusion", the volume available for pathological materials is physically limited, so determining higher concentration and colloidal osmotic pressure, a fact which strongly influences pre-capillary lymph drainage (Granger et al. 1984) and reduces the interstitial protein content (Reed et al. 1989). This is a major reason for the higher "restriction" of macromolecular traffic in HA solutions (also compared with water). As a consequence, the importance of HA in microcirculatory exchange is secondary to its action at the level of interstitial connective tissue, and when HA concentrations increase, both inflammatory cells and macromolecule diffusion decrease in asthmatics as well as in chronic bronchitics.

## 3.6 Hydration and Osmolarity in the Pathogenesis of Obstruction

The "classic" Editorial "*Is asthma an epithelial disease?*" (Hogg and Eggleston 1984) starts with the following statement: "*Years ago Allegra and Bianco reported bronchial hyper-reactivity in asthmatics to inhaled distilled water aerosol*", a phenomenon which was absent in asthmatics subjected to aerosols of iso-osmolar solutions. At the end of the 1960s, some observations were published regarding broncho-constrictive responses to hypo-osmolar solutions aerosolised during anesthaesia in *chronic bronchitics* (Cheney and Butler 1968, 1970).

The above studies were followed by Allegra and Biancos' observations (1974, 1980) on the evidence, in *asthmatics*, of broncho-constrictive events during foggy days in Milano (a city where fog was usual in winter!), which prompted them to propose a novel elective bronchial provocation test, the ultrasonically nebulised distilled water (UNDW) test (or "fog" test). This test was considered to be more "physiological" and safer than others (Allegra and Dal Negro 1993) and was later employed by other groups, mainly in Australia (Schoeffel et al. 1981; Anderson et al. 1983; Anderson 1985), Vancouver (Hogg and Eggleston 1984; Elwood et al. 1982) and San Francisco (Sheppard et al. 1983; Eschenbacher et al. 1984).

Evidence obtained using bronchial provocation induced by aerosols of hyper-osmolar solutions were linked with those with hypo-osmolar solutions (Cade and Pain 1972; Magyar et al. 1983; Anderson and Brannan 2004).

The links between hydration and osmolarity and reciprocal regulation in the tissues are strongly dependent on the physical properties of HA. Disturbances of osmolarity-regulating homeostasis and water/ion movements alter the bronchial tone in obstructive patients. The correct hydrating mechanisms of epithelium and underlying connective tissue are central to the functional regulation of the bronchial patency.

The conclusion of the above-mentioned Editorial by Hogg and Eggleston (1984) was the following: "*Thus, data from several different experimental approaches suggest that the asthmatic airway adapts poorly to osmotic stress. . .A basic defect responsible for the asthmatic state could be an inability to control the osmolarity and ion concentration of the fluid lining the airway surface*".

The tests employing hypo/hyper-osmolar aerosols contributed to clarify the mechanism of another important non specific provocation: the commonly used exercise test, the positivity of which in asthmatics was originally suggested to be due to cooling of the airways (McFadden et al. 1986). However, such effects are attributed to loss of water (evaporation) during effort with consequent osmolarity disturbances at the bronchial surface level (Anderson et al 1982; Sheppard and Eschenbacher 1984).

The pathological effects of disturbed water movement interferes furthermore with the admirable defence mechanism constituted by muco-ciliary clearance: not only in relation to the efficiency of bronchial cilia (which beat into an iso-osmolar "sol" layer and transport upon their tip the relevant load of the "gel" mucus layer) but also in relation to the rheology of bronchial secretions (Marchette and Daviskas 1985; Daviskas and Anderson 2006), ~95% of which is constituted by iso-osmolar water (Braga and Allegra 1988).

In conclusion, both hydration-dependent tissue osmolarity changes and phenomena due to hypo/hyper-osmolarity of aerosols (Anderson and Brannan 2004) induce vago-mediated broncho-constrictive responses in patients with chronic obstructive lung diseases (Singleton et al. 1986; Potter et al 1991; Valerio et al 2007).

The "unique" characteristic of HA, related to its variable functions, is hydration (up to ~6,000 molecules $H_2O$/molecule of HA). Also in the lung "*the concentration of HA determines its water content....HA's major role is most likely the regulation of the fluid balance in the interstitium*", as demonstrated in normal animal lungs (Turino 2003).

## 3.7 *Chronic Obstructive Bronchopulmonary Diseases*

Recent reports on bronchopulmonary obstructive diseases have provided evidence that such diseases cannot be solely characterised and explained on the basis of an altered immune response and that their chronicity can be explained on the basis of the so-called "remodelling". It is worth remembering that molecules belonging to GAG family, and HA in particular, are deeply involved in matrix homeostasis and function, malfunction of structure-forming cells and disturbed homeostasis of bronchial and pulmonary extracellular matrix, and it is increasingly recognised that their qualitative- or quantitative insufficiency significantly contributes to the pathology of such diseases and reflects airway and parenchyma pathological remodelling. Tissue remodelling describes the structural alterations in bronchi and lung due to chronic inflammation and involves changes not only in cell density, basal membrane and sub-mucosal architecture, but also in composition and function of extracellular matrix, with consequent impairment of lung mechanical functions, for example airway resistance and lung elastance (Johnson et al. 2001; Postma and Timens 2006; Bush 2008).

Regarding matrix disturbances due to HA functional or structural impairment, recent interesting and promising observations on HA content in human airway smooth muscle cells (HASMC) have been published (Klagas et al. 2009) by a group of scientists from Universities in Sydney, Australia, Thessaloniki, Greece, and Basel, Switzerland. Thus, in asthmatic and COPD patients (a) secretion of HA and (b) the expression of HA receptor CD44 were significantly decreased, (c) expression of HAS-1 and HAS-2 was also significantly reduced, (d) the RHAMM receptor was not expressed, (e) hyaluronidase-1 was significantly increased. Such observations suggest the use of HA in such patients.

### 3.7.1 HA in Experimental and Human Obstructive Diseases: Recent History

In the last 15 years, HA has been proposed and used successfully in the treatment of a number of lung diseases:

- In 1995, the Columbia University team (New York, NY, USA) led by G.J. Turino demonstrated protection from experimental emphysema in hamsters following intratracheal HA (Cantor et al. 1995)
- In 1996, the Aarhus University team (Denmark) led by R. Dahl observed that subcutaneously administered HA offered significant protection of COPD patients from exacerbations (Venge et al. 1996)
- From 2000 on, we have demonstrated protective and therapeutic effects of HA administered by aerosol in human asthma (Petrigni and Allegra 2006; Allegra et al. 2006, 2008) and COPD
- In 2001, the pharmacologists R. Lever and C. Page (London, UK) postulated the clinical use of glycosaminoglycans in order to correct bronchial hyperresponsiveness and airways inflammation (Lever and Page 2001) and with the present author went onto to demonstrate that SC enoxaparin added to conventional therapy for COPD, conferred additional improvement in lung function (Brown et al. 2006)
- From 2001 on, the Miami Mount Sinai Medical Center team (Fl, USA) led by W.M. Abraham, demonstrated beneficial effects of aerosolised HA in experimentally broncho-constricted sheep (Scuri et al. 2001; Scuri and Abraham 2003

### 3.7.2 Data on Human Use

Specifically:

1. The single administration of aerosolised HA to subjects with asthma, but only at predetermined (relatively high) molecular weight, concentration (0.3%), and dose (see below), prevents bronchoconstriction induced by:
   (a) Muscular exercise (Petrigni and Allegra 2006)
   (b) Inhalation of UNDW (Allegra et al. 2006)
   (c) Inhalation of methacholine (Mch) (Allegra et al. 2008)
2. Furthermore, it has now been demonstrated that repeated administration of aerosolised HA determines functional improvement in *patients suffering from COPD* (preliminary unpublished data).

In order to demonstrate such effects, HA (iso-osmolar, buffered) colloidal preparations possessing the above characteristics were aerosolised to patients with asthma according to tests A and B (see below). HA was represented by a mixture of $mw_s$ ranging from 400 to 4,000 kDa, while in the study utilising the test C, as well as in the study on the effect of repeated administrations in COPD, HA was represented by a preparation with mw ~1,000 kDa (mixture of mw ranging from 800 to 1,200 kDa).

### 3.7.3 Protective Activity of Aerosolised HA (Single Dose) in Asthmatics Undergoing Bronchial Provocation

Protection from bronchial provocation has been observed in three studies on asthmatics submitted to three different bronchoprovocation tests, two indirect (test A: exercise, test B: UNDW) and one direct (test C: progressively increasing nebulised doses of Mch).

The patients selected for such studies were 43 asymptomatic non-smokers, suffering from mild-to-moderate persistent asthma (in most of them, of allergic nature), characterised by forced expiratory volume in the first second ($FEV_1$) $\geq$80% (predicted) on enrollment: 14 were challenged with exercise (mean age 22 years), 15 with UNDW (mean age 33 years) 14 with Mch (mean age 36 years).

The study designs were randomised, placebo controlled (saline: P), cross over, single blind.

In *Tests A and B*: 4 ml of the above-mentioned P or buffered HA (iso-osmolar solution containing 0.3% of the above-mentioned colloidal preparations of HA in saline) was administered by aerosol to each patient in two non-consecutive days using a nebuliser, 30′ prior to the beginning of the tests (respectively consisting of a free run of 10′ or 5′ exposure to UNDW). Clinical and spirographic data have been evaluated before and 5′ following the end of exercise or UNDW. $FEV_1$ was the functional variable considered.

In *Test C*: 3 ml of the above-mentioned P or buffered HA was administered to each patient by aerosol, according to the American Thoracic Society Guidelines (2000): $PD_{20}FEV_1$ (dose of Mch inducing a fall of $FEV_1 \geq 20\%$) was the variable considered and was calculated as the cumulative dose of Mch nebulised using a dosimeter, with an initial dose of 15 μg followed by doubled successive doses at a distance of 2′ each, with $FEV_1$ measured 90″ after each Mch inhalation. Bronchial hyper-reactive response was demonstrated by a previous challenge with Mch. Once such $FEV_1$ reduction was reached, the test was stopped. In order to reverse the induced bronchoconstriction, salbutamol (200 μg) was administered immediately at the end of the test.

The results of the studies with Test A, B and C are represented in Table 1.

### 3.7.4 Therapeutic Effect of HA in COPD (Repeated Administrations)

Nineteen ex-smokers (mean age 69.6, range 61–79), suffering from COPD (with mild-to-moderate severity) were selected for the study (April to June 2009). All patients were symptomatic and at least four weeks apart from last exacerbation. They were all treated with different bronchodilators, plus supportive therapy when necessary; 7 of them were also receiving inhaled glucocorticosteroids; none was

**Table 1** Bronchoconstriction expressed as loss of $FEV_1$ or increased $PD_{20}FEV_1$ in patients submitted to provocation tests (exercise, UNDW, Mch)

| | | | | |
|---|---|---|---|---|
| Exercise test ($FEV_1$) | Post-P vs. Pre-P | −36% | Post HA vs. Pre-HA | −12%*** |
| UNDW test ($FEV_1$) | Post-P vs. Pre-P | −36% | Post-HA vs. Pre-HA | −24%*** |
| Mch test ($PD^{20}FEV_1$, μg) | Post-P | 263 | Post-HA | 413 (+57%**) |

**$p < 0.01$; ***$p < 0.001$

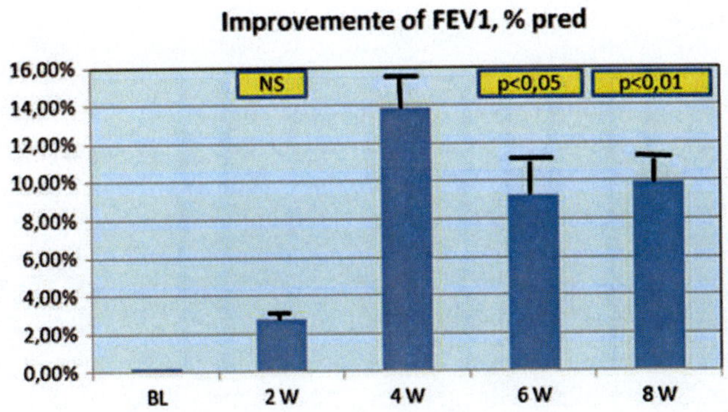


**Fig. 1** Functional improvement evaluated every 2 weeks during 8 weeks of daily treatment with aerosolised HA in COPD patients.To be noted that $FEV_1$ increase was particularly significant after 4 wks treatment ($p \leq 0.001$). *Vertical axis*: % increase of $FEV_1$ (mean ± s.e.). *Horizontal axis*: BL = baseline; w = weeks

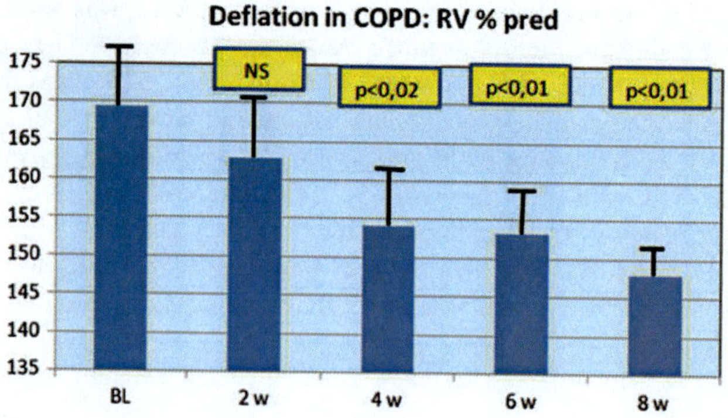


**Fig. 2** Progressive deflation of residual volume (RV) evaluated every 2 weeks during 8 weeks of daily treatment with HA in COPD patients. *Vertical axis*: % RV (mean ± s.e.). *Horizontal axis*: BL = baseline; w = weeks

receiving treatment with antibiotics; none suffered from exacerbation during the study, so that said therapies remained unchanged for the entire period.

As add-on daily treatment they received once-a-day during 8 consecutive weeks aerosolised HA (mw ~1,000 kDa, 3 ml, 0.3%). A complete plethysmographic study was performed the day before said treatment, and every two weeks for the entire period.

$FEV_1$ and residual volume (RV) at each control are reported in Figs. 1 and 2, respectively.

# 4 Conclusions

## *4.1 Rationale for Clinical Use of HA in Respiratory Diseases*

There is now clear evidence of a pleiotropic protective effect of HA in the lung. Its unique capacity to link and retain water molecules in the interfibrillar spaces (Laurent 1970; Keity et al. 1992), via osmotic pressure and effects on consequent resistance to airflow, contributes to the "structure" of the amorphous colloidal matrix which, in the connective tissue, glues together cells and fibres (Laurent and Reed 1991; Fraser et al. 1997; Toole 2004). This provides HA with the ability to hydrate and control solute transport and microcirculatory exchanges, due to its influence on interstitial volume, hydraulic conductibility and macromolecule diffusion (Bert and Pierce 1984; Granger et al. 1984; Bhattacharya et al. 1989). Other physiological functions of HA include exclusion effects (a sort of barrier effect), stabilisation of extracellular matrix structure by electrostatic interactions, lubrication on account of its rheological properties (Laurent and Reed 1991; Reed and Laurent 1992; Fraser et al. 1997), increased muco-ciliary clearance (Marchette et al 1985;

Scuri et al. 2007), retention of homeostatic enzymes at the apical surface (Klagas 2009), prevention of elastase-induced elastin degradation by a mechanism of providing a protective coating against the actions of elastin (Baccarani-Contri 1990; Turino 2003), stabilisation of lung surfactant (Baccarani-Contri 1990), stabilisation of proteoglycans in the extracellular matrix (Granger 1981), contribution to tissue repair (King et al. 1991; Goldberg 1993), inhibition of migration, chemotaxis and aggregation of polymorphonuclear leucocytes and macrophages (Smedrson 1990; Toole 2004).

Reduced levels or altered function of HA could constitute one of the hallmarks of chronic obstructive lung diseases. Data on the possibility that administration of (relatively) high mw HA (either by aerosol or through different ways of administration) results in clinical improvements in asthma and COPD need further work, but the work to date is optimistic. Clinical properties such as regulation of correct hydration/osmosis, antagonism with inflammation/oxidation, restoration of pathologically remodelled lung tissues (at least partial, when possible) may constitute the objectives of novel possibilities of treatment in such diseases: to date no therapeutic interventions have addressed the damaged extracellular matrix of bronchial and pulmonary tissues.

The potential of high mw HA in the regulation of lung matrix properties can be summarised from a recent (2008) statement by Dahl: "*It's time to change the focus to connective tissue and extracellular matrix substances (such as HA) in order to prevent and treat the chronic lung diseases (cancer, asthma, COPD and others)*".

## 4.2 *Potential Therapeutic Effects of HA in Respiratory Diseases and Disorders*

According to Dahl's (2008) hypothesis, HA could be of clinical interest in some lung diseases due to a number of evidences and possible mechanisms.

*In bronchial asthma*:

- Reduction of airway hyper-responsiveness to direct challenges (Mch).
- Reduction of airway hyper-responsiveness to indirect challenges (exercise, 403 UNDW).
- HA blocks airway obstruction by kallikrein.

*In COPD and pulmonary emphysema*:

- Prevention of elastolysis by barrier function
- Decreased chemotaxis through decreased elastin fragmentation.
- Blocking elastase secretion by neutrophils and macrophages.
- Blocking airway obstruction by kallikrein.
- Restoring hyaluronan content (reduced in human COPD and emphysema).
- Restoring HA (degraded by smoke in vitro and in vivo) in respiratory tissues.

*In cystic fibrosis*:

- Prevention of elastolysis in the airways.
- Decreased chemotaxis through decreased elastin fragmentation.
- Blockade of elastase secretion by neutrophils and macrophages.

*And furthermore. . .in pharyngeal streptococcal infection?*

- Blocking epithelial cell CD44 HA-receptor prevents streptococcal colonisation.

## References

Ahmed T, Abraham WM, D'Brot J (1992) Effects of inhaled heparin on immunologic and non-immunologic bronchoconstriction responses in sheep. Am Rev Respir Dis 145:566–570

Ahmed T, Garrigo J, Danta I (1993) Preventing bronchoconsriction in exercise-induced asthma with inhaled heparin. N Engl J Med 329:90–95

Ahmed T, Gonzales BJ, Danta I (1999) Prevention of exercise-induced bronchoconstriction by inhaled low molecular weight heparin. Am J Respir Crit Care Med 160:576–581

Ahmed T, Hallmon J, Abraham WM (1991) The effect of heparin on immunologically-induced tracheal smooth muscle contraction in allergic sheep. Am Rev Respir Dis 143(Suppl):A360, abstract

Ahmed T, Syriste T, Mendelssohn R, Sorace D, Mansour E, Lansing M, Abraham WM, Robinson MJ (1994) Heparin prevents antigen-induced airway hyperresponsiveness: interference with !P3-mediated mast cell degranulation? J Appl Physiol 76:893–901

Allegra L, Abraham WM, Fasano V, Petrigni G (2008) Methacholine challenge in asthmatics is protected by aerosolised hyaluronan at high (1,000 kDa) but not low (300 kDa) molecular weight. Ital J Chest Dis 62:297–301

Allegra L, Bianco S, Petrigni G, Robuschi M (1974) Lo sforzo muscolare e la nebulizzazione ultrasonica di $H_2O$ come test di provocazione aspecifica del broncospasmo. In: Pasargiklian M, Bocca E (eds) Progressi in Medicina Respiratoria. Terme di Sirmione, Sirmione, pp 81–86

Allegra L, Bianco S (1980) Non-specific bronchoreactivity obtained with an ultrasonic aerosol of distilled water. Eur J Respir Dis 61(Suppl 106):41–49

Allegra L, Dal Negro R (1993) Physical agents in bronchial provocation tests: nebulised hypoosmolar aerosol. In: Allegra L, Braga PC, Dal Negro R (eds) Methods in asthmology. Springer, Berlin, pp 219–225

Allegra L, Della Patrona S, Comi A, Fasano V, Longo L, Terzano C, Petrigni G (2006) Aerosols of hyaluronan partially prevent bronchoconstriction induced by ultrasonically nebulised distilled water in asthmatics. Ital J Chest Dis 60:243–248

American Thoracic Society (2000) Guidelines for methacholine and exercise challenge testing 1999. Am J Respir Crit Care Med 161:309–329

Anderson SD (1985) Bronchial challenge by ultrasonically nebulized aerosols. Clin Rev Allergy 3:427–439

Anderson SD, Brannan JD (2004) Methods for "indirect" challenge tests including exercise, eucapnic voluntary hyperpnea, and hypertonic aerosols. Paediatr Drugs 6:161–175

Anderson SD, Schoeffel RE, Finney M (1983) Evaluation of ultrasonic distilled water for provocation testing in patients with asthma. Thorax 38:284–291

Anderson SD, Schoeffel RE, Follet R, Perry CP, Daviskas E, Kendall M (1982) Sensitivity to heat and water loss at rest and during exercise in asthmatic patients. Eur J Respir Dis 63:459–471

Baccarani-Contri M, Vincenzi D, Cicchetti F, Mori G, Pasquali-Ronchetti I (1990) Immunocytochemical localization of proteoglycans within normal elastin fibers. Eur J Cell Biol 53:305–312

Balasz A, Laurent TC, Jeanlo RW (1986) Nomenclature of hyaluronic acid. Biochem J 235:903, abstract
Bert JL, Pierce RH (1984) The interstitium and microvascular exchange. In: Renkin EM (ed) Handbook of physiology. Section 2: the cardiovascular system, Vol. IV, Pt 1: Microcirculation. American Physiological Society, Bethesda, pp 521–547
Bhattacharya J, Cruz T, Bhattacharya S, Bray BA (1989) Hyaluronan affects extravascular water in lungs of unanesthetized rabbits. J Appl Physiol 66:2595–2599
Blackwood RA, Cantor JO, Moret J, Mandl I, Turino GM (1983) Glycosaminoglycan synthesis in endotoxin-induced lung injury. Proc Soc Exp Biol Med 174:343–349
Bothner H, Wyk O (1987) Rheology of hyaluronate. Acta Otorinolaryngol 442(suppl):25–30
Braya PC, Allegra L. (1988) Methods in Bronchial Mucology, Raven Press, New York
Brown RA, Allegra L, Matera MG, Page CP, Cazzola M (2006) Additional clinical benefits of enoxaparin in COPD patients receiving salmeterol and fluticasone propionate in combination. Pulm Pharmacol Ther 19:419–424
Bush A (2008) How early do airway inflammation and remodelling occur? Allergol Int 57:11–19
Cade JF, Pain MCF (1972) Lung function in provoked asthma: response to inhaled urea, methacholine and isoprenaline. Clin Sci 43:759–769
Cantor JO, Cerreta JM, Keller S, Turino GM (1995) Modulation of airspace enlargement in elastase-induced emphysema by intratracheal instillment of hyaluronidase and hyaluronic acid. Exp Lung Res 21:423–436
Cheney FW jr, Butler J (1968) The effect of ultrasonically produced aerosols on airway resistance in man. Anesthesiology 29:1099–1106
Cheney FW jr, Butler J (1970) The effects of ultrasonica aerosols on the total respiratory resistance in the intubated patient. Anesthesiology 32:456–458
Cleland RL, Wang JL (1970) Tonic polysaccharides. III: dilute solution properties of hyaluronic acid fraction. Biopolymers 9:799–810
Dahl R (2008) Glycosaminoglycans: therapeutic opportunities for hyaluronan in respiratory diseases. Ital J Chest Dis 62:289–296
Daviskas E, Anderson SD (2006) Hyperosmolar agents and clearance of mucus in the diseased airway. J Aerosol Med 19:100–109
Dentener MA, Vernooy JH, Hendriks S et al (2005) Enhanced levels of hyaluronan in lungs of patients with COPD: relationship with lung function and local inflammation. Thorax 60:114–119
Elwood RK, Hogg JC, Pare PD (1982) Airway response to osmolar challenge in asthma. Am Rev Respir Dis 61(Suppl):125, abstract
Eschenbacher WL, Boushey HA, Sheppard D (1984) Alteration in osmolarità of inhaled aerosols cause bronchoconstriction and cough, but of a permanent anion causes cough alone. Am Rev Respir Dis 129:211–215
Fraser JR, Laurent TC, Laurent UB (1997) Hyaluronan: its nature, distribution, function and turnover. J Intern Med 242:27–33
Fraser JRE, Kimpton WG, Laurent TC, Cahill RN, Vakakis N (1988) Uptake and degradation of hyaluronan in lymphatic tissue. Biochem J 256:153–158
Fryer A, Huang YC, Rao G, Jacobi D, Mancilla E, Whorton R et al (1997) Selective O- desulfation produces nonanticoagulant heparin that retains pharmacological activity in the lung. J Pharmacol Exp Ther 282:208–219
Garrigo J, Danta I, Ahmed T (1996) Time course of the protective effect of inhaled heparin on exercise-induced asthma. Am J Respir Crit Care Med 153:1702–1707
Goh LT (1998) Effect of colture conditions on rates of intrinsic hyaluronic acid production by *Streptococcus equi* spp *Zooepidermicus*. Biotechnol Lett 16:507–512
Goldberg EP, Burns JW, Jacobi Y (1993) Prevention of postoperative adhesion to precoating tissues with diluted sodium hyaluronate solutions. Prog Clin Biol Res 381:191–204
Granger HJ, Laine GA, Barnes JE, Lewis RE (1984) Dynamics and control of transmicrovascular fluid exchange. In: Staub NC et al (eds) Edema. Raven, New York, pp 198–228

Granger HJ (1981) Physicochemical properties of the extracellular matrix. In: Hargens AR (ed) Tissues fluid pressure and composition. Williams & Wilkins, Baltimore, pp 43–61
Hallgren R, Eklund A, Engstrom-Laurent A et al (1985) Hyaluronate in bronchoalveolar lavage fluid: a new marker in sarcoidosis reflecting pulmonary disease. Br Med J 290:1778–1781
Haynes BF, Hua-Xin L, Patton KL (1991) The transmembrane hyaluronate receptor (CD44): multiple functions, multiple forms. Cancer Cells 3:347–350
Hogg JC, Eggleston PA (1984) Is asthma an epithelial disease? Am Rev Respir Dis 129:207–208
Itano N, Sawai T, Yoshida M et al (1999) Three isoforms of mammalian hyaluronan synthases have distinct enzymatic properties. J Biol Chem 274:25085–25092
Johnson PR, Roth M, Tamm M et al (2001) Airway smooth muscle cell proliferation is increased in asthma. Am J Respir Crit Care Med 164:474–477
Keity CM, Whittaker SP, Grant MF, Shuttleworth CA (1992) Type VI collagen microfibrils: evidence for a structural association with hyaluronan. J Cell Biol 118:979–990
King SR, Hickerson WL, Proctor KG (1991) Beneficial actions of exogenous hyaluronic acid on wound healing. Surgery 109:76–84
Klagas I, Goulet S, Karakiulakis G, Zhong J, Baraket M, Black JL, Papakonstantinou E, Royh M (2009) Decreased hyaluronan in airway smooth muscle cells from patients with asthma and COPD. Eur Respir J 34:616–628
Kunz LIZ, van Rengen ELJ, Sterk PJ (2006) Inhaled hyaluronic acid against exercise-induced broncho-constriction in asthma. Pulm Pharmacol Ther 19:286–291
Laurent TC (1970) Structure of hyaluronic acid. In: Balasz EA (ed) Chemistry and molecular biology of the intercellular matrix. Academic, London, pp 703–732
Laurent UGB, Reed RK (1991) Turnover of hyaluronan in the tissue. Adv Drug Deliv Rev 7:237–256
Lever R, Page C (2001) Glycosaminoglycans, airways inflammation and bronchial inflammation. Pulm Pharmacol Ther 14:249–254
Magyar P, Dervaderics M, Tòth A, Lantos A (1983) Inhalation of hypertonic potassium chloride solution: a specific bronchial challenge for asthma. Ital J Chest Dis 37:29–34
Marchette LC, Marchette BE, Abraham WM, Wanner A (1985) The effect of systemic hydration on normal and impaired mucociliary function. Pediatr Pulmonol 1:107–111
McFadden ER, Lenner KAM, Strohl KP (1986) Postexertional airway rewarming and thermally induced asthma. New insights into pathophysiology and possible pathogenesis. J Clin Invest 78:18–25
Meyer K, Palmer JW (1934) The polysaccharide in vitreous humour. J Biol Chem 107:629–634
Monzon ME, Casalino-Matsuda SM, Forteza RM (2006) Identification of glycosaminoglycans in human airway secretions. Am J Respir Cell Mol Biol 34:135–141
Papakonstantinou E, Karakiulakis G, Eickelberg O et al (1998) A 340 kDa hyaluronic acid secreted by human vascular smooth muscle cells regulates their proliferation and migration. Glycobiology 8:821–830
Petrigni G, Allegra L (2006) Aerosolised hyaluronic acid prevents exercise-induced bronchoconstriction, suggesting novel hypotheses on the correction of matrix defects in asthma. Pulm Pharmacol Ther 19:166–171
Poole AR, Dieppe P (1994) Biological markers in rheumatoid arthritis. Semin Arthritis Rheum 23:17–31
Postma DS, Timens W (2006) Remodelling in asthma and chronic obstructive pulmonary disease. Proc Am Thor Soc 3:434–439
Potter PC, Klein M, Weinberg EG (1991) Hydration in severe acute asthma. Arch Dis Child 66:216–299
Reed RK, Bowen BD, Bert JL (1989) Microvascular exchange and interstitial volume regulation in the rat: implication of the model. Am J Physiol 257H:2081–2091
Reed RK, Laurent UGB (1992) Turnover of hyaluronan in the microcirculation. Am Rev Respir Dis 146:S37–S39
Schmid K, Grundboeck-Jusco J, Kimura A, Tschopp FA, Zollinger R, Binette JP, Lewis W, Hayashi S (1982) The distribution of the glycosaminoglycans in the anatomic components of

the lung and the changes in concentration of these macromolecules during development and aging. Biochem Biophys Acta 716:178–187
Schoeffel RE, Anderson SD, Altounyan RE (1981) Bronchial hyperreactity in response to inhalation of ultrasonically nebulized solutions of distilled water and saline. Br Med J 283:1285–1287
Scuri M, Abraham WM, Botvinnikova Y, Forteza R (2001) Hyaluronic acid blocks porcine pancreatic elastase (PPE)-induced broncho-constriction in sheep. Am J Crit Care Med 164:1855–1859
Scuri M, Abraham WM (2003) Hyaluronan blocks human neutrophil elastase (HNE)-induced air way responses in sheep. Pulm Pharmacol Ther 16:335–340
Scuri M, Sabater J, Abraham WM (2007) Hyaluronan blocks porcine pancreatic elastase-induced mucociliary dysfunction in allergic sheep. J Appl Physiol 102:2324–2331
Shaw RJ, Anderson SD, Durham SR, Taylor KM, Schoeffel RE, Green W, Torzillo P, Kay AB (1985) Mediators of hypersensitivity and "fog"-induced asthma. Allergy 40:48–57
Sheppard D, Rtzk N, Boushey HA (1983) Mechanism of cough and bronchoconstriction induced by distilled water aerosol. Am Rev Respir Dis 127:691–694
Sheppard D, Eschenbacher WL (1984) Respiratory water loss as a stimulus to exercise-induced bronchoconstriction. J Allergy Clin Immunol 73:640–642
Singleton R, Moel DI, Cohn RA (1986) Preliminary observation of impaired water excretion in treated status asthmaticus. Am J Dis Child 140:59–61
Smedrson B, Perthof H, Gustavson S, Laurent TC (1990) Scavenger function of the liver endothelial cell. Biochem J 266:313–327
Toole BP (1990) Hyaluronan and its binding proteins, hyaladerins. Curr Opin Cell Biol 2:839–844
Toole BP (1991) Glycosaminoglycans and morphogenesis. In: Hay ED (ed) Cell biology and extracellular M2trix. Plenum, New York, pp 259–294
Toole BP (2004) Hyaluronan: from extracellular glue to pericellular cue. Nat Rev Cancer 4:528–539
Tranfa CM, Vatrella A, Parrella R, Pelaia G, Bariffi F, Marsico S (2001) Effect of inhaled heparin on water-induced bronchoconstriction in allergic asthmatics. Eur J Clin Pharmacol 57:59
Turino GM (1986) The pulmonary parenchyma: a dynamic matrix (J. Burns Amberson lecture). Am Rev Respir Dis 132:1324–1334
Turino GM (2003) Hyaluronan in respiratory injury and repair. Am J Respir Crit Care Med 167:1169–1175
Underhill CB (1992) CD44: the hyaluronan receptor. J Cell Sci 103:193–198
Valerio G, Salerno FG, Bracciale P (2007) The i.v. infusion of mannitol decreases air way responsiveness to methacholine in asthma. Respir Physiol Neurobiol 156:374–377
Venge P, Pedersen B, Hakansson L, Hallgren R, Lindblad G, Dahl R (1996) Subcutaneous administration of hyaluronan reduces the number of infectious exacerbations in patients with chronic bronchitis. Am J Respir Crit Care Med 153:312–316
Vitanzo PC jr, Sennett BJ (2006) Hyaluronans: is clinical effectiveness dependent on molecular weight? Am J Orthop 35:421–428
Yang B, Zhang L, Turley EA (1993) Identification of two hyaluronan-binding domains in the hyaluronan receptor RHAMM. J Biol Chem 268:8617–8623

# Semi-synthetic Heparinoids

**P. Oreste and G. Zoppetti**

## Contents



**Abstract** New chemical-enzymatic technology based on the modification of the bacterial polysaccharide K5 from *Escherichia coli* leads to the synthesis of a number of heparin/heparan sulfate-like molecules with different biological activities.

With this technology, two families of sulfated compounds were synthesized, which differ in their uronic acid content. The first group contains only glucuronic acid, whereas the second group contains about 50% iduronic acid following epimerization by immobilized recombinant C5 epimerase.

This has led to the development of various anticoagulant and nonanticoagulant K5 derivatives endowed with different – and sometimes highly specific – antitumor, antiviral, and/or anti-inflammatory activities.

**Keywords** AIDS • Angiogenesis • Cell proliferation • Heparin from K5 polysaccharide • Herpes simplex virus • HIV-1 • Inflammation • K5 polysaccharide • K5 polysaccharide derivatives • Papillomavirus

P. Oreste (✉)
Glycores 2000 S.r.l., Milan, Italy
e-mail: oreste@glycores.it

R. Lever et al. (eds.), *Heparin - A Century of Progress*,
Handbook of Experimental Pharmacology 207,
DOI 10.1007/978-3-642-23056-1_18, 

# 1 Heparin-like Molecules

## 1.1 Introduction

Heparin has been used for decades as an anticoagulant after surgery and as an antithrombotic drug for patients requiring short-and long-term antithrombotic therapy. For many years, it was produced from beef organs (lung and mucosa) mainly by American and European factories. Since the extraction process is possible only where large amounts of animal organs are available, heparin production has been developed close to large slaughter houses for the meat market, present in North and South America and in Europe. The extraction requires large amounts of solvents and chemicals. This is one of the major drawbacks in the extraction of heparin.

The appearance of Creutzfeldt–Jacob disease has raised the important issue of finding alternative sources of heparin to beef and has led to increased use of pig tissues. For this reason, the production of the raw material was moved from South America to China. However, the total number of pigs available was just enough to cover the worldwide demand for heparin and not enough for any further development of the heparin market as was recently observed (see Chess et al. 2011, for a case study). Porcine heparin produced by the extractive process requires close supervision by regulatory authorities, and the risk of biological contamination remains, including possible admixture of bovine heparin.

Furthermore, the pollution caused by the use of large amounts of solvents and chemicals and the need to avoid the presence of potential biological contaminants such as prions is a further problem with the current extraction of heparin. Thus, there is a real need to consider new methodologies to produce heparin.

In spite of the aforementioned problems, the market for heparin is developing, as illustrated by the extended use profile of and range of available low-molecular-weight heparins (LMWHs) over recent decades. Moreover the discovery of the utility of heparin-like molecules in some other therapeutic fields such as cancer, AIDS, and diabetes should result in an increased demand for heparin.

In addition to the binding of heparin to factors in the coagulation cascade, heparin is also able to interact with a variety of other proteins, enzymes, cytokines, and viral proteins. This capacity depends on the presence of specific sequences along the polysaccharide chain some of which have been discovered and published. The elucidation of these sequences together with the knowledge of the structure of heparan sulfates and their physiological roles is the basis for the synthesis of new molecules of non-animal origin to mimic some of the useful actions of heparin.

## 1.2 Synthesis of Heparin from K5 Polysaccharide

To overcome the problems in the production of commercial heparin, in 1989, the possibility to set up a new process of production of non animal derived-heparin with anticoagulant and antithrombotic properties was studied. At that time, the steps of the biosynthesis of heparin were under study by Lindahl et al. (1989), the structure and properties of the active pentasaccharide were already known (Lindahl et al. 1980; Choay et al. 1980; Petitou et al. 1988) and also the requirement of the regular sequences for the expression of the anticoagulant activity had been ascertained (Choay 1989; Barrowcliffe 1989).

On the contrary, it was already published that some strains of *Escherichia coli* biosynthesize natural polysaccharides in their capsule involved in the protection of these microorganisms against the non-specific host defense in the preimmune phase of infection (Ørskov et al. 1977). In particular, it was discovered that the strain 010: K5:H4 is able to synthesize K5 polysaccharide (K5), which has the same structure as the natural biosynthetic precursor of heparin, N-acetylheparosan (Vann et al. 1981, see also Jann and Jann 1990). In fact, K5 is formed by the repetition of a disaccharide composed by glucuronic acid (GlcA) α 1–4 linked to N-acetylglucosamine (GlcNAc) (Fig. 1), as found by Vann et al. who prepared a purified K5 polysaccharide by fermentation and elucidated its structure (Vann et al. 1981).

The fermentation of K5 polysaccharide was studied and modified by Cavazzoni et al. (Cavazzoni et al. 1992; Manzoni et al. 1993, 1996) and the purification was optimized. A very pure K5 polysaccharide free from lipophilic substances was obtained with the use of isopropanol in the presence of high concentrations of NaCl in the last step of purification (Oreste and Zoppetti 2002a, b). A further improvement in the fermentation step and yield was achieved by Viskov et al. (2006).

Two ways were possible to produce a non-animal derived-heparin starting from K5 polysaccharide: an enzymatic process involving the use of the enzymes acting in the biosynthesis of heparin or a mixture of chemical-enzymatic steps. The enzymatic approach was not applicable because the enzymes were not yet cloned and expressed and little was known about the possibility to make them work in a concerted action like that occurring in natural biosynthesis. The chemo-enzymatic approach was chosen as the only possible way of making heparin from K5 polysaccharide.

Thus, in 1991 Kusche et al. demonstrated that the addition of N-deacetylated N-sulfated K5 (N-sulfate K5) to a homogenate of solubilized enzymes from a heparin-producing mouse mastocytoma in the presence of the sulfate donor adenosine 3′-phosphate 5′-phospho[$^{35}$S]sulfate (PAPS) results in the synthesis of a heparin-like molecule with affinity for antithrombin (AT) (Kusche et al. 1991).

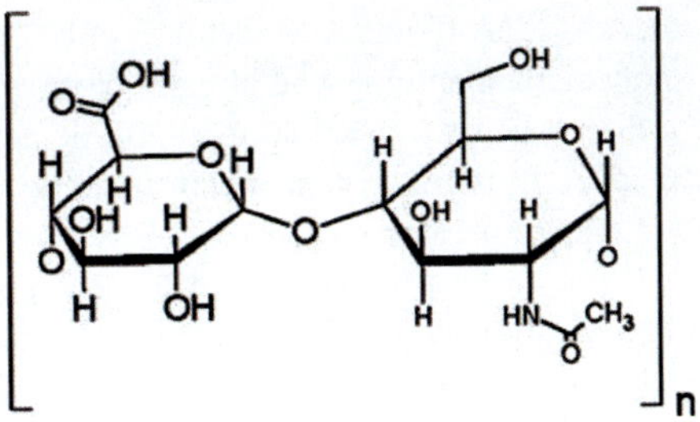

Fig. 1 Structure of the disaccharide repeating unit of K5 polysaccharide

**Fig. 2** Scheme of synthesis of heparin from K5 polysaccharide

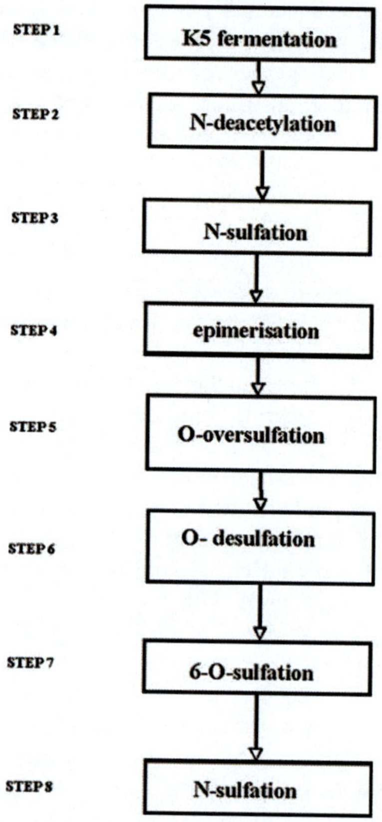


Subsequently, a synthetic route to heparin-like K5 derivatives, consisting of a combined enzymatic-chemical step system involving the C5 epimerization of an N-sulfated K5 polysaccharide followed by sulfation steps, was put into practice.

According to the synthetic scheme described (Fig. 2), the purified K5 polysaccharide underwent an N-deacetylation with NaOH and an N-sulfation with the sulfo trioxide pyridine adduct (pyr.$SO_3$) in water, to reach 100% N-sulfation and a heparin-like molecule (Casu et al. 1994a).

The N-sulfate K5 intermediate (which resembles N-sulfoheparosan) is the natural substrate for C5 epimerase, the biosynthetic enzyme, which converts GlcA to IdoA (Hagner-McWhirter et al. 2000). A non-epimerized compound with anticoagulant activity had already been synthesized, but its activity was only 50% of that of heparin (Razi et al. 1995), demonstrating that the epimerization of GlcA to IdoA is necessary to reach an anticoagulant activity comparable to that of commercial heparin.

Some further evidence had to be taken into account during the setup of the epimerization step:

1. The epimerization reaction by purified C5 epimerase is freely reversible and the maximum degree of conversion affordable was approximately 30% (Campbell et al. 1994).
2. In the aortic heparan sulfate the minimum degree of epimerization is 50% (Maccarana et al. 1996).
3. The reaction of extraction of $^3H$ in the medium from the carboxyl group of the substrate is not influenced by metal ions (Malmström et al. 1980).
4. Heparin anticoagulant activity needs the regular sequence 2-sulfate IdoA-N,6-disulfate GlcN to exert its activity against thrombin (IIa) (Choay 1989; Barrowcliffe 1989).

The problem of the degree of epimerization was approached by studying the formation of IdoA by $^1H$-NMR (Casu et al. 1994a). In fact, in $^1H$-NMR the chemical shift of the anomeric signals of IdoA is easily distinguishable from those of GlcA. In Fig. 3, the $^1H$-NMR of the N-sulfate K5 shows the anomeric signals attributable to N-sulfate GlcN at 5.6 ppm and to GlcA at 4.55 ppm. After epimerization under standard conditions, in the presence of 50 mM KCl, three new signals appear (Casu et al. 1994a). They are attributable to the H-1 of N-sulfate

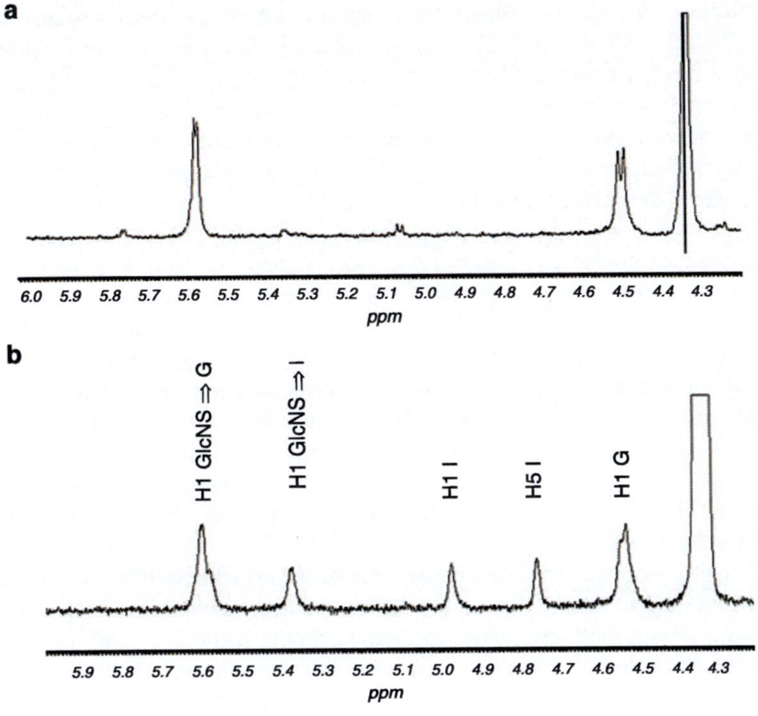


**Fig. 3** The $^1H$-NMR spectrum of the N-sulfate K5 polysaccharide (**a**) is compared with the one of the epimerized product, (**b**) containing 26% IdoA. Both were recorded on a 500 MHz instrument in $D_2O$

GlcN linked to IdoA at 5.38 ppm, the H-1 of IdoA at 4.98 ppm, and H-5 of IdoA at 4.73 ppm. It is possible to quantitate the amount of IdoA formed calculating the area of these peaks.

To set up the epimerization step, first, the standard conditions in the presence of HEPES buffer were used (Casu et al. 1994a) and 30% of epimerization was reached. Surprisingly, in contrast with the previous literature, after addition of divalent cations, such as calcium, barium, magnesium, and manganese, the degree of epimerization was increased and reached 50% in the presence of 50 mM $CaCl_2$ (Oreste and Zoppetti 2000). None of the metal ions had an effect on tritium release from radiolabeled substrate (Malmström et al. 1980), thus indicating that the metal ions do not directly affect the reaction rate of the enzyme.

Moreover, previous observations had shown that calcium ions strongly bind IdoA residues in the monosaccharide form more than GlcA residues (Whitfield and Sakar 1991). IdoA residues can be found in the spatial conformation in three different forms $^4C_1$, $^1C_4$, and $^2S_0$. $^1C_4$ is the preferred conformation for calcium binding (Angulo et al. 2000). Also, the carboxyl group of IdoA and the N-sulfate group of glucosamine are essential for calcium binding in modified heparin (Liang et al. 1982; Ayotte and Perlin 1986). Contiguous nonsulfated IdoA residues in an oligo- or polysaccharide chain are substantially in the $^1C_4$ form (Petitou et al. 1987; Van Boeckel et al. 1987). Again, the spectrum of 50% epimerized N-sulfate K5 shows that the NMR correlation constant $J_{H\text{–}H}$ and $^1J_{H\text{–}H}$ of IdoA are mantained less than 2 Hz also after removal of divalent cations by EDTA, in agreement with the $^1C_4$ conformation (Ferro et al. 1990). Thus, it seems possible that during the epimerization reaction, calcium ions strongly bind to newly formed IdoA residues preventing their back epimerization to GlcA and the equilibrium of this reaction is then pushed toward IdoA formation.

C5 epimerase was sequenced and cloned (Li et al. 1997) and the best conditions of reactions of the recombinant enzyme were studied. In solution in the presence of HEPES buffer containing $CaCl_2$, the recombinant enzyme reached 50% epimerization at pH 7.0 and 30°C. The same enzyme was also immobilized on an inert support and again the best conditions of epimerization were studied. In its immobilized form, the maximum level of conversion was achieved again at pH 7.0 and 30°C. It was also noted that the chemical shift of the signals in the anomeric region of the 50% epimerized K5 indicates that the residues are not organized in clusters, but in an alternate structure favoring the possibility of synthesis of the backbone structure of the active pentasaccharide (Oreste and Zoppetti 2003a, c).

The other problem faced during the setup of the synthetic scheme was the need to introduce a 3-O sulfate group into GlcN to obtain an active pentasaccharide able to bind AT with high specificity. In fact, according to Petitou et al. (1988), the active pentasaccharide contains a disaccharide formed by 6,3-O disulfated GlcN linked with a GlcA that is necessary to obtain the Anti-Xa activity. Moreover, in commercial heparin, 30% of the chains contain the active binding site for AT, while the remaining 70% contain the same pentasaccharide, but without 3-O sulfation (Björk and Lindahl 1982).

Experiments of direct sulfation performed on O-desulfated heparin (Naggi et al. 2001) resulted in an oversulfated compound lacking N-sulfate groups, but that was rich in 3-O sulfated GlcN.

Following the results with O-desulfated heparin, to favor the sulfation of position 3 of GlcN, the epi N-sulfate K5 intermediate was oversulfated according to process C in Casu et al. (1994b). Under these conditions most of the hydroxyl groups are O-sulfated, while the sulfate group in position 2 of GlcN is lost. Due to the very high degree of sulfation, this oversulfated intermediate is devoid of anticoagulant activity. During the oversulfation step some sulfate groups translocate from position 2 to position 3 of GlcN (Naggi et al. 2001; Ogamo et al. 1989).

The excess of sulfation was eliminated by a controlled desulfation in the presence of dimethylsulfoxide/methanol. The anticoagulant activity of the final product was retained only when the reaction was performed from 2 to 4 h at 45°C. According to Naggi et al. (2001) there is a sequence of elimination of the O-sulfate groups along the chain and the 3-O sulfate groups on GlcN are the most resistant, while the 6-O sulfate groups are the easiest to remove. It turns out that during the desulfation reaction all the 6-O sulfate groups on GlcN and most of the sulfate groups on uronic acids are eliminated while most of the 3-O sulfate groups remain in the chain.

It was demonstrated that in the active pentasaccharide of heparin, 6-O and N-sulfate groups on GlcN are essential for the anticoagulant activity (Petitou et al. 1988; Thunberg et al. 1982). To replace 6-O sulfate groups lost during the selective desulfation step a reaction of sulfation at 4°C was performed, while, as the last step, the lost N-sulfate groups were restored with a sulfation reaction in water (Casu et al. 1994a).

### *1.3 Anticoagulant Activity*

The high-molecular-weight heparin-like compounds obtained with the above-outlined process have the characteristics indicated in Tables 1 and 2, sample A. In particular, they show structural features similar to commercial heparin apart from the presence of 3-O sulfated GlcA residues and higher proportions of nonsulfated uronic acid and 3-O sulfated GlcN (Table 1) (Oreste and Zoppetti 2002).

If compared with commercial heparin, they show an increase in the anti-IIa activity and a decrease in the aPTT potentiation (Table 2).

To evaluate the affinity for AT of these compounds, some of them were passed through a column of immobilized AT and the results are shown in Fig. 4.

All the products are characterized by the presence of an affine and a nonaffine fraction whose proportion is very close to that of commercial heparin. The same experiments performed in the presence of a linear gradient of NaCl demonstrated

**Table 1** Structural characteristics of heparins from K5 polysaccharide

| | A | B | C |
|---|---|---|---|
| Molecular weight | 12,700 | 7,400 | 9,700 |
| $SO_3^-/COO^-$ | 2.7 | 2.55 | 2.83 |
| % IdoA | 54 | 54 | 52 |
| % Glc NS | >90 | >90 | >90 |
| % 6-OS GlcN | 90 | 90 | 80 |
| % 3-OS GlcN | 20 | 20 | 50 |
| % 2 OS IdoA | 25 | 25 | 20 |
| % 3-OS GlcA | 30 | 30 | 40 |
| Nonsulfated UA | 55 | 55 | nd |

**Table 2** In vitro anticoagulant activity of heparins from K5 polysaccharide

| SAMPLE | Anti-Xa % | aPTT % | Anti-IIa % | HCII % | AT binding % |
|---|---|---|---|---|---|
| Hep IV Int.Std | 100 | 100 | 100 | 100 | 34.2 |
| A | 157 | 78 | 373 | 161 | 54.8 |
| B | 99 | 52 | 203 | 108 | 33.8 |
| C | 152 | n.d. | 75 | n.d. | n.d. |

that the strength of binding is of the same order of magnitude that of commercial heparin.

Mostly, the commercial heparin used in clinics is a LMWH obtained with different chemical or enzymatic methods. To obtain LMWH from K5 polysaccharide, the deamination reaction in the presence of nitrous acid was chosen. Two possible steps of the synthetic scheme were identified for the deaminative cleavage: either just after the epimerization or at the end of the process. The deaminative cleavage at the end of the process was first applied (Rusnati et al. 2005) and the compound shown in Tables 1 and 2, sample B was obtained.

According to Jacobsson et al.(1984) a decrease in the efficiency of the epimerization reaction is observed with the decrease of the molecular weight of the substrate, so the deaminative cleavage on N-sulfate K5 was not applicable before the epimerization reaction with C5 epimerase.

The nitrous acid deamination applied on the epimerized N-sulfate K5 intermediate is proven to be very clean and reproducible because it is not influenced by the O-sulfate groups. After oversulfation, selective desulfation, 6-O resulfation, and re-N-sulfation of the low-molecular-weight epimerized compound the product in Tables 1 and 2 sample C was obtained. Different from the low-molecular-weight compounds obtained by carring out the deamination reaction at the end of the process, these new molecules show a higher amount of 3-O sulfation on GlcN, in part, due to the presence of a new 3-O sulfate 2,5 anhydromannitol residue at the reducing end of the chain. Moreover the profile of the anticoagulant activity is closer to that of commercial LMWH but with a higher anti-Xa activity (Oreste and Zoppetti 2003b).

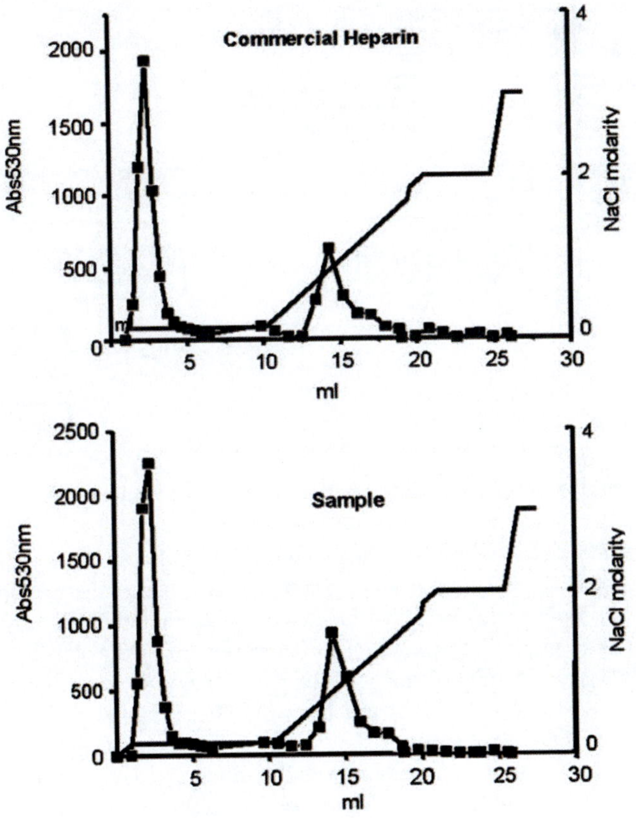


**Fig. 4** Affinity for AT of heparins from K5 polysaccharide

# 2 Nonanticoagulant Molecules

## *2.1 Synthesis*

With the same process scheme setup to obtain the anticoagulant heparin-like molecules from K5 polysaccharide, non-epimerized and epimerized products devoid of activity in the coagulation cascade have been obtained.

K5 polysaccharide was chemically sulfated under mild (Casu et al. 1994a) or strong (Casu et al. 1994b, Method C) conditions either directly in the O-position to obtain O-sulfated K5 polysaccharide with a low (K5OS(L)) and, respectively, high (K5OS(H)) degree of sulfation, or through a previous N-deacetylation followed by an N-sulfation both to obtain N,O-sulfated derivatives with low (K5N,OS(L)) and, respectively, high (K5N,OS(H)) sulfate content. In the same way, epimerized compounds containing 50% IdoA starting from the epiN-sulfate intermediate were synthesized (Fig. 5).

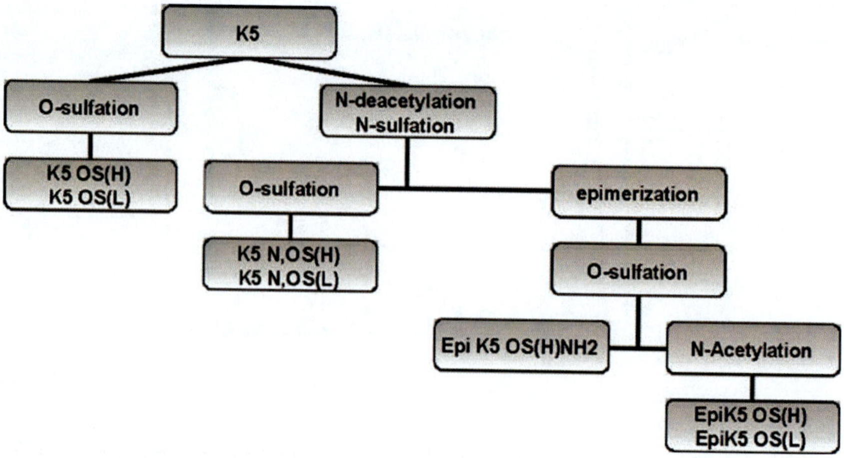


**Fig. 5** Synthetic scheme of nonanticoagulant compounds from K5 polysaccharide

The sulfation under strong reaction conditions (oversulfation) was also improved in order to obtain the maximal degree of sulfation and, hence, the maximal anionic power for the epimerized or non-epimerized, O-oversulfated K5-polysaccharide having its amino group in free form (K5-(epi or non-epi)-OS(H)NH2) (Oreste and Zoppetti 2003a, c). The improvement of the oversulfation reaction according to the Method C of Casu et al. (1994b) has been introduced in the preparation of the tetrabutylammonium (TBA) salt of the epimerized or non-epimerized, N-sulfated K5 polysaccharide before the treatment with the sulfating agent. Instead of adjusting the pH to 5.5 and freeze-drying, as described in Casu et al. (1994b), the solution was brought to pH 7, and this pH value was maintained by addition of further TBA until stabilization (30–60 min), before isolating the TBA salt by freeze-drying. By operating under these conditions, compounds with a degree of sulfation higher than 3.5 were synthesized.

## 2.2 *Antiviral Activity*

One of the crucial events in viral infection is their binding to the cell surface through a series of negatively charged receptors such as heparan sulfate proteoglycans (HSPGs). HSPGs modulate a number of biological events by storing and releasing proteins, enzymes, and growth factors. Due to their negative charges, they are able to selectively bind to the surface proteins of viruses that cause the infection (Vives et al. 2005; Adamiak et al. 2007; Knappe et al. 2007). These are usually basic proteins, or they contain clusters of basic domains, able to interact with the negative charges of the sulfated domains of HSPGs located on the cell surface. HSPGs can be the initial receptors for the virus at the beginning of the infection but the entry into the cell can be mediated by other factors like, such as,

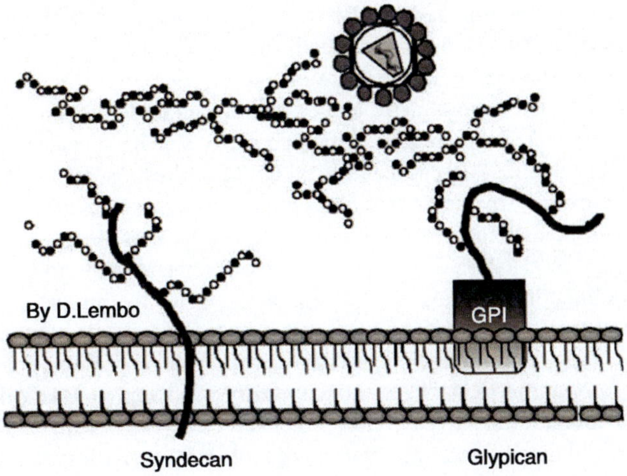


**Fig. 6** Model of the interference of free sulfated GAGs to the binding of viruses to HSPGs (with the permission of Dr. D. Lembo)

tumor necrosis factor (TNF) in the case of herpes simplex virus (HSV) (Shukla and Spear 2001).

Free sulfated polysaccharides can interfere with the attachment of the virus to the host cell, competing for the binding of viral determinants to surface HSPGs. Heparin in particular has shown activity against HIV, the human immunodeficiency virus (Harrop et al. 1994; Rider et al. 1994; Rider 1997), HSV (Herold et al. 1996), and human papilloma virus (HPV) (Knappe et al. 2007) (Fig. 6).

### 2.2.1 Anti-HIV Activity

The infection of the immune cells by HIV is the result of a multistep process in which the virus binds to the surface HSPGs with its gp120 protein. Glycoprotein120 engages CD4 producing a conformational change that causes its binding to the chemokine co-receptors CCR5 (R5 viruses) and CXCR4 (X4 viruses), located on the cell surface, and the internalization of the complex (Moore and Stevenson 2000). The infection results into the progressive destruction of the immune system of the host and the increase in the vulnerability towards other infections and diseases.

It is known that several polyanions inhibit HIV infection (Rider 1997; De Clercq 1989), being more effective against T-cell-tropic HIV variants (X4) with respect to the macrophage-tropic R5 variants. Relevant to this point, R5 viruses are chiefly responsible for the HIV/AIDS epidemic in underdeveloped countries such as sub-Saharan Africa and Southeast Asia.

Among K5 derivatives the highest sulfated molecules, K5OS(H) and K5N,OS (H) together with K5N,OS(L), inhibit the binding of gp120, present on the HIV surface, to heparin immobilized in a BIACORE sensor system, with a dissociation constant (Kd) similar to that calculated for heparin-gp120 interaction ($ID_{50}$ 1–10 nM) (M. Rusnati personal communication).

However, different from heparin and other polysulfated compounds, K5OS(H) and K5N,OS(H) exert the highest activity on both X4 and R5 species (Vicenzi et al. 2003; Rusnati et al. 2009). In particular K5N,OS(H) is the most effective against the so-called dual-tropic virus R5X4 that binds to both the types of chemokines and that appears during the development of the pathology in 50% of the cases.

K5N,OS(H) was also tested on ectocervical tissue explants and showed no toxicity up to concentrations of 4 μg/ml. No significant induction of pro-inflammatory cytokines or chemokines by K5N,OS(H) was detected in a whole blood assay up to 10 μg/ml in comparison with lipopolysaccharides (E. Vicenzi personal comunication)

The clinical features of AIDS cannot be ascribed to simple CD4+ cell infection by HIV. HIV-infected cells release viral proteins such as Tat, able to act on different types of uninfected cells, exerting a variety of biological activities related to AIDS-associated central and peripheral neuropathies (Dewhurst et al. 1996), immune suppression (Caputo et al. 1999; Gatignol and Jeang 2000), and tumorigenesis (Caputo et al. 1999), in particular Kaposi's sarcoma. To exert its various biological activities, Tat needs to interact with the different receptors expressed on target cells among which are HSPGs (Rusnati et al. 1998)

Tat protein is formed by a basic domain, which is able to interact with the polysulfated compounds and HSPGs (Rusnati et al. 1998; Goldstein 1996); in particular, heparin has shown to be effective as Tat inhibitor (Rusnati et al. 1998, 1999; Tyagi et al. 2001).

Some non-epimerized K5 derivatives with different degrees of sulfation have been tested for their capacity to bind to Tat protein in two different assays and to inhibit different pathological effects exerted by Tat in target cells.

The results demonstrated that the most sulfated compounds, K5OS(H) and K5N, OS(H), bind to Tat with an affinity similar to that of heparin. However, in comparison with heparin, K5OS(H) and K5N,OS(H) were able to inhibit a wider array of Tat pathological effects implicated in the appearance of Kaposi's sarcoma (Urbinati et al. 2004).

### 2.2.2 Anti-Herpes Activity

Herpes simplex virus (HSV) causes a number of diseases of the lips, eyes, and genitals, particularly in immunocompromised individuals (Whitley and Roizman 2001). Two different species of HSV are known: HSV-1 responsible for the infections on the lips, (Arduino and Porter 2008) and HSV-2, which causes lesions in the genitals (Corey and Wald 1999). Both species induce the production of pro-inflammatory cytokines and chemokines that disrupt part of the epithelia. As a consequence, a close correlation between HIV and HSV infection exists since very often infection by HSV-2 enhances the possibility of transmission of AIDS (Freeman et al. 2006). However, the immunosuppression caused by HIV infection increases both the risk and the morbidity of HSV infection (McGrath and Newman 1994).

Polyanions are known to bind to the envelope of HSV and to inhibit its entry into target cells, and a number of sulfated polysaccharides have demonstrated anti-HSV activity (Cheshenko et al. 2004).

Both epimerized and non-epimerized compounds with different degrees of sulfation have been tested against HSV-1 and 2 (Rusnati et al. 2005; Pinna et al. 2008). The highest sulfated compounds show the best activity against both the viruses. In particular, K5N,OS(H) and the epimerized oversulfated N-acetylated compound epiK5OS(H) are more effective than heparin in the inhibition of HSV-1, while K5N,OS(H) was as active as heparin, and epiK5OS(H) was about ten times more active on HSV-2 than on HSV-1. Both the compounds limited the cell–cell spread that is the most effective way of propagation of the infection and are also active in the postentry step.

When the compounds are tested against two different viruses, HIV and HSV-1, K5N,OS(H) shows almost the same IC50 for both the viruses while epi K5 has a high specificity for HSV-1 (Vicenzi personal communication).

### 2.2.3 Anti-papilloma Activity

Mucosotropic HPV mainly infects the genital area and is responsible for the appearance of cervical cancer causing about 250,000 deaths of women worldwide per year (Bosch and de Sanjose 2003).

Genital HPVs are classified according to their association with cervical cancer. Infections with low-risk types (primarily types 6 and 11) can cause benign or low-grade cervical cell changes and genital warts, but are not associated with cervical cancer. Infection with high-risk types (primarily types 16, 18, 31, and 45) can cause low-grade and high-grade cervical cell abnormalities that are precursors to cancer (Lowy and Howley 2001).

Again, the binding of HPV virions to HSPGs is the basis of the infection with the virus (Giroglou et al. 2001; Joyce et al. 1999; Shafti-Keramat et al. 2003).

Some non-epimerized K5 derivatives with different degrees of sulfation have been tested for their capacity to prevent HPV infection.

K5N,OS(H) and K5OS(H) were the most effective compounds in the inhibition of infection with both high-risk and low-risk HPV types, together with the N,O-sulfated low K5 derivative (K5N,OS(L)), which showed 10 fold higher activity than heparin, (Rusnati et al. 2009; Lembo et al. 2008). Relevant to this point, heparin and K5N,OS(L) have similar degrees of sulfation, but differ in the uronic acid contents, mainly IdoA in heparin and only GlcA in K5N,OS(L). This suggests a very important role of the uronic acid conformation in the backbone.

All the active compounds also exert post attachment activity. In particular, K5OS(H) retains its postattachment activity over time, disrupting the already estabilished binding between the virions and cellular HSPGs. Moreover, in the

BIACORE assay, it was very effective in inhibiting heparin-HPV-16 PSV pseudovirus (PsV) interaction and in detaching HPV-16 PsV from the BIACORE heparin surface.

## 2.3 *Antitumor Activity*

### 2.3.1 Antiangiogenic activity

The higher risk of thromboembolic complications in cancer patients has led to clinical trials in which heparin has shown a direct effect on cancer independently of its anticoagulant and antithrombotic activity. (Lebeau et al. 1994; von Tempelhoff and Heilmann 2000; Zacharski et al. 2000). This may be due to different mechanisms of action. Indeed, heparin interacts with various growth factors, leading to the inhibition of tumor growth and neovascularization. Also, heparin inhibits heparanase activity, thus inhibiting cell migration and invasion, and it affects cell–cell and cell–matrix interactions mediated by selectins, integrins, and chemokines, resulting in the inhibition of the metastatic process (Zacharski et al. 2000; Engelberg 1999).

Fibroblast growth factors (FGFs) are involved in tumor growth and neovasularization through their interaction with high affinity receptors (FGFRs) (Engelberg 1999) and low-affinity HSPGs receptors (Schlessinger et al. 1995). Heparin is able to inhibit the interaction between fibroblast growth factor-2 (FGF2) and FGFR when the cells bear HSPG (Ishihara et al. 1993), probably competing with the proteoglycans for the interaction with the growth factor.

The non-epimerized K5 derivatives have been tested as antiangiogenic molecules in a number of in vitro assays, including: competition with immobilized heparin for $^{125}$I-FGF-2 interaction; inhibition of the formation of the HSPG/FGF2/FGFR ternary complex; inhibition of the binding of $^{125}$I-FGF2 to HSPGs and FGFR on the endothelial cell surface; inhibition of FGF2-mediated endothelial cell proliferation; inhibition of endothelial cell sprouting and morphogenesis (Rusnati et al. 2005; Leali et al. 2001). Again, the highly sulfated K5OS(H) and K5N,OS(H) were the most effective compounds, but only K5N,OS(H) was able to inhibit angiogenesis triggered by FGF2 when tested in vivo on the chick embryo chorioallantoic membrane assay. This in vivo antiangiogenic activity is maintained by a low-molecular-weight derivative (MW 4,250 D) obtained by nitrous acid deamination of K5N,OS(H) (Presta et al. 2005).

It should be noted that, at variance with K5N,OS(H), heparin has no activity when tested in vitro on the endothelial cell sprouting assay nor in vivo on the chick embryo chorioallantoic membrane assay. Apart from the uronic acid conformation, the structural difference of K5N,OS(H) with respect to heparin is the sulfation on GlcA in positions 2 and 3 and the high degree of 3-O sulfation on GlcN.

### 2.3.2 Anti-proliferative Activity

HSPGs are involved in the regulation of FGF signaling. FGF signaling is thought to stimulate tumor cell proliferation and some malignant tumors have been demonstrated to overexpress FGFs and/or their receptors (Basilico and Moscatelli 1992; McKeehan et al. 1998).

For instance, FGF8 is overexpressed in breast (Marsh et al. 1999) and prostate cancer (Dorkin et al. 1999). Experiments have been performed on S115 breast cancer cells stimulated by testosterone (Borgenstrom et al. 2003). Under these conditions, S115 cells overexpress FGF8b, which stimulates cell proliferation and transformation. When treated with K5OS(H), these cells normalize the malignant phenotype induced by testosterone, thus suggesting that K5OS(H) may interfere with the malignant growth of breast cancer cells by binding FGF8b with high affinity and preventing the binding to its receptors.

## 2.4 *Anti-inflammatory Activity*

It is well known that heparin, apart from its anticoagulant activity, is also effective at inhibiting proinflammatory cytokines (Attanasio et al. 1998).

K5 derivatives were added to mononuclear cells stimulated with lipopolysaccharides (LPS), which activates the production of Interleukin 1β (IL-1β), Interleukin-6 (IL-6), Interleukin-10 (IL-10), and tumor necrosis factor (TNF-α). Among all the K5 derivatives tested, only K5OS(H) and epiK5N,OS(H) were able to inhibit both the production of IL-1β, IL-8 and TNF-α at a concentration of 5 and 10 μg/ml without affecting the production of the antiinflammatory cytokine IL-10. These data indicate that highly specific structural features are required for the anti-inflammatory activity of K5 derivatives (Gori et al. 2004).

# 3 Conclusions

The use of K5 polysaccharides as precursors for the synthesis of new molecules also results in tools for the synthesis of non-epimerized and epimerized derivatives that resemble the structure and activity of the naturally occurring heparin/heparan sulfate. These products differ in their degree of sulfation and distribution of sulfate groups along their molecular backbone, leading to novel compounds with greater affinity and specificity than commercial heparin.

The first important characteristic of this technology is the flexibility of the synthetic scheme which, by the modulation of the sulfation profile of epimerized or non epimerized K5 polysaccharide chains, results in nonanticoagulant compounds but also in heparin-like molecules. The use of solvents is minimal and no reagents of animal origin are involved. Moreover, the production of K5 derivatives is independent of a requirement for animals and a complete coverage of the need of heparin market can be envisaged.

As demonstrated by the results shown in this chapter, even if the most sulfated molecules have unspecific activity in a variety of tests, some others display greater selectivity in their spectrum of activities, and it will be in the future possible to design yet more specific, novel active molecules without undesired side effects. Also, since they are formed from saccharide sequences that mimic the natural occurring heparin/heparan sulfate, minimal toxicity is expected.

**Acknowledgments** The authors thank all the participants in the research and, in particular, Dr. Crisafulli, Dr. Lembo, Dr. Rusnati, and Dr. Presta for the critical revision of the manuscript.

## References

Adamiak B, Ekblad M, Bergström T, Ferro V, Trybala E (2007) Herpes simplex virus type 2 glycoprotein G is targeted by the sulfated oligo- and polysaccharide inhibitors of virus attachment to cells. J Virol 81:13424–13434

Angulo J, dePaz J-L, Nieto PM, Martin Lomas M (2000) Interaction of heparin with $Ca^{2+}$: a model study with a synthetic heparin-like hexasaccharide. Isr J Chem 40:289–299

Arduino PG, Porter SR (2008) Herpes simplex virus type 1 infection: overview on relevant clinico-pathological features. J Oral Pathol Med 37:107–121

Attanasio M, Gori AM, Giusti B, Pepe C, Comeglio P, Brunelli T, Prisco D, Abbate R, Gensini GF, Neri Serneri GG (1998) Cytokines gene expression in human LPS- and IFNg-stimulated mononuclear cells is inhibited by heparin. Thromb Haemost 79:959–962

Ayotte L, Perlin AS (1986) N.M.R. spectroscopic observations related to the function of sulfate groups in heparin. Calcium binding vs. biological activity. Carbohydr Res 145:267–277

Barrowcliffe TW (1989) The anticoagulant activity of heparin: measurement and relationship to chemical structure. J Pharm Biomed Anal 7:217–226

Basilico C, Moscatelli D (1992) The FGF family of growth factors and oncogenes. Adv Cancer Res 59:115–165

Bjõrk I, Lindahl U (1982) Mechanism of the anticoagulant action of heparin. Mol Cell Biochem 48:161–182

Borgenström M, Jalkanen M, Salmivirta M (2003) Sulfated derivatives of *Escherichia coli* K5 polysaccharide as modulators of fibroblast growth factor signalling. J Biol Chem 278: 49882–49889

Bosch FX, de Sanjose S (2003) Human papillomavirus and cervical cancer-burden and assessment of causality. J Natl Cancer Inst Monogr 31:3–13

Campbell P, Hannessonn HH, Sandbäck D, Rodén L, Lindahl U, Li J-P (1994) Biosynthesis of heparin/heparan sulfate. Purification of the D-glucuronyl C-5 epimerase from bovine liver. J Biol Chem 268:26953–26958

Caputo A, Boarini BM, Mantovani I, Corallini A, Barbanti-Brodano G (1999) Multiple functions of human immunodeficiency virus type 1 Tat protein in the pathogenesis of AIDS. Virology 1:753–778

Casu B, Grazioli G, Hannessonn HH, Jann B, Jann K, Lindahl U, Naggi A, Oreste P, Razi N, Torri G, Tursi F, Zoppetti G (1994a) Biologically active, heparan sulfate-like species by combined chemical and enzymic modification of the *Escherichia coli* polysaccharide K5. Carbohydr Lett 1:107–114

Casu B, Grazioli G, Razi N, Guerrini M, Naggi A, Torri G, Oreste P, Tursi F, Zoppetti G, Lindahl U (1994b) Heparin-like compounds prepared by chemical modification of capsular polysaccharide from *Escherichia coli* K5. Carbohydr Res 263:271–284

Cavazzoni V, Manzoni M, Bergomi S (1992) Production and characterization of the *Escherichia coli* K5 polysaccharide in the extracellular form. Ann Microbiol Enzymol 42:101–110

Cheshenko N, Keller MJ, MasCasullo VG, Jarvis A, Cheng H, John M et al (2004) Candidate topical microbicides bind herpes simplex virus glycoprotein B and prevent viral entry and cell-to-cell spread. Antimicrob Agents Chemother 48:2025–2036

Chess EK et al (2011) Case study: contamination of heparin with oversulfated chondroitin sulfate. In: Lever R, Mulloy B, Page CP (eds) Heparin – a century of progress. Springer, Heidelberg

Choay J (1989) Structure and activity of heparin and its fragments: an overview. Semin Thromb Hemost 15:359–364

Choay J, Lormeau J-C, Petitou M, Sinay P, Casu B, Oreste P, Torri G, Gatti G (1980) Anti-Xa active heparin oligosaccharides. Thromb Res 18:573–578

Corey L, Wald A (1999) Genital herpes. In: Holmes KK, Sparling PF, Mardh PA et al (eds) Sexually transmitted diseases. McGraw-Hill, New York, pp 285–312

De Clercq E (1989) Potential drugs for treatment of AIDS. J Antimicrob Chemoter 23(Suppl A): 35–46

Dewhurst S, Gelbard HA, Fine SM (1996) Neuropathogenesis of AIDS. Mol Med Today 2:16–23

Dorkin TJ, Robinson MC, Marsh C, Bjartell A, Neal DE, Leung HY (1999) FGF8 over-expression in prostate cancer is associated with decreased patient survival and persists in androgen independent disease. Oncogene 18:2755–2761

Engelberg H (1999) Actions of heparin that may affect the malignant process. Cancer 85:257–272

Ferro DR, Provasoli A, Ragazzi M, Casu B, Torri G, Bossenec V, Perly B, Sinay P, Petitou M, Choay J (1990) Conformer populations of L-Iduronic acid residues in glycosaminoglycans sequences. Carbohydr Res 195:157–167

Freeman EE, Weiss HA, Glynn JR, Cross PL, Withworth JA, Hayes RJ (2006) Herpes symplex virus 2 infection increases HIV acquisition in men and women: systematic review and meta-analysis of longitudinal studies. AIDS 20:73–83

Gatignol A, Jeang KT (2000) Tat as a transcriptional activator and a potential therapeutic target for HIV-1. In: Jeang KT (ed) Advances in pharmacology, vol 48. Academic Press, San Diego, CA, pp 209–227

Giroglou T, Florin L, Shäfer F, Streek RE, Sapp M (2001) Human papillomavirus infection requires cell surface heparan sulphate. J Virol 75:1565–1570

Goldstein G (1996) HIV-1 Tat protein as a potential AIDS vaccine. Nat Med 2:960–964

Gori AM, Attanasio M, Gazzini A, Rossi L, Lucarini L, Miletti S, Chini J, Manoni M, Abbate R, Genuini GF (2004) Cytokine gene expression and production by human LPS-stimulated mononuclear cells are inhibited by sulfated heparin-like semi-synthetic derivatives. J Thromb Haemost 2:1657–1662

Hagner-McWhirter A, Hannessonn HH, Campbell P, Westley J, Rodén L, Lindahl U, Li J-P (2000) Biosynthesis of heparin/heparan sulfate: kinetic studies of the glucuronyl C5-epimerase with N-sulfated derivatives of the *Escherichia coli* K5 capsular polysaccharide as substrates. Glycobiology 10:159–171

Harrop HA, Coombe DR, Rider CC (1994) Heparin specifically inhibits binding of V3 loop antibodies to HIV-1 gp120, an effect potentiated by CD4 binding. AIDS 8:183–192

Herold BC, Gerber SI, Belval BJ, Siston AM, Shulman N (1996) Differences in the susceptibility of herpes simplex virus types 1 and 2 to modified heparin compounds suggest serotype differences in viral entry. J Virol 70:3461–3469

Ishihara M, Tyrrell DJ, Stauber GB, Brown S, Counses LS, Stack RJ (1993) Preparation of affinity-fractionated, heparin-derived oligosaccharides and their effects on selected biological activities mediated by basic fibroblast growth factor. J Biol Chem 268:4675–4683

Jacobsson I, Lindahl U, Jensen JW, Rodén L, Prihar H, Feingold DS (1984) Biosynthesis of heparin – substrate specificity of heparosan N-sulfate d-glucuronosyl 5-epimerase. J Biol Chem 259:1056–1063

Jann B, Jann K (1990) Structure and biosynthesis of the capsular antigens of *Escherichia coli*, vol 150, Current topics in microbiology and immunology. Springer, Heidelberg, pp 19–42

Joyce JG, Tung JS, Przysiecki CT, Cook JC, Lehman ED, Sands JA, Jansen KU, Keller PM (1999) HPV: from infection to cancer. J Biol Chem 274:5810–5822

Knappe M, Bodevin S, Selinka HC, Spillmann D, Streeck RE, Chen XS (2007) Surface-exposed aminoacids of HPV16L1 protein mediating interaction with cell surface heparan sulphate. J Biol Chem 282:27913–27922

Kusche M, Hannesson H, Lindahl U (1991) Biosynthesis of heparin. Use of *Escherichia coli* K5 capsular polysaccharide as a model substrate in enzymic polymer-modification reactions. Biochem J 275:151–158

Leali D, Belleri M, Urbinati C, Coltrini D, Oreste P, Zoppetti G, Ribatti D, Rusnati M, Presta M (2001) Fibroblast growth factor-2 antagonist activity and angiostatic capacity of sulfated *Escherichia coli* K5 polysaccharide derivatives. J Biol Chem 276:37900–37908

Lebeau B, Chastang C, Brechot JM, Capron F, Dautzenberg B, Delaisements D et al (1994) Subcutaneous heparin treatment increases survival in small cell lung cancer. Cancer 74:38–45

Lembo D, Donalisio M, Rusnati M, Bugatti A, Cornaglia M, Cappello P, Giovarelli M, Oreste P, Landolfo S (2008) Sulfated K5 *Escherichia Coli* polysaccharide derivatives as wide-range inhibitors of genital types of human papillomavirus. Antimicrob Agents Chemother 52: 1374–1381

Li J-P, Hagner-McWhirter Å, Kjellen L, Palgi J, Jalkanen M, Lindahl U (1997) Biosynthesis of heparin/heparan sulfate. cDNA cloning and expression of D-glucuronyl C5 epimerase from bovine lung. J Biol Chem 272:28158–28163

Liang JN, Chakrabarti B, Ayotte L, Perlin AS (1982) An essential role for the 2-sulfamino group in the interaction of calcium ion with heparin. Carbohydr Res 106:101–109

Lindahl U, Bäckström G, Thunberg L, Leder IG (1980) Evidence for a 3-O-sulfated D-glucosamine residue in the antithrombin-binding sequence of heparin. Proc Natl Acad Sci USA 77:6551–6555

Lindahl U, Kusche M, Lidholt K, Oscarsson L-G (1989) Biosynthesis of heparin and heparan sulphate. Ann NY Acad Sci 556:36–50

Lowry DR, Howley PM (2001) Papillomaviruses. In: Knipe DM, Howley PM (eds) Fields virology. Lippincott Williams & Wilkins, Philadelphia, pp 2231–2264

Maccarana M, Tawada A, Yoshida K, Lindahl U (1996) Domain structure of heparan sulfates from bovine organs. J Biol Chem 271:17804–17810

Malmström A, Rodén L, Feingold DS, Jacobsson I, Bäckström G, Lindahl U (1980) Biosynthesis of heparin. Partial purification of the uronosyl C-5 epimerase. J Biol Chem 255:3878–3883

Manzoni M, Bergomi S, Cavazzoni V (1993) Extracellular K5 polysaccharide of *Escherichia coli*: production and characterization. J Bioact Compat Polym 8:251–259

Manzoni M, Bergomi S, Cavazzoni V (1996) Production of K5 polysaccharides of different molecular weight by *Escherichia coli*. J Bioact Compat Polym 11:301–311

Marsh SK, Bansal GS, Zammit C, Barnard R, Coope R, Roberts-Clarke D (1999) Increased expression of fibroblast growth factor 8 in human breast cancer. Oncogene 18:1053–1060

McGrath BJ, Newman CI (1994) Genital herpes simplex infections in patients with the acquired immunodeficiency syndrome. Pharmacotherapy 14:529–542

McKeehan WL, Wang F, Kan M (1998) The heparan sulfate-fibroblast growth factor family: diversity of structure and function. Prog Nucleic Acids Res Mol Biol 59:135–176

Moore JP, Stevenson M (2000) New targets for inhibitors of HIV-1 replication. Rev Mol Cell Biol 1: 40–49

Naggi A, Torri G, Casu B, Oreste P, Zoppetti G, Li J-P, Lindahl U (2001) Toward a biotechnological heparin through combined chemical and enzymatic modification of the *Escherichia coli* K5 polysaccharide. Semin Thromb Hemost 27:437–443

Ogamo A, Metori A, Ukiyama N (1989) Reactivity toward chemical sulfation of hydroxyl groups of heparin. Carbohydr Res 193:165–172

Oreste P, Zoppetti G (2000) Glycosaminoglycans derived from K5 polysaccharide having high antithrombin activity and process for their preparation. EP01271394.7

Oreste P, Zoppetti G (2002a) Glycosaminoglycans derived from K5 polysaccharide having high anticoagulant and antithrombotic activity and process for their preparation. US 2002/0062019

Oreste P, Zoppetti G (2002b) Highly sulfated derivatives of K5 polysaccharide and their preparation. EP1366082

Oreste P, Zoppetti G (2003a) Epimerized derivatives of K5 polysaccharide with a very high degree of sulfation. EP1513880

Oreste P, Zoppetti G (2003b) Low molecular polysaccharides having antithrombin activity. EP1694714

Oreste P, Zoppetti G (2003c) Process for the manufactiurer of N-acyl-(epi)K5-amine-O-sulfate-derivatives and product thus obtained. US 12/120,167

Ørskov I, Ørskov F, Jann B, Jann K (1977) Serology, chemistry, and genetics of O and K antigens of *Escherichia coli*. Bacteriol Rev 41(3):667–710

Petitou M, Duchaussoy P, Lederman I, Choay J, Jacquinet JC, Sinay P, Torri G (1987) Synthesis of heparin fragments: a methyl-α pentaoside with high affinity for antithrombin III. Carbohydr Res 167:67–75

Petitou M, Lormeau JC, Choay J (1988) Interaction of heparin and antithrombin III. The role of O-sulfate groups. Eur J Biochem 176:637–640

Pinna D, Oreste P, Coradin T, Kajaste-Rudnisti A, Ghezzi S, Zoppetti G, Rotola A, Argnani R, Poli G, Manservigi R, Vicenzi E (2008) Inhibition of herpes simplex virus types 1 and 2 in vitro infection by sulfated derivatives of *Escherichia Coli* K5 polysaccharide. Antimicrob Agents Chemother 52:3078–3084

Presta M, Oreste P, Zoppetti G, Belleri M, Vanghetti E, Leali D, Urbinati C, Bugatti A, Ronca R, Vicoli S, Moroni E, Stabile H, Camozzi M, Hernandez GA, Mitola S, Dell'Era P, Rusnati M, Ribatti D (2005) Antiangiogenetic activity of semisynthetic biotechnological heparins. Low molecular weight-sulfated *Escherichia coli* polysaccharide derivatives as fibroblast growth factor antagonists. Arterioscler Thromb Vasc Biol 25:71–76

Razi N, Feyzi E, Björk I, Naggi A, Casu B, Lindahl U (1995) Structural and functional properties of heparin analogs obtained by chemical sulphation of *E. coli* K5 capsular polysaccharide. Biochem J 270:11267–11275

Rider CC (1997) The potential for heparin and its derivatives in the therapy and prevention of HIV-1 infection. Glycoconj J 14:639–642

Rider CC, Coombe DR, Harrop HA, Hounsell EF, Bauer C, Feeney J et al (1994) Anti-HIV-1 activity of chemically modified heparins: correlation between binding to the V3 loop of gp120 and inhibition of cellular HIV-1 infection in vitro. Biochemistry 33:6974–6980

Rusnati M, Tulipano G, Urbinati C, Tanghetti E, Giuliani R, Giacca M, Ciompi M, Corallini A, Presta M (1998) The basic domain in HIV-1 Tat protein as a target for polysulfonated heparin-mimicking extracellular Tat antagonists. J Biol Chem 273:16027–16037

Rusnati M, Tulipano G, Spillmann D, Vanghetti E, Oreste P, Zoppetti G, Giacca M, Presta M (1999) Multiple interaction of HIV-1 Tat protein with sized-defined heparin oligosaccharides. J Biol Chem 274:28198–29205

Rusnati M, Oreste P, Zoppetti G, Presta M (2005) Biotechnological engineering of heparin/heparan sulphate: a novel area of multi-target drug discovery. Curr Pharm Des 11:2489–2499

Rusnati M, Vicenzi E, Donalisio M, Oreste P, Landolfo S, Lembo D (2009) Sulfated K5 *Escherichia Coli* polysaccharide derivatives: a novel class of candidate antiviral microbicides. Pharmacol Ther 123:310–322

Schlessinger J, Lax I, Lemmon M (1995) Regulation of growth factor activation by proteoglycans: What is the role of the low affinity receptors? Cell 83:357–360

Shafti-Keramat S, Handisurya A, Kriehuber E, Menguzzi G, Slupetzky K, Kirnbauer R (2003) Different heparan sulfate proteoglycans serve as cellular receptors for human papillomaviruses. J Virol 77:13125–13135

Shukla D, Spear PG (2001) Herpesviruses and heparan sulfate: an intimate relationship in aid of viral entry. J Clin Invest 108:503–510

Thunberg L, Bäckström G, Lindahl U (1982) Further characterization of the antithrombin-binding sequence in heparin. Carbohydr Res 100:393–410

Tyagi M, Rusnati M, Presta M, Giacca M (2001) Internalization of HIV-1 Tat requires cell surface heparan sulfate proteoglycans. J Biol Chem 276:3254–3261

Urbinati C, Bugatti A, Oreste P, Zoppetti G, Waltenberger J, Mitola S, Ribatti D, Presta M, Rusnati M (2004) Chemically sulfated *Escherichia coli* K5 polysaccharide derivatives as extracellular HIV-1 Tat protein antagonists. FEBS Lett 568:171–177

Van Boeckel CAA, van Aelst SF, Wanegaars GN, Mellema JR (1987) Conformational analysis of synthetic heparin-like oligosaccharides containing α-L-idopyranosiluronic acid. Recl Trav Chim Pays Bas 106:19–29

Vann F, Schmidt MA, Jann B, Jann K (1981) The structure of the capsular polysaccharide (K5 antigen) of urinary-tract-infective *Escherichia coli* 010:K5:H4. Eur J Biochem 116:359–364

Vicenzi E, Gatti A, Ghezzi S, Oreste P, Zoppetti G, Poli G (2003) Broad spectrum inhibition of HIV-1 infection by sulfated K5 *Escherichia Coli* polysaccharide derivatives. AIDS 17: 177–181

Viskov C, Lux F, Gervier, R, Colas G (2006) Method for producing K5 polysaccharide. WO 2006/030099

Vives RR, Imberty A, Sattentau Q, Lortat-Jacob H (2005) Heparan sulfate targets the HIV-1 envelope glycoprotein gp 120 coreceptor binding site. J Biol Chem 22:21353–21357

von Tempelhoff GF, Heilmann L (2000) Antithrombotic therapy in gynecologic surgery and gynecologic oncology. Hematol Oncol Clin North Am 14:1151–1169

Whitfield DM, Sakar B (1991) Metal binding to heparin monosaccharides: D-glucosamine-6-sulphate, D-glucuronic acid and L-iduronic acid. J Inorg Biochem 41:157–170

Whitley RJ, Roizman B (2001) Herpes simplex virus infections. Lancet 357:1513–1518

Zacharski LR, Ornstein DL, Mamourian AC (2000) Low-molecular-weight heparin and cancer. Semin Thromb Hemost 26:69–77

# Heparin-like Entities from Marine Organisms

**S. Colliec-Jouault, C. Bavington, and C. Delbarre-Ladrat**

**Contents**



**Abstract** Polysaccharides are ubiquitous in animals and plant cells where they play a significant role in a number of physiological situations e.g. hydration, mechanical properties of cell walls and ionic regulation. This review concentrates on heparin-like entities from marine procaryotes and eukaryotes. Carbohydrates from marine prokaryotes offer a significant structural chemodiversity with novel material and biological properties. Cyanobacteria are Gram-negative photosynthetic prokaryotes considered as a rich source of novel molecules, and marine bacteria are a rich source of polysaccharides with novel structures, which may be a good starting point

S. Colliec-Jouault (✉)
Laboratoire de Biotechnologie et Molécules Marines, Ifremer, Rue de l'Ile d'Yeu, BP 21105, 44311 Nantes Cedex 3, France
e-mail: Sylvia.Colliec.Jouault@ifremer.fr

R. Lever et al. (eds.), *Heparin - A Century of Progress*,
Handbook of Experimental Pharmacology 207,
DOI 10.1007/978-3-642-23056-1_19, 

from which to synthesise heparinoid molecules. For example, some sulphated polysaccharides have been isolated from gamma-proteobacteria such as *Alteromonas* and *Pseudoalteromonas* sp. In contrast to marine bacteria, all marine algae contain sulphated wall polysaccharides, whereas such polymers are not found in terrestrial plants. In their native form, or after chemical modifications, a range of polysaccharides isolated from marine organisms have been described that have anticoagulant, anti-thrombotic, anti-tumour, anti-proliferative, anti-viral or anti-inflammatory activities.

In spite of the enormous potential of sulphated oligosaccharides from marine sources, their technical and pharmaceutical usage is still limited because of the high complexity of these molecules. Thus, the production of tailor-made oligo- and polysaccharidic structures by biocatalysis is also a growing field of interest in biotechnology.

**Keywords** Anticoagulant activity • Anti-inflammatory activity • Anti-thrombotic activity • Chemical modification • Cyanobacteria • Derivatives • Exopolysaccharides • Heparinoids • Heparin-like entities • Marine algae • Marine bacteria • Marine fungi • Structure • Sulphated polysaccharides

## 1 Heparin-like Entities from Cyanobacteria

Cyanobacteria (blue-green algae) are Gram-negative prokaryotic organisms that are highly structurally organised and morphologically differentiated. They are among the most primitive forms of life on earth. During the two billion years they have flourished on the earth, they have virtually not changed morphologically and physiologically (Sergeev et al. 2002).

Cyanobacteria include edible and toxic species. *Nostoc*, *Spirulina* and *Aphanizomenon* are the main edible cyanobacteria. Many of the commercially important natural products that are derived from cyanobacteria are complex organic compounds that possess unique structures and stereochemistry. Rastogi and Sinha (2009) reviewed the innovative pharmacologically active compounds derived from cyanobacteria showing biological activities as antibiotics, immunosuppressants, anticancer, anti-viral, anti-inflammatory and protease inhibitors (Rastogi and Sinha 2009). Cyanobacteria produce a wide variety of toxins and other bioactive compounds, which include lipopeptides, amino acids, fatty acids, macrolides and amides (Singh et al. 2005).

Despite the importance of marine cyanobacteria as a source of bioactive secondary metabolites (Tan 2007), very few marine cyanobacteria have been studied for the polysaccharides they produce. Polysaccharides are mainly present as capsules and/or slimes in cyanobacteria. A small portion of them are usually released as water-soluble polymers (RPS) (De Philippis et al. 2001) and are complex anionic heteropolysaccharides, usually containing uronic acids. Some of these polysaccharides are also sulphated (De Philippis et al. 2001; Parikh and Madamwar 2006).

### 1.1 *Spirulina*

Spirulan, existing as an ionic form (calcium or sodium), is a sulphated polysaccharide isolated from *Arthrospira platensis* (formerly *Spirulina platensis*) which consists of two types of disaccharide repeating units, [→3)-α-L-Rha(1→2)-α-L-Aco-(1→], where Aco (acofriose) is 3-O-methyl-Rha with sulphate groups and O-hexuronosyl-rhamnose. It also contains trace amounts of xylose, glucuronic acid and galacturonic acid (Lee et al. 2000). The molecular weight of spirulan is about 200,000 with between 5 and 20% sulphation depending on the source (Kaji et al. 2004; Majdoub et al. 2009).

Spirulan and spirulan-like substances have been widely studied for anti-viral activity without any cytotoxic effects (Rechter et al. 2006; Hayashi et al. 1996).

The ultrafiltrated spirulan is also endowed with an anticoagulant activity, only five times lower than standard unfractionated heparin in the APTT and TT assays (Majdoub et al. 2009). Calcium spirulan also exhibits anti-thrombin activity by activation of heparin cofactor II, an inhibitor of thrombin, thus by a mechanism that is different from that of heparin (Hayakawa et al. 1996, 2000). Lee et al. (2007) suggested that sodium spirulan might have beneficial effects as an anticoagulant agent on the blood coagulation fibrinolytic system. This biological effect is dependent on molecular weight and/or sulphate content. Sodium spirulan can also prevent atherosclerosis by inhibiting the proliferation of the arterial smooth muscle cells, as with heparin for which the effect is not dependent on the anticoagulant activity, without exhibiting any toxic effects on the vascular endothelial cell layers (Yamamoto et al. 2006). Calcium spirulan has similar biological activity to sodium spirulan. However, removal of the ion or desulphation reduces activity; thus, the effect on the prevention of atherosclerosis requires a molecular mass of 14,700 or more, the presence of sulphate groups and sodium or calcium ions (Yamamoto et al. 2006).

### 1.2 *Other Cyanobacteria*

Cyanobacteria of the genera *Aphanocapsa*, *Cyanothece*, *Gloeothece*, *Synechocystis*, *Phormidium*, *Anabaena* and *Nostoc* are able to produce sulphated polysaccharides containing uronic acids (De Philippis et al. 2001; Parikh and Madamwar 2006). Applications of cyanobacterial polysaccharides have been poorly investigated in the biomedical field except as anti-viral agents (Rechter et al. 2006; Lee et al. 2007; Zheng et al. 2006).

## 2 Heparin-like Entities from Marine Bacteria

A number of microbial extracellular polysaccharides are produced on an industrial scale (for a review Rehm 2009). Xanthan (from *Xanthomonas campestris*) and gellan (*Sphingomonas paucimobilis*) are both widely used in food applications.

Other well-known polysaccharides produced by bacteria, all non-marine, include cellulose, dextran, curdlan, alginates, succinoglycans, and hyaluronic acid (Sutherland 1998; Kumar and Mody 2009; De Angelis 2002). Although no microbial strain produces heparin, a strain of *Escherichia coli* serotype K5 does form a capsular polysaccharide in which the disaccharide repeat unit is essentially a form of desulphatoheparin or N-acetyl heparosan (Leali et al. 2001). Chitin, the most abundant marine polysaccharide, has been extensively studied, as has its derivative (chitosan), for its biomedical applications, in particular, as an anticoagulant after chemical sulphation (Tharanathan and Kittur 2003; Jayakumar et al. 2007; Nishimura et al. 1986). It is widely distributed in crustaceans, insects, fungi and yeast, but it is not produced by prokaryotes and is therefore not within the scope of this chapter. In the course of the discovery of novel polysaccharides of biotechnological interest, the marine environment and especially, deep-sea hydrothermal vents have provided a relevant source of a variety of new micro-organisms and novel polysaccharides.

## 2.1 Marine Biodiversity

The marine environment covers more than 70% of the earth's surface, that is 361 million km$^2$ with an average depth of 3,800 m. Therefore, it represents a large reservoir of micro-organisms (Whitman et al. 1998). Still, the marine biodiversity is a largely underexplored field (Boeuf and Kornprobst 2009) offering great opportunities in terms of chemodiversity (Bourguet-Kondracki and Banaigs 2009). This makes marine micro-organisms an attractive area in the search for new biomolecules.

A great variety of habitats exists all over the ocean, depending on environmental conditions such as water temperature, pressure and organic and mineral composition. The deep sea is not only the largest habitat on earth, but it is also the most difficult environment in which to survive because of the extreme conditions. Deep-sea environments are characterised by low temperature (1–2°C), high pressure (1 MPa more for every 100 m depth), high-salt and low-nutrient conditions. They were once considered as a biological desert.

In contrast, deep-sea hydrothermal vents, which represent only a small portion of the whole ocean, are real "oases". They were discovered on oceanic geological ridges such as those of the Galapagos and the Pacific East (2,500 m of depth), as well as the mid-Atlantic ridge, but also at the level of the oceanic basins where there are tectonic activities, e.g. the Guaymas (−2,000 m) and the North-Fijian (−2,000 m) basins. Because of the high temperature of the salt water in the contact of the magma, waters which go out of these hydrothermal springs are extremely mineralised; insoluble metal salts form chimneys called smokers. The stream which goes out can border 350°C while some centimetres farther, the temperature of the water is close to 2°C. These ecosytems are characterised by the development around the smokers of a dense population of invertebrates based on heterotrophic and autotrophic bacterial communities. Micro-organisms can exist freely in the water column or by colonising animal and mineral surfaces.

It has been postulated that life on Earth originated at a deep sea vent (Pace 1991; Baross and Hoffman 1985). The phylogeny and the metabolism of organisms isolated from marine hydrothermal springs can be new and very diverse (Brittany Culture Collection: http://www.ifremer.fr/souchotheque). The ceaselessly increasing number of these newly described micro-organisms, as well as the evidence provided by molecular analytical methods, of new phyla of not previously cultured micro-organisms, show the archaeal and bacterial diversity in the deep oceanic environments (Miroshnichenko and Bonch-Osmolovskaya 2006). These bacteria have created a large biotechnological interest for the isolation of novel bio-molecules because of the particular properties of their cellular machinery. In metagenomic studies (reviewed by Siezen and Wilson 2009), it has been reported that deep-sea microbial communities are enriched in genes, among others, encoding polysaccharide biosynthesis. Most of the sequenced culturable micro-organisms from the deep-sea, excluding hydrothermal vents, are Alteromonadales from the Gammaproteobacteria (Siezen and Wilson 2009). Among mesophilic strains from deep-sea hydrothermal vents, Alteromonadales (in particular *Alteromonas* and *Pseudoalteromonas* genus) (Raguenes et al. 1996, 1997a) and a *Vibrio* strain, (*Vibrio diabolicus*) have been identified, *Vibrio* genus is also widely distributed in other marine environments (Raguenes et al. 1997b).

From these microbial taxonomic groups, glycopolymer biosynthesis has been demonstrated. Bacterial polysaccharides (BPS) are either present in the cellular wall as essential constituent of lipopolysaccharides (LPS), or as capsular material (CPS) that closely surrounds the producing microbial cell and bound outside of the cell or as material that is released more widely into the surrounding environment as a dispersed slime, exopolysaccharide (EPS) (Guo et al. 2008). Their role has been reviewed by Mancuso Nichols et al. (2004). They play an important role in the interaction between bacteria and their environment, participating in the cellular attachment and adhesion to surfaces, increasing survival compared with growth in an unattached state. Polysaccharides form a layer that protects cells against toxic compounds or against digestion by other organisms. BPS may also prevent cells from desiccation or damage. The ultrastructural network protects cells and facilitates cellular interactions. Polysaccharides from marine bacteria living in extreme condition, usually show peculiar chemical features as a consequence of their adaptation to their environment and large amounts of the starting polymer can be obtained by biotechnological methods.

## 2.2 *Sulphated Polysaccharides are Produced by Some Marine Bacteria*

Anticoagulant activities of polysaccharides have been described to depend mainly on the sulphate groups present within the molecule, even if some other structural characteristics such as the polyanionic features or the molecular weight influence the biological activity (Shanmugam and Mody 2000a).

Although no marine micro-organism produces heparin, some of them synthesise polysaccharides, sometimes sulphated, with neutral or hexosamine sugar and uronic acids. Most of the EPS-producing marine bacteria belong to the genus *Vibrio*, *Flavobacterium*, *Pseudomonas*, and *Alteromonas* or *Pseudoalteromonas* (Bramhachari and Dubey 2006). Although common in animal cells, sulphated carbohydrates are rare in prokaryotes, having been reported so far in rhizobia i.e. *Sinorhizobium meliloti* (Lerouge et al. 1990) or *Mesorhizobium loti* (Townsend et al. 2006) as well as *Azospirillum brasilense* Sp7 (Vanbleu et al. 2005) and *Mycobacterium* (Rivera-Marrero et al. 2002). As far as marine bacteria are concerned, sulphated polysaccharides have been described in *Pseudomonas* species (Matsuda et al. 2003) and in some marine *Alteromonas* strains or *Pseudoalteromonas* species (Rougeaux et al. 1999a; Roger et al. 2004; Mata et al. 2008). This Alteromonadaceae family seems rich in sulphated EPSs. However, Ivanova et al. (1994), Nazarenko et al. (1993), Zubkov et al. (1995), Perepelov et al. (2005), and Saravanan and Jayachandran (2008) have described polysaccharides of some marine *Alteromonas* or *Pseudoalteromonas* species composed of different neutral sugars, as well as hexosamine and uronic acid residues, but there was no evidence of sulphate groups. Some of these polysaccharides contain novel sugar residues emphasising the chemodiversity of marine micro-organisms (Zubkov et al. 1995; Perepelov et al. 2005). The diversity of the structures of polysaccharides from *Pseudoalteromonas* and *Shewanella* sp., both belonging to the Alteromonadaceae family, has been reviewed by Nazarenko et al. (2003). Some sulphates have also been detected in *Marinobacter* sp. extracellular polymeric substances, but there is no evidence of the sulphation status of the polysaccharides which compose these exopolymeric substances (Bhaskar et al. 2005), even if sulphated polysaccharides have also been described in a *Marinobacter* strain (Bramhachari and Dubey 2006).

Our studies of numerous isolates from deep-sea hydrothermal vents revealed a few polymers with interesting properties. They are high-molecular-weight carbohydrate polymers, either linear (Raguenes et al. 1997b; Rougeaux et al. 1999b) or highly branched (Rougeaux et al. 1998, 1999a; Roger et al. 2004). Most of them have high uronic acid content, and bear different substitutions (sulphate, pyruvate, lactate) (Rougeaux et al. 1996; Guezennec et al. 1994). *Pseudoalteromonas* Strain HYD721, *Alteromonas infernus* (Raguenes et al. 1997a) and *Alteromonas macleodii* subsp. *fijiensis biovar deepsane* (Cambon-Bonavita et al. 2002) can produce sulphated polysaccharides (HYD721, GY785 and HYD657, respectively) (Fig. 1) (Rougeaux et al. 1999a; Roger et al. 2004). GY785 is a water soluble acidic heteropolysaccharide composed of glucose, galactose, glucuronic and galacturonic acids (1:1:0.7:0.4) and 3% (w/w) sulphur content corresponding to 9% sulphate groups (Roger et al. 2004). The structure of polysaccharide HYD657 has not been elucidated yet, but this is currently in progress in Ifremer; however this polysaccharide has shown a level of 9% sulphate content (w/w).

EPS composed of neutral sugar, hexosamine and/or uronic acid residues have also been described in *Vibrio* strains such as *V. harveyi* (Bramhachari and Dubey 2006), *V. diabolicus* (Rougeaux et al. 1999b), *V. furnissii* (Bramhachari et al. 2007) and *V. alginolyticus* (Muralidharan and Jayachandran 2003). They do not bear

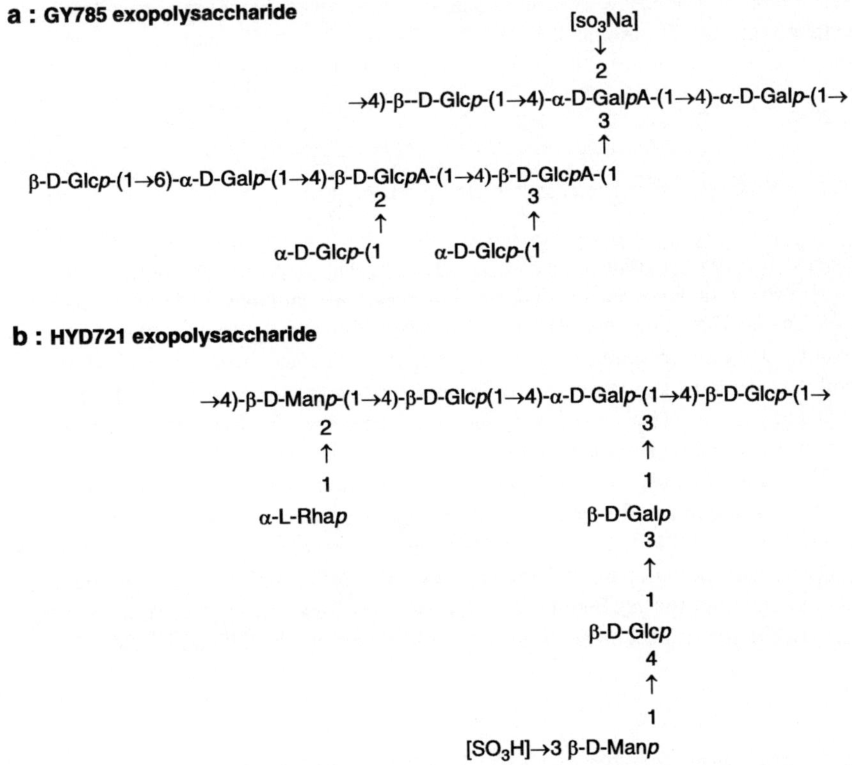


**Fig. 1** Structures of the repeating unit of the main exopolysaccharides produced by the marine microorganims *Alteromonas infernus* (**a**) and *Pseudoalteromonas* strain HYD721 (**b**)

HE800

→3)-β-D-GlcpNAc-(1→4)-β-D-GlcpA-(1→4)-β-D-GlcpA-(1→4)-α-D-GalpNAc-(1→

Hyaluronic acid

→3)-β-D-GlcpNAc-(1→4)-β-D-GlcpA-(1→

**Fig. 2** Osidic sequence of the repeating unit of the polysaccharides HE800 and hyaluronic acid

any sulphate groups but their structure may show some homology with glycosaminoglycans (GAG), especially hyaluronic acid (Fig. 2). The first species of *Vibrio* to be isolated from a vent sample was a mesophile that secretes an innovative EPS of potential medical interest for its chemical resemblance to heparin. The EPS HE800 secreted by *Vibrio diabolicus* has a linear repetitive unit constituted by four residues: two of glucuronic acid, one N-acetyl-glucosamine and one N-acetyl-galactosamine. It is a structural analogue of heparan sulphate or heparin with the succession of glucuronic acid and hexosamine residues; however, it does not have sulphate groups. Its molecular weight is about $10^6$ g mol$^{-1}$ and varies from one production lot to another (Raguenes et al. 1997b; Rougeaux et al. 1999b).

## 2.3 Native Polysaccharides are Bioactive

Only a few marine bacterial polysaccharides having GAG-like biological activities have been reported. Native EPS exhibit interesting biological activities such as efficient bone-healing with HE800. This EPS secreted by *Vibrio diabolicus* was evaluated on the restoration of bone integrity in an experimental animal model and was demonstrated to be a strong bone-healing material without inducing any inflammatory reaction. Moreover, the new bone was histologically normal and the degree of new vascularisation was significant (Guezennec 2002; Zanchetta et al. 2003; Colliec-Jouault et al. 2004). However, high-molecular-weight polysaccharides, even when sulphated, do not possess anticoagulant properties like heparin (Colliec-Jouault et al. 2001). Thus, these high-molecular-weight polysaccharides could function as a material for the regeneration of a wide variety of tissues for both wound care and the regeneration of damaged or diseased organs. Another *Vibrio* polysaccharide has been shown to have antitumor activity (Okutani 1984).

## 2.4 Polysaccharides May Be Modified to Obtain Heparin-like Entities

Many polysaccharides exhibit useful properties when they undergo structural modifications and thus may find varied applications in the food, pharmaceutical and other industries. Structural derivatives of extracellular polymers (EPS) with unusual structures produced by marine bacteria isolated from hydrothermal deep-sea vents have been prepared by chemical modification to design compounds with better activity and specificity. Marine polysaccharides are suitable as starting materials to synthesise heparin-like drugs and sulphated or over-sulphated derivatives have been prepared. Modification processes giving GAG-like entities (semi-synthetic sulphated polysaccharides) are up to now chemical, e.g. acid hydrolysis (Guezennec et al. 1998), radical depolymerisation (Nardella et al. 1996), N-deacetylation with sodium hydroxide (Zou et al. 1998) or sulphation (Nishino and Nagumo 1992).

Heparin-like derivatives were obtained from chemical modifications of the native GY785 EPS secreted by *Alteromonas infernus* (Colliec-Jouault et al. 2001). This EPS is naturally sulphated (9% sulphate groups w/w). After chemical over-sulphation, only the free OH groups can bear a sulphate group. Two depolymerization processes were used to obtain homogeneous low-molecular-weight (LMW) and over-sulphated derivatives (20–30 $10^3$ g/mol and sulphate content 20–40%). The compounds generated by radical processes were more homogeneous than those obtained by acid hydrolysis with respect to their molecular mass. The derivatives obtained after over-sulphation and depolymerisation were compared with heparin and anticoagulant activity was detected in over-sulphated derivatives, but not in the native EPS. The free radical depolymerised and over-sulphated derivative inhibited thrombin generation in both contact-activated and thromboplastin-activated plasma, showing a prolonged lag phase only in the contact-activated assay. Affinity co-electrophoresis studies suggested that a single population of polysaccharide chains binds to anti-thrombin and that only a subpopulation strongly interacts with heparin cofactor. The preparation of new heparinoids or heparin-like entities from other EPS secreted by strains isolated from deep-sea vents have been undertaken. These LMW over-sulphated derivatives presenting differences in structural features were endowed with original anticoagulant properties compared to heparin. They presented a lower anticoagulant activity than heparin and so could be promising new anti-thrombotic drugs without a major bleeding risk (Colliec-Jouault et al. 2003a).

## 3 Heparin-like Entities from Marine Eukaryotes

All marine algae and microalgae contain sulphated wall polysaccharides. The proportion of highly acidic polysaccharides is greater in the outer regions of the cell wall and in the outer cellular layers of the thallus. Heparin-like entities, with biological properties similar to heparin rather than similarity of structure, have been extracted from marine algae and have been well described over the last 60 years. Sulphated polysaccharides from the three major divisions of marine algae, Rhodophyta, Phaeophyta and Chlorophyta have been studied to explore their potential as a cheap and safe source of new types of heparinoids or heparin-like entities. Among the numerous algal polysaccharides, the fucoidan family, a minor matrix component in brown algae – the by-products of alginate production in food and cosmetic industries – can be considered as the “marine heparin” and has been most widely studied. This algal sulphated polysaccharide family with complex, heterogeneous structures shares a lot of biological properties with heparin, especially low-molecular-weight homogeneous fucoidan preparations. The low-molecular-weight fucoidan with a high arterial anti-thrombotic activity presents both low anticoagulant activity and therefore a low risk of haemorrhage. This compound is a promising anti-thrombotic drug, which could be of interest in preventing restenosis or potentiating neovascularization of ischaemic areas.

### *3.1 Heparin-like Entities from Macroalgae*

Algal sulphated polysaccharides are complex and high molecular mass molecules (molecular mass ranges between $10^5$ and $10^6$) with no clear evidence of repeating units in their structure. These anionic macromolecules are essentially found in three major divisions of marine algae: Rhodophyta (red algae), Phaeophyta (brown algae) and Chlorophyta (green algae). The sulphated polysaccharides extracted from red algae are galactans consisting entirely of galactose units (carrageenans and agars), and they are mainly linear sulphated homopolysaccharides. The sulphated polysaccharides found in brown algae are more complex being highly branched heteropolysaccharides. According to the species of brown algae, the sulphated polysaccharides are more or less complex and often described in three groups. The first group, fucoidans or homofucans, is highly branched and primarily composed of L-fucose units with small contents of D-xylose, D-galactose, D-mannose and uronic acid units. The second group, ascophyllans, is xylofucoglycuronans with large proportions of D-xylose and uronic residues. The third group contains glycuronogalactofucans or glycuronofucogalactans in which galactose becomes preponderant over fucose and uronic acid units. Often fucoidan fractions contain a small amount of proteins, probably covalently attached, and suggesting that the brown algal sulphated polysaccharides exist in vivo as proteoglycans. The sulphated polysaccharides in Chlorophyta are highly branched heteropolysaccharides made up of D-xylose, L-rhamnose, galactose and glucuronic acid with three main groups glucuronoxylorhamnans, glucuronoxylorhamnogalactans and xyloarabinogalactans. Some of the green algal sulphated polysaccharides are found covalently attached to a protein and they are characterised as proteoglycans (Kloareg and Quatrano 1988; Shanmugam and Mody 2000b; Pomin and Mourao 2008).

### 3.1.1 The Galactan Family of Polysaccharides from Rhodophyceae

The composition and structure of the sulphated galactans found in Rhodophyta vary according to the algal genus and are known commercially as agar and carrageenan. Galactans are usually extracted for food application (e.g. jelly candies and canned meats for agars and frozen dessert stabilisers for carrageenans) and other industrial applications such as important life science applications (e.g. chromatography and microbial and cell cuture media for agars and agaroses derived from agars). Sulphated polysaccharides from red algae have been well described by many workers for their anti-viral activities against certain viruses responsible for human infection diseases (Bourgougnon et al. 1993; Damonte et al. 1994; Witvrouw and De Clercq 1997). However, more than 40 species of red algae have also been studied for their anticoagulant activities. A considerable structural variation in the red agal sulphated galactans occurs among different species from different environments. Consequently, it is often very difficult to class them as agar-type or carrageenan-type.

The agars are mainly extracted from *Gelidium*, *Gracilaria*, *Pterocladia*, *Gracilariopsis* and *Porphyra*. They are low sulphated galactans (from 2 to 5% of sulphate groups) and the disaccharidic repeating unit is a 1,4-linked α-D-galactose and 3,6-anhydro-α-L-galactose (agarobiose). Consequently, very few studies on biological activities and especially anticoagulant properties of native agars are described due to their low sulphate content.

A sulphated galactan from *Gelidium crinale* was studied by Pereira et al. (2005) According to its algal origin, this sulphated galactan can be considered as agar, it shows a very heterogeneous sulphation pattern and presents a very low anticoagulant activity (65 IU/mg compared with a heparin standard found at 180 IU/mg in the APTT clotting assays). This low anticoagulant activity is probably due to its low proportion of sulphate groups more than their distribution on the osidic backbone.

Carrageenans mainly extracted from *Chondrus*, *Gigartina*, *Euchema* and *Hypnea* are highly sulphated galactans (from 20 to 40% of sulphate groups). Carrageenans are divided into three groups with regard to differences of solubility and also the content and the position of sulphate groups on the galactose unit. The kappa- and iota-carrageenans have a disaccharidic repeating unit alternating 1,3-linked α-D-galactose and 1,4 linked 3,6-anhydro-β-D-galactose (carrabiose) substituted with varying positions and percentages of sulphate groups. For kappa, the galactose unit is 4-sulphated (25% of sulphate groups) and for iota, the galactose unit is also 4-sulphated as well as the 3,6-anhydrogalactose unit is 2-sulphated (32% of sulphate groups). The lambda-carrrageenan disaccharidic repeating unit is composed of alterning 2-sulphated 1,3 linked α-D-galactose and 2,6 disulphated 1,4-linked β-D-galactose (35% of sulphate groups). Among the different groups of carrageenans, the lambda is the most hydrosoluble. Contrary to the kappa- and iota-, the lambda-carrageenan is a non-gelling polysaccharide and presents a proportion of sulphate groups close to that found in glycosaminoglycans. The lambda-carrageenan is described as more toxic than the kappa, although this toxicity seems dependent upon the molecular weight.

Hawkins and Leonard (1962) decided to study carefully the anticoagulant activities of the lambda- and kappa-types of carrageenins or carrageenans extracted from *Chondrus crispus* with the intention to clarify the previous studies. Both in vitro and in vivo assays were performed. For the in vivo effect, the products were administered intravenously into dogs at different doses and blood collected at different times after the injection (from 0.5 to 2 h). At 5 mg/kg of lambda-carrageenan and 0.5 mg/kg of heparin, the same prolongations in both clotting times and thrombin times were observed ½ h after the injection. Kappa-carrageenan had no effect. In vitro anticoagulant assays performed in human and dog plasma showed a similar effect for lambda-carrageenin and heparin on thrombin time at 10 μg/mL and 1 μg/mL, respectively. This study showed an anticoagulant property of the lambda-carrageenan tenfold lower than heparin. Anderson and Duncan (1965) confirmed by in vivo experiments in the rabbit and in vitro assays (prothrombin time and thrombin time) that the lambda-carrageenan is more anticoagulant than the corresponding kappa-carrageenan. The lambda-carrageenan is also more toxic than the kappa-carrageenan.

Guven et al. (1991), compared different carrageenan types extracted from *Grateloupia dichotoma*; the highest anticoagulant activity measured in vitro by clotting assays was found with the lambda-type.

Carlucci et al. (1997) described both anti-viral and anticoagulant activities of carrageenans extracted from *Gigartina skottsbergii*. Both the kappa- and lambda types have a high molecular weight ($10^5$). The lambda-type has a higher content of sulphate groups (40%) than the kappa-type (30%). These compounds were not toxic and their anticoagulant activities were determined in vitro by the thrombin time (TT). The kappa- and lambda-carrageenans can prolong the TT. The concentrations required to double the control time (15.6 s) were around 5 μg/ml for the lambda-type and 200 μg/ml for the kappa-type. The lambda-type presents the highest anticoagulant effect and at 50 μg/ml the TT is above 180 s. A recent publication described also the anticoagulant activity of a native lambda-carrageenan extracted from *Tichocarpus crinitus* (Baranova et al. 2008). The anticoagulant activity of the non-gelling fraction was measured in vitro using the activated partial thromboplastin time (APTT). At 100 μg/ml, the APTT was prolonged and the coagulation time measured at 343 s (58.7 s for the control time).

Others have isolated homogeneous low molecular fractions in order to isolate a better characterised product and improve its specificity. The group of Yamada et al. (1997) modified chemically carrageenans in order to obtain lower molecular weight fractions and tested their in vitro anticoagulant activity using the APTT. The native lambda-type is more anticoagulant than the kappa- and iota-types and a decrease of the anticoagulant activity was observed with the decrease of the molecular weight. The over-sulphation of the depolymerised carrageenans can increase or maintain the anticoagulant activity. Opoku et al. (2006) showed that the chemical over-sulphation of a kappa-carrageenan can increase its anticoagulant activity (30-fold), whilst the over-sulphated kappa-carrageenan was tenfold less effective than unfractionated heparin with the same percentage of sulphate as heparin.

The mechanism of anticoagulant activity of carrageenans has been suggested to be via the inhibition of thrombin through the catalysis of anti-thrombin (Shanmugam and Mody 2000).

The anticoagulant activity of a low-molecular-weight fraction obtained by free radical depolymerisation of the sulphated galactan extracted from *Schizymenia binderi* was studied by Zuniga et al. (2006). In this study, the very low-molecular-weight (LMW) fraction (MW 8,500 sulphate 25% and a mixture of agaran and carrageenan) was less active than the corresponding native polysaccharide and chemical sulphation increased the activity. The concentrations required to observe the same prolongation were 1.5 μg/ml of heparin, 50 μg/mL of native polysaccharide, 150 μg/mL of LMW fraction and 50 μg/mL of over-sulphated LMW fraction. Pushpamali et al. (2008) attempted to hydrolyse a sulphated polysaccharide from the red seaweed *Lomentaria catenata* by fermentation of seaweeds. After fermentation, the molecular weight of the sulphated polysaccharides remains high (from $10^5$ to $5 \times 10^5$) and its sulphate content is close to 22%. Recently, Mourao's group performed an extensive study on the anticoagulant and anti-thrombotic properties of algal sulphated galactans from the red alga *Botryocladia occidentalis* (Fonseca et al. 2008; Melo et al. 2008). The analysis of the native product showed an unusual dual effect in a rat model of venous thrombosis. In order to avoid this dual effect, low-molecular-weight derivatives were prepared by acid hydrolysis.

A very low-molecular-weight fraction (MW 5 × 10$^3$) lost this dual effect, and its effect was similar to unfractionated heparin (UFH) in venous thrombosis, with a very weak anticoagulant effect. This product had little activity in arterial thrombosis in rats demonstrating that improved or dose-related effects can be obtained by the preparation of LMW derivatives from heterogeneous products.

### 3.1.2 The Fucoidan family from Phaeophyceae

Fucoidan, first described by Killing (1913), was originally a member of the sulphated polysaccharides found in Phaeophyceae for which the general name of fucans was given; a family of compounds including fucoidin, ascophyllan, sargassan and glucuronoxylofucan. Fucoidan is a highly sulphated polysaccharide (30–40%) like heparin, but in contrast to heparin, fucoidan contains neither N-acetylated nor N-sulphated groups. Fucoidan is primarily composed of 2-sulphated, 1,3, and 1,4 linked α-L-fucose with branching or sulphate group at position 4 (Fig. 3).

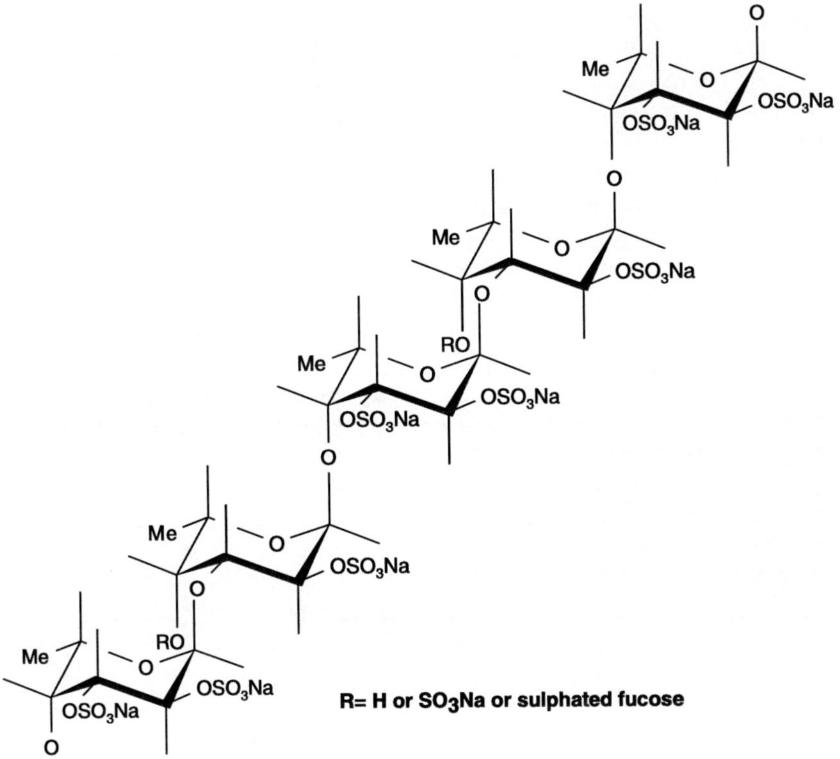


**Fig. 3** Structure of sulphated oligofucans constitutive of the low-molecular-weight fucoidan isolated from *Ascophyllum nodosum* (Chevolot et al. 2001)

The biological activities of fucoidan depend on the structure of the starting material, the purification process, its structural heterogeneity and molecular mass dispersion. Fucoidan has a wide spectrum of biological activities (Berteau and Mulloy 2003).

## Anticoagulant Activity (or Anti-thrombin Activity) of Fucoidan

In 1957 and 1962, the anticoagulant activity of heterogeneous extracts rich in fucoidan was reported (Springer et al. 1957 and Bernadi and Springer 1962). Furthermore, different extracts partially purified from brown algae presenting variable molecular mass have been studied for their biological activities. In 1985, Deaconsmith et al. studied heterogeneous algal extracts from different genera of Phaeophyceae (*Laminaria, Fucus, Ascophyllum*). They showed an inhibiting effect on the thrombin time performed with plasma, and also pure fibrinogen, and suggested an immediate inhibition of thrombin activity. In 1989, different research teams published the anticoagulant activity of fucose-containing sulphated polysaccharides from various brown algae. First, Nishino et al. (1989) isolated different sulphated polysaccharidic fractions obtained first by hot-water extraction from *Ecklonia kurome* followed by successive fractionation steps using anion-exchange and gel filtration chromatographies. A low-molecular-weight (21,000 g/mol) homogeneous fraction was obtained, containing mainly fucose units and ester sulphate groups. This highly purified fucoidan possessed a high anticoagulant activity especially in APTT and no anti-factor Xa activity was detected. Then, Grauffel et al. (1989) prepared a relatively LMW fucan fraction (50,000 g/mol) from different species (*Pelvetia canaliculata*, *Fucus vesiculosus*, *Laminaria digitata*, *Sargassum muticum* and *Ascophyllum nodosum*) and investigated the anticoagulation mechanism by checking the involvement of the anti-thrombin III (AT) or other plasma inhibitors. According to the species, the anticoagulant activity was found to be highly related to chemical composition and extraction conditions. Direct interaction between fucan and thrombin seems largely responsible for the kinetic effect of thrombin inactivation and inactivation of thrombin was accelerated when an LMW fucan fraction was incubated with AT (the predominant inhibitor of thrombin in plasma). This kinetic effect of thrombin inactivation by AT was less important with fucoidan than with heparin. No neutralisation of factor Xa was observed in the presence of this fucan fraction. Finally, Church et al. (1989) studied the interaction of fucoidan with heparin cofactor II (HC II), AT and thrombin using a commercial heterogeneous HMW fucoidan extracted from *Fucus vesiculosus* (100,000 g/mol). This group showed that the anti-thrombin action of fucoidan (ex vivo in a plasma system) is mediated through HC II, and not through AT, and this correlates well with the anti-thrombin effect of fucoidan observed in vitro.

Colliec et al. (1991) confirmed these results using a homogeneous LMW fucoidan extracted from *Pelvetia canaliculata* (20,000 g/mol). This LMW fucoidan was anticoagulant in vitro and as potent on the prolongation of APTT and TT as heparin, at a concentration 50 times higher than heparin (on a weight to weight basis). Fucoidan was a weak inhibitor of thrombin generation (100 times lower than heparin). In comparison with Church's data (using a HMW fucoidan) (Church et al. 1989), Colliec et al. (1991) found roughly a same enhancement on the rate of thrombin inhibition by HC II ($1.5 \times 10^8$ $M^{-1}$ $min^{-1}$ and $3.8 \times 10^8$ $M^{-1}$ $min^{-1}$ with 10 μg/mL of HMW fucoidan and LMW fucoidan, respectively), but a stronger inhibition mediated by AT ($3 \times 10^8$ $M^{-1}$ $min^{-1}$ with 10 μg/mL of LMW fucoidan and $5.7 \times 10^7$ $M^{-1}$ $min^{-1}$ with 30 μg/mL of HMW fucoidan). On the contrary, Church et al. (1989) observed a weak factor Xa inhibition by AT in the presence of the HMW fucoidan at very high concentration (500 μg/mL), while at this high concentration the LMW fucoidan is not able to produce noticeable inhibition. In conclusion, this LMW fucoidan exerts its anticoagulant activity by enhancing thrombin inhibition in the presence of either AT or HC II. Among the known anticoagulant sulphated polysaccharides, this LMW fucoidan appears to be the only one which is as potent on the formation of an AT-thrombin complex as it is on the HC II-thrombin complex. In fact, heparin requires much higher concentrations to enhance the activity of HC II than for influencing AT. On the contrary, pentosan polysulphate and dermatan sulphate are mainly active via an HC II pathway. Moreover, the absence of factor Xa inactivation by this LMW fucoidan, correlated with its low viscosity (compared to HMW fucoidan) and its high thrombin inactivation, could be an important feature for further clinical use, according to some studies. Indeed, it has been demonstrated that LMW heparin fragments, with high factor Xa inhibition and negligible activity on thrombin inhibition, are poor anti-thrombotic agents.

The same year, Nishino et al. (1991) showed the relationship between both molecular weight and sulphate content of fucoidan and its anti-thrombin effect. Its anti-thrombin activity in the presence of HC II was improved with increase in its molecular weight and reduced with decrease in its sulphate content.

In 1992 and 1993, Soeda et al. reported that fucoidan in vitro stimulated tissue plasminogen activator (t-PA) catalysed plasminogen activation and prevented the formation of fibrin polymer according to the degree of sulphation. This group also reported that the in vitro abilities of over-sulphated fucoidan to stimulate t-PA, catalyse plasminogen activation and to potentiate thrombin inhibition by AT or HC II decreased with a decrease in molecular size. For the first time, the potential therapeutic benefit of fucoidan for the prevention of thrombus formation was described in endotoxin-induced hepatic vein thrombosis in hyperlipemic rats.

## Anti-thrombotic Activity of Fucoidan

Following on from Soeda's publication, the anti-thrombotic activity of fucoidan has been widely described in different animal models. In 1995, Mauray et al. showed the venous anti-thrombotic and anticoagulant activities of a homogeneous LMW fucoidan extracted from *Ascophyllum nodosum* (20,000 g/mol). In a Wessler model of venous thrombosis, the LMW fucoidan injected intravenously to rabbits exhibited anti-thrombotic activity and the dose which inhibited 80% of mean thrombus weight ($ED_{80}$) was 1.8 mg/kg, compared to a heparin $ED_{80}$ of 0.1 mg/kg. At this

$ED_{80}$, the anti-thrombotic effect of the LMW fucoidan persisted longer than that of heparin (30 min vs. 15 min), whilst the anticoagulant effect measured ex vivo was related to an haemorrhagic risk. Millet et al. (1999) reported the anti-thrombotic and anticoagulant activities of a very LMW fucoidan from *Ascophyllum nodosum* (8,000 g/mol) by the subcutaneous route. In the Wessler model, this LMW fucoidan exhibited dose-related venous anti-thrombotic activity, with an ED80 of about 20 mg/kg, two hours after a single subcutaneous injection. At the same anti-thrombotic activity, LMW fucoidan exhibited a lower effect on ex vivo coagulation tests, and a lower prolongation of the bleeding time than the LMW heparin (dalteparin), which corresponded to a weaker haemorrhagic effect.

Then, Colliec-Jouault et al. (2003) showed that LMW fucoidan (injected intravenously or subcutaneously) exhibited arterial anti-thrombotic properties in rabbits and rats at the same doses than in a venous thrombosis model (1.8–2 mg/kg intravenous and 10–20 mg/kg subcutaneous). In the same arterial thrombosis models, both unfractionated heparin and LMW heparin had to be used at higher doses than in the Wessler venous thombosis model. Thus, the anticoagulant effect, the prolongation of the bleeding time and the haemorrhagic risk are much more pronounced, in these arterial thrombosis models, with heparins than with fucoidan. These results were confirmed recently by Durand, Helley et al. (2008) in another rabbit model of arterial thrombosis. Thrombosis was induced in femoral arteries by in situ induction of endothelial apoptosis and the animals treated by subcutaneous injection of 15 mg/kg of LMW fucoidan from *Ascophyllum nodosum* (8,000 g/mol) and 2.5 mg/kg of LMW heparin (enoxaparin). LMW fucoidan appeared to be more effective than LMW heparin for preventing arterial thrombosis in this experimental model. LMW fucoidan also had lower haemorrhagic risk than LMW heparin. The plasma concentration of tissue factor pathway inhibitor (TFPI) was significantly increased after LMW fucoidan injection, whereas no change was observed after LMW heparin treatment.

This effect on TFPI was previously described for heparin, and by Girault et al. (1998a) for fucoidan who actually showed that fucoidan induces TFPI release from cultured human umbilical vein endothelial cells, which may contribute to its anti-thrombotic effect. In summary, the anti-thrombotic effect of LMW fucoidan may in part be explained by the observed effect on the tissue factor pathway. Previously, Tholarius et al. reported that a heterogeneous HMW fucoidan prevented microvascular thrombus formation induced by endothelial damage in arterioles and venules in vivo. This protective effect of fucoidan is not attributable to inhbition of P- and L-selectin function, but may be related to the anticoagulant capacity of fucoidan (Thorlacius et al. 2000).

## Endothelial Cell Interactions and Fucoidan

Endothelial wound repair is a crucial step to prevent rethrombosis and restenosis of a damaged arterial vessel wall. Besides its anticoagulant and anti-thrombotic effects, fucoidan can induce angiogenesis in vitro by modulating the proangiogenic properties of heparin-binding growth factors such as fibroblast growth factor-2 (FGF-2). In 1983, Glabe et al. described a reversible disruption of cultured endothelial monolayers and found that fucoidan appears to bind at two distinct sites on endothelial monolayers. One site is inhibitable by heparin, while the other site seems to be specific for fucoidan. These authors suggested that sulphated fucose-containing glycoconjugates may play a role in the adhesive interactions of endothelial cells. The effect of a LMW fucoidan from *Ascophyllum nodosum* on the growth and migration of human umbilical vein endothelial cells was compared to heparin in the presence of human growth factors. Fucoidan modulated FGF-2 induced cell proliferation, which was dependent on FGF-2 concentration, whereas unfractionated heparin had an inhibitory effect (Giraux et al. 1998b). It has also been described that LMW fucoidan from *Ascophyllum nodosum* can enhance FGF-2 induced tube formation of endothelial cells through $\alpha_6$ overexpression that was heparan sulphate dependent (Matou et al. 2002; Chabut et al. 2003).

Other properties of LMW fucoidan from *Ascophyllum nodosum* could also be of interest against arterial thrombosis, as these molecules are able to enhance vascular tube formation as described above and to inhibit smooth cell proliferation and neointimal hyperplasia. LMW fucoidan and heparin share some similar mechanisms of action, such as smooth muscle cell growth inhibition, binding and internalisation (Logeart et al. 1997). This fucoidan with high affinity for smooth muscle cells reduced intimal hyperplasia in rabbit iliac artery in-stent restenosis model and may be potentially relevant for the treatment of in-stent restenosis (Deux et al. 2002). Moreover, in a rat model of critical limb ischemia, LMW fucoidan promoted therapeutic revascularization induced by FGF-2. LMW fucoidan promotes FGF-2 effects in vivo, suggesting its potential interest for use in vascular tissue repair and angiogenesis (Luyt et al. 2003). The inhibition of platelet-neutrophil interactions by a HMW heterogeneous fucoidan was also demonstrated and this inhibition reduces adhesion and vasoconstriction after acute arterial injury by angioplasty in pigs (Chauvet et al. 1999). In a rat cardiac allograft model, the LMW fucoidan from *Ascophyllum nodosum* appeared very effective to prevent arterial and parenchymal lesions occurring in response to alloimmune injury. However this protective effect does not appear to depend on mobilisation of bone marrow-derived cells (Alkhatib et al. 2006).

### Neoangiogenesis Induced by Endothelial Progenitor Cells with fucoidan

Another effect of fucoidan is the ability to promote progenitor stem cell mobilisation via the release of stromal-derived factor-1 (SDF-1) from heparan sulphate sites. This effect was previously described by Sweeney et al. for HMW heterogeneous fucoidan and other glycosaminoglycans, such as dextran sulphates and chondroitin sulphate. Sweeney et al. also indicated that plasma metalloproteinase MMP-9 significantly increases in response to intravenous injection of HMW fucoidan (Sweeney et al. 2002).

In constrast, LMW fucoidan from *Ascophyllum nodosum* did not induce an increase in MMP-9 level in vivo (Luyt et al. 2003). These results suggest that sulphated polysaccharides from the same family may exhibit different properties depending on their molecular weight. Boisson-Vidal et al. (2007) reported that LMW fucoidan from *Ascophyllum nodosum* enhances the proangiogenic properties of endothelial progenitor cells and can mobilise bone marrow progenitor cells in peripheral blood, enhancing their recruitment to sites of active angiogenesis and increasing blood vessel formation (Zemani et al. 2005; Boisson-Vidal et al. 2007).

#### Anti-inflammatory Activity of fucoidan

Like proteoglycans, fucoidan interacts with a wide range of proteins and thus has pleiotropic properties including anti-inflammatory activity (Tissot et al. 2003; Angstwurm et al. 1995). More recently, Cumashi et al. (2007) studied the biological properties of fucoidans obtained from nine species of brown algae. All fucoidans inhibited leucocyte recruitement in an inflammation model in rats. In 2008, Medeiros et al. reported that fucoidan from *Lobophora variegata* (Phaeophyceae, Dictyotales) inhibits leucocyte migration to an inflammatory site. Ear swelling caused by croton oil was also inhibited when polysaccharides form *Fucus vesiculosus* and *Lobophora variegata* were used. The polysaccharides studied may have therapeutic potential in a range of inflammatory disorders.

### 3.1.3 The Rhamnan Family and Arabinan Family from Chlorophyceae

In Clorophyta, the major polysaccharides are polydisperse and highly branched sulphated heteropolysacccharides rich in rhamnose, galactose and arabinose sugars. The marine green algae represent an important biomass that is still little used compared to red and brown algae. Also compared to Rhodophyta, the reports on biological activities and anticoagulant activity of green algal polysaccharides are less abundant. Sulphated polysaccharides from two orders (Uvales and Bryopsidales) were mainly studied for their anticoagulant properties. Polysaccharides from Ulvales are glucuronoxylorhamnans rich in rhamnose, they are sulphated (22% of ester sulphalte groups) and carboxylated (20% of uronic acid sugars). The most frequent repeating sequence is D-glucuronosyl 1,4 linked α-L-rhamnosyl-2sulphate 1,3 linked β-D-glucuronic acid with ramifications formed by 1,4 linked D-xylose. In Bryopsidales, the sulphated polysaccharides are xylogalactoarabinans or xyloarabinogalactans, they are rich in arabinose or galactose respectively. The backbone consists of 1,4 linked L-arabinose blocks separated by D-galactose residues. All D-xylose residues and part of galactose residues are in terminal positions and they contain about 17% of ester sulphate groups.

In 1991, Maeda et al. compared the anticoagulant properties of different hot-water extracts from Ulvales (Monostromataceae and Ulvaceae) and Bryopsidales (Codiaceae, Caulerpaceae and Bryopsidaceae) (Maeda et al. 1991). The yields of crude polysaccharides from dry algae were from 5 to 20%, and the sulphate ester contents from crude polysaccharides were from 5 to 25%. The most sulphated polysaccharides were found in *Monostroma nitidum* (Ulvales). After purification steps, the sulphated high rhamnose-containing polysaccharide (65% of L-rhamnose, 6% of D-glucose, 5% of glucuronic acid and 23% of ester sulphate) was sixfold more anticoagulant than standard heparin measured by in vitro clotting assays.

The others studies reported the anticoagulant properties of sulphated polysaccharides isolated from different species of the genus *Codium* (Bryopsidales). Jurd described in 1995 (Jurd et al. 1995) the anticoagulant properties of sulphated polysaccharides from *Codium fragile* ssp. *atlanticum*. After extraction and different purification steps (size exclusion and ion exchange chromatographies), different product was isolated and studied: a high-molecular-weight sulphated (18%) proteoglycan and two lower molecular weight sulphated (7 and 10% of ester sulphate groups, respectively) polysaccharides. The highest anticoagulant activity detected using in vitro clotting assays (APTT, TT and PT) was found with the proteoglycan. In APTT, the concentrations required to double the control clotting time with proteoglycan, 7% sulphated polysaccharide and 10% sulphated polysaccharide were 4, 30 and 250 μg/mL, respectively. The proteoglycan isolated from *Codium fragile* ssp. *Atlanticum* inhibited thrombin and factor Xa through the catalysis of anti-thrombin III, whereas the two polysaccharides inhibited only thrombin via heparin cofactor II catalysis. No direct activity on thrombin and factor Xa was demonstrated. The anticoagulant effect is correlated, with not only the degree of sulphate but also with the increase of the molecular weight.

In 1999, Siddhanta et al. isolated from the green marine alga *Codium dwarkense* Boergs. (Siddhanta et al. 1999). (Bryopsidales), two sulphated polysaccharides: one arabinan and one arabinogalactan. The very high-molecular-weight arabinan sulphate (estimated MW was $3 \times 10^6$ and containing only arabinose sugars and 40% of ester sulphate groups) exhibited stronger anticoagulant activity than the lower molecular weight arabinogalactan sulphate (estimated MW was $3 \times 10^5$ and containing arabinose and galactose residues and 32% of ester sulphate groups). The anticoagulant activity is proportional to the arabinose and sulphate contents and inversely proportional to the protein and uronic acid contents, but also probably proportional to the molecular weight. The highly sulphated arabinan is only composed of α-L-arabinofuranose. It prolongs APTT and TT, in APTT the same anticoagulant effect was obtained for the arabinan sulphate and heparin (140.3 units/mg) at 15 and 4 μg/ml, respectively.

Recently, Ciancia et al. (2007) compared two species of the genus *Codium* (Bryopsidales), the crude extract isolated from *Codium vermilara* was more sulphated than the one from *Codium fragile*, 30 and 20%, of ester sulphate groups, respectively. The two extracts were sulphated arabinogalactans, so the major sugars were galactose (62% from *C. fragile* and 50% from *C. vermilara*) and arabinose (23% from *C. fragile* and 45% from *C. vermilara*). The molecular weight of the arabinogalactan from *C. vermilara* was higher than the one from *C. fragile* ($66 \times 10^3$ and $11 \times 10^3$, respectively). The sulphated arabinogalactan from *C. vermilara* was the most active in clotting assays (in APTT, a same prolongation was observed at 20 μg/ml for the extract from *C. vermilara* and 100 μg/ml for the other one from *C. fragile*). As previously described above, the anticoagulant activity is proportional to the arabinose and sulphate contents but probably to the molecular weight.

### 3.2 *Heparin-like Entities from Marine Fungi*

Reports of sulphated polysaccharides from marine fungi are extremely rare, probably because the native polysaccharides are not sulphated. In 2005, Chen et al. studied the antiangiogenic activities of polysaccharides isolated from terrestrial medicinal fungi. They are very high molecular weight neutral polysaccharides, the most active are rich in fucose, glucose and mannose; so Chen et al. suggested that these monosaccharides may play a role in the inhibitory effect of these fungi on endothelial tube formation. In 2005, a polysaccharide YCP from a marine filamentous fungus *Keissleriella* sp., YS4108 was chemically sulphated. The YCP sulphate significantly prolonged clotting times (APTT, TT and PT) and the anticoagulant activity improved with the increasing degree of sulphation and decreased molecular weight (Han et al 2005).

## 4 Concluding Remarks and Future Directions

Oligosaccharides showing heparin-like activities have been obtained from polysaccharides produced by marine bacteria. Relationships between the structure and the anticoagulant activities of marine polysaccharides and their derivatives remain to be established in order to fully control the production of therapeutic drugs.

The demand for clean environmentally friendly processes is increasing significantly; biotechnology and in particular “white” biotechnology (which aims to avoid harmful substances by application of enzymes as catalysts and utilisation of renewable raw materials) may propose new processes for sustainable industry. Enzymes are ideal biocatalysts to assist in the synthesis of various compounds by offering catalysis with stereo- and regio-selectivity, under mild conditions, in aqueous solutions. The use of an enzymatic step in processes involving sulphation may eliminate the need for solvents and multiple protection and de-protection steps, increasing the final yield and lowering the duration of production. Sulfotransferases able to sulphate these EPS should be of high interest in the aim to produce bioactive derivatives. Marine bacteria producing sulphated polysaccharides would likely be a relevant source of new sulfotransferase enzymes to be active on these molecules. Although chemical synthesis was the major route to obtain structurally defined heparin oligosaccharides, the features of enzymatic methods to get oligosaccharides of biological relevance meet well the needs of better control of targeted modifications and of environmentally safer processing steps.

Furthermore, knowledge of the biosynthesis of EPS would facilitate the development of novel approaches, either by enzymes or by metabolic engineering, to synthesise heparinoids from marine bacterial EPS and to produce new molecules with high specificity for biological targets.

## References

Alkhatib B, Freguin-Bouilland C, Lallemand F et al (2006) Low molecular weight fucan prevents transplant coronaropathy in rat cardiac allograft model. Transpl Immunol 16(1):14–19

Anderson W, Duncan JGC (1965) Anticoagulant activity of carrageenan. J Pharm Pharmacol 17(10):647–654

Angstwurm K, Weber JR, Segert A et al (1995) Fucoidin, a polysaccharide inhibiting leukocyte rolling, attenuates inflammatory responses in experimental pneumococcal meningitis in rats. Neurosci Lett 191(1–2):1–4

Barabanova A, Shashkov A, Glazunov V et al (2008) Structure and properties of carrageenan-like polysaccharide from the red alga *Tichocarpus crinitus* (Gmel.) Rupr. (Rhodophyta, Tichocarpaceae). J Appl Phycol 20(6):1013–1020

Baross JA, Hoffman SE (1985) Submarine hydrothermal vents and associated gradient environments as sites for the origin and evolution of life. Orig Life Evol Biosph 15(4):327–345

Bernardi G, Springer GF (1962) Properties of highly purified fucan. J Biol Chem 237(1):75–80

Berteau O, Mulloy B (2003) Sulfated fucans, fresh perspectives: structures, functions, biological properties of sulfated fucans and overview of enzymes active towards this class of polysaccharide. Glycobiology 13(6):29R–40R

Bhaskar PV, Grossart H-P, Bhosle NB et al (2005) Production of macroaggregates from dissolved exopolymeric substances (EPS) of bacterial and diatom origin. FEMS Microbiol Ecol 53(2): 255–264

Boeuf G, Kornprobst J-M (2009) Biodiversité et chimiodiversité marines. Biofutur 301:28–32

Boisson-Vidal C, Zemani F, Caligiuri G et al (2007) Neoangiogenesis induced by progenitor endothelial cells: effect of fucoidan from marine algae. Cardiovasc Hematol Agents Med Chem 5(1):67–77

Bourgougnon N, Lahaye M, Chermann J-C et al (1993) Composition and antiviral activities of a sulfated polysaccharide from schizymenia dubyi (rhodophyta, gigartinales). Bioorg Med Chem Lett 3(6):1141–1146

Bourguet-Kondracki M-L, Banaigs B (2009) Des métabolites pas si secondaires. Biofutur 301: 34–38

Bramhachari PV, Dubey SK (2006) Isolation and characterization of exopolysaccharide produced by *Vibrio harveyi* strain VB23. Lett Appl Microbiol 43(5):571–577

Bramhachari PV, Kavi Kishor PB, Ramadevi R et al (2007) Isolation and characterization of mucous exopolysaccharide (EPS) produced by *Vibrio furnissii* strain VB0S3. J Microbiol Biotechnol 17(1):44–51

Cambon-Bonavita MA, Raguenes G, Jean J et al (2002) A novel polymer produced by a bacterium isolated from a deep-sea hydrothermal vent polychaete annelid. J Appl Microbiol 93(2): 310–315

Carlucci MJ, Pujol CA, Ciancia M et al (1997) Antiherpetic and anticoagulant properties of carrageenans from the red seaweed *Gigartina skottsbergii* and their cyclized derivatives: correlation between structure and biological activity. Int J Biol Macromol 20(2):97–105
Chabut D, Fischer AM, Colliec-Jouault S et al (2003) Low molecular weight fucoidan and heparin enhance the basic fibroblast growth factor-induced tube formation of endothelial cells through heparan sulfate-dependent alpha 6 overexpression. Mol Pharmacol 64(3):696–702
Chabut D, Fischer AM, Helley D et al (2004) Low molecular weight fucoidan promotes FGF-2-induced vascular tube formation by human endothelial cells, with decreased PAI-1 release and ICAM-1 downregulation. Thromb Res 113(1):93–95
Chauvet P, Bienvenu JG, Theoret JF et al (1999) Inhibition of platelet-neutrophil interactions by fucoidan reduces adhesion and vasoconstriction after acute arterial injury by angioplasty in pigs. J Cardiovasc Pharmacol 34(4):597–603
Chen SC, Lu MK, Cheng JJ et al (2005) Antiangiogenic activities of polysaccharides isolated from medicinal fungi. FEMS Microbiol Lett 249(2):247–254
Chevolot L, Mulloy B, Ratiskol J et al (2001) A disaccharide repeat unit is the major structure in fucoidans from two species of brown algae. Carbohydr Res 330(4):529–535
Church FC, Meade JB, Treanor RE et al (1989) Antithrombin activity of fucoidan. The interaction of fucoidan with heparin cofactor II, antithrombin III, and thrombin. J Biol Chem 264(6):3618–3623
Ciancia M, Quintana I, Vizcarguenaga MI et al (2007) Polysaccharides from the green seaweeds *Codium fragile* and *C. vermilara* with controversial effects on hemostasis. Int J Biol Macromol 41(5):641–649
Colliec S, Fischer AM, Taponbretaudiere J et al (1991) Anticoagulant properties of a fucoidan fraction. Thromb Res 64(2):143–154
Colliec-Jouault S, Chevolot L, Helley D et al (2001) Characterization, chemical modifications and in vitro anticoagulant properties of an exopolysaccharide produced by Alteromonas infernus. Biochim Biophys Acta 1528:141–151
Colliec-Jouault S, Roger O, Ratiskol J et al (2003a) Bioactive bacterial exopolysaccharides: modification, characterization and preliminary results on biological activity. In: Colliec S (ed) Marine biotechnology: an overview of leading field, pp 141–147 Plouzané: Ifremer Publications.
Colliec-Jouault S, Millet J, Helley D et al (2003b) Effect of low-molecular-weight fucoidan on experimental arterial thrombosis in the rabbit and rat. J Thromb Haemost 1(5):1114–1115
Colliec-Jouault S, Zanchetta P, Helley D et al (2004) Les polysaccharides microbiens d'origine marine et leur potentiel en thérapeutique humaine. Pathol Biol 52(3):127–130
Cumashi A, Ushakova NA, Preobrazhenskaya ME et al (2007) A comparative study of the anti-inflammatory, anticoagulant, antiangiogenic, and antiadhesive activities of nine different fucoidans from brown seaweeds. Glycobiology 17(5):541–552
Damonte E, Neyts J, Pujol CA et al (1994) Antiviral activity of a sulphated polysaccharide from the red seaweed nothogenia fastigiata. Biochem Pharmacol 47(12):2187–2192
De Angelis PL (2002) Microbial glycosaminoglycan glycosyltransferases. Glycobiology 12(1):9R–16R
De Philippis R, Sili C, Paperi R et al (2001) Exopolysaccharide-producing cyanobacteria and their possible exploitation: a review. J Appl Phycol 13:293–299
Deaconsmith RA, Leepotter JP, Rogers DJ (1985) Anticoagulant activity in extracts of British marine-algae. Bot Mar 28(8):333–338
Deux JF, Meddahi-Pelle A, Le Blanche AF et al (2002) Low molecular weight fucoidan prevents neointimal hyperplasia in rabbit iliac artery in-stent restenosis model. Arterioscler Thromb Vasc Biol 22(10):1604–1609
Durand E, Helley D, Zen AAH et al (2008) Effect of low molecular weight fucoidan and low molecular weight heparin in a rabbit model of arterial thrombosis. J Vasc Res 45(6):529–537
Fonseca RJ, Oliveira SN, Melo FR et al (2008) Slight differences in sulfation of algal galactans account for differences in their anticoagulant and venous antithrombotic activities. Thromb Haemost 99(3):539–545

Giraux JL, Tapon Bretaudiere J, Matou S et al (1998a) Fucoidan, as heparin, induces tissue factor pathway inhibitor release from cultured human endothelial cells. J Thromb Haemost 80(4): 692–695

Giraux JL, Matou S, Bros A et al (1998b) Modulation of human endothelial cell proliferation and migration by fucoidan and heparin. Eur J Cell Biol 77:352–359

Glabe CG, Yednock T, Rosen SD (1983) Reversible disruption of cultured endothelial monolayers by sulfated fucans. J Cell Sci 61(May):475–490

Grauffel V, Kloareg B, Mabeau S et al (1989) New natural polysaccharides with potent antithrombic activity: fucans from brown algae. Biomaterials 10(6):363–368

Guezennec J (2002) Deep-sea hydrothermal vents: a new source of innovative bacterial exopolysaccharides of biotechnological interest? J Ind Microbiol Biotechnol 29:204–208

Guezennec JG, Pignet P, Raguenes G et al (1994) Preliminary chemical characterization of unusual eubacterial exopolysaccharides of deep-sea origin. Carbohydr Polym 24(4):287–294

Guezennec J, Pignet P, Lijour Y et al (1998) Sulphation and depolymerization of a bacterial exopolysaccharide of hydrothermal origin. Carbohydr Polym 37:19–24

Guo H, Yi W, Song JK et al (2008) Current understanding on biosynthesis of microbial polysaccharides. Curr Top Med Chem 8:141–151

Guven KC, Ozsoy Y, Ulutin ON (1991) Anticoagulant, fibrinolytic and antiaggregant activity of carrageenans and alginic acid. Bot Mar 34(5):429–432

Han F, Yao W, Yang X et al (2005) Experimental study on anticoagulant and antiplatelet aggregation activity of a chemically sulfated marine polysaccharide YCP. Int J Biol Macromol 36(4):201–207

Hawkins WW, Leonard VG (1962) Antipeptic and antithrombic properties of carrageenin. J Lab Clin Med 60:641–648

Hayakawa Y, Hayashi T, Hayashi K et al (1996) Heparin cofactor II-dependent antithrombin activity of calcium spirulan. Blood Coagul Fibrinolysis 7(5):554–560

Hayakawa Y, Hayashi T, Lee J-B et al (2000) Activation of heparin cofactor II by calcium spirulan. J Biol Chem 275(15):11379–11382

Hayashi T, Hayashi K, Maeda M et al (1996) Calcium spirulan, an inhibitor of enveloped virus replication, from a blue-green alga *Spirulina platensis*. J Nat Prod 59(1):83–87

Ivanova EP, Gorshkova RP, Mikhailov VV et al (1994) Capsular polysaccharides of marine-bacteria of the genus *Alteromonas*. Microbiology 63(2):120–123

Jayakumar R, Nwe N, Tokura S et al (2007) Sulphated chitin and chitosan as novel biomaterials. Int J Biol Macromol 40(3):175–181

Jurd KM, Rogers DJ, Blunden G et al (1995) Anticoagulant properties of sulphated polysaccharides and a proteoglycan from *Codium fragile* ssp. *atlanticum*. J Appl Phycol 7:339–345

Kaji T, Okabe M, Shimada S et al (2004) Sodium spirulan as a potent inhibitor of arterial smooth muscle cell proliferation in vitro. Life Sci 74(19):2431–2439

Killing, H. (1913) Zur biochemie der Meersalgen. Z. Physiol. Chem., 83, 171–197

Kloareg B, Quatrano RS (1988) Structure of the cell walls of marine algae and ecophysiological functions of the matrix polysaccharides. Oceanogr Mar Biol Annu Rev 26:259–315

Kumar AS, Mody K (2009) Microbial exopolysaccharides: variety and potential applications. In: Rehm BHA (ed) Microbial production of biopolymers and polymer precursors. Applications and Perspectives, pp 229–253 Caister Academic Press

Leali D, Belleri M, Urbinati C et al (2001) Fibroblast growth factor-2 antagonist activity and angiostatic capacity of sulphated *Escherichia coli* K5 polysaccharide derivatives. J Biol Chem 276(41):37900–37908

Lee J-B, Hayashi T, Hayashi K et al (2000) Structural analysis of calcium spirulan (Ca-SP)-derived oligosaccharides using electrospray ionization mass spectrometry. J Nat Prod 63(1): 136–138

Lee JB, Hou XL, Hayashi K et al (2007) Effect of partial desulphation and oversulphation of sodium spirulan on the potency of anti-herpetic activities. Carbohydr Polym 69(4):651–658

Lerouge P, Roche P, Faucher C et al (1990) Symbiotic host-specificty of *Rhizobium meliloti* is determined by a sulphated and acylated glucosamine oligosaccharide signal. Nature 344: 781–784

Logeart D, Prigent-Richard S, Boisson-Vidal C et al (1997) Fucans, sulfated polysaccharides extracted from brown seaweeds, inhibit vascular smooth muscle cell proliferation. II. Degradation and molecular weight effect. Eur J Cell Biol 74(4):385–390

Luyt CE, Meddahi-Pelle A, Ho-Tin-Noe B et al (2003) Low-molecular-weight fucoidan promotes therapeutic revascularization in a rat model of critical hindlimb ischemia. J Pharmacol Exp Ther 305(1):24–30

Maeda M, Uehara T, Harada N et al (1991) Heparinoid-active sulphated polysaccharides from Monostroma nitidum and their distribution in the chlorophyta. Phytochemistry 30(11): 3611–3614

Majdoub H, Mansour MB, Chaubet F et al (2009) Anticoagulant activity of a sulphated polysaccharide from the green alga *Arthrospira platensis*. Biochim Biophys Acta 1790(10):1377–1381

Mancuso Nichols CA, Garon S, Bowman JP et al (2004) Production of exopolysaccharides by Antarctic marine bacterial isolates. J Appl Microbiol 96:1057–1066

Mata JA, Béjar V, Bressollier P et al (2008) Characterization of exopolysaccharides produced by three moderately halophilic bacteria belonging to the family Alteromonadaceae. J Appl Microbiol 105(2):521–528

Matou S, Helley D, Chabut D et al (2002) Effect of fucoidan on fibroblast growth factor-2-induced angiogenesis in vitro. Thromb Res 106(4–5):213

Matsuda M, Yamori T, Naitoh M et al (2003) Structural revision of sulphated polysaccharide B-1 isolated from a marine *Pseudomonas* species and its cytotoxic activity against human cancer cell lines. Mar Biotechnol 5:13–19

Mauray S, Sternberg C, Theveniaux J et al (1995) Venous antithrombotic and anticoagulant activities of a fucoidan fraction. J Thromb Haemost 74:1280–1285

Medeiros VP, Queiroz KCS, Cardoso ML et al (2008) Sulfated galactofucan from *Lobophora variegata*: anticoagulant and anti-inflammatory properties. Biochemistry 73(9):1018–1024

Melo FR, Mourao PAS (2008) An algal sulfated galactan has an unusual dual effect on venous thrombosis due to activation of factor XII and inhibition of the coagulation proteases. Thromb Haemost 99(3):531–538

Millet J, Jouault SC, Mauray S et al (1999) Antithrombotic and anticoagulant activities of a low molecular weight fucoidan by the subcutaneous route. J Thromb Haemost 81(3):391–395

Miroshnichenko M, Bonch-Osmolovskaya E (2006) Recent developments in the thermophilic microbiology of deep-sea hydrothermal vents. Extremophiles 10(2):85–96

Muralidharan J, Jayachandran S (2003) Physicochemical analyses of the exopolysaccharides produced by a marine biofouling bacterium, *Vibrio alginolyticus*. Process Biochem 38(6): 841–847

Nardella A, Chaubet F, Boisson-Vidal C et al (1996) Anticoagulant low molecular weight fucans produced by radical process and ion exchange chromatography of high molecular weight fucans extracted from the brown seaweed *Ascophyllum nodosum*. Carbohydr Res 289:201–208

Nazarenko EL, Gorshkova RP, Zubkov VA et al (1993) Structure of the repeating unit of the acidic polysaccharide from *Alteromonas* sp 4MC17. Bioorganicheskaya Khimiya 19(7):733–739

Nazarenko EL, Komandrova NA, Gorshkova RP et al (2003) Structures of polysaccharides and oligosaccharides of some Gram-negative marine Proteobacteria. Carbohydr Res 338(23): 2449–2457

Nishimura S-I, Nishi N, Tokura S et al (1986) Inhibition of the hydrolytic activity of thrombin by chitin heparinoids. Carbohydr Res 156:286–292

Nishino T, Nagumo T (1992) Anticoagulant and antithrombin activities of oversulphated fucans. Carbohydr Res 229(2):355–362

Nishino T, Yokoyama G, Dobashi K et al (1989) Isolation, purification, and characterization of fucose-containing sulfated polysaccharides from the brown seaweed *Ecklonia kurome* and their blood-anticoagulant activities. Carbohydr Res 186(1):119–129

Nishino T, Aizu Y, Nagumo T (1991) The influence of sulfate content and molecular weight of a fucan sulfate from the brown seaweed *Ecklonia kurome* on its antithrombin activity. Thromb Res 64(6):723–731

Okutani K (1984) Antitumor and immunostimulant activities of polysaccharide produced by a marine bacterium of the genus *Vibrio*. Bull Jap Soc Sci Fish 50(6):1035–1037

Opoku G, Qiu X, Doctor V (2006) Effect of oversulfation on the chemical and biological properties of kappa carrageenan. Carbohydr Polym 65(2):134–138

Pace NR (1991) Origin of life-facing up to the physical setting. Cell 65(4):531–533

Parikh A, Madamwar D (2006) Partial characterization of extracellular polysaccharides from cyanobacteria. Bioresour Technol 97(15):1822–1827

Pereira MG, Benevides NMB, Melo MRS et al (2005) Structure and anticoagulant activity of a sulfated galactan from the red alga, Gelidium crinale. Is there a specific structural requirement for the anticoagulant action? Carbohydr Res 340(12):2015–2023

Perepelov AV, Shashkov AS, Torgov VI et al (2005) Structure of an acidic polysaccharide from the agar-decomposing marine bacterium *Pseudoalteromonas atlantica* strain IAM 14165 containing 5,7-diacetamido-3,5,7,9-tetradeoxy-l-glycero-l-manno-non-2-ulosonic acid. Carbohydr Res 340(1):69–74

Pomin VH, Mourao PAS (2008) Structure, biology, evolution, and medical importance of sulfated fucans and galactans. Glycobiology 18(12):1016–1027

Pushpamali WA, Nikapitiya C, De Zoysa M et al (2008) Isolation and purification of an anticoagulant from fermented red seaweed *Lomentaria catenata*. Carbohydr Polym 73(2): 274–279

Raguenes G, Pignet P, Gauthier G et al (1996) Description of a new polymer-secreting bacterium from a deep-sea hydrothermal vent, *Alteromonas macleodii* subsp. *fijiensis*, and preliminary characterization of the polymer. Appl Environ Microbiol 62(1):67–73

Raguenes GH, Peres A, Ruimy R et al (1997a) *Alteromonas infernus* sp. nov., a new polysaccharide-producing bacterium isolated from a deep-sea hydrothermal vent. J Appl Microbiol 82(4): 422–430

Raguenes G, Christen R, Guezennec J et al (1997b) *Vibrio diabolicus* sp. nov., a new polysaccharide-secreting organism isolated from a deep-sea hydrothermal vent polychaete annelid, *Alvinella pompejana*. Int J Syst Bacteriol 47(4):989–995

Rastogi RP, Sinha RP (2009) Biotechnological and industrial significance of cyanobacterial secondary metabolites. Biotechnol Adv 27(4):521–539

Rechter S, Konig T, Auerochs S et al (2006) Antiviral activity of *Arthrospira*-derived spirulan-like substances. Antiviral Res 72(3):197–206

Rehm BHA (2009) Microbial production of biopolymers and polymer precursors. Applications and perspectives. Caister Academic, Norfolk, p 293

Rivera-Marrero CA, Ritzenthaler JD, Newburn SA et al (2002) Molecular cloning and expression of a novel glycolipid sulfotransferase in *Mycobacterium tuberculosis*. Microbiology 148(3): 783–792

Roger O, Kervarec N, Ratiskol J et al (2004) Structural studies of the main exopolysaccharide produced by the deep-sea bacterium *Alteromonas infernus*. Carbohydr Res 339(14): 2371–2380

Rougeaux H, Pichon R, Kervarec N et al (1996) Novel bacterial exopolysaccharides from deep-sea hydrothermal vents. Carbohydr Polym 31(4):237–242

Rougeaux H, Talaga P, Carlson RW et al (1998) Structural studies of an exopolysaccharide produced by *Alteromonas macleodii* subsp. *fijiensis* originating from a deep-sea hydrothermal vent. Carbohydr Res 312(1–2):53–59

Rougeaux H, Guezennec J, Carlson RW et al (1999a) Structural determination of the exopolysaccharide of *Pseudoalteromonas* strain HYD721 isolated from a deep-sea hydrothermal vent. Carbohydr Res 315(3–4):273–285

Rougeaux H, Kervarec N, Pichon R et al (1999b) Structure of the exopolysaccharide of *Vibrio diabolicus* isolated from a deep-sea hydrothermal vent. Carbohydr Res 322(1–2): 40–45

Saravanan P, Jayachandran S (2008) Preliminary characterization of exopolysaccharides produced by a marine biofilm-forming bacterium *Pseudoalteromonas ruthenica* (SBT 033). Lett Appl Microbiol 46(1):1–6

Schaeffer DJ, Krylov VS (2000) Anti-HIV activity of extracts and compounds from algae and cyanobacteria. Ecotoxicol Environ Saf 45(3):208–227

Sergeev VN, Gerasimenko LM, Zavarzin GA (2002) The proterozoic history and present state of cyanobacteria. Microbiology 71(6):623–637

Shanmugam M, Mody KH (2000) Heparinoid-active sulphated polysaccharides from marine algae as potential blood anticoagulant agents. Curr Sci 79(12):1672–1683

Siddhanta AK, Shanmugam M, Mody KH et al (1999) Sulphated polysaccharides of Codium dwarkense Boergs. from the west coast of India: chemical composition and blood anticoagulant activity. Int J Biol Macromol 26(2–3):151–154

Siezen RJ, Wilson G (2009) Genomics of deep-sea and sub-seafloor microbes. Microb Biotechnol 2(2):157–163

Singh S, Kate BN, Banerjee UC (2005) Bioactive compounds from cyanobacteria and microalgae: an overview. Crit Rev Biotechnol 25(3):73–95

Soeda S, Sakaguchi S, Shimeno H et al (1992) Fibrinolytic and anticoagulant activities of highly sulfated fucoidan. Biochem Pharmacol 43(8):1853–1858

Soeda S, Ohmagari Y, Shimeno H et al (1993) Preparation of oversulfated fucoidan fragments and evaluation of their antithrombotic activities. Thromb Res 72(3):247–256

Springer GF, Wurzel HA, Mc Neal GM et al (1957) Isolation of anticoagulant fractions from crude fucoidin. Proc Soc Exp Biol Med 94:404–408

Sutherland IW (1998) Novel and established applications of microbial polysaccharides. Trends Biotechnol 16(1):41–46

Sweeney EA, Lortat-Jacob H, Priestley GV et al (2002) Sulfated polysaccharides increase plasma levels of SDF-1 in monkeys and mice: involvement in mobilization of stem/progenitor cells. Blood 99(1):44–51

Tan LT (2007) Bioactive natural products from marine cyanobacteria for drug discovery. Phytochemistry 68(7):954–979

Tharanathan RN, Kittur FS (2003) Chitin – the undisputed biomolecule of great potential. Crit Rev Food Sci Nutr 43(1):61–87

Thorlacius H, Vollmar B, Seyfert UT et al (2000) The polysaccharide fucoidan inhibits microvascular thrombus formation independently from P- and L-selectin function in vivo. Eur J Clin Invest 30(9):804–810

Tissot B, Montdargent B, Chevolot L et al (2003) Interaction of fucoidan with the proteins of the complement classical pathway. Biochim Biophys Acta 1651(1–2):5–16

Townsend GE II, Forsberg LS, Keating DH (2006) *Mesorhizobium loti* produces nodPQ-dependent sulphated cell surface polysaccharides. J Bacteriol 188(24):8560–8572

Vanbleu E, Choudhury BP, Carlson RW et al (2005) The nodPQ genes in *Azospirillum brasilense* Sp7 are involved in sulphation of lipopolysaccharides. Environ Microbiol 7(11): 1769–1774

Whitman WB, Coleman DC, Wiebe WJ (1998) Prokaryotes: the unseen majority. Proc Natl Acad Sci USA 95(12):6578–6583

Witvrouw M, De Clercq E (1997) Sulfated polysaccharides extracted from sea algae as potential antiviral drugs. Gen Pharmacol 29(4):497–511

Yamada T, Ogamo A, Saito T et al (1997) Preparation and anti-HIV activity of low-molecular-weight carrageenans and their sulfated derivatives. Carbohydr Polym 32(1):51–55

Yamamoto C, Fujiwara Y, Tg K (2006) The biological effects of depolymerized sodium spirulan and sulphated colominic acid on vascular cells are beneficial in preventing atherosclerosis. J Health Sci 52(3):205–210

Zanchetta P, Lagarde N, Guezennec J (2003) A new bone-healing material: a hyaluronic acid-like bacterial exopolysaccharide. Calcif Tissue Int 72(1):74–79

Zemani F, Benisvy D, Galy-Fauroux I et al (2005) Low-molecular-weight fucoidan enhances the proangiogenic phenotype of endothelial progenitor cells. Biochem Pharmacol 70(8): 1167–1175

Zheng WF, Chen CF, Cheng QP et al (2006) Oral administration of exopolysaccharide from *Aphanothece halophytica* (Chroococcales) significantly inhibits influenza virus (H1N1)-induced pneumonia in mice. Int Immunopharmacol 6(7):1093–1099

Zou W, Laferriere CA, Jennings HJ (1998) Oligosaccharide fragments of the type III group B streptococcal polysaccharide derived from *S. pneumoniae* type 14 capsular polysaccharide by a chemoenzymatic method. Carbohydr Res 309:297–301

Zubkov VA, Nazarenko EL, Gorshkova RP et al (1995) Structure of the capsular polysaccharide from *Alteromonas* sp. CMM 155. Carbohydr Res 275:147–154

Zuniga EA, Matsuhiro B, Mejias E (2006) Preparation of a low-molecular weight fraction by free radical depolymerization of the sulfated galactan from *Schizymenia binderi* (Gigartinales, Rhodophyta) and its anticoagulant activity. Carbohydr Polym 66(2):208–215

# Index




R. Lever et al. (eds.), *Heparin - A Century of Progress*,
Handbook of Experimental Pharmacology 207,
DOI 10.1007/978-3-642-23056-1, © Springer-Verlag Berlin Heidelberg 2012

## G



## H